Margot Van Voorhies
The Art of Mexican Enamelwork

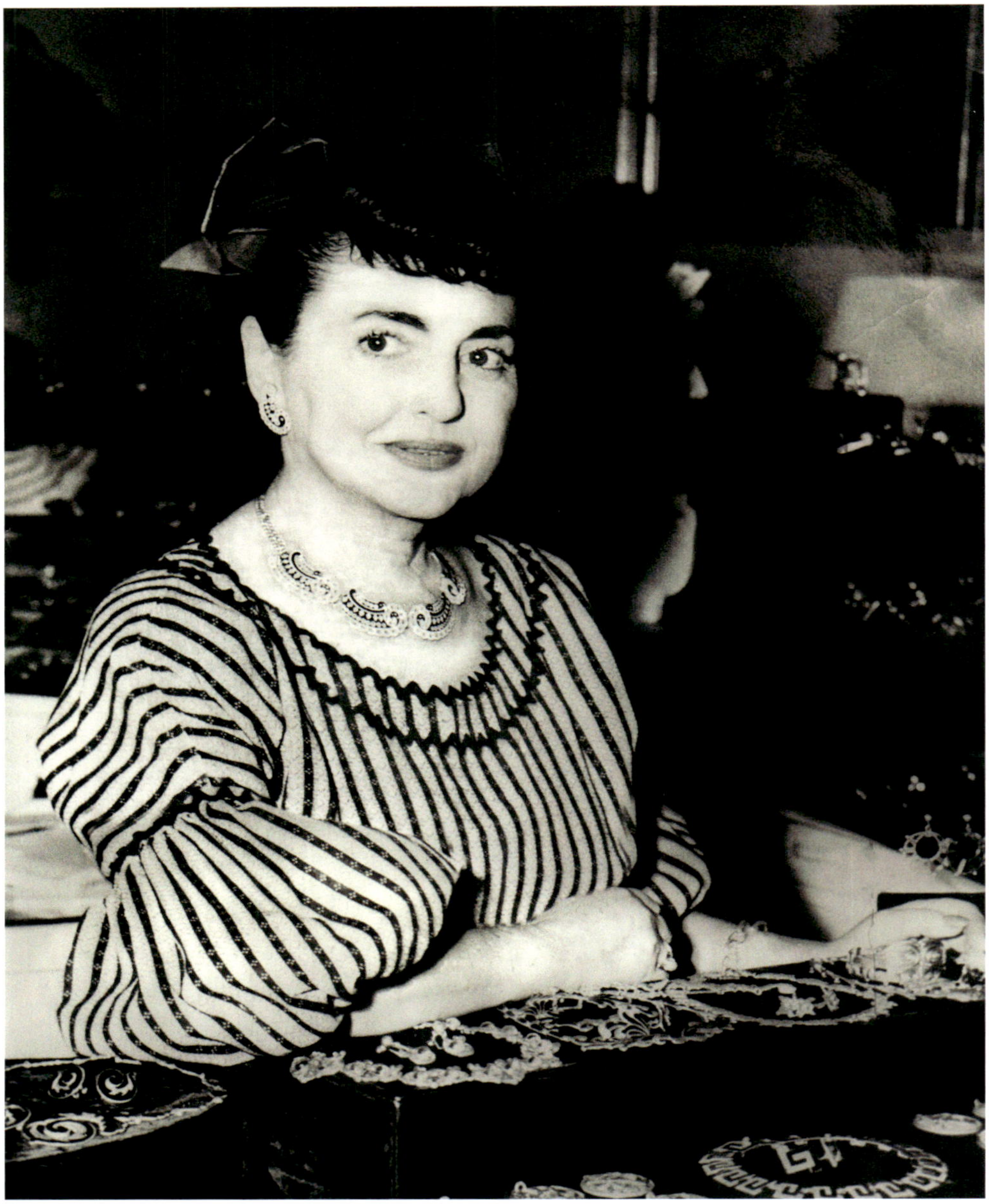

Margot Van Voorhies
The Art of Mexican Enamelwork

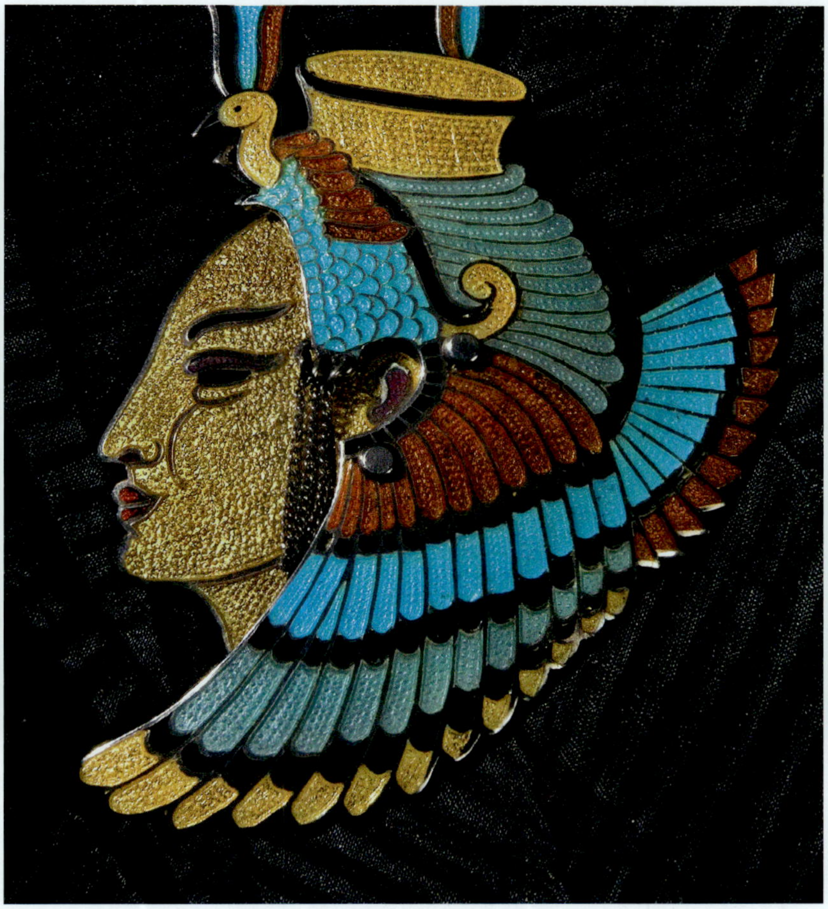

Penny C. Morrill

Photographs by Luisa Di Pietro

Elizabeth Finch, Research Assistant

Schiffer Publishing Ltd

4880 Lower Valley Road, Atglen, PA 19310

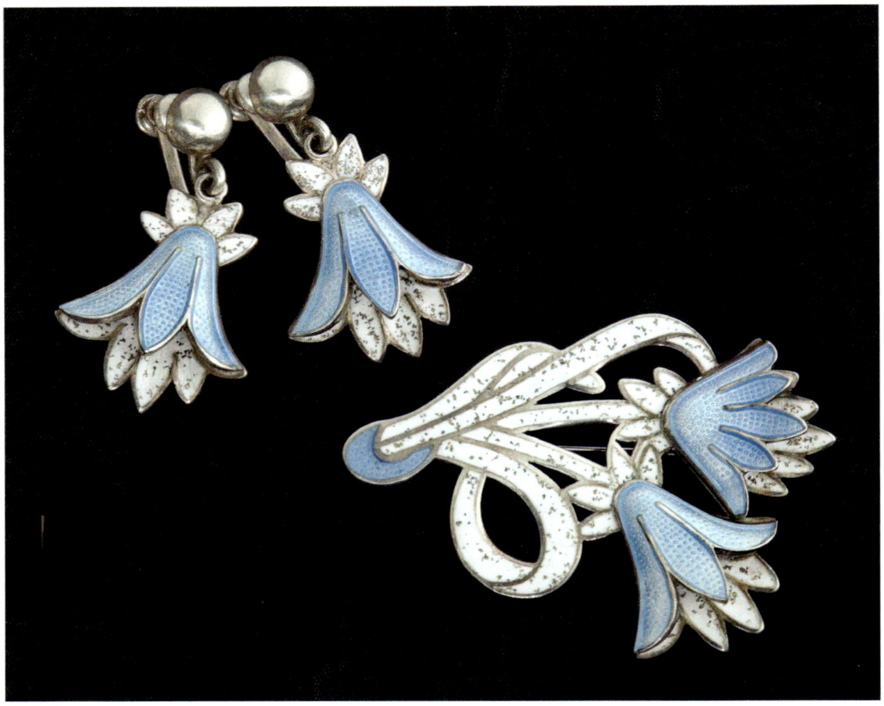

Library of Congress Control Number: 2010937141

FRONT COVER: Margot de Taxco 5644. *Fish in the Waves Set* in blue. bracelet 7 x 1.25 in.; brooch 1.75 x 3.5 in. Stella and Fred Krieger Collection
SPINE: Margot de Taxco 5914. *Z* brooch/pendant 2.25 x 1.88 in. J. P. Mather Collection
BACK COVER: Margot de Taxco 5554. *Serpent Set* in green mélange. necklace 16 x 0.63 in.; bracelet 7.5 in. d. Private Collection
Margot de Taxco 5560. *Eagle Knight Cigarette Box* 1.5 x 5 x 3.88 in. photograph provided by collectors. Marc Navarro and Gunther Cohn Collection
FRONTISPIECE: Margot van Voorhies in her store, photograph c. 1955, Rodríguez Family Collection
TITLE PAGE: Detail of 5242
ABOVE: Margot de Taxco 5837. *Blue Flower Set*. brooch/pendant 2.125 x 1.38 in.; earrings 1/5 x o.88 in. J. P. Mather Collection [1/4]
CONTENTS PAGE: Margot de Taxco 5628. *Butterfly of Many Colors* matching brooches 1.88 x 2.13 in. Elizabeth Finch Collection

Designed by John P. Cheek
Cover design by Bruce Waters
Type set in Benguiat Bk BT/New Baskerville BT
ISBN: 978-0-7643-3549-5
Printed in China

Schiffer Books are available at special discounts for bulk purchases for sales promotions or premiums. Special editions, including personalized covers, corporate imprints, and excerpts can be created in large quantities for special needs. For more information contact the publisher:

Published by Schiffer Publishing Ltd.
4880 Lower Valley Road
Atglen, PA 19310
Phone: (610) 593-1777; Fax: (610) 593-2002
E-mail: Info@schifferbooks.com

For the largest selection of fine reference books on this and related subjects, please visit our web site at
www.schifferbooks.com
We are always looking for people to write books on new and related subjects. If you have an idea for a book please contact us at the above address.

This book may be purchased from the publisher.
Include $5.00 for shipping.
Please try your bookstore first.
You may write for a free catalog.

In Europe, Schiffer books are distributed by
Bushwood Books
6 Marksbury Ave.
Kew Gardens
Surrey TW9 4JF England
Phone: 44 (0) 20 8392 8585; Fax: 44 (0) 20 8392 9876
E-mail: info@bushwoodbooks.co.uk
Website: www.bushwoodbooks.co.uk

Contents

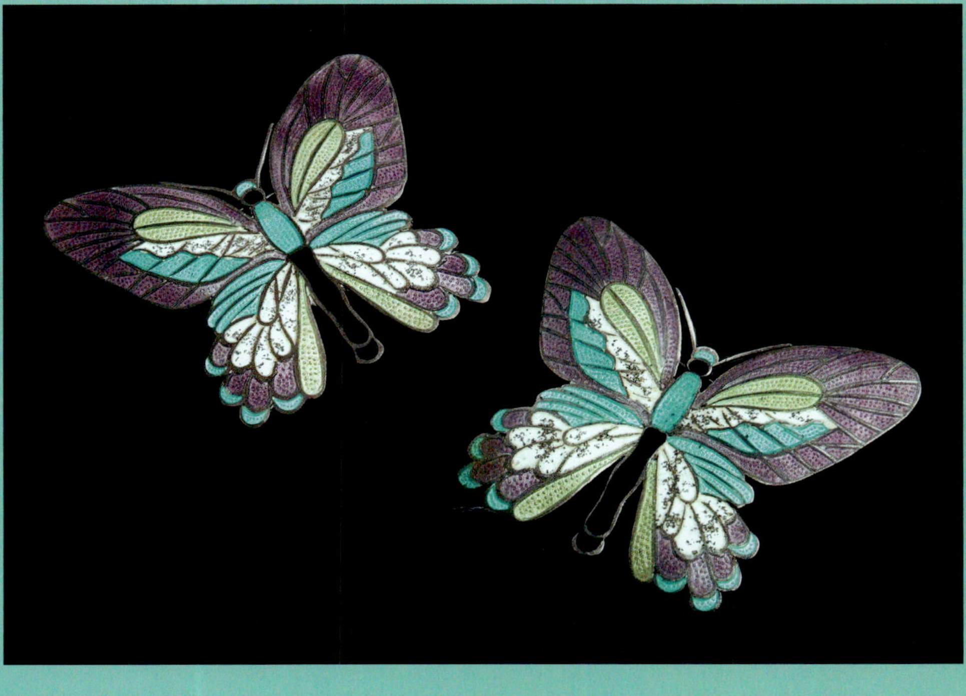

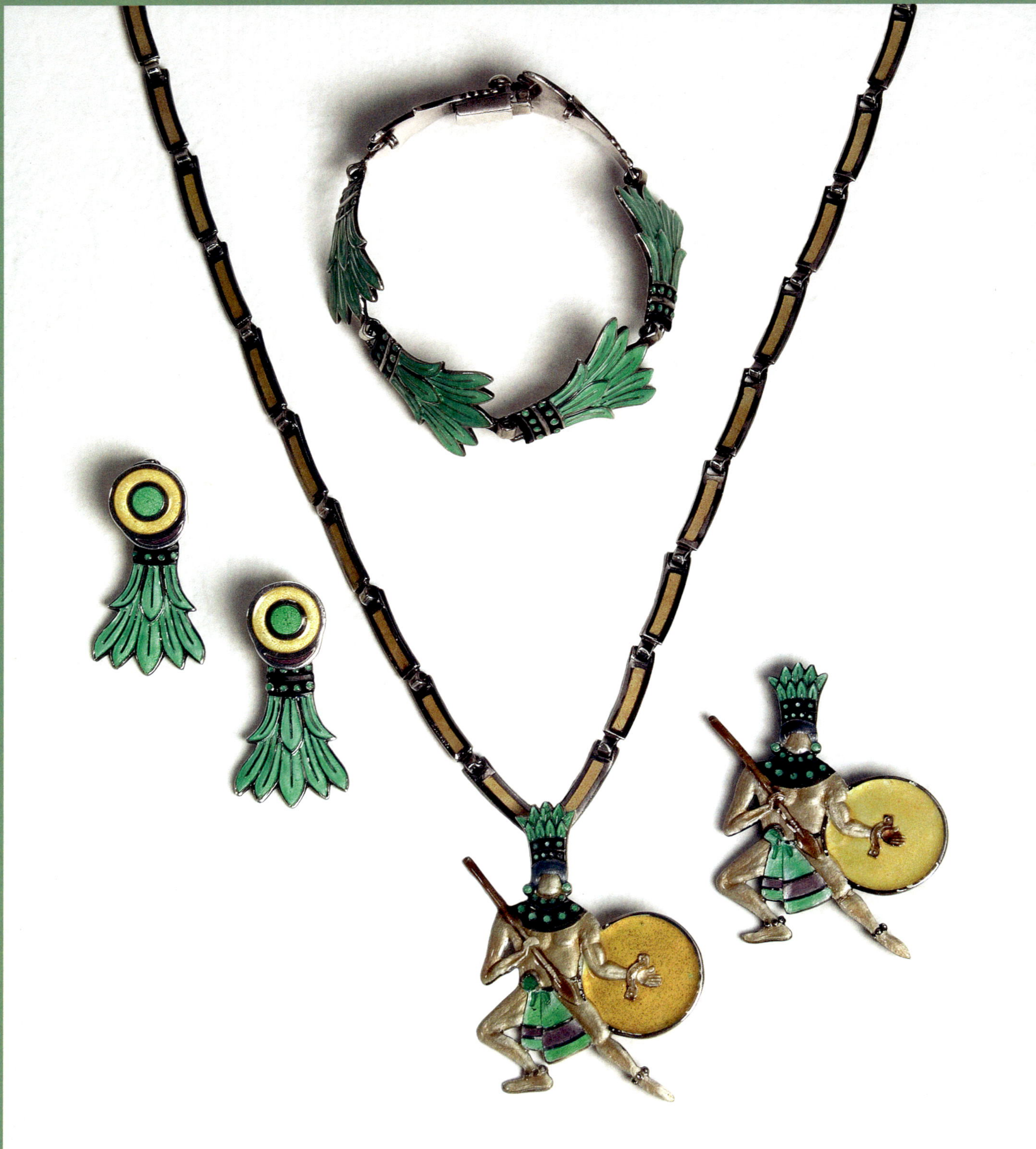

F-1 Margot de Taxco 5764. *Aztec Warrior Set*. necklace 20 x 0.12 in.; pendant/brooch 8.38 x 2 in.; bracelet 7.25 x 1.6 in.; earrings 0.6 x 0.88 in. Things Finer in Santa Fe Collection

Foreword

Author's note: The work of silver designer Margot van Voorhies has defied the passage of time. For collectors, Margot's jewelry designs will remain current because they are a reflection of her life story and of the depth of her aesthetic experience. While Margot drew from the art of Asia and the Americas, her oeuvre has a personal quality that is recognizable. Below are two short essays by collectors who greatly admire the artist, Margot de Taxco.

I have loved antique jewelry all my life. This wearable art reflects the spirit of the maker, the owners, and the time of its creation. I have a particular love of enamel work since, in my mind, this is more a reflection of the artist than stones can ever be. Enameled jewelry has been made for thousands of years, and all over the world. It is both fragile and easy to crack yet durable, since the material itself does not decay. About fifteen years ago, I walked into a shop and saw a bracelet with some of the most beautiful enamel work I had seen. The green with small patches of rainbow colors was unique and perfectly set off by the touches of matte black and the silver setting itself (5396, Plate 12). Then, when I tried it on, it was wonderful to wear, flexible and soft feeling. It was clear that the maker actually wore jewelry as well as designed it. This piece was my introduction to Margot de Taxco. Once introduced, I started looking for everything I could find about her and her work.

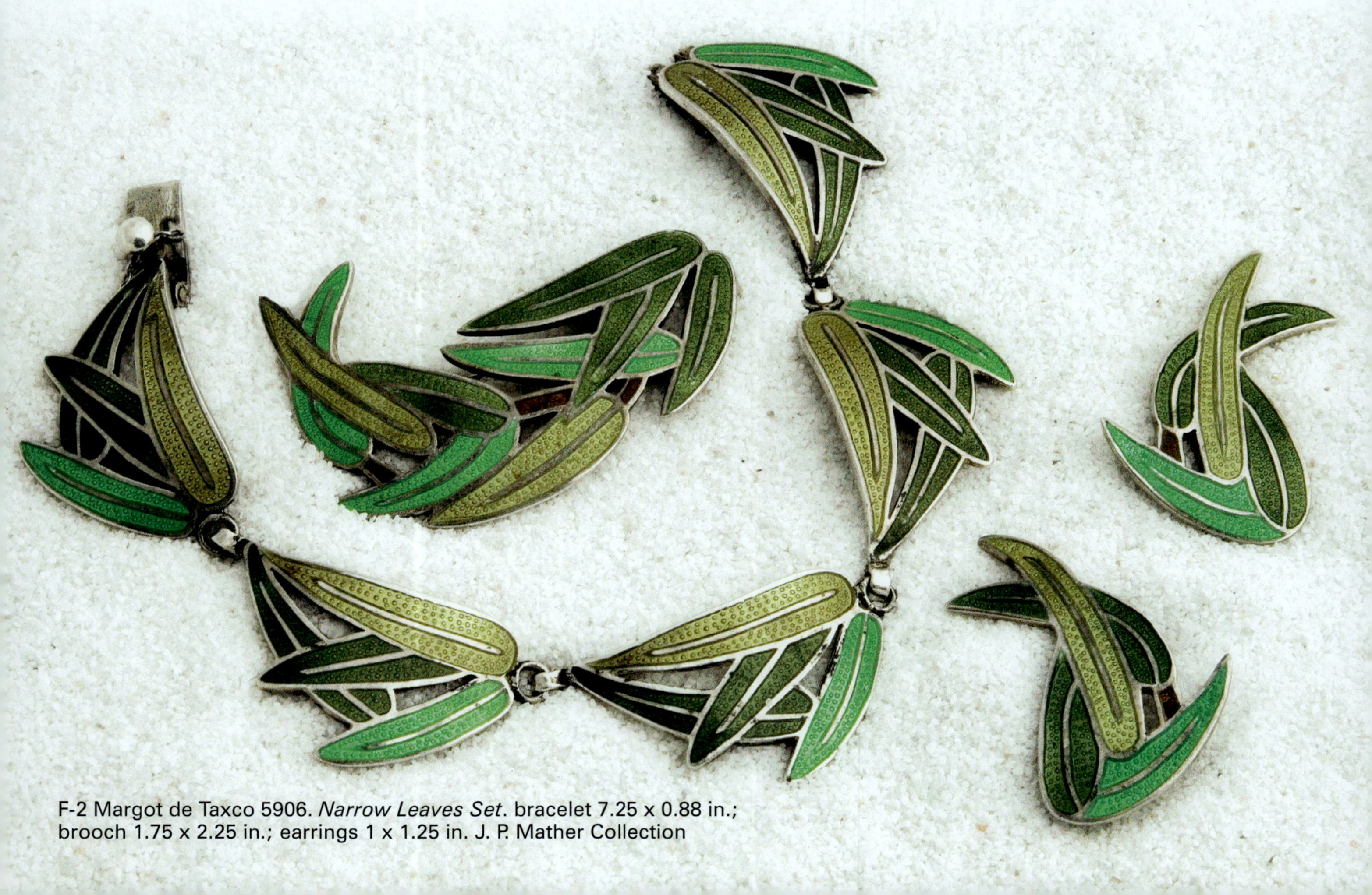

F-2 Margot de Taxco 5906. *Narrow Leaves Set*. bracelet 7.25 x 0.88 in.; brooch 1.75 x 2.25 in.; earrings 1 x 1.25 in. J. P. Mather Collection

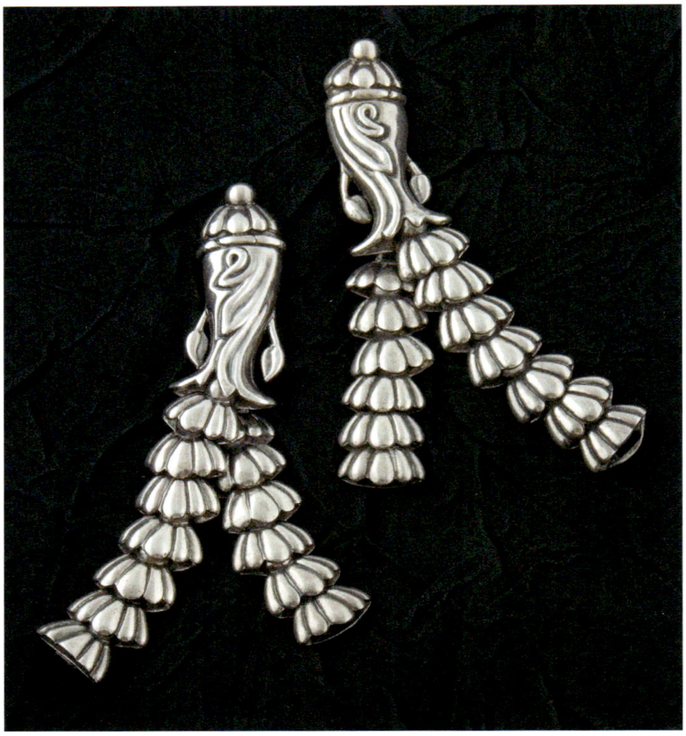

F-3 Los Castillo 181. *Pendulous Flower* brooches 4 in. Stella and Fred Krieger Collection

This, of course, led me to the story of the Taxco artists and Penny Morrill's other books on those master silver workers. I found, in looking through my jewelry, that a bracelet I had bought many years before in a pawnshop was by Fred Davis. My "Mexican Silver Collection" was started, and continues to grow with pieces from most of the Taxco Masters.

I still feel, however, that Margot's jewelry has a special place even in this very talented company. There is the obvious outstanding nature of her enamels. The colors and vibrancy are uniquely hers. But it is also the silver work itself that is special. The beautiful designs reflect a familiarity with a broad range of styles from Aztec and Maya forms to Art Nouveau and Art Deco — all transformed into timeless, wearable pieces. She clearly was a woman who combined aesthetic sensibility and creative talent with a real spirit of adventure. She obviously loved trying new things, materials, colors, and styles. I can personally relate to that spirit and it makes collecting her work exciting. I am very glad that this book is being written to celebrate this remarkable artist and help promote a broader appreciation of her work.

J. P. Mather

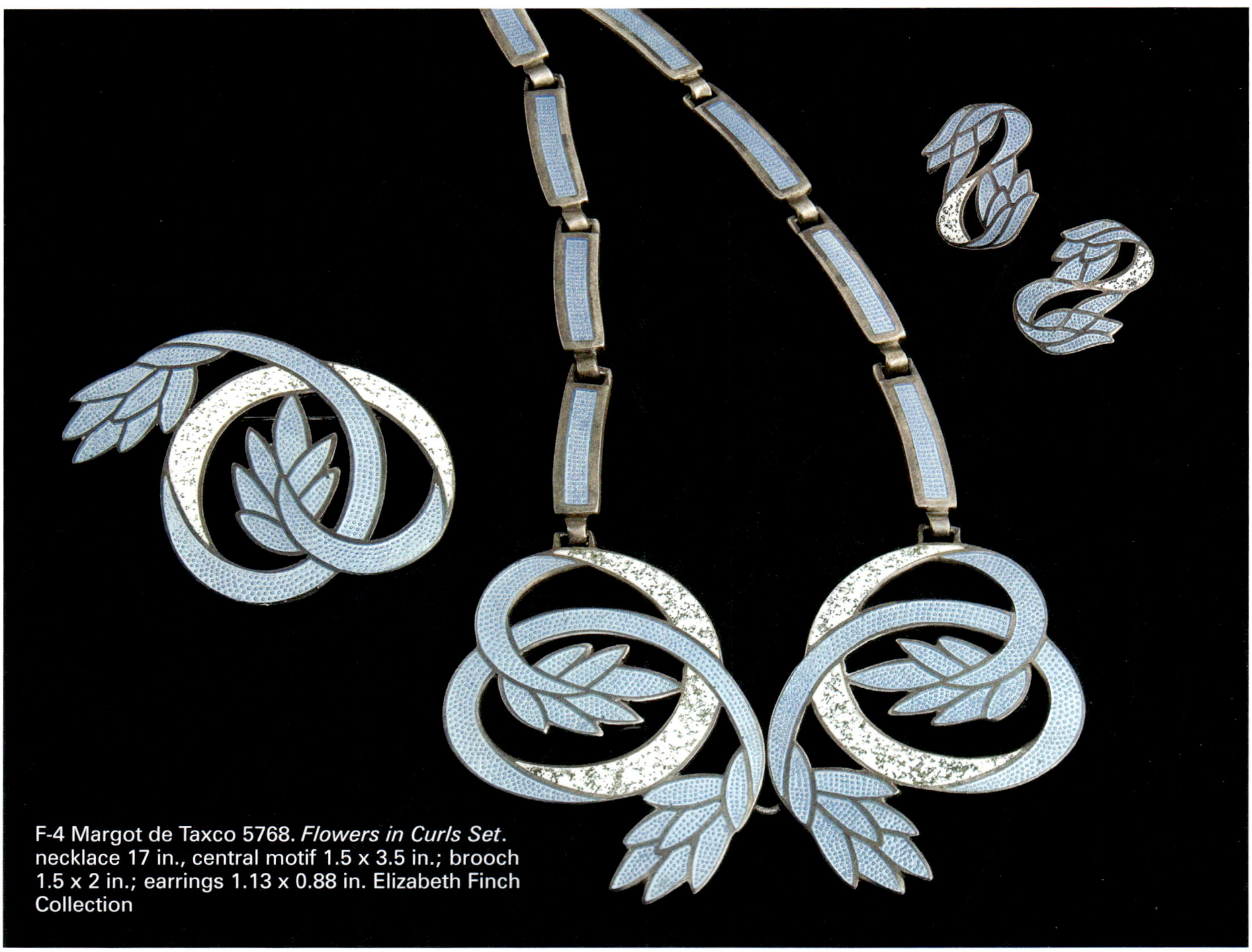

F-4 Margot de Taxco 5768. *Flowers in Curls Set*. necklace 17 in., central motif 1.5 x 3.5 in.; brooch 1.5 x 2 in.; earrings 1.13 x 0.88 in. Elizabeth Finch Collection

Overpowered by habitual passion. . . attracted by rarity, seduced by example, and inflamed by competition.

—Samuel Johnson described collecting perfectly in 1759.

The works of Margot de Taxco are especially entertaining and fun to collect and to wear. I bought my first piece of Margot at an antique show in Florida twenty years ago. It was a fish necklace, and I eventually found the earrings and matching bracelet. Later, I placed the set with another collector and used the funds to acquire another set that I wear more often. Sometimes the thrill is in the chase, and sometimes in the capture.

Margot created suites: parures, demi-parures, sets, suites, whatever you call them, collectively, the whole outfit. Women who decry "matchy-matchy" as outmoded and déclassé miss the central reason for buying a suite: we don't have to wear the pieces all at once, but we can choose to wear a selection, or mix and match with other sets, and the pieces will always work together. Sets always look coordinated, intentional, and stylish. The lure of the top to toe outfit can still be heady.

Lana Turner, Linda Darnell, and other Hollywood fashion icons of the forties and fifties accessorized their jackets and dresses with clusters of brooches, cuff bracelets, stylish collars and earrings. Animals, figures, roman numerals, Japanese, zodiac, and floral designs in silver and enamel display attention to detail and a sense of whimsy that distinguish Margot's work from that of others. Her revolutionary employment of women to hand fire the layers of enamel and gold champlevé marks Margot as a visionary and courageously independent thinker.

By today's standards the earrings are often a bit small, and the bracelets and brooches are large and striking. The generous cuffs make hands look more graceful and delicate; the brooches and asymmetry add interest to any jacket and display both humor and authority. A brooch changes a basic jacket from humdrum to snazzy in a single action.

So, if wearing Margot de Taxco jewelry is fabulous, can the thrill of acquiring it be far behind? Amassing a matching suite is like reuniting a family. And when reunited, the collective is far more compelling than the individual pieces (and more valuable). Worn with stark black, denim, a chic suit, or a simple linen top, Margot completes outfits with authority, humor, and style.

Elizabeth Pettus

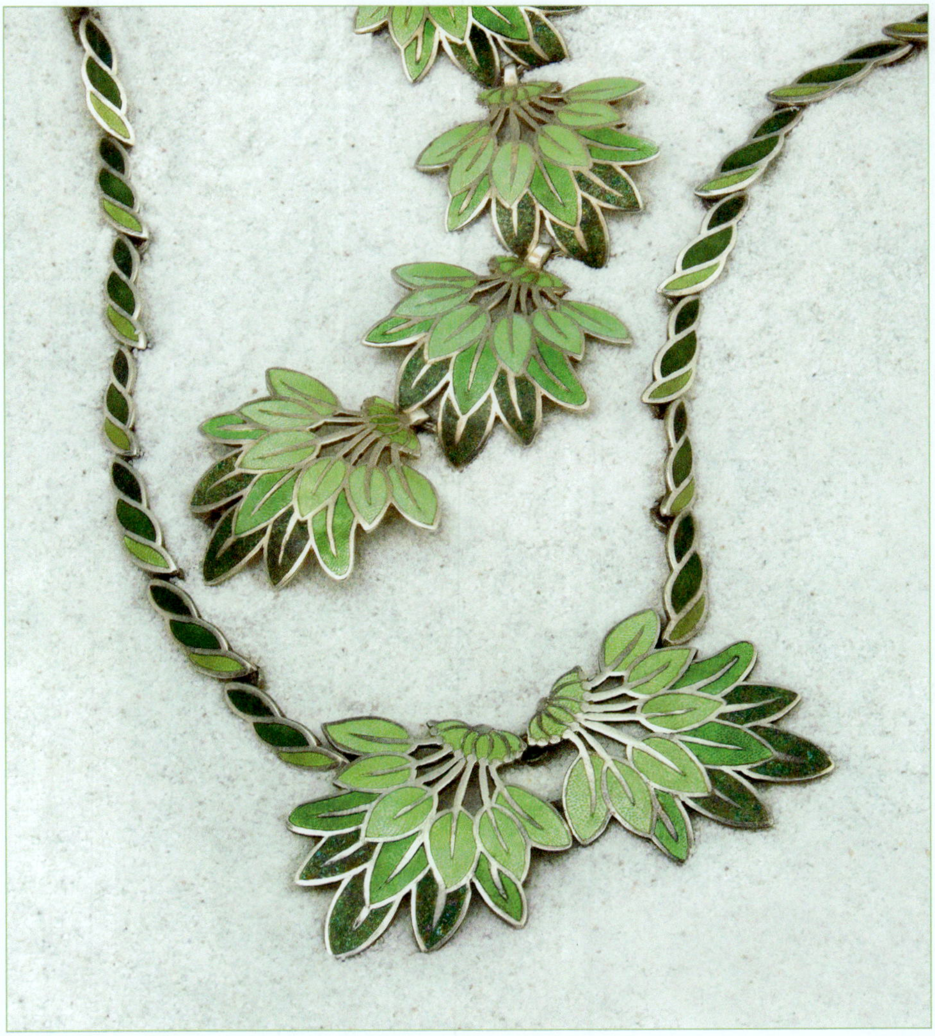

F-5 Margot de Taxco 5405. *Green Leaf Set*. necklace 16 in., central motif 2.25 x 3.75 in.; bracelet 7.5 x 2 in. J. P. Mather Collection

Ack-1 Margot de Taxco 5755. *Quetzal Set.* necklace 21 in.; pendant brooch 2 in.; bracelet 7 x 1 in.; earrings 1.5 x 0.5 in. Elizabeth Finch Collection

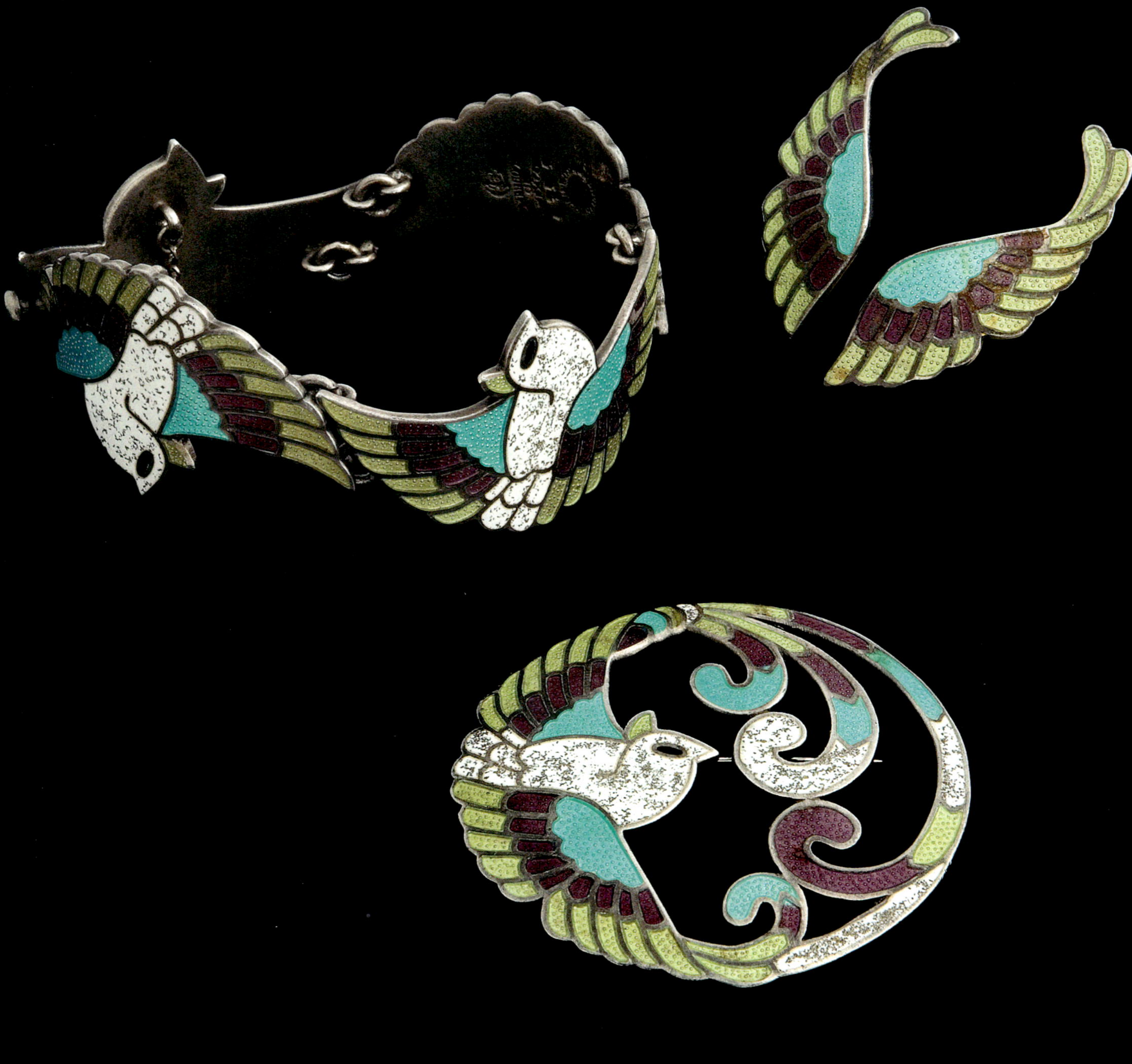

Acknowledgments

Accompanying Margot Van Voorhies along her life's path was a complex, enlightening, and at times, a very sad journey. There was much to discover about this strong woman, who was creative, generous, and brave when the moment demanded any of these responses. How fortunate, then, to have been able to share this adventure with several people who provided immeasurable assistance and support.

Through great effort and remarkable research skills, Elizabeth Finch has shed light on the world Margot inhabited before her arrival in Mexico. Nothing could have prepared us for what we discovered. Elizabeth's insights were of great value in interpreting the archival material and in telling the story. Elizabeth also appreciates and has much of consequence to say about Margot's artistry. Collectors and those with a general interest in Mexican silver are in her debt for her wonderful contributions.

Luisa Di Pietro is more than an accomplished photographer; she has remained a creative and enthusiastic partner in this project. Luisa is a perfectionist and an artist. With assistant Andrew Rivers at her side, she has produced the beautiful photographs that give life to Margot's designs.

Jaime Quiróz Duque, Siguifrido Pineda, María Pineda, and Melecio Rodríguez knew and admired Margot. They have told her story, each from a different perspective. We have been the recipients of a gift, for they have all interwoven their own life stories in the telling of Margot's. Additionally, I would like to extend thanks to Claudia Quiróz Millán and Melecio Villarreal, family members who were always willing to give their time.

Christie Romero has been of great assistance with questions concerning technique and materials used by Margot. Her recent death has been a personal loss for me, for she was a friend. Christie admired Margot's work, and the enthusiasm and support she shared with her friends and colleagues have now become a part of this book.

Several institutions played a vital role in developing Margot's biography. Perhaps the most important has been Tulane University's Latin American Library (LAL), where Margot's enamel design drawings now reside. I am deeply indebted to the director, Dr. Hortensia Calvo, for her friendship and support. I also want to thank David Dressing and Verónica Sánchez at LAL for their invaluable assistance, and Margarita Becerra Cano, graduate student at Tulane University.

Margot's life in San Francisco would have remained a mystery without the efforts of two stalwart researchers at the California Genealogical Society: Nancy Peterson and Jane Knowles Lindsey. Their interaction with Elizabeth Finch made it possible to go far beyond the original boundaries of Margot's biography. Irene Wainwright at the Louisiana City Archives division of the New Orleans Public Library was most helpful concerning research into the Voorhies family, as was Judy Bethea, researcher in New Orleans (sadly, no longer with us), and Voorhies family historians, Ann Nunn and Florence Christoph. Elizabeth Freeman, a cousin of Margot's, took an ac-

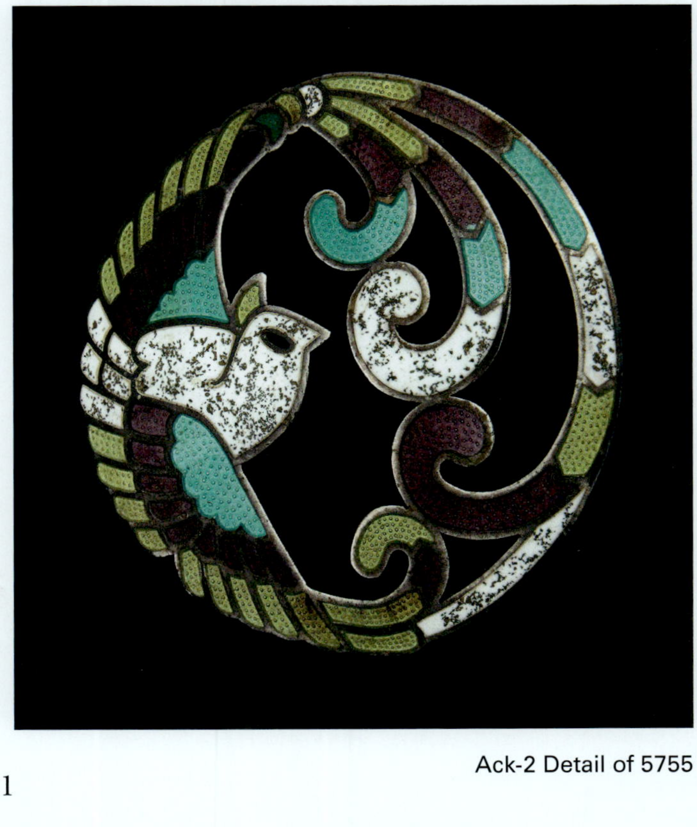

Ack-2 Detail of 5755

tive interest in the story, and brought enlightenment concerning the Chabot family. Assistance was also provided by the staffs at the University of Louisiana Lafayette Archives, the National Archives (especially Elizabeth Gray), the Bancroft Library, University of California, Berkeley, California, and at the Library of Congress Microfilm Division and the Hispanic Reading Room.

Authorities in several fields were able to assist in the identification of elements in Margot's work that required their expertise. Many thanks to Dr. Robert De Caroli, Director of Art History at George Mason University, and James T. Ulak, Deputy Director of the Freer Gallery of Art and the Arthur Sackler Gallery, Smithsonian Institution, Washington, D.C. for help with the Asian ivory figurines. I also want to thank Patricia H. Svoboda, Research Coordinator, and Linda Thrift, Director CEROS, Center for Electronic Research and Outreach Services, National Portrait Gallery, Smithsonian Institution, for their interpretation of the lovely portrait miniature (5780A, Plate 1). Melanie A. Marts, G.G. (G.I.A.) was most helpful with the identification of stones in Margot's jewelry. Carol Leake, Assistant Professor of painting and drawing, College of Music and Fine Arts, Loyola University, New Orleans, Louisiana, studied the Margot design drawings and identified them as gouaches. My thanks also go to Nancy Schiffer, a supportive editor.

I owe a special debt of gratitude to those individuals and families who have lent works to be photographed for the book. It has been a wonderful experience to meet and collaborate with people

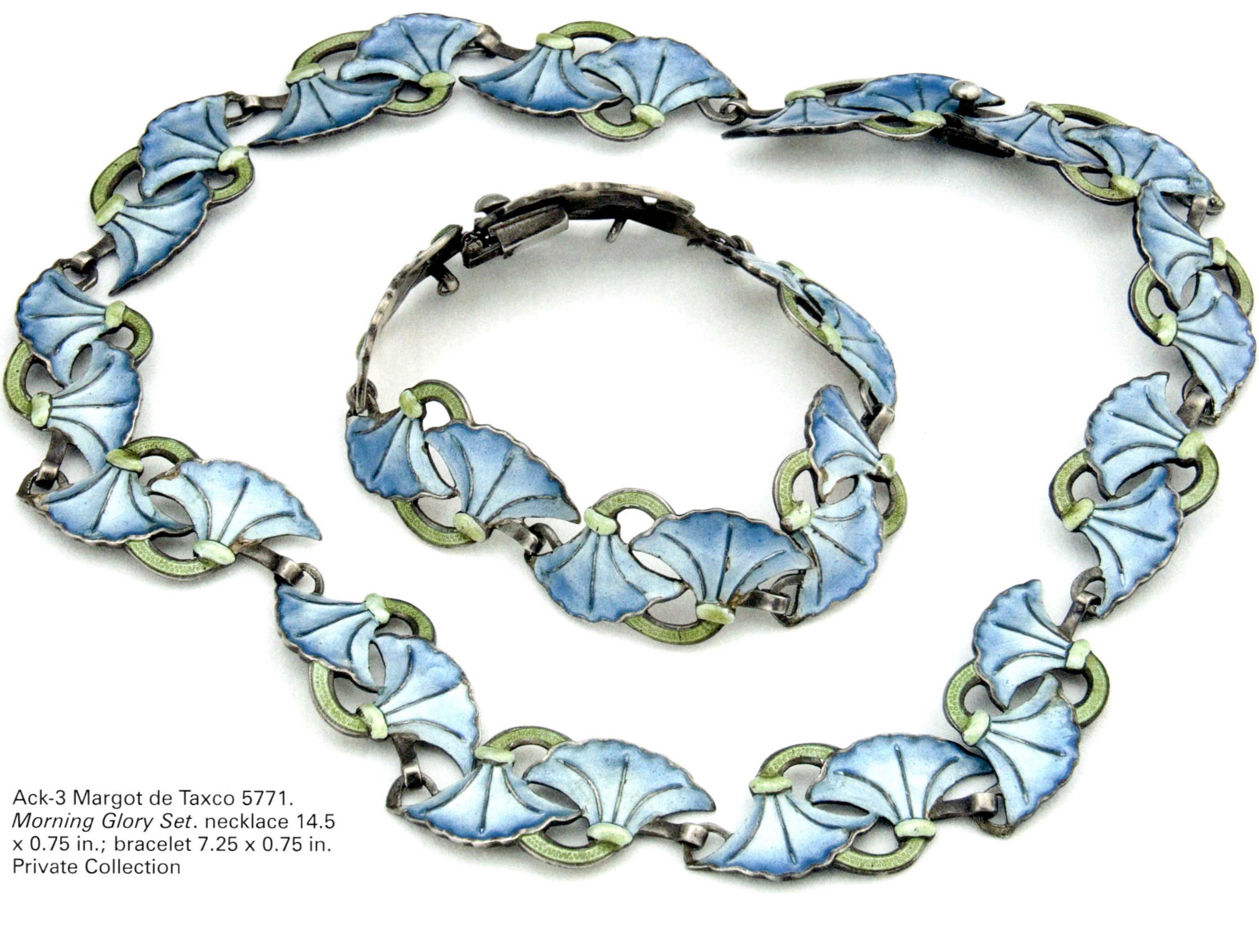

Ack-3 Margot de Taxco 5771. *Morning Glory Set*. necklace 14.5 x 0.75 in.; bracelet 7.25 x 0.75 in. Private Collection

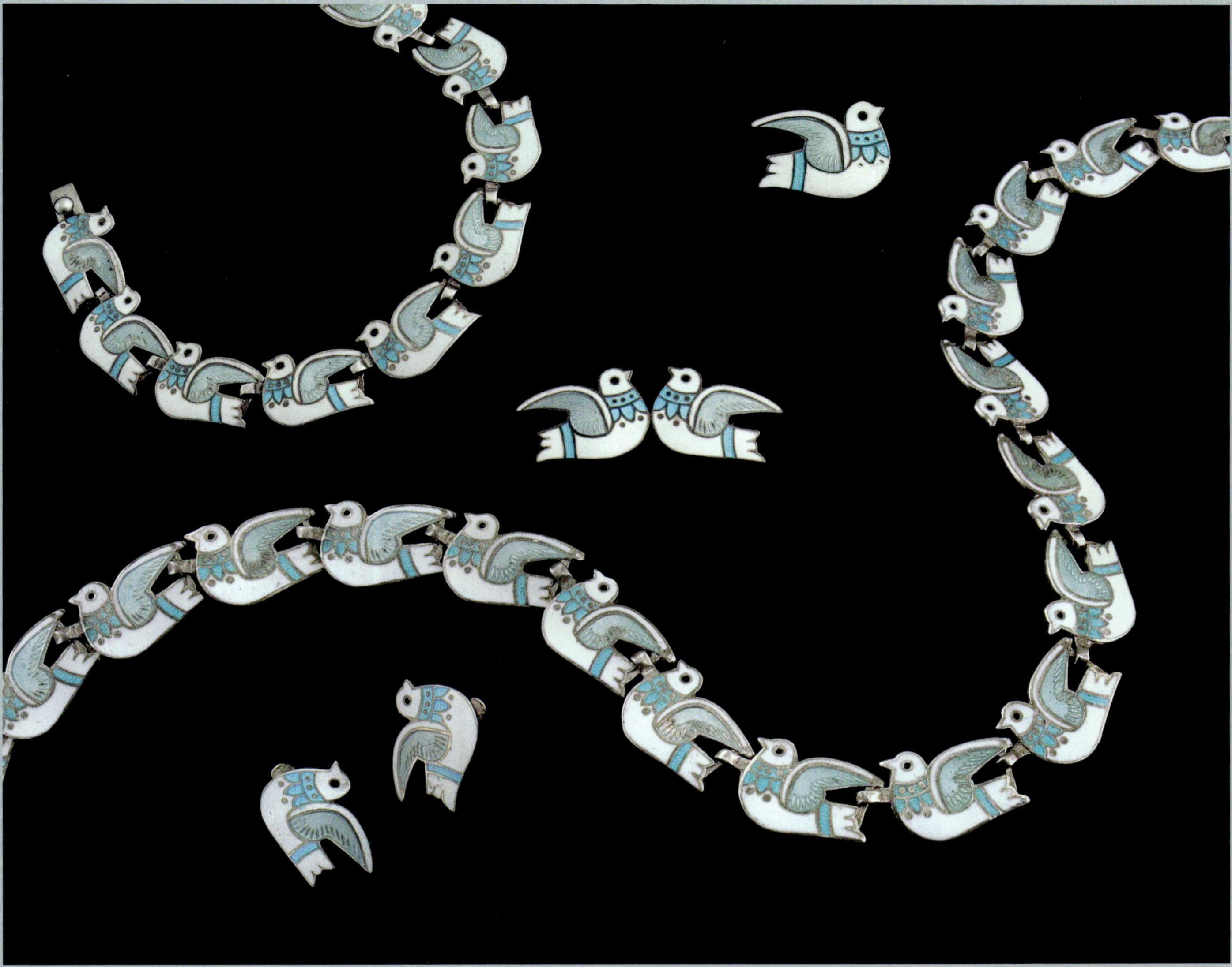

Ack-4 Margot de Taxco 5868. *Folk Art Doves Set*. necklace 14 in.; bracelet 7.75 x 0.63 in.; double bird brooch 5/6 x 1.75 in.; earrings and single bird brooch 5/6 x 7.8 in. Elizabeth Finch and Private Collections

who are remarkable in their own right, and who also happen to be enthusiastic collectors of Margot's work. I offer my thanks to Linda Affleck, Katherine Armitre and Stephen Etre, Ron Belkin, Carole Berk, Lona Bjornsen, James Black and James Tracy, Adrian Brun and his daughter Aline Brun, Priscila Canales, Byron and Jill Crawford, Luisa Di Pietro, Andrea Fatell, Elizabeth Finch, Mary Jane Fletcher, Elizabeth Freeman, Susan Toomey Frost, Germán Fuentes Flores and Graciela Jácopo de Fuentes, Rocio Sierra-Honigmann, Patrick Kapty, Stella and Fred Krieger, J. P. Mather, Susan McCloskey, Marc Navarro and Gunther Cohn, Sheila Pamfiloff, María Pineda, Miguel Pineda, Ellen Poss, Jaime Quiróz Duque and the Quiróz family, Orlando Rigono, Melecio Rodríguez, Stella Rubin, Ilon Specht, Judy P. Steele, Gabrielle Stodd, Elizabeth and Robert Pettus, Pam Tournier, Tulane University Latin American Library's Spratling-Taxco Collection, Violante and Consuelo Ulrich, Marlene and James Vitanza, Betsy Vought, and the Zamora/Carbajál family.

To my family, Jackson, Nicole, Julia, and Jesse, and to my two sisters, Susan and Mary Jane, many thanks for your enthusiasm and support. The biggest cheerleader has been my husband Jim. His favorite designer is Margot, so he was delighted when I made the decision to write her biography. He has shared expenses, taken phone messages, and done a bit of editing. I am most appreciative.

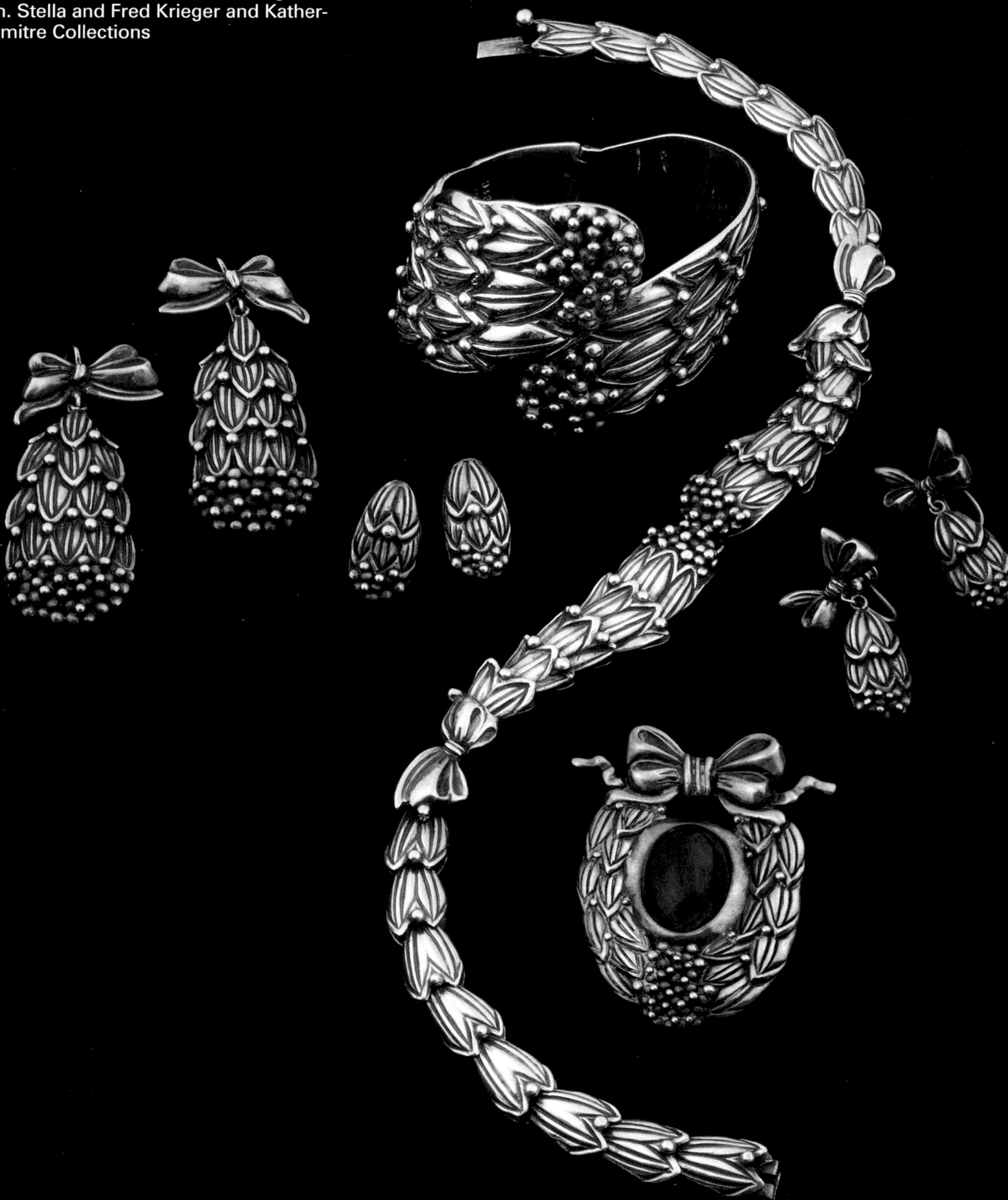

Margot de Taxco 5316. *Corn Flower Set*. necklace 15.5 x 0.6 in.; cuff bracelet 6.5 x 2.5 in.; brooch with amethyst 3.25 x 1.5 in.; large earrings 2.75 x 1 in.; small earrings with bows 1.5 x 0.75 in.; small earrings 1 x 0.75 in. Stella and Fred Krieger and Katherine Armitre Collections

Infatuated with Color: Twenty Designs

Margot van Voorhies was a celebrated jewelry designer in the mid-twentieth century. One of the early protagonists in the Taxco, Mexico silver business, she and her husband Antonio Castillo built upon William Spratling's initial success to create Los Castillo in 1939. This workshop and storefront remained synonymous with innovation for more than fifty years. Margot instituted a second workshop in 1947, "Margot de Taxco," where she developed her unique designs in *champlevé*. While collaborating with Antonio and then later on her own, Margot provided employment for hundreds of artisans and highly skilled *maestros* [master silversmiths]. Additionally, Margot de Taxco was one of the only workshops in Taxco to employ women.

While Margot strived for and achieved success, the reality of her life is barely credible. The challenges she faced were so difficult they impelled her to drastically alter the direction of her life not once but several times. These dramatic events drove her forward and defined her, while her strong will and sense of purpose sustained her in the worst of times.

Margot is quoted as having said, "El día que llueva sopa, nos va a tocar tenedores a tí y a mi." ["The day it rains soup, you and I will end up with forks."] Margot must have felt the wearying presence of misfortune, and yet her life was not characterized by defeat. Instead, Margot's story is compelling because she remained optimistic and continuously looking forward toward the future. Her special qualities and strengths kept her from degenerating into self-focus. While she had this marvelous strength for survival, Margot did allow herself, unfortunately, to be trusting of others to the point of vulnerability, a virtue that became a tragic flaw.

Margot was primarily an artist who was equally accomplished as a creative entrepreneur and businesswoman in a foreign and male-dominated world. She instituted innovative approaches that enhanced the business of producing silver jewelry. Her creativity and that of her contemporaries, William Spratling, Valentín Vidaurreta, the Castillo brothers, Salvador Terán, and Siguifrido Pineda, defined what was within the realm of the possible. Unbounded by tradition, their daring approach to design led to the development of an industry in which experimentation became the means of success.

Margot can be considered one of the most productive silver artists of Taxco, for her designs number in the hundreds. Many of the necklaces and bracelets produced in the early years with Los Castillo continued in production for decades, long after she had left the company. Serious collectors study and admire her work for its beauty and diversity. How fortunate then that Margot's artistry in silver and enamel can be a validation of her triumphs and remain as her legacy.

Twenty designs have been chosen as particularly revealing of Margot's creative genius. Each of these special pieces is accompanied by the original design drawings and related works that amplify an appreciation and understanding of Margot's design process.

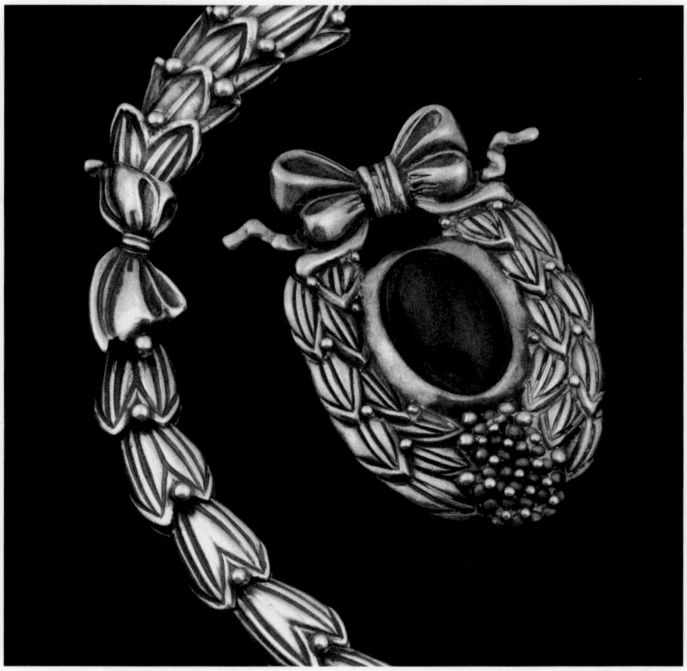

Detail of 5316

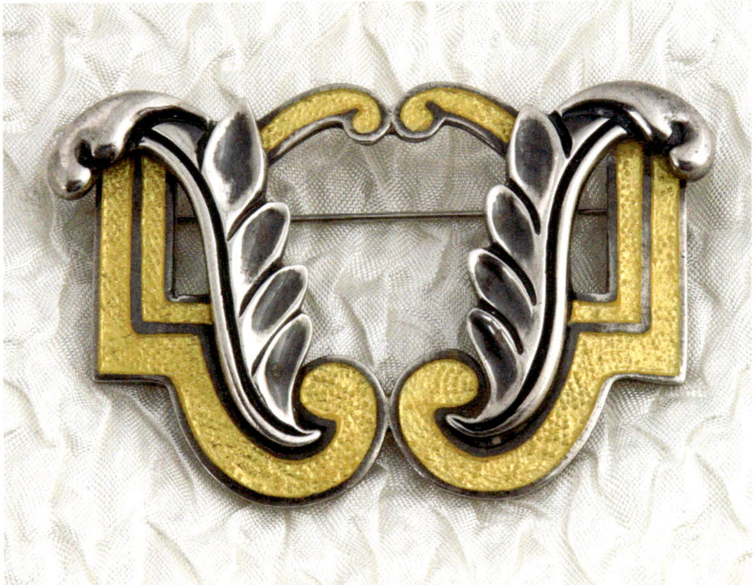

Plate 1 Margot de Taxco 5389. *Leaf & Scroll* in a square brooch 1.5 x 2.25 in. Elizabeth Finch Collection

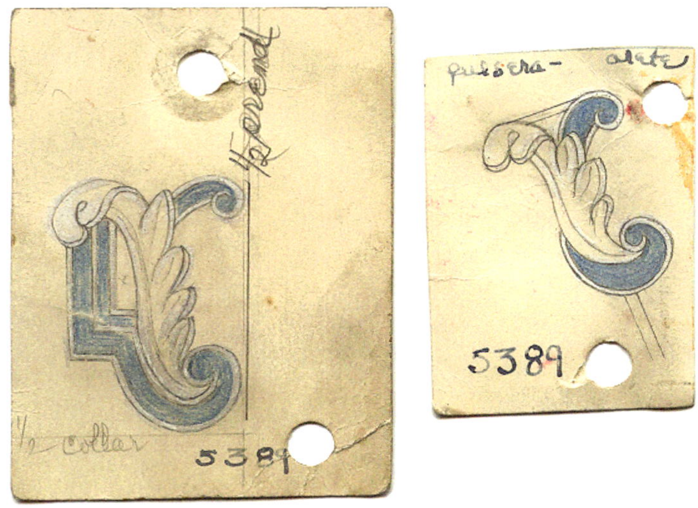

Drawings 5389 *Leaf & Scroll*

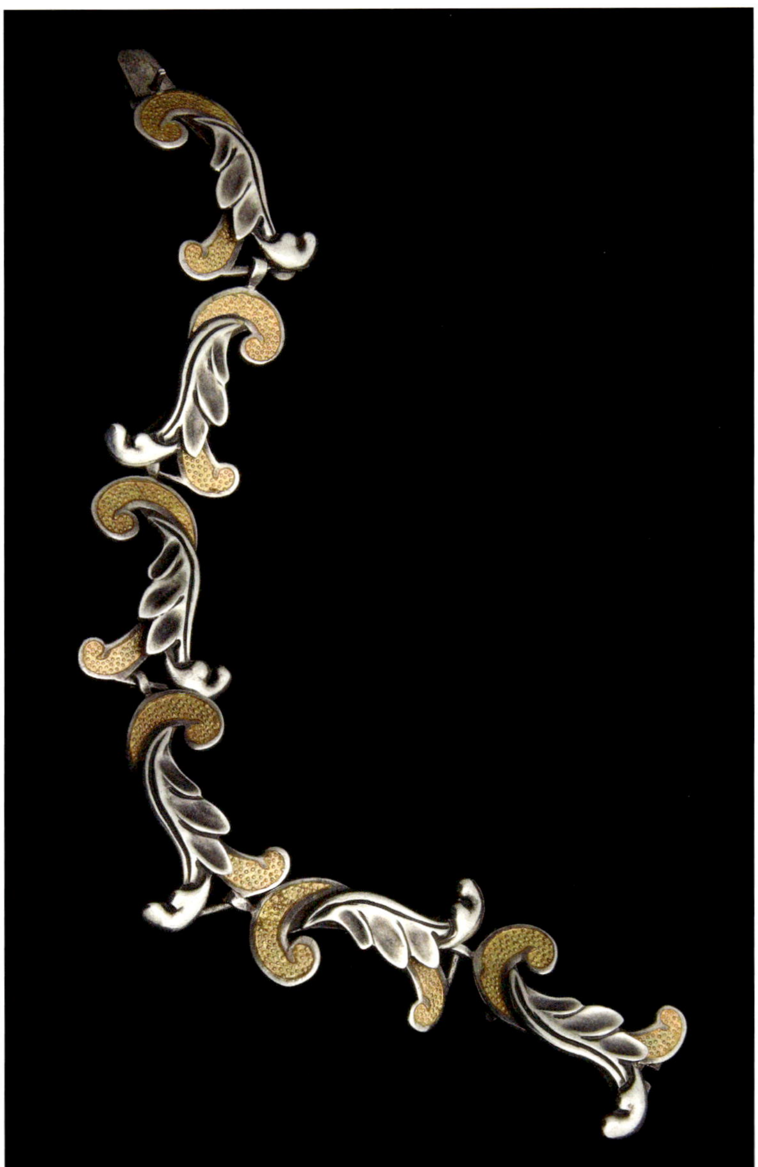

Plate 2 Margot de Taxco 5389. *Leaf & Scroll* in yellow bracelet 7.5 x 1 in. J. P. Mather Collection

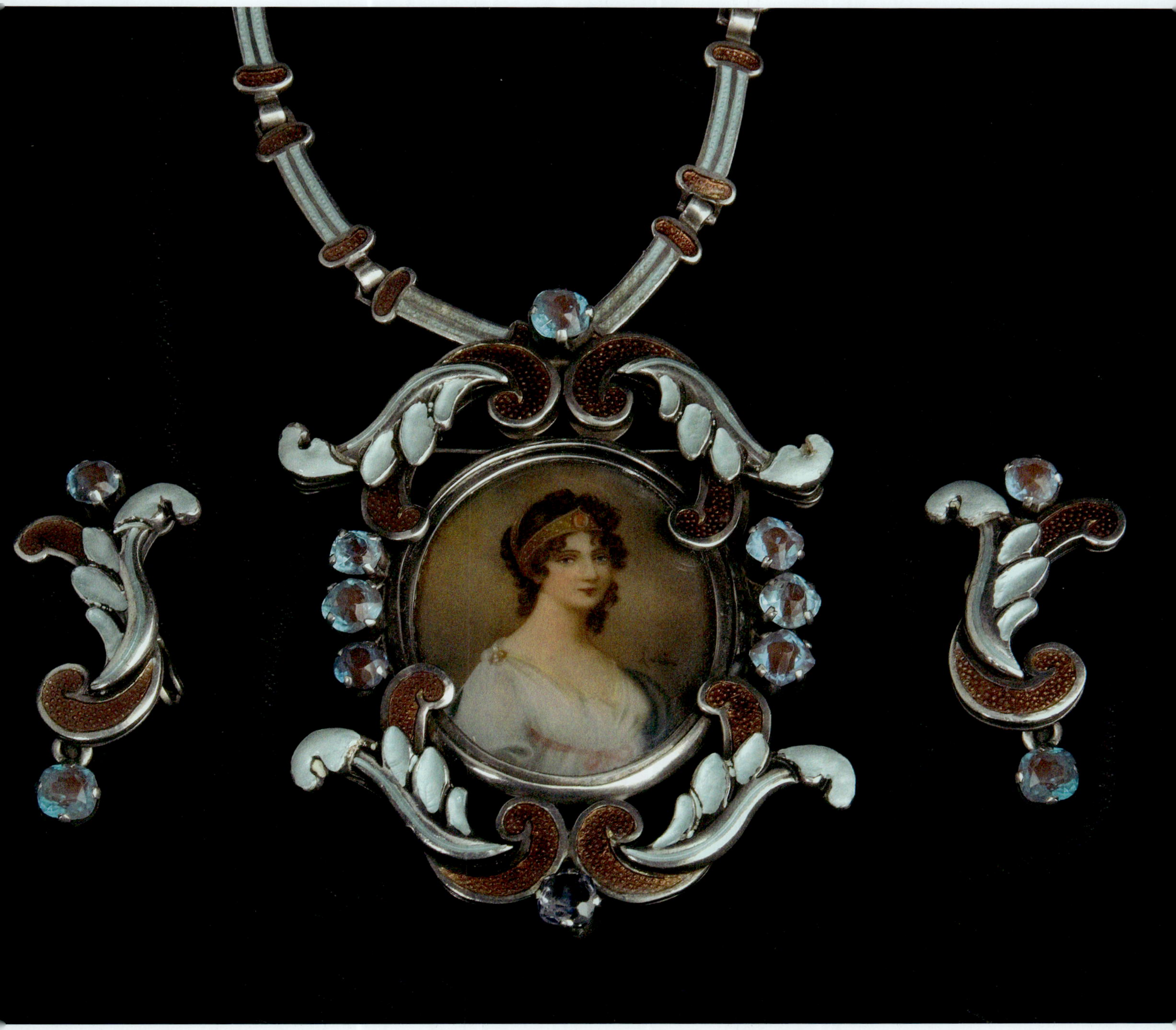

Plate 3 Margot de Taxco 5780A. *Portrait Set*. necklace 18.5 in., central motif 3.25 x 2.5 in.; earrings 1.75 x 1 in. J. P. Mather Collection

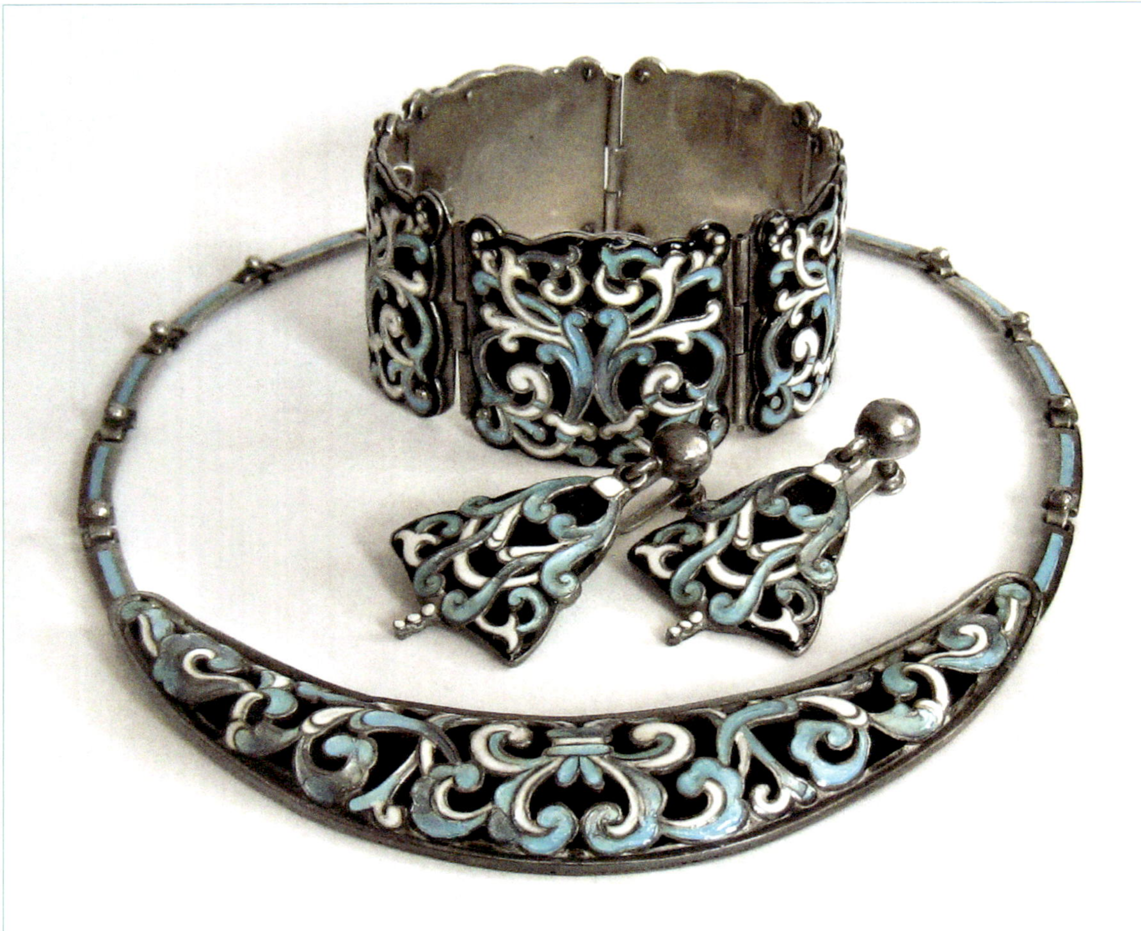

Plate 4 Margot de Taxco 5583. *Classical Scrolls Set.* necklace 15 in.; bracelet; earrings. photograph provided by collector. Gabrielle Stodd Collection.

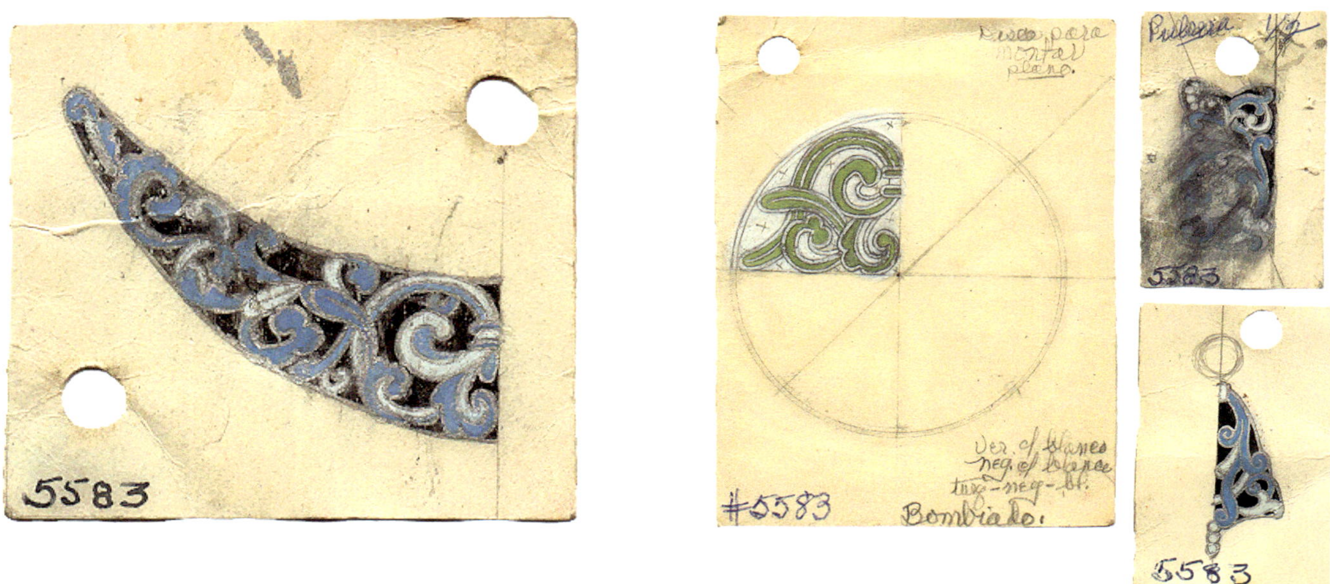

Drawings 5583 *Leaf Curls* and *Scrolls In Circle*

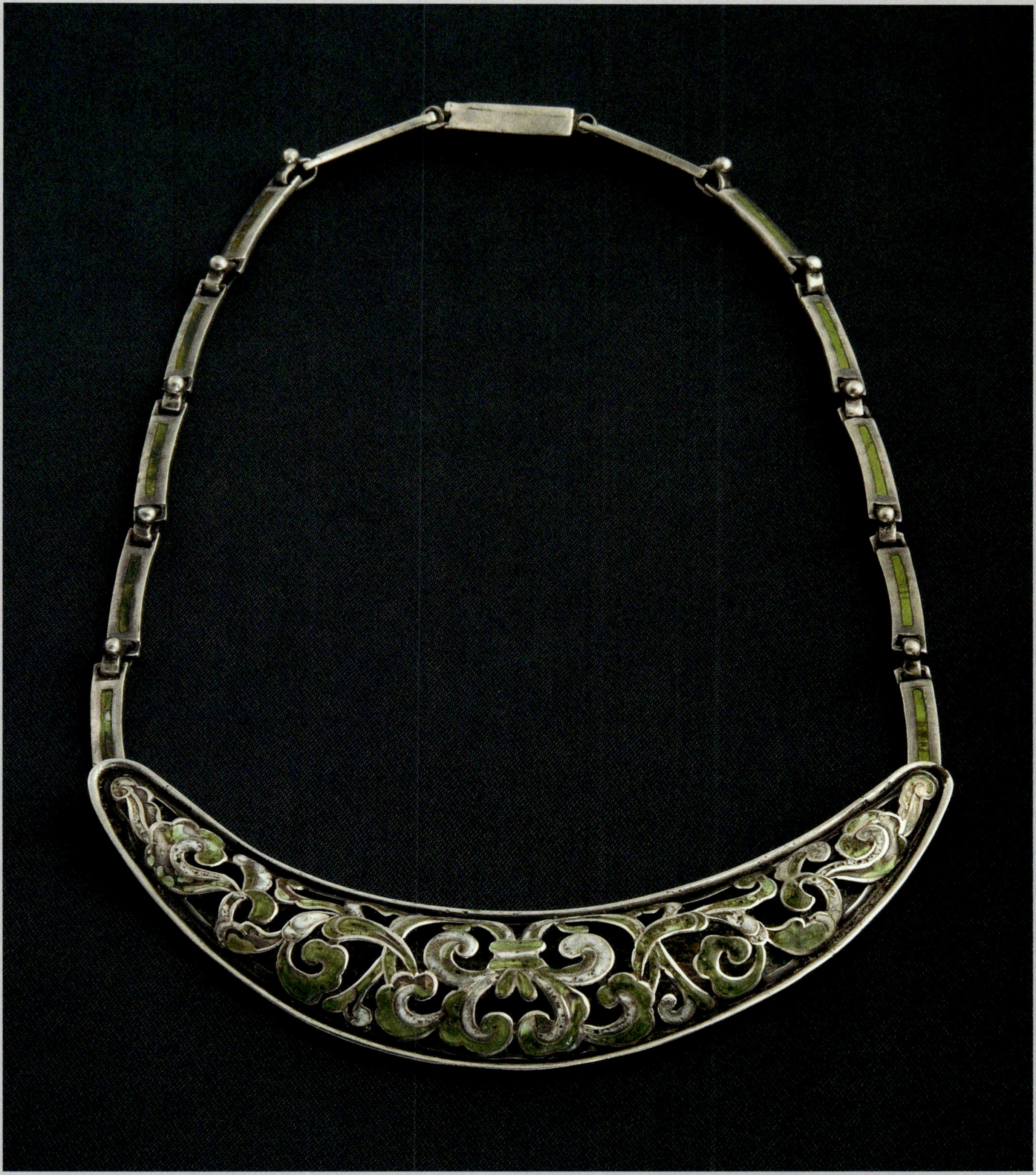

Plate 5 Margot de Taxco 5583. *Classical Scrolls* necklace 15 in., central motif 0.88 x 4 in. Elizabeth Finch Collection

Plate 6 Margot de Taxco 5685. *Chrysanthemum Sets*. necklaces 15 in., central motifs 3 x 1.63 in.; bracelet 7.75 x 1.63 in.; earrings 1.5 x 1 in. Margot de Taxco 5851. earrings 1.25 x 1 in. J. P. Mather Collection

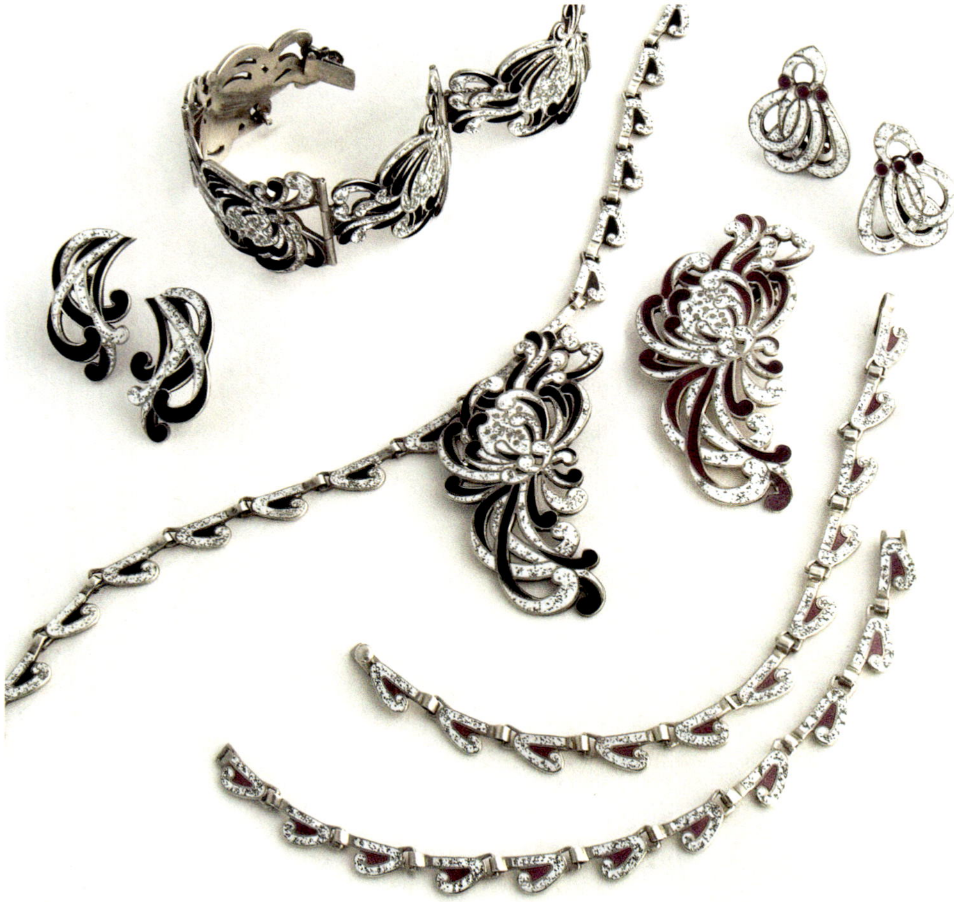

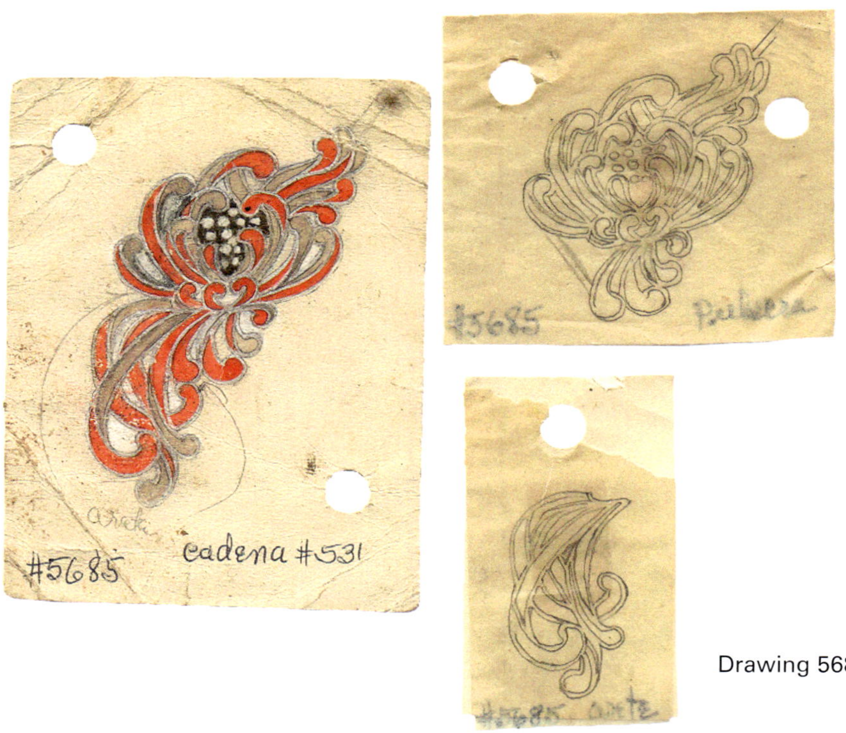

Drawing 5685 *Chrysanthemum*

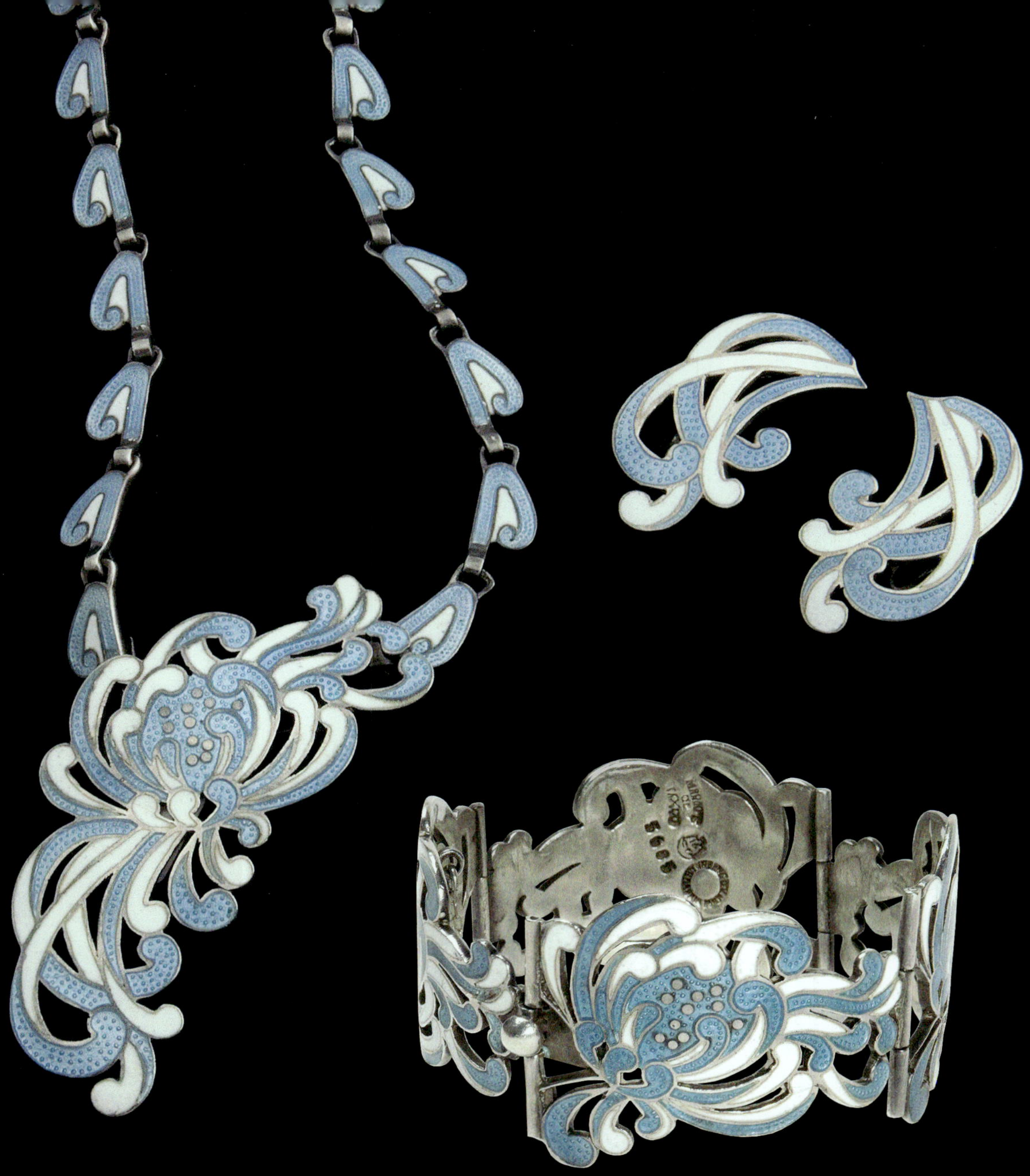

Plate 7 Margot de Taxco 5685. *Chrysanthemum Set* in blue and white. necklace 15 in., central motif 3 x 1.75 in.; bracelet 7 x 1.5 in.; earrings 1.5 x 0.75 in. Betsy Vought Collection

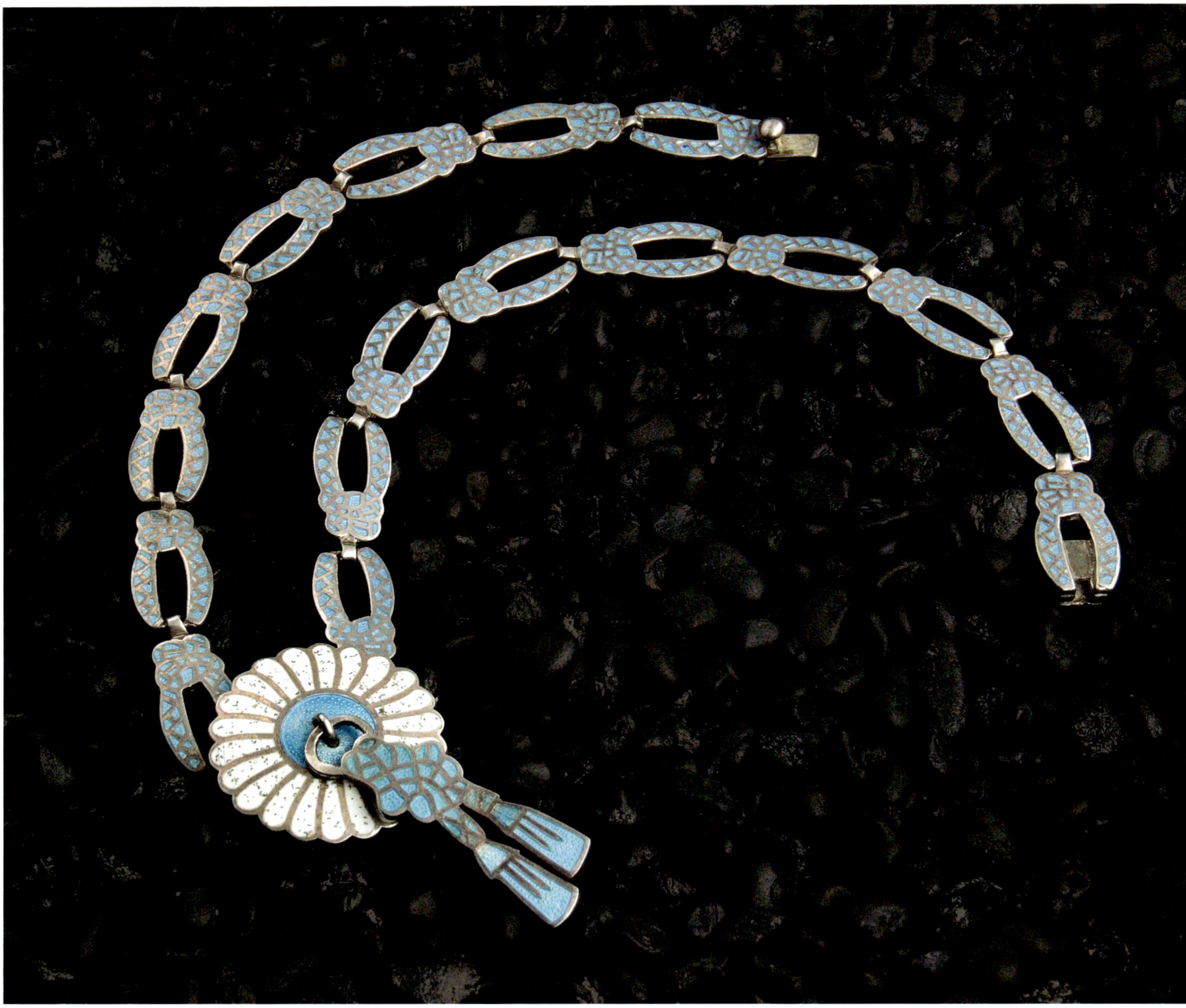

Plate 8 Margot de Taxco 5774. *Flower with Tassel* (flower from 5669) necklace 15 in.; central motif 2.25 x 1.25 in. Private Collection

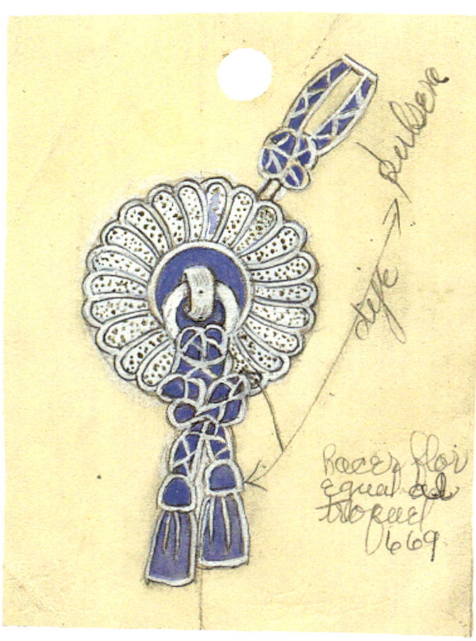

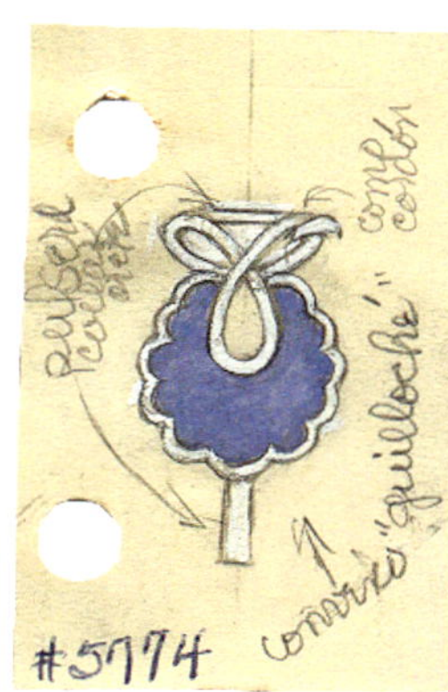

Drawing 5774 *Tassels*

Plate 9 Margot de Taxco 5774. *Delicate Textures* in yellow. necklace 16.5 x 1 in.; bracelet 8.25 x 0.75 in. Elizabeth Finch Collection

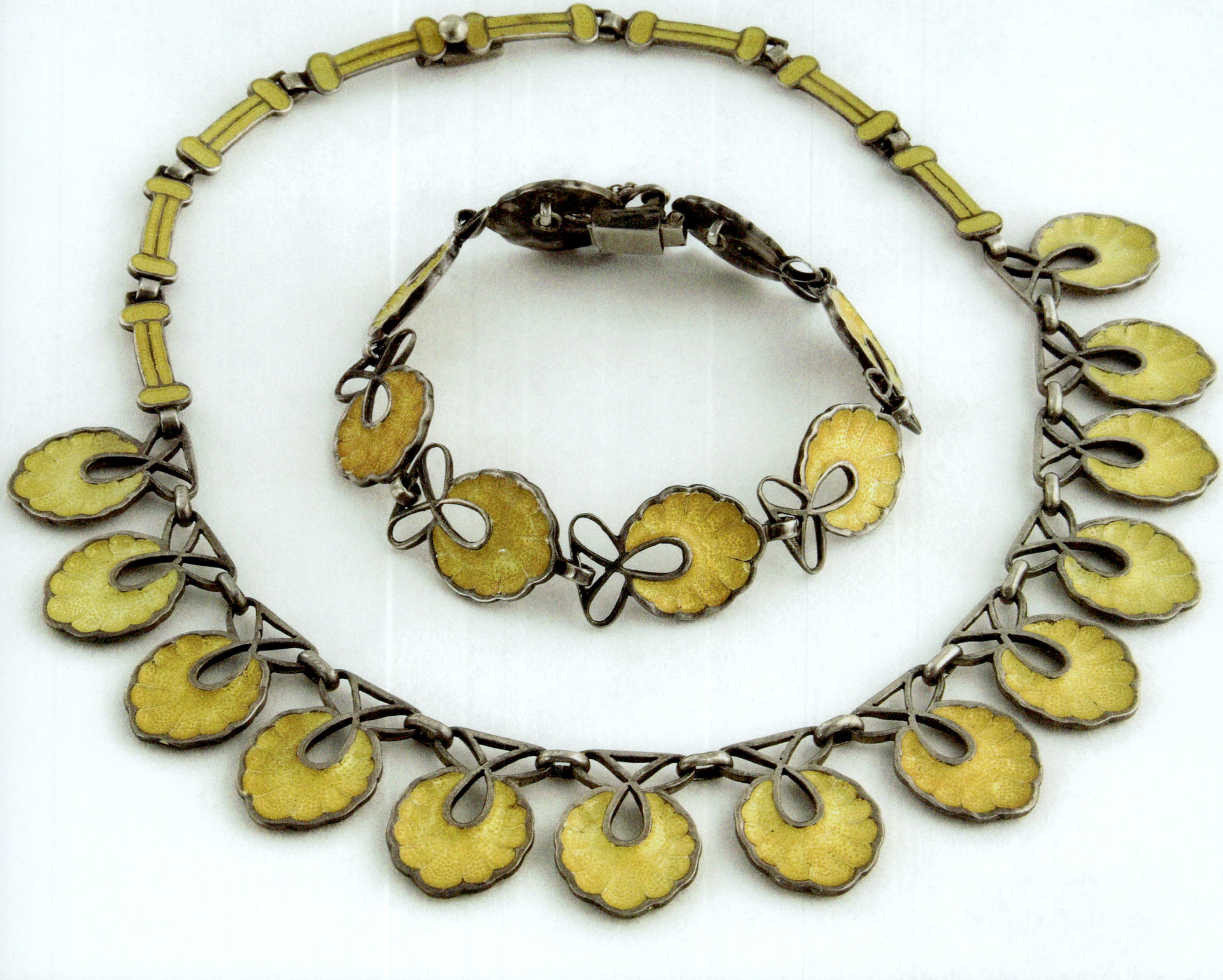

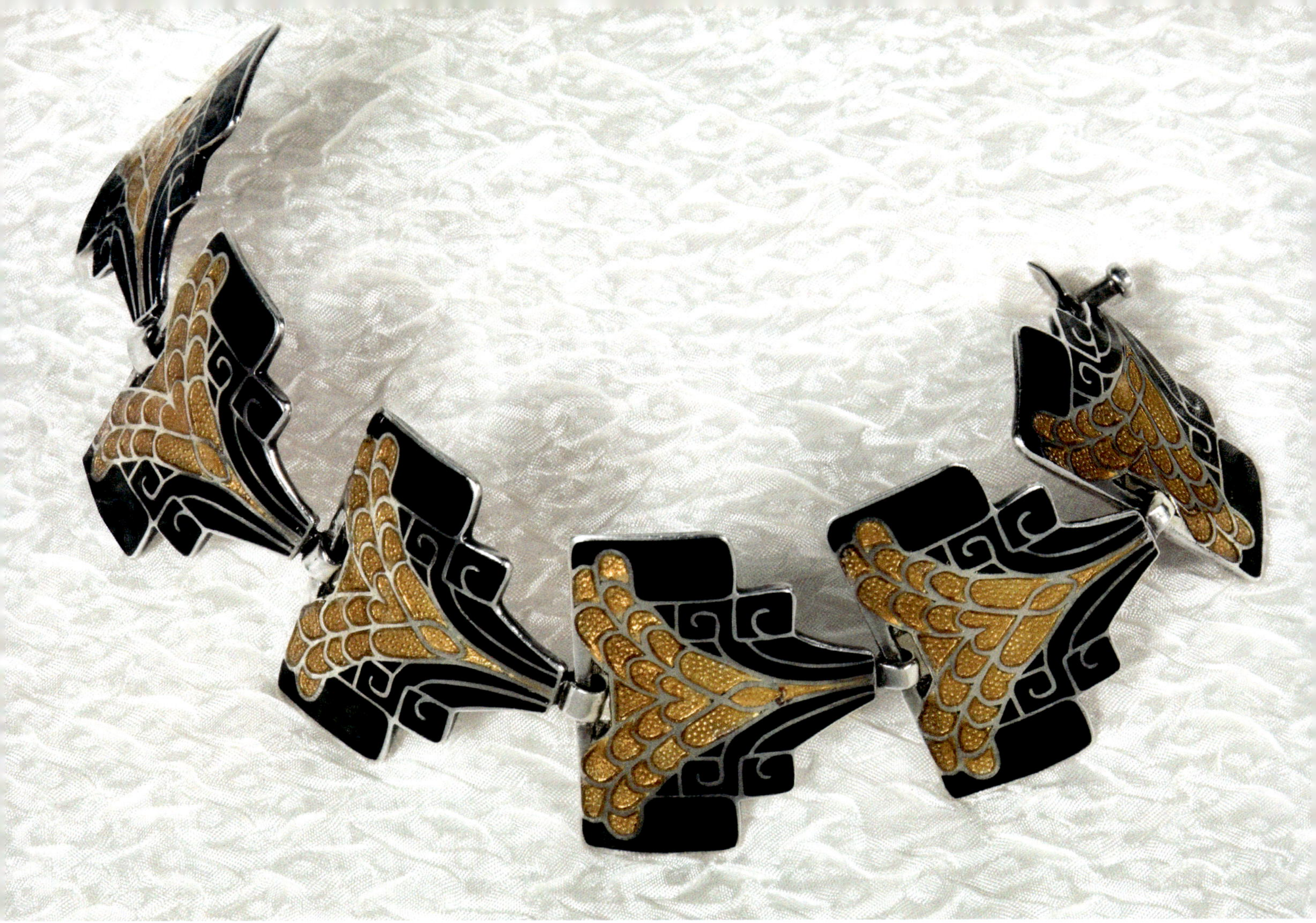

Plate 10 Margot de Taxco 5905. *Black and Gold Dragon's Scales* bracelet 7 x 1.5 in. Maestros de Taxco Collection

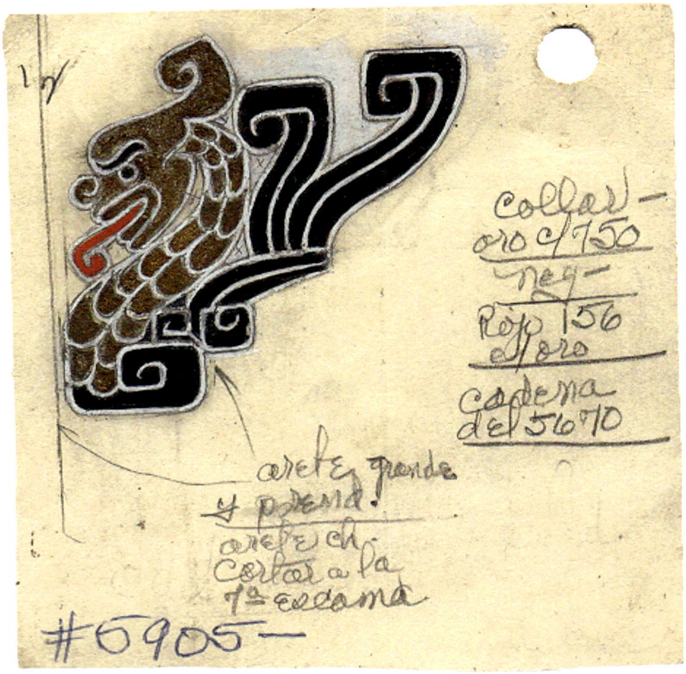

Drawing for 5905 *Dragon*

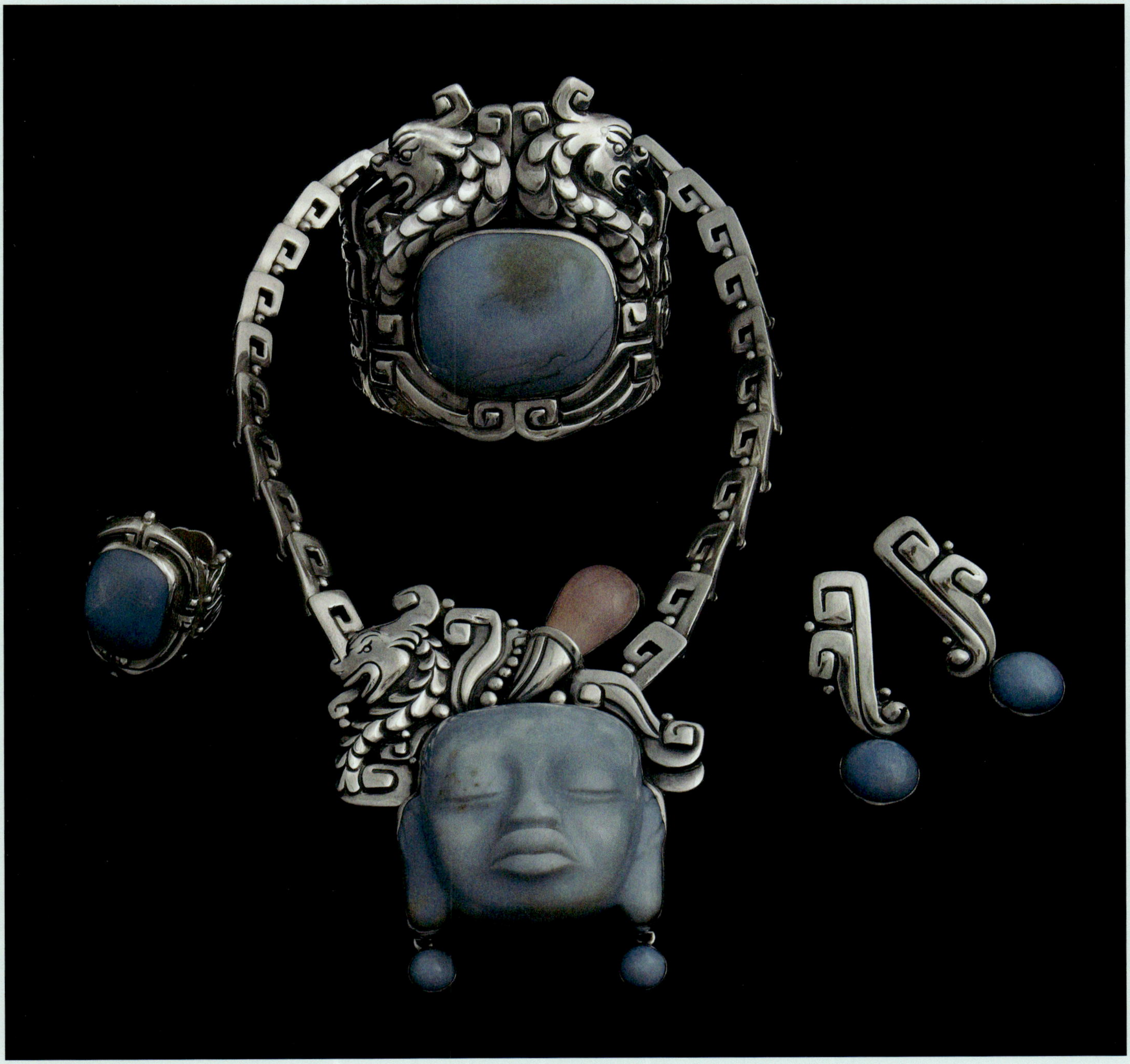

Plate 11 Margot de Taxco 5255. *Silver Dragons, Potch Blue Opals, and Pink Quartzite Ensemble.* brooch/pendant 3.5 x 3 in.; necklace 15.6 in.; cuff bracelet 3 x 7 in.; earrings 2 in.; ring 1.5 x 3.75 in. Linda Affleck Collection

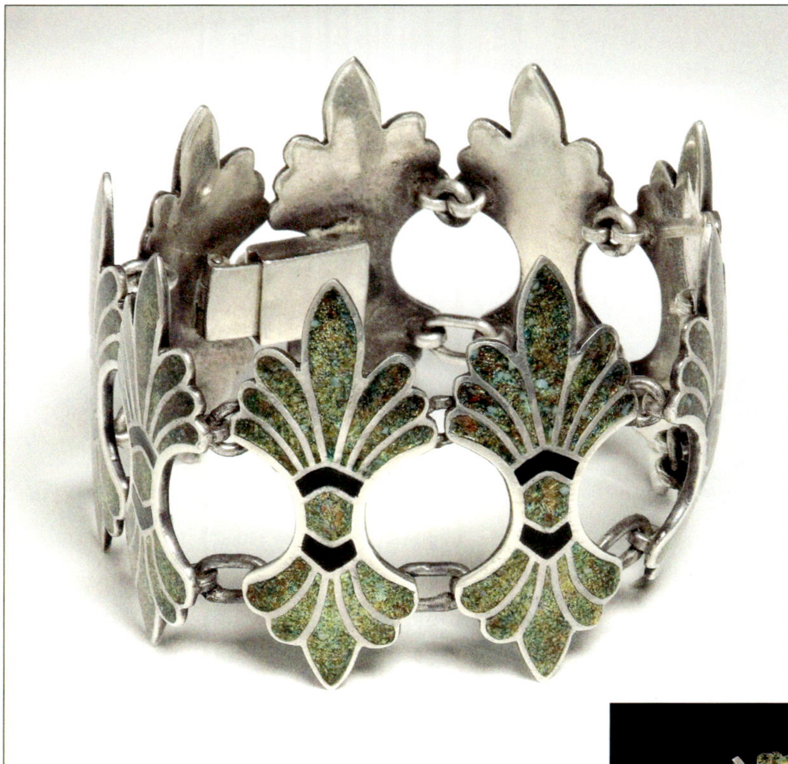

Details

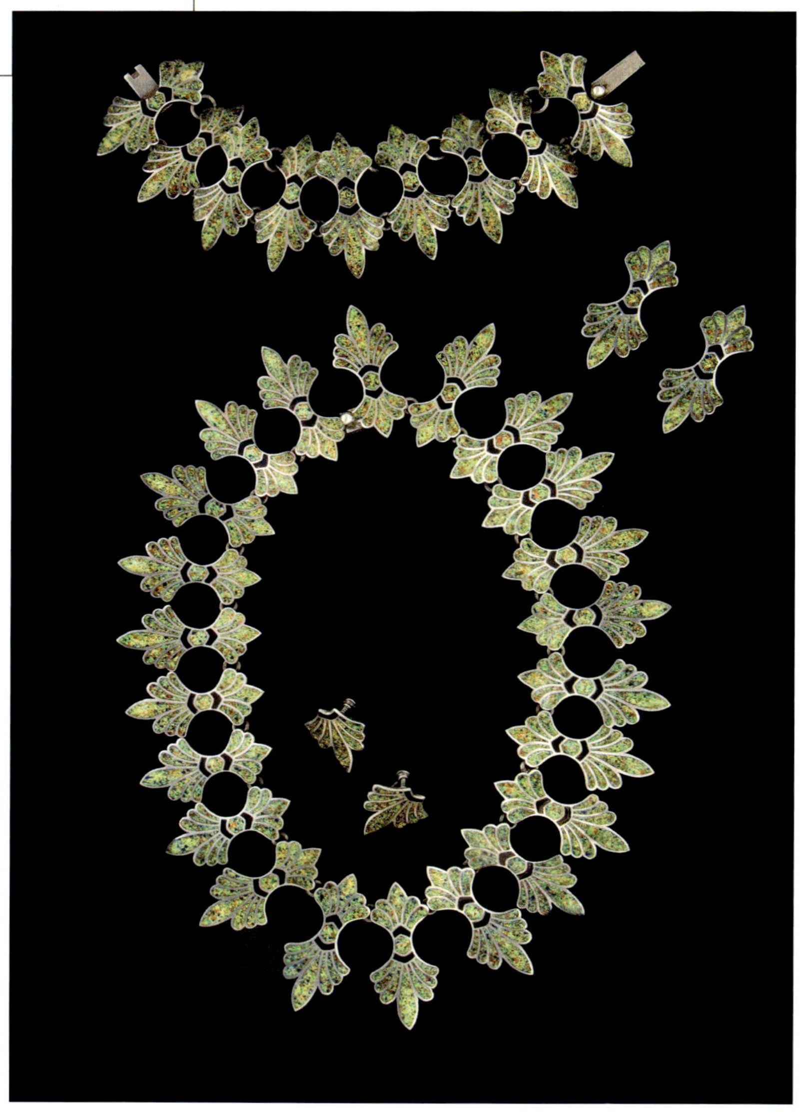

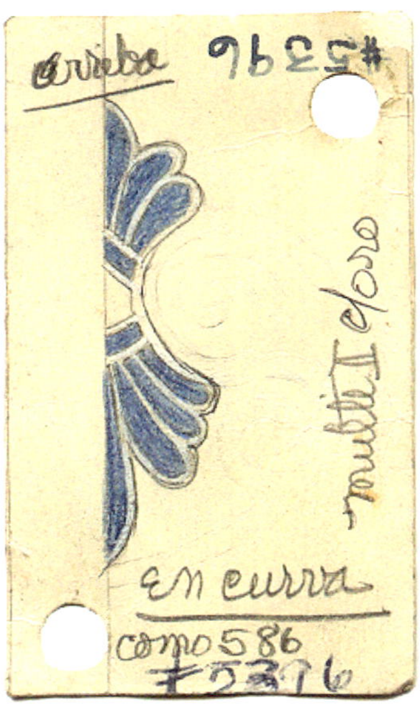

Drawing 5396 *Flared Bow*

Plate 12 Margot de Taxco 5396. *Gathered Leaf Pattern Set.* necklace 16 x 2 in.; bracelet 7.75 x 1.75 in.; small earrings 1 x 0.88 in.; large earrings 1.75 x 0.75 in.
J. P. Mather Collection

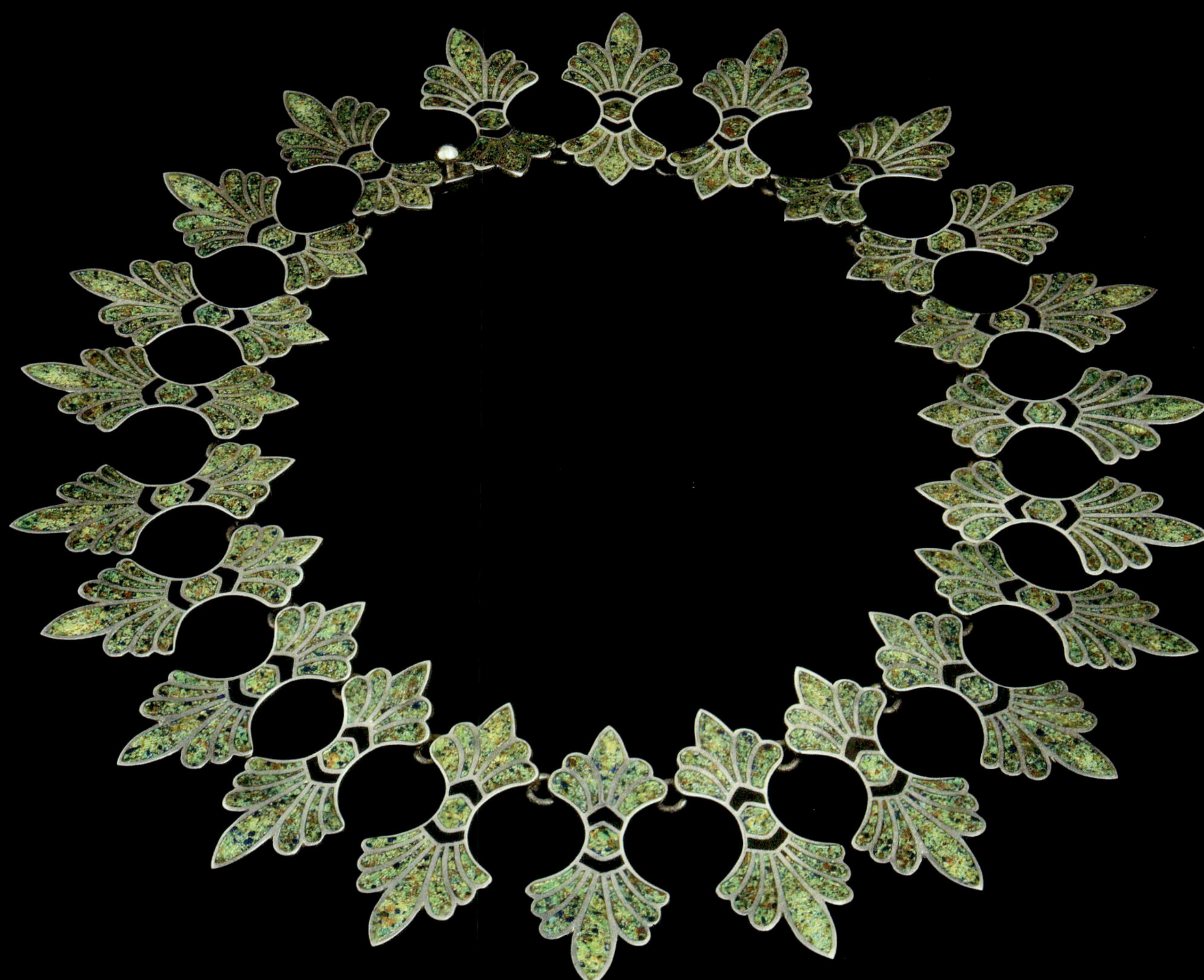

Drawing 5696 *Scalloping*

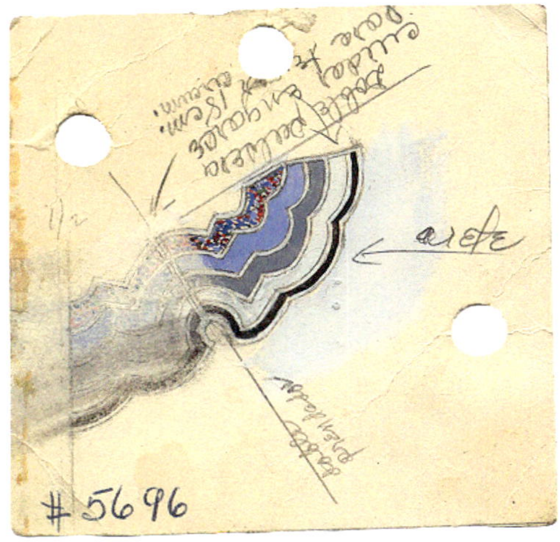

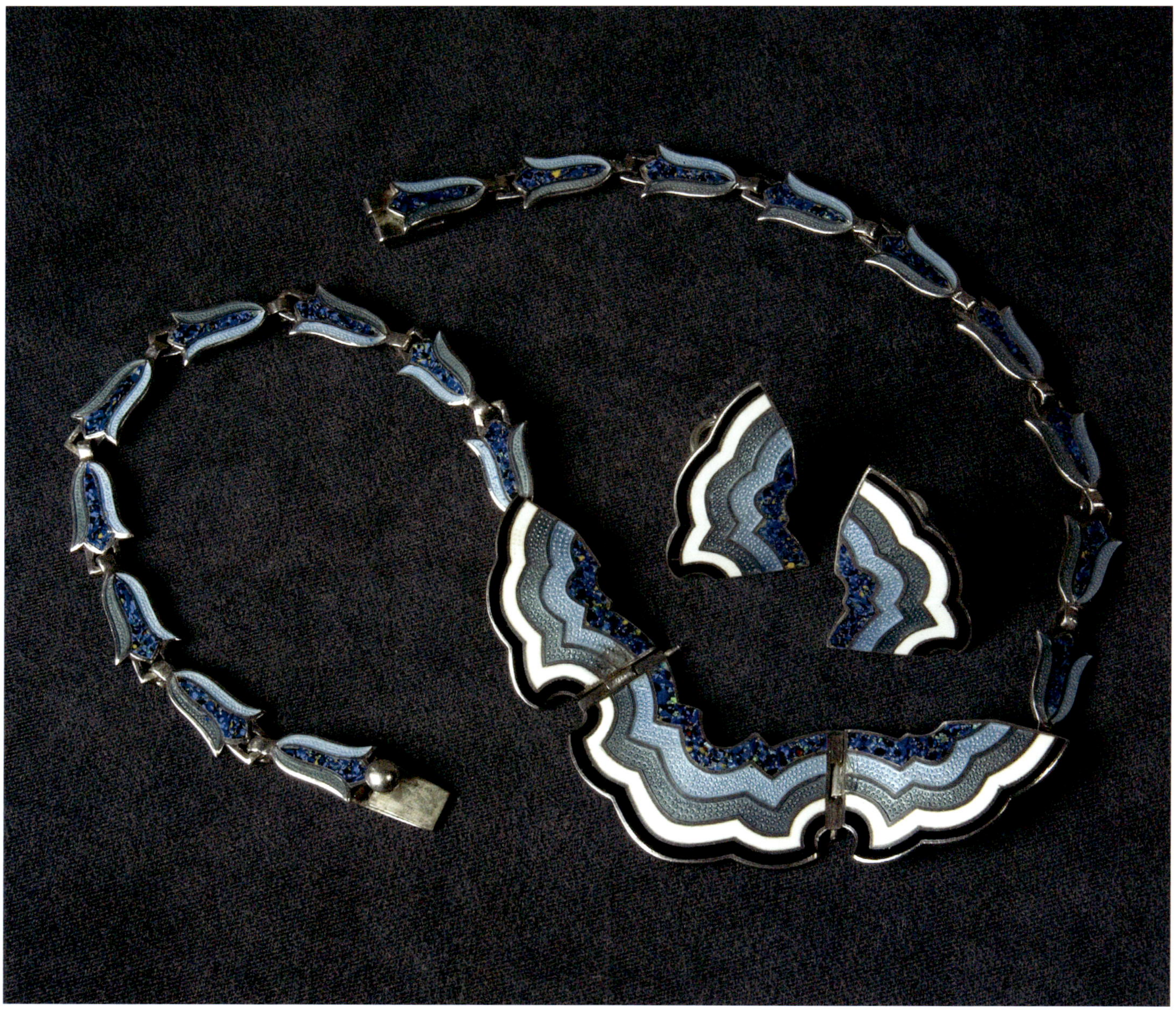

Plate 13 Margot de Taxco 5696. *Elegant Geological Enamel Set*. necklace 15 x 1 in.; earrings 1.25 x 0.75 in. Katherine Armitre Collection

Plate 14 Margot de Taxco 5744. *White Flower* pendant/brooch 1.5 x 3 in. Elizabeth Finch Collection

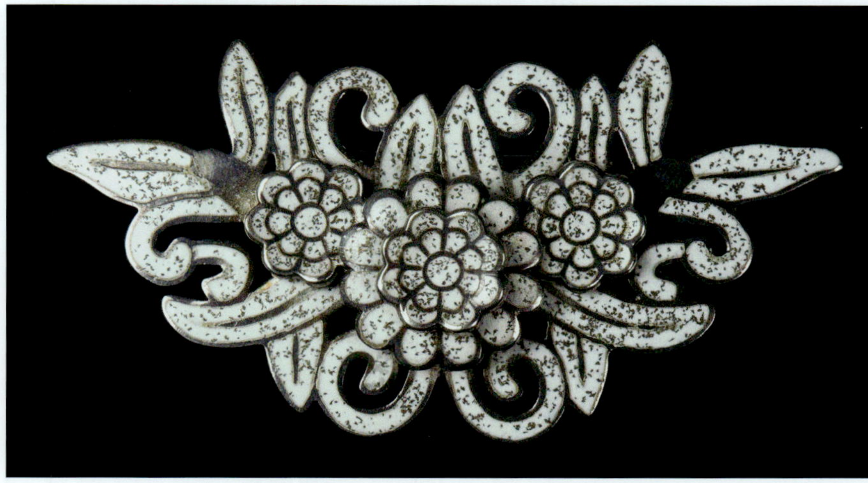

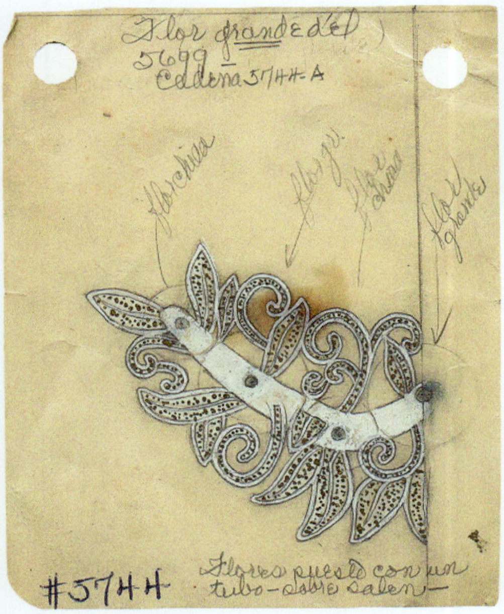

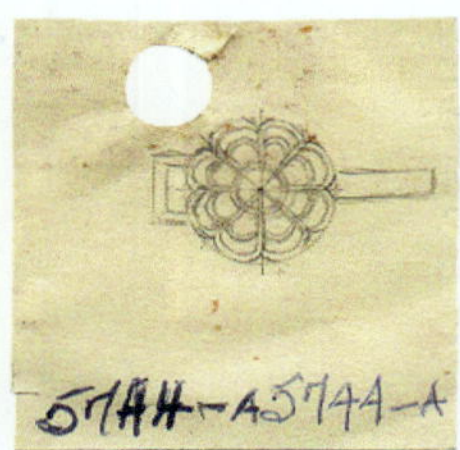

Drawings 5744 *Wild Leaves With Flowers*

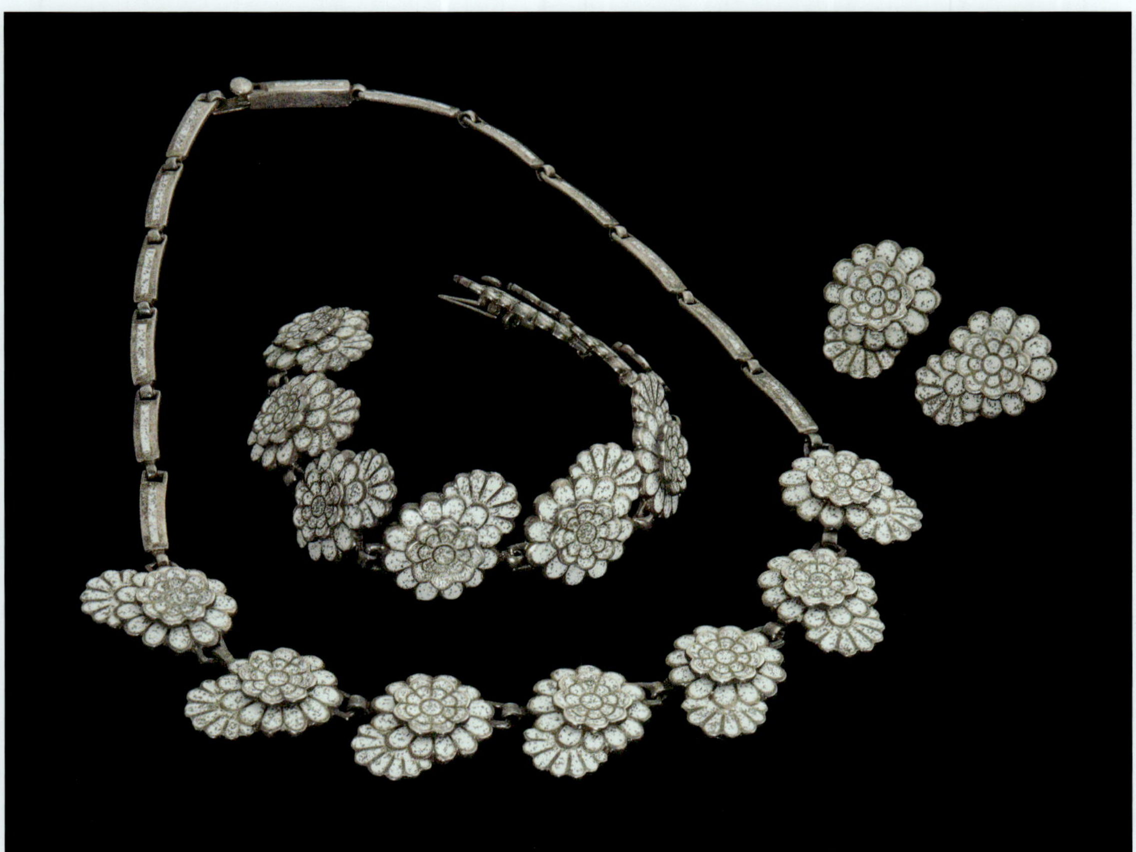

Plate 15 Margot de Taxco 5744. *White Flowers Set*. necklace 16.5 x 1 in.; bracelet 6.5 in. x 1 in.; earrings 1.13 x 0.75 in. James Black and James Tracy Collection

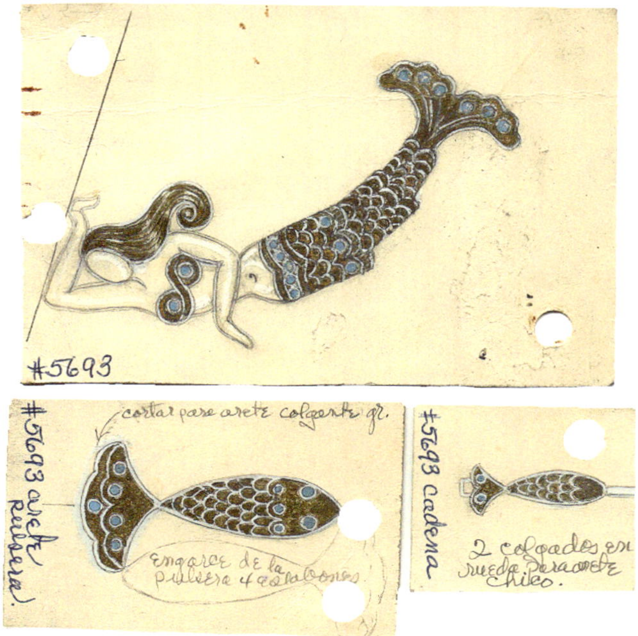

Drawings 5693
Mermaid & Fish

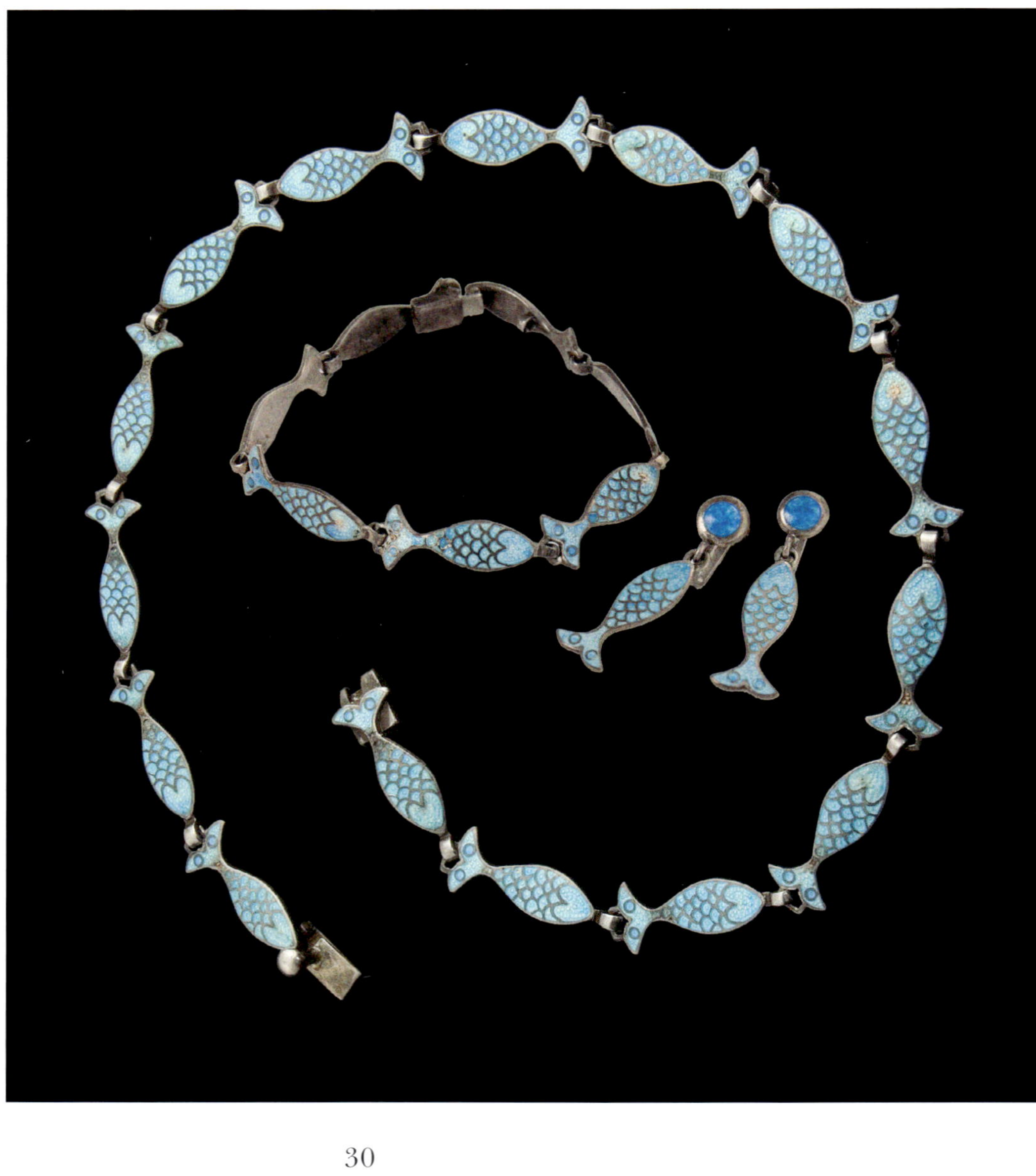

Plate 16 Margot de Taxco 5693. *Small Blue Fish Set*. necklace 17 in.; bracelet 7 in.; earrings 1.25 in. Elizabeth Finch and Private Collections

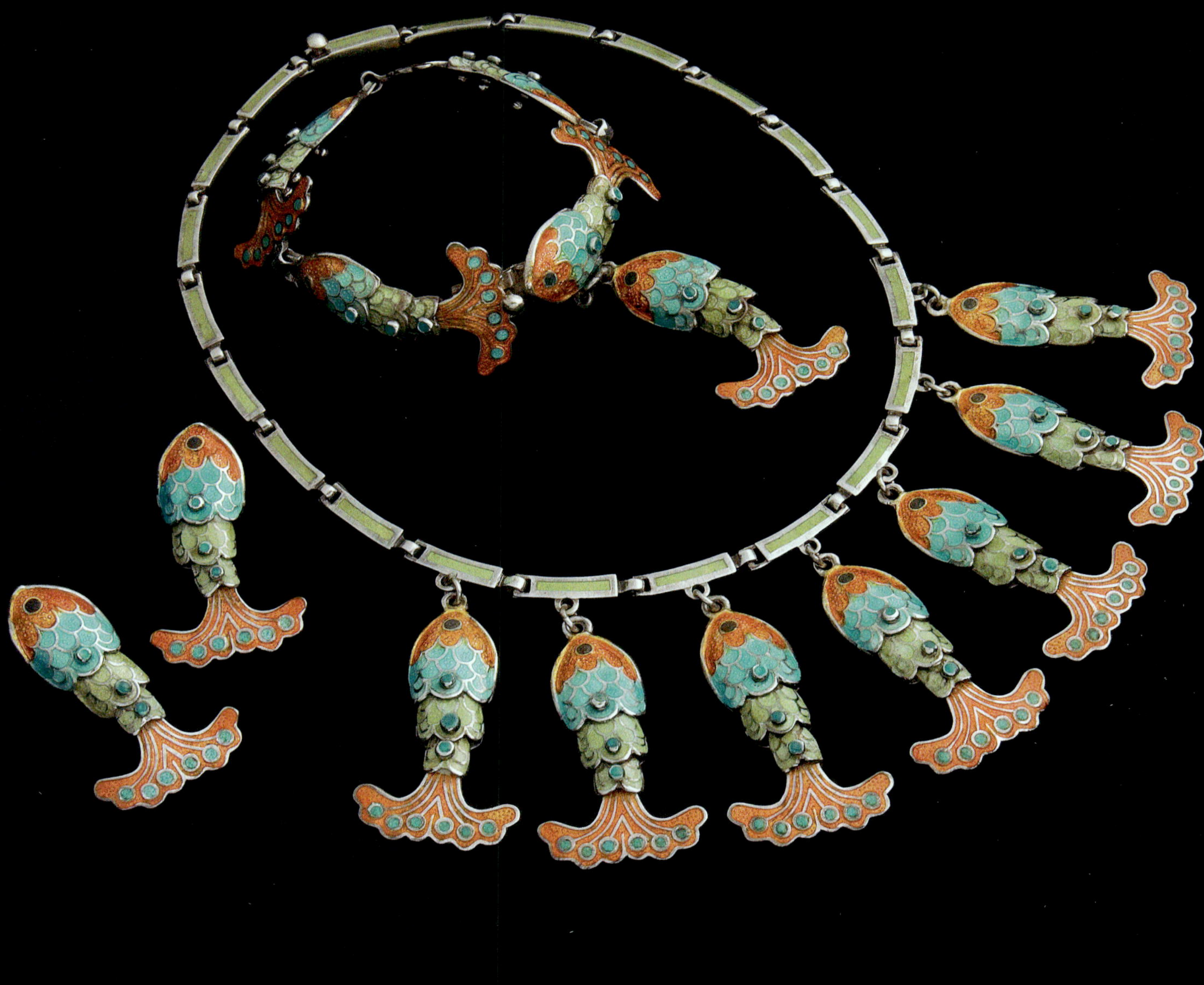

Plate 17 Margot de Taxco 5693. *Polychrome Fish Set*. necklace 17 75 in., fish pendants 2 x 1 25 in.; bracelet 8 x 1.13 in.; earrings 2 x 1.13 in. Betsy Vought Collection

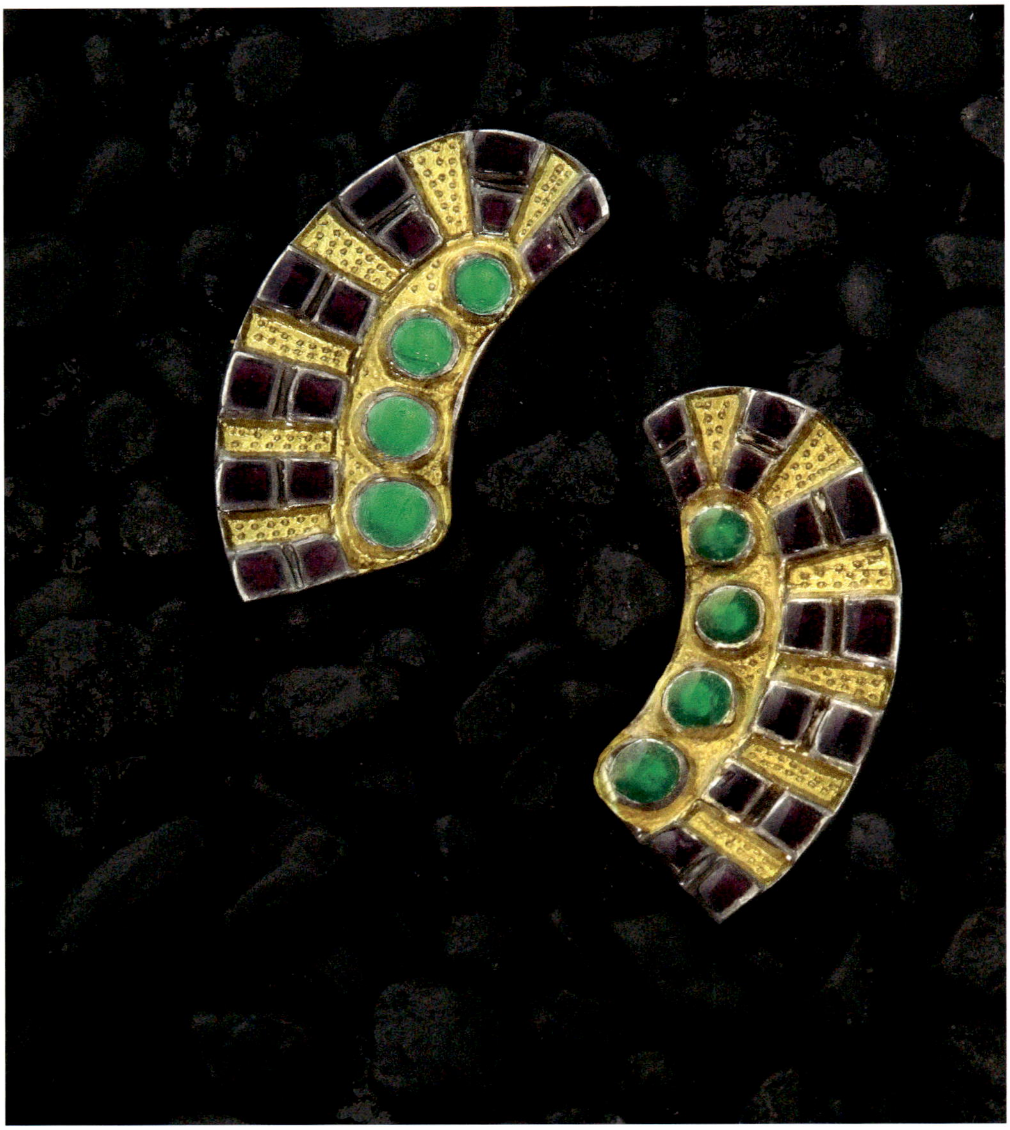

Plate 18 Margot de Taxco 5603. *Jeweled Circles and Squares* earrings 1.5 x 5/8 in. Stella and Fred Krieger Collection

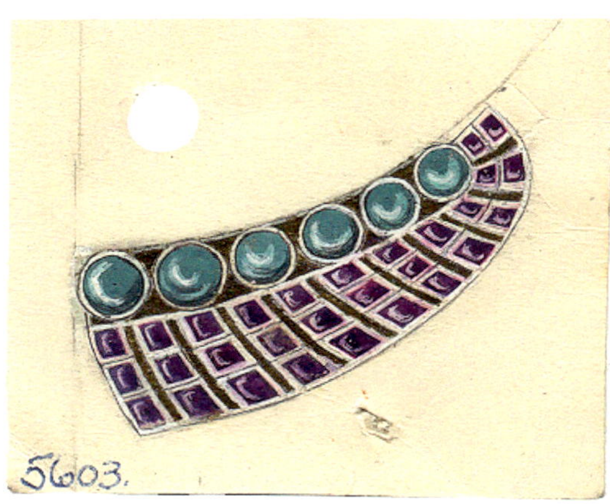

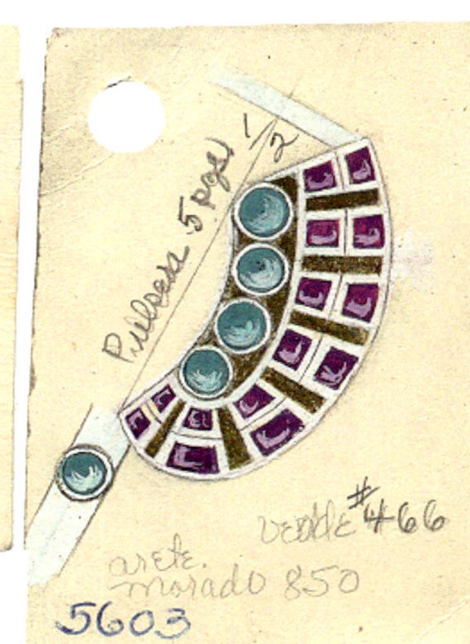

5603 Jeweled Circles & Squares

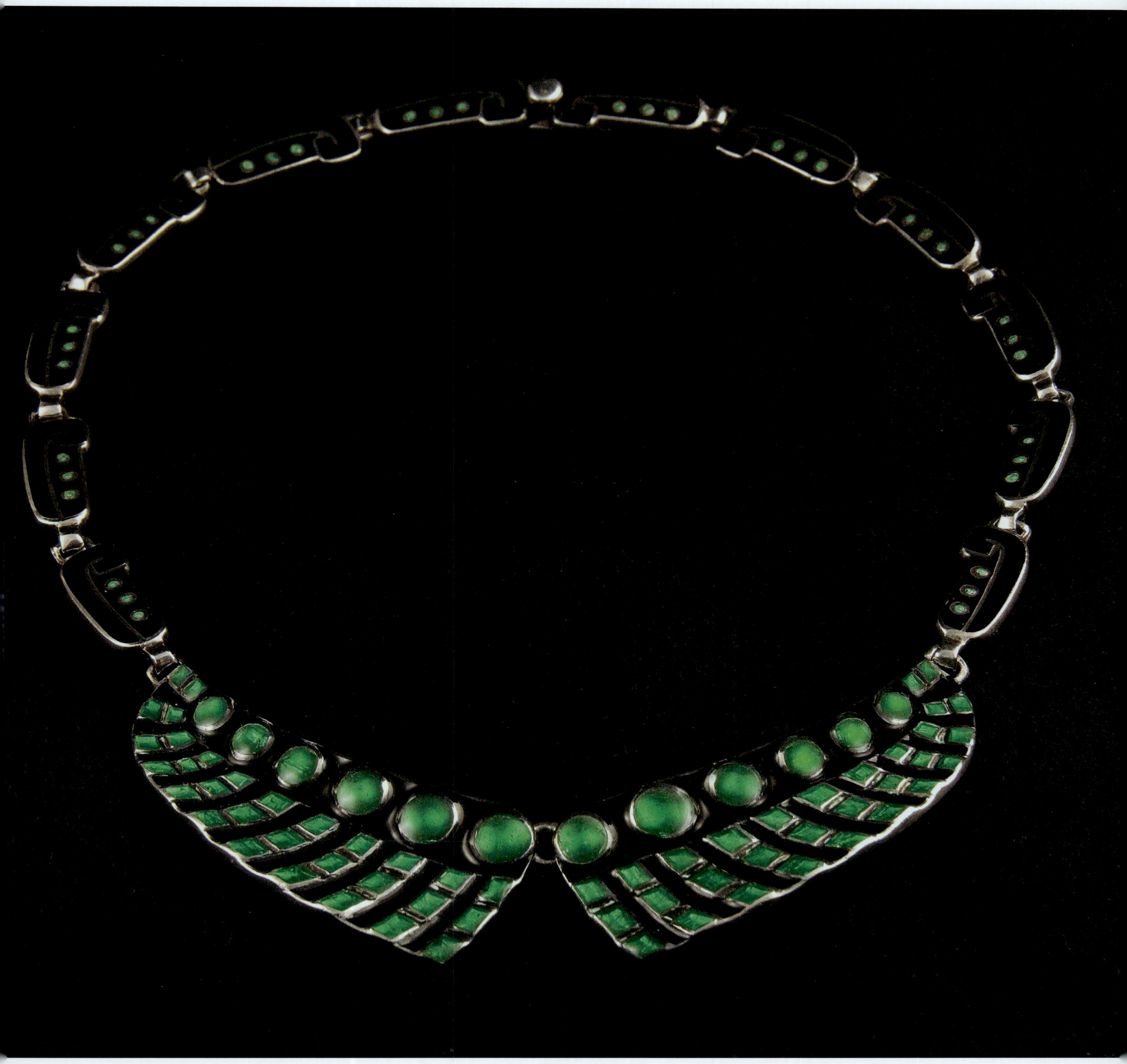

Plate 19 Margot de Taxco 5603. *Jeweled Circles and Squares* in green necklace 14.5, central motif 0.88 x 3.75 in. Andrea Fatell Collection

Plate 20 Margot de Taxco 5758. *Egyptian Scroll* earrings 1.5 x 1.25 in. Private Collection

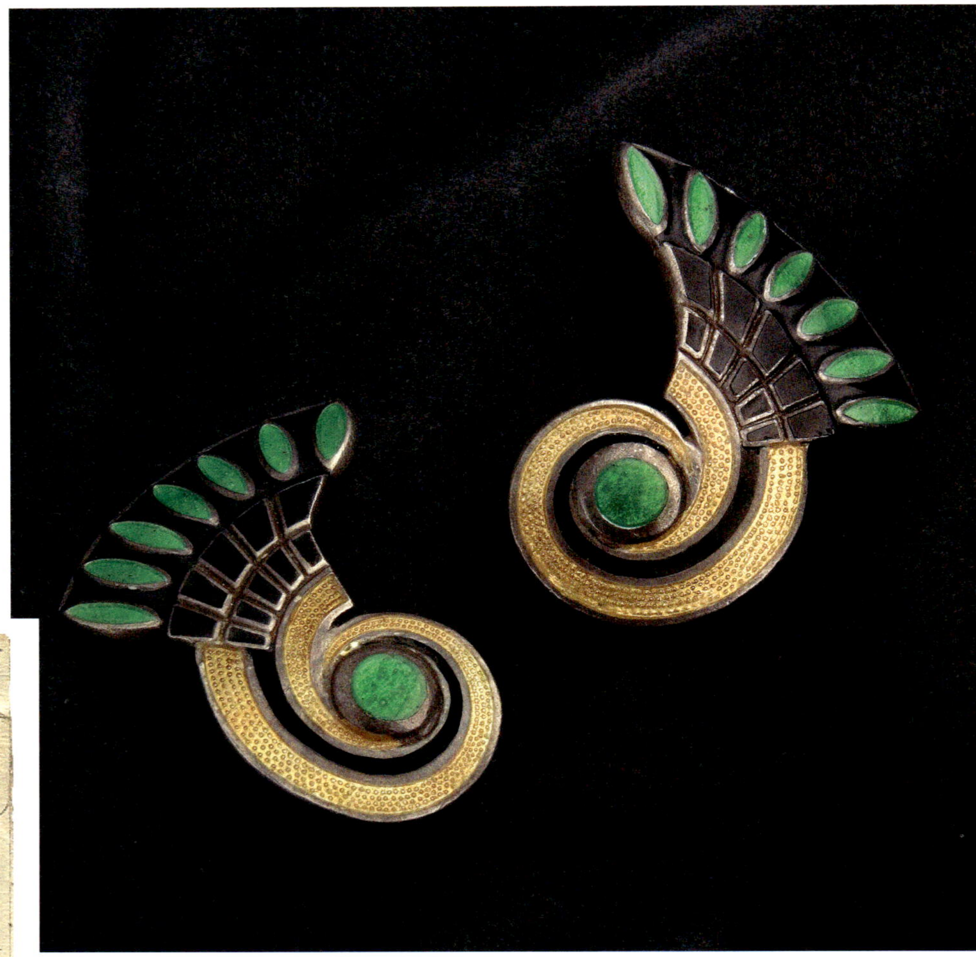

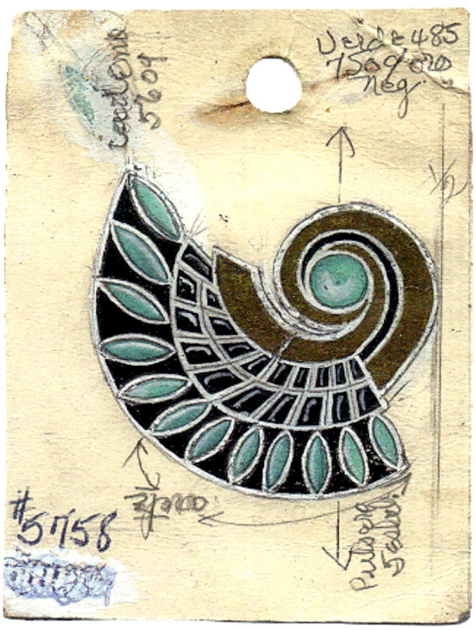

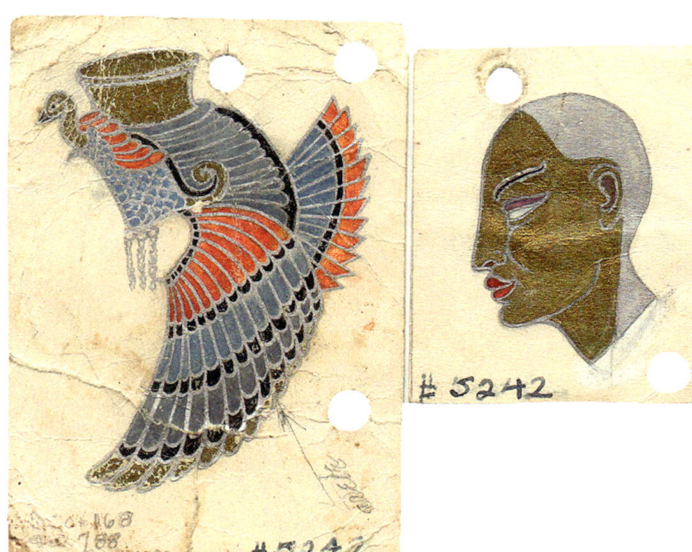

Drawings 5242 *Cleopatra* and 5758 *Egyptian Scroll*

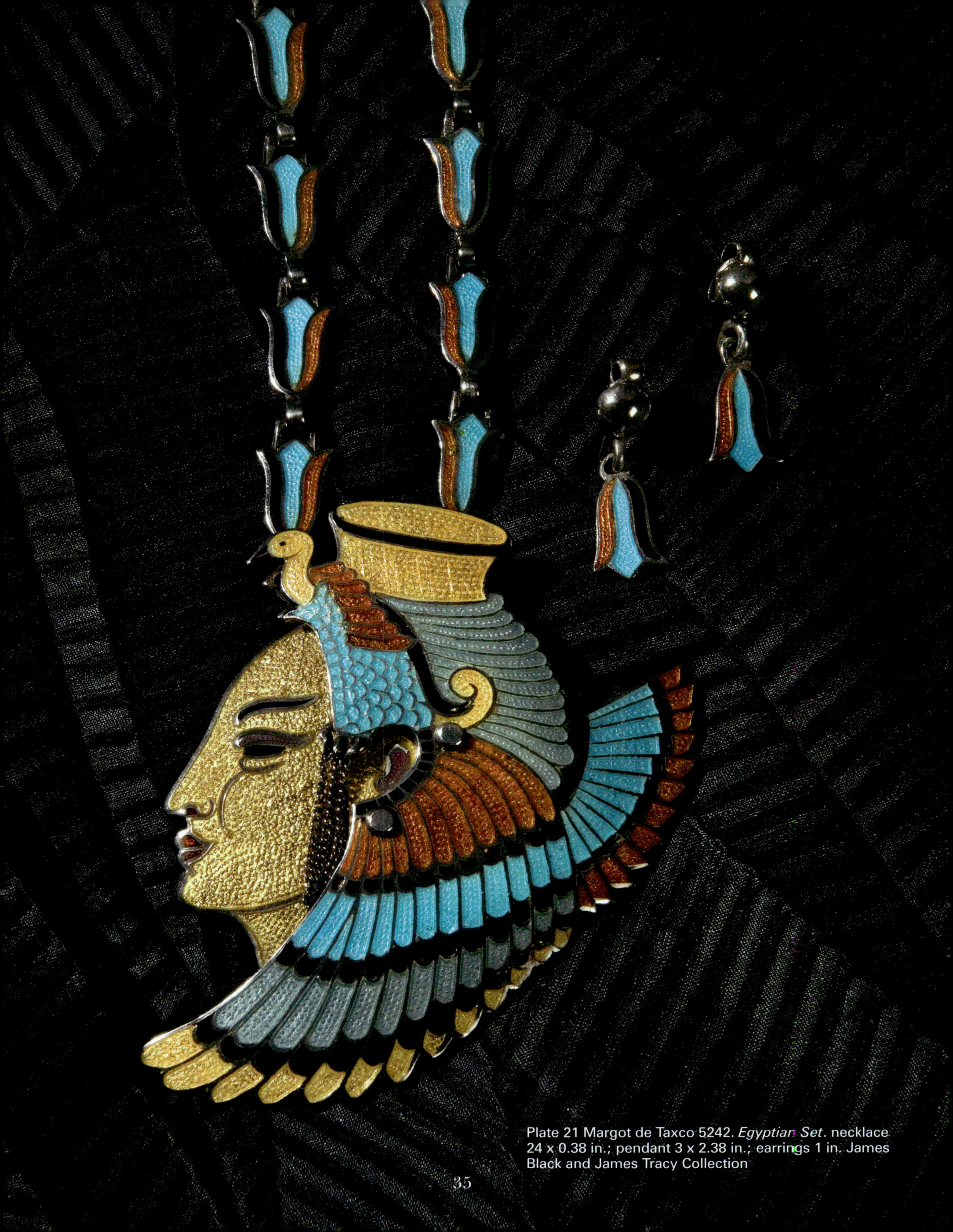

Plate 21 Margot de Taxco 5242. *Egyptian Set*. necklace 24 x 0.38 in.; pendant 3 x 2.38 in.; earrings 1 in. James Black and James Tracy Collection

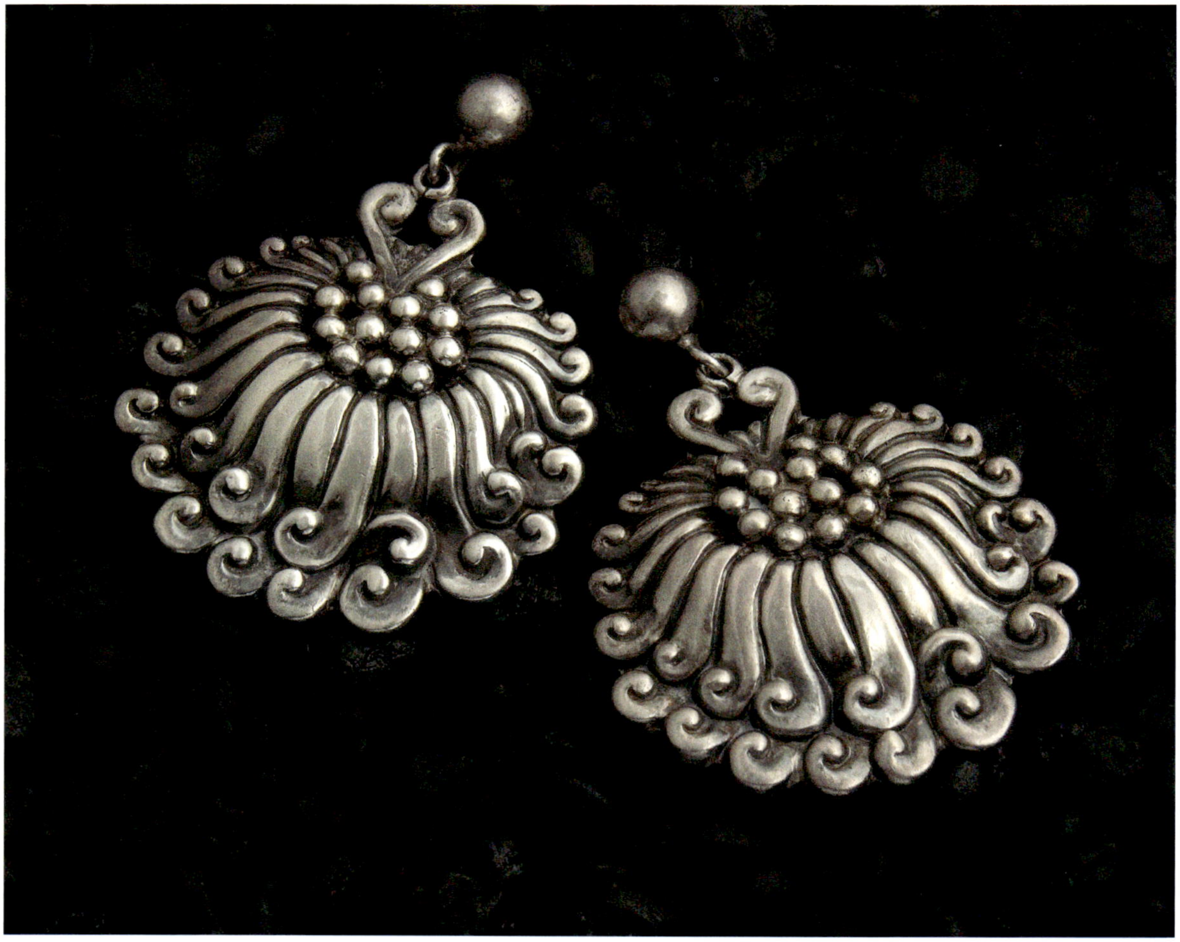

Plate 22 Margot de Taxco. *Floral* silver earrings 2 x 1.25 in. Private Collection

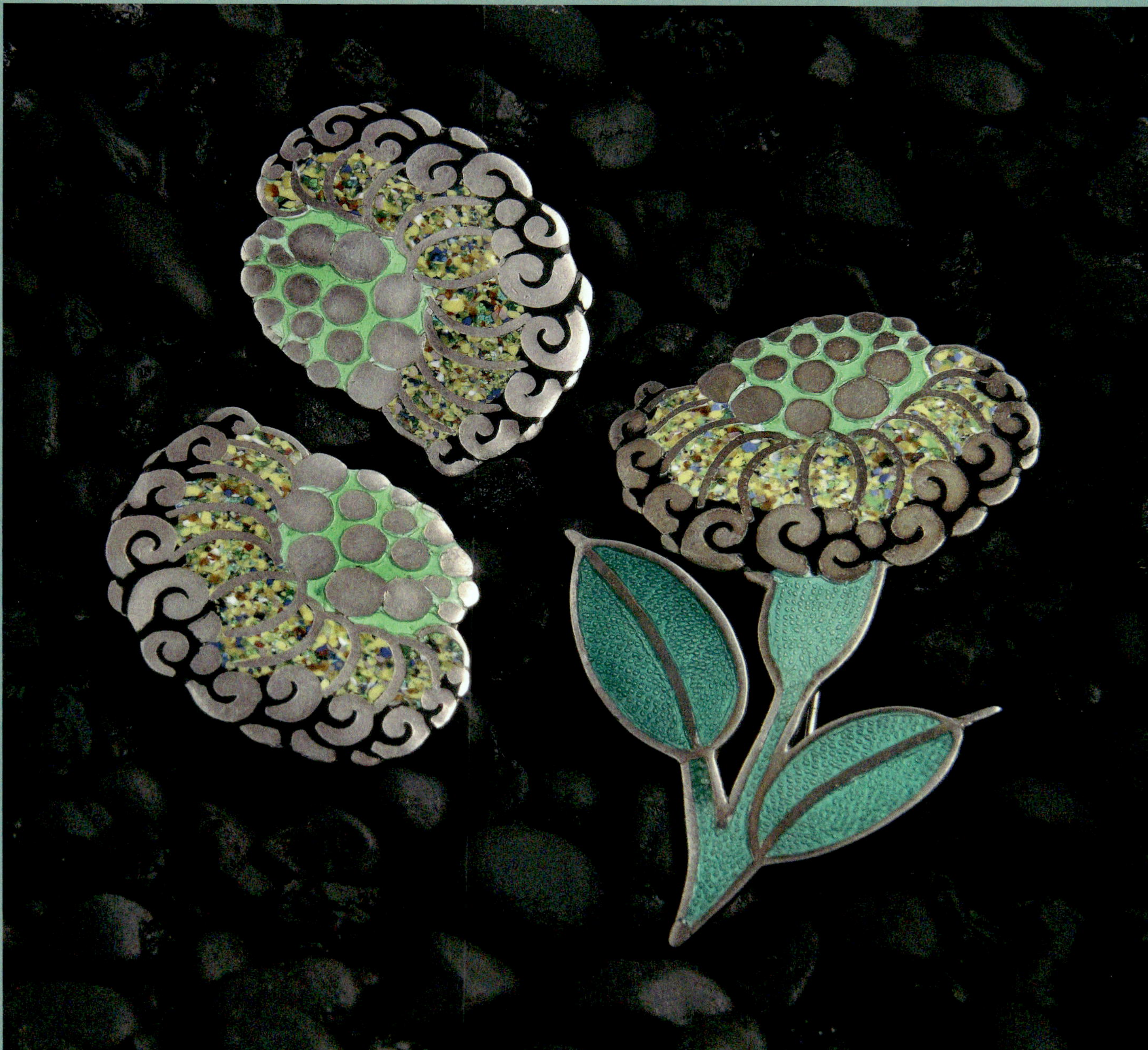

Plate 23 Margot de Taxco 5408. *Polychrome Flower Set*. brooch 2 x 1 in.; earrings 1.25 x 1 in. Private Collection

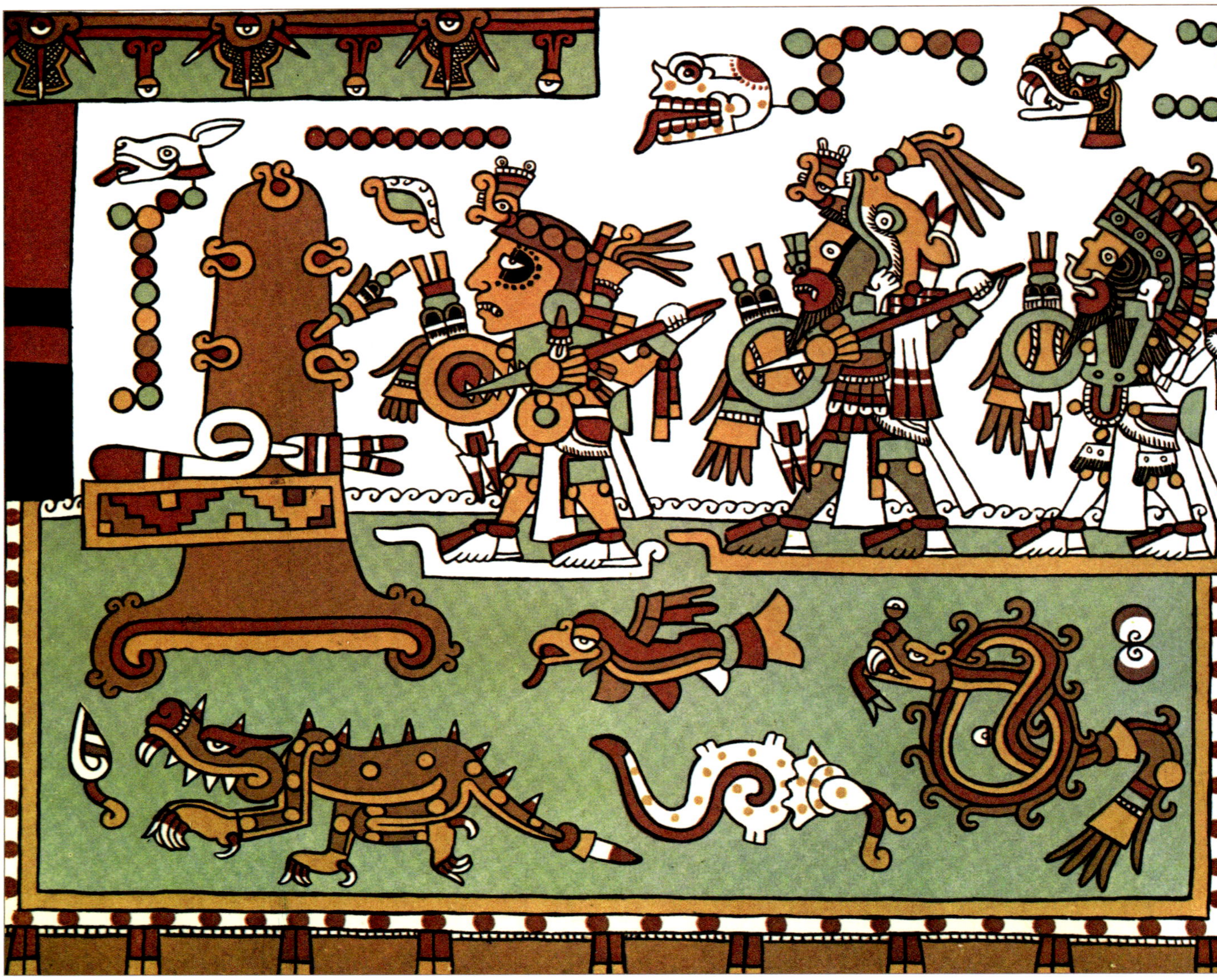

Plate 24 *Codex Nuttall*. Folio 75.

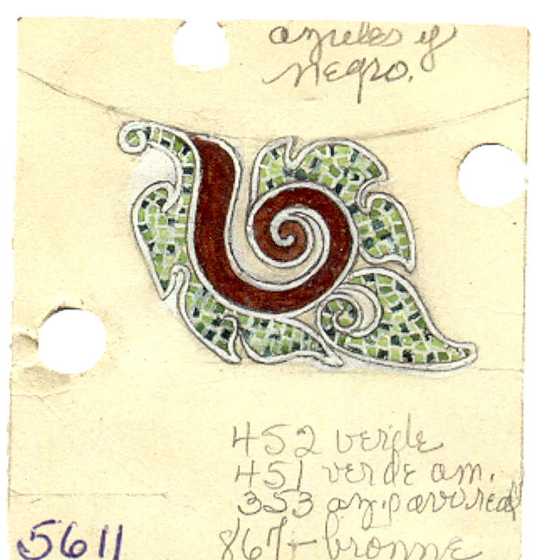

Drawing 5611 *Leaf Shell*

Plate 25 Margot de Taxco 5611. *Codex Nuttall Shell Set* in blue and black. necklace 16.38 x 1 in.; bracelet 7.75 x 1.13 in.; earrings 2 x 1.13 in. Betsy Vought Collection

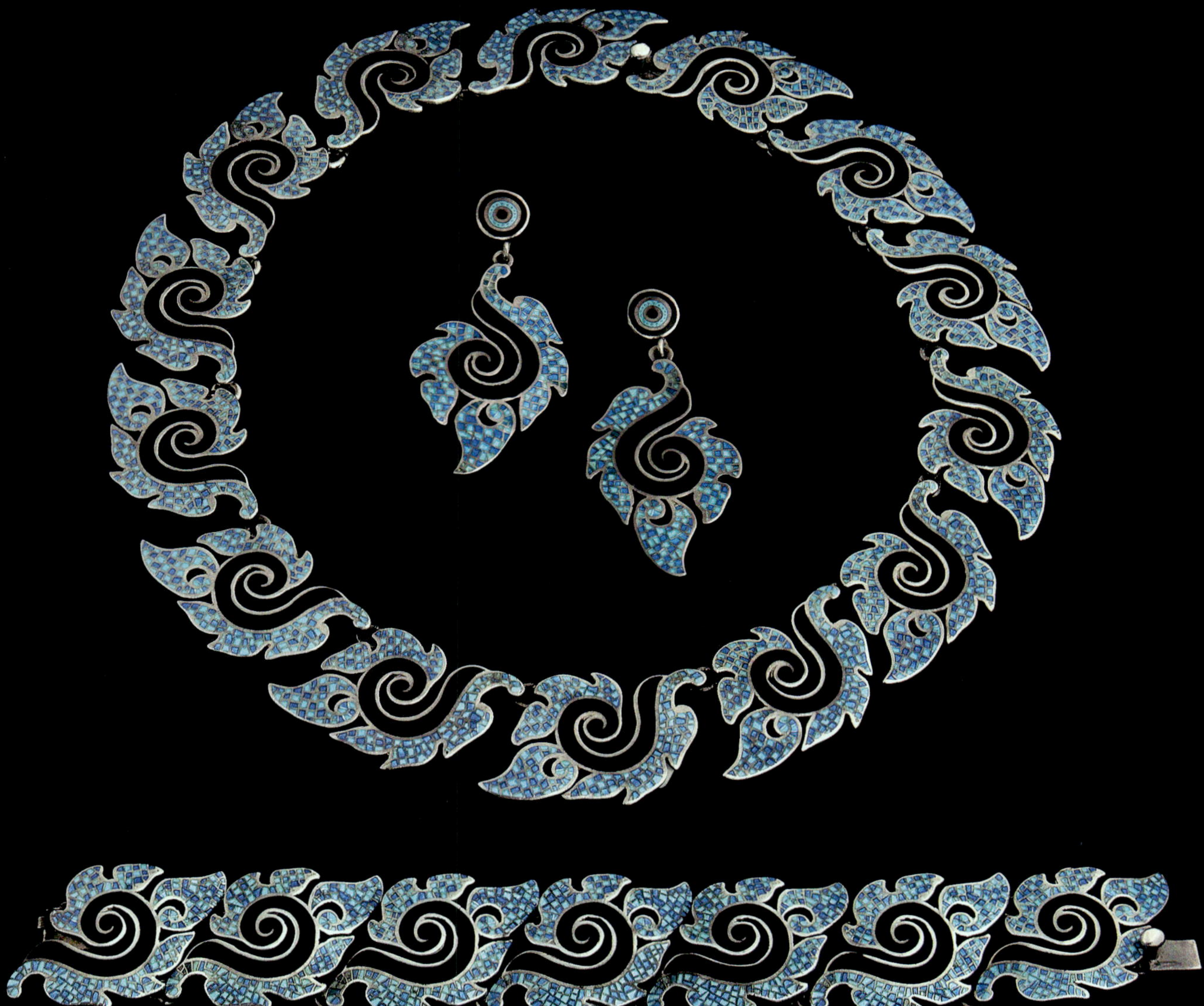

Plate 26 Margot de Taxco 5531. *Arabesque* cuff bracelet 7.25 x 1.25 in. Margot de Taxco 5651. *Gathered Swirls* bracelet 7.25 x 1 in. Private Collection

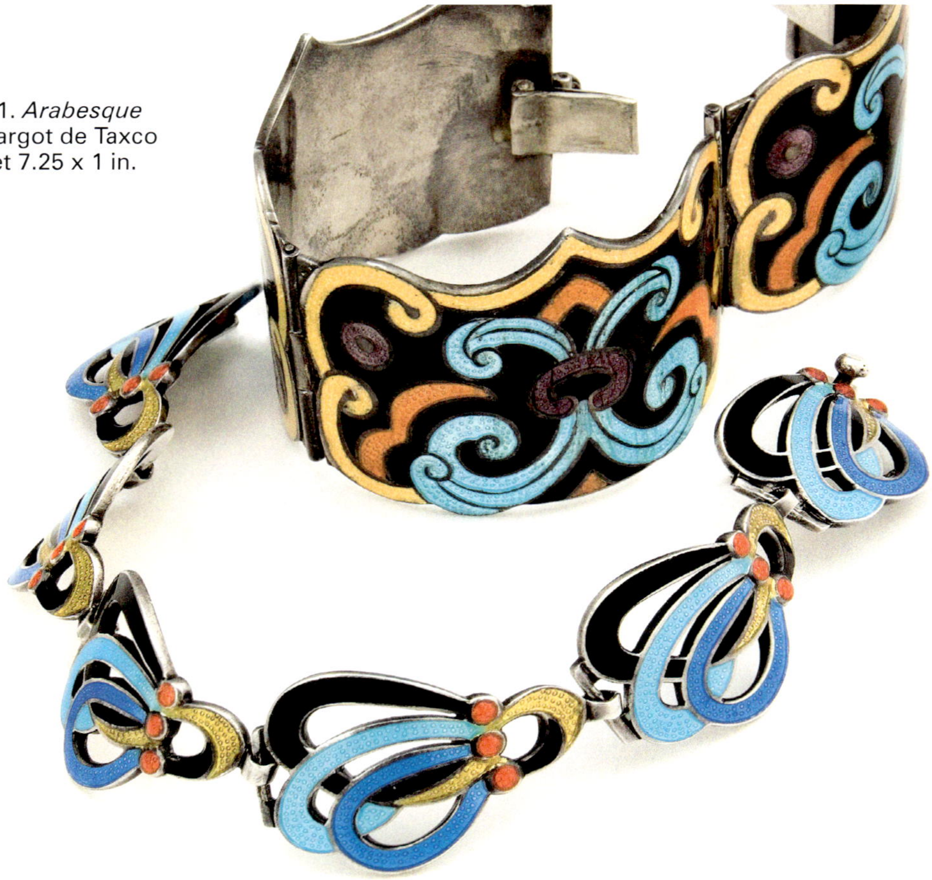

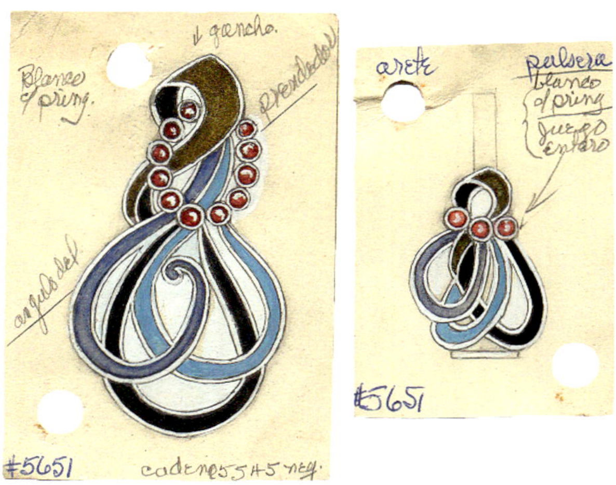

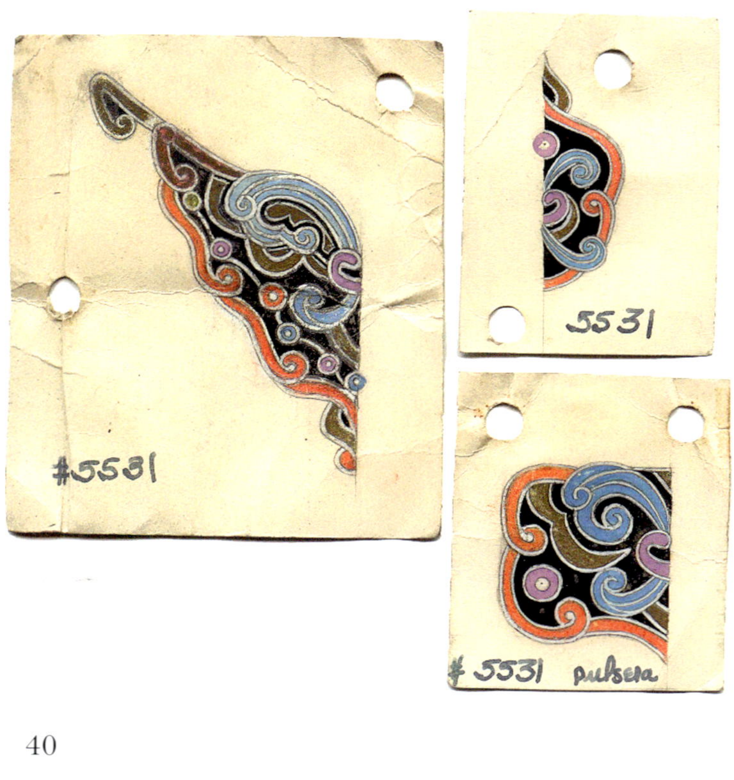

Drawings, 5651 *Gathered Swirls* and 5531 *Arabesque*

Plate 27 Margot de Taxco 5531. *Arabesque* necklace 15 in., pendant 3 x 3 in.; earrings 2.25 x 1 in. Silver Huntress Collection

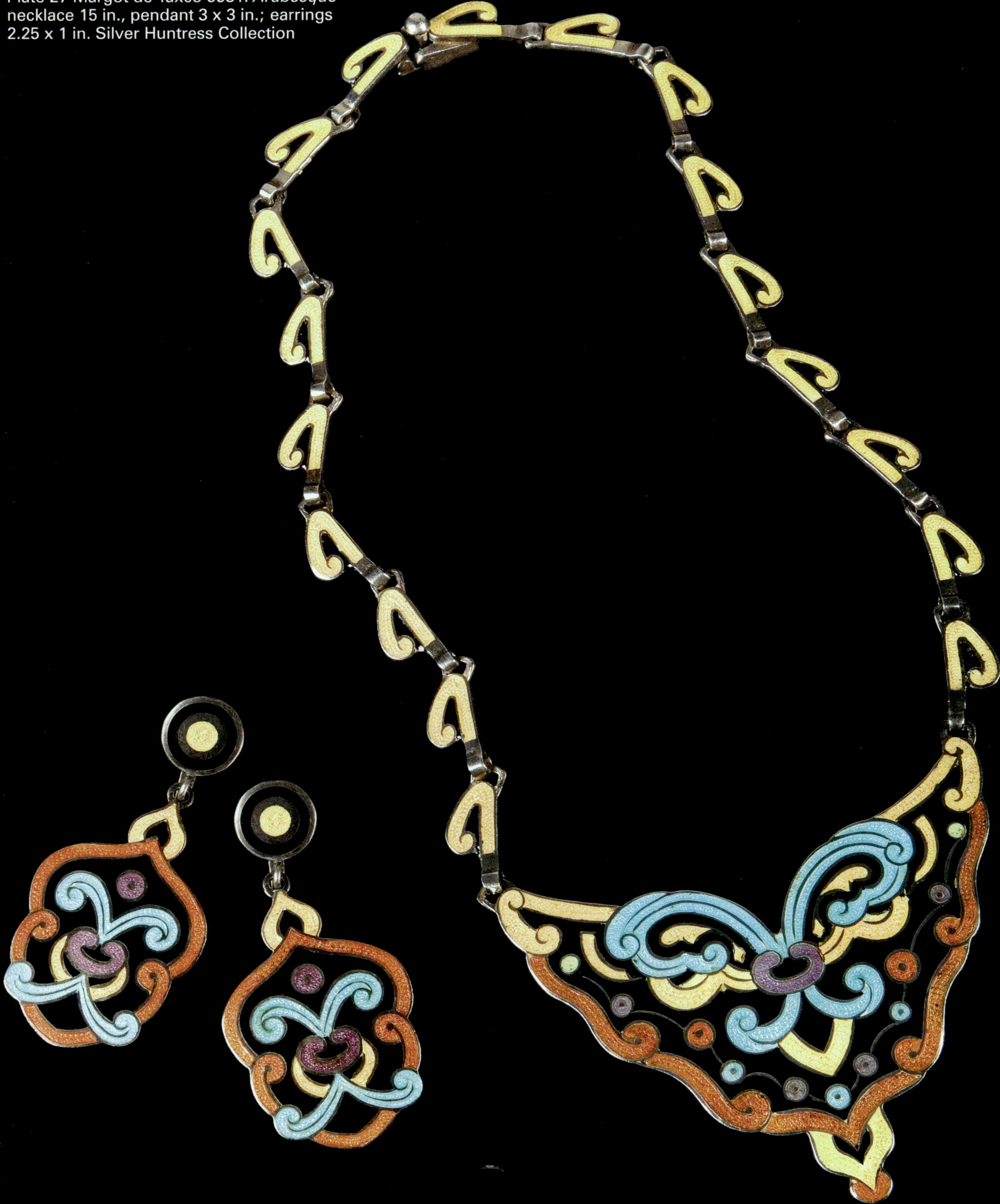

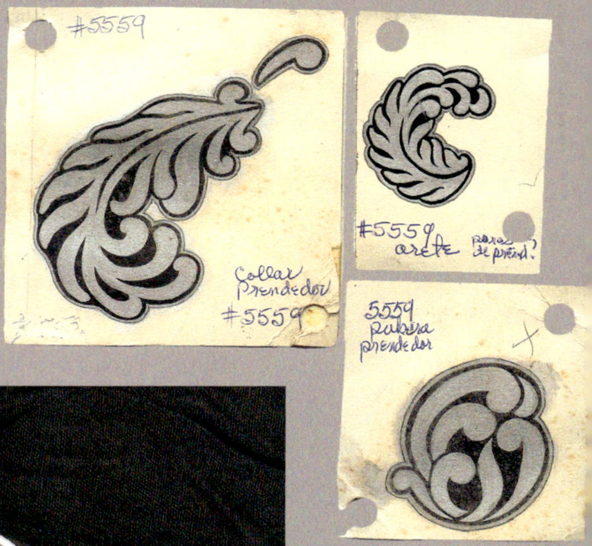

Drawings 5559 *Feathery Leaf*

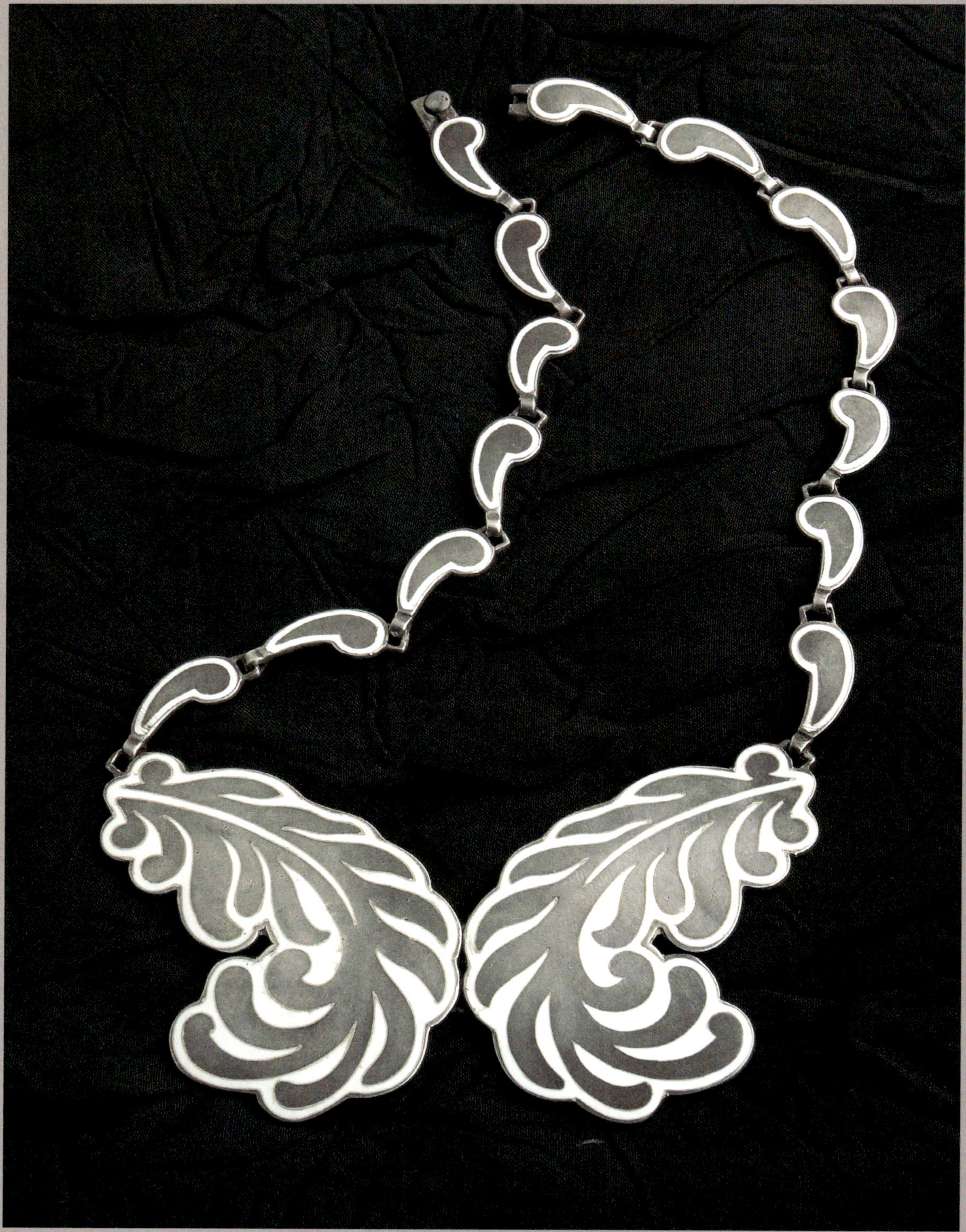

Plate 28 Margot de Taxco 5559. *Large Facing Leaves* necklace 16 in., central motif 2.38 x 4 in. J. P. Mather Collection

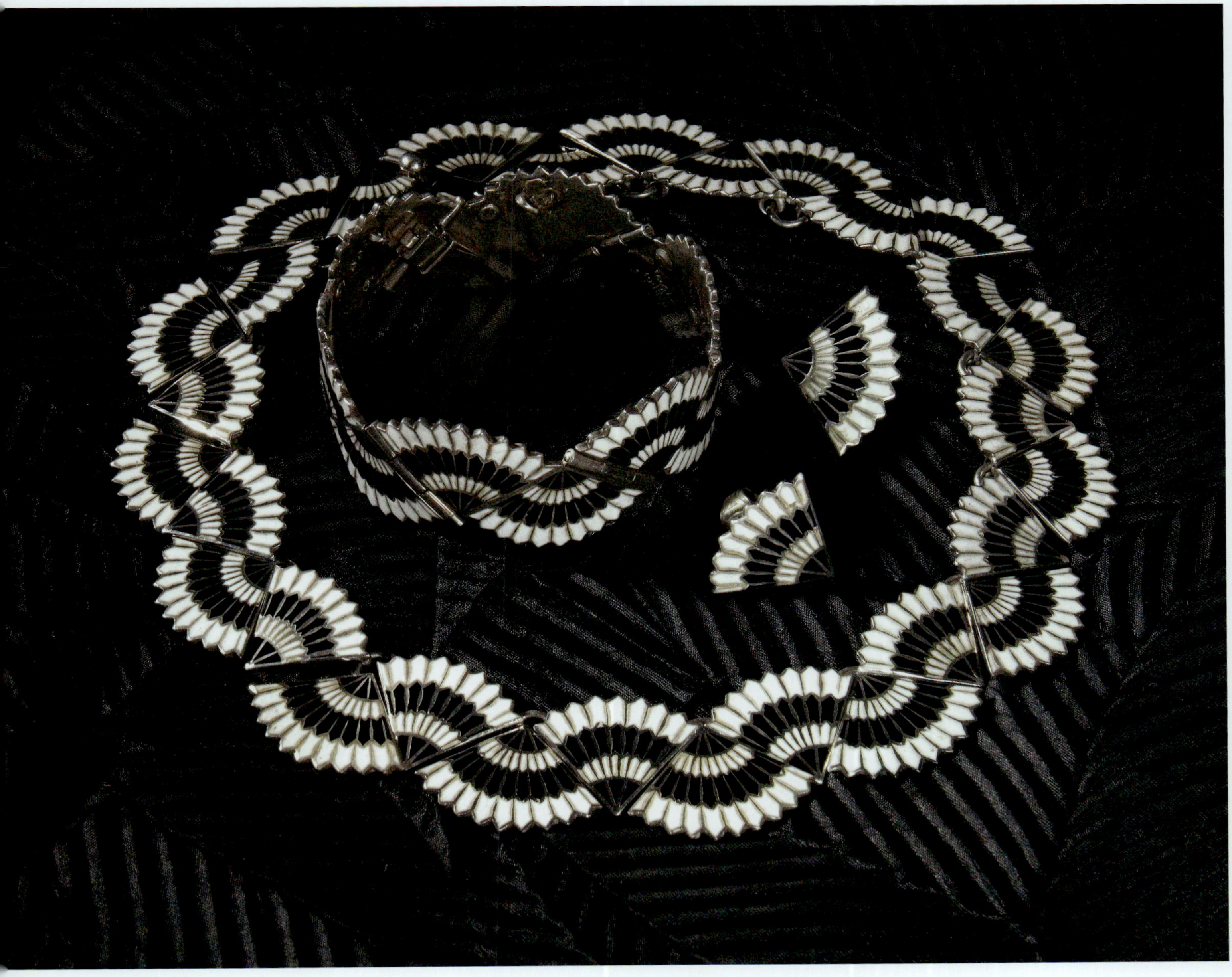

Plate 29 Margot de Taxco 5677. *Alternating Fan Set* in black and white. necklace 15 x 1 in., bracelet 1.13 x 6.5 in., earrings 0.75 in. James Black and James Tracy Collection

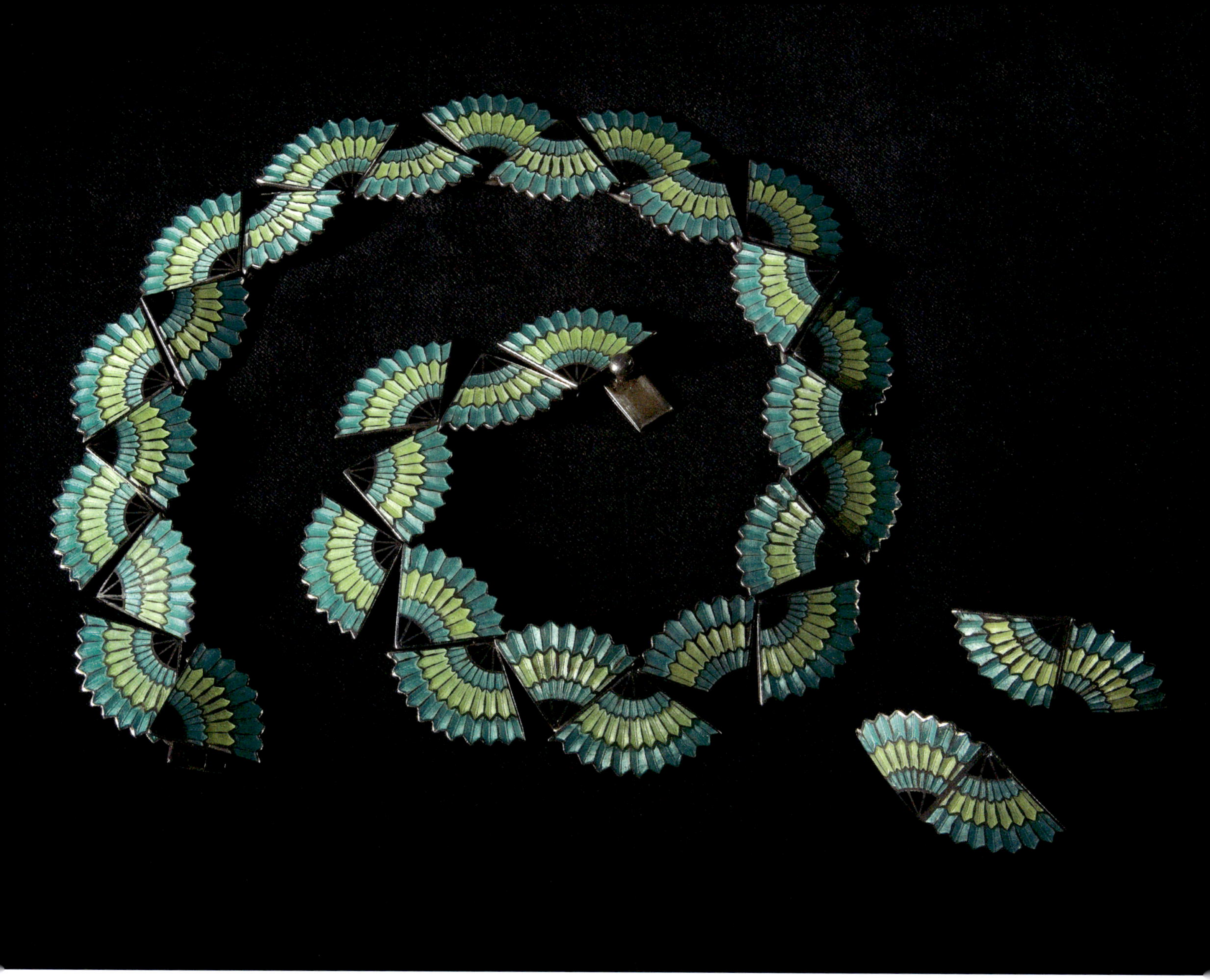

Plate 30 Margot de Taxco 5677. *Alternating Fans Set* in shades of green. necklace 15 x 1 in.; earrings 1/5 x 0.88 in. Katherine Armitre Collection

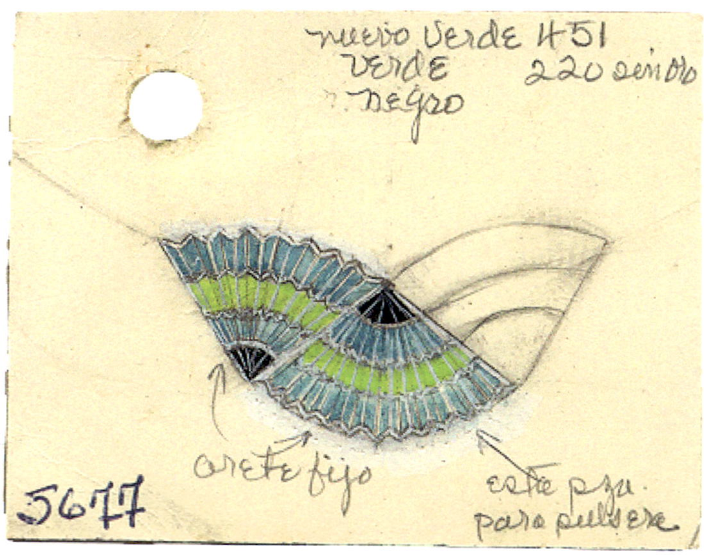

Drawing 5677 *Alternating Fans*

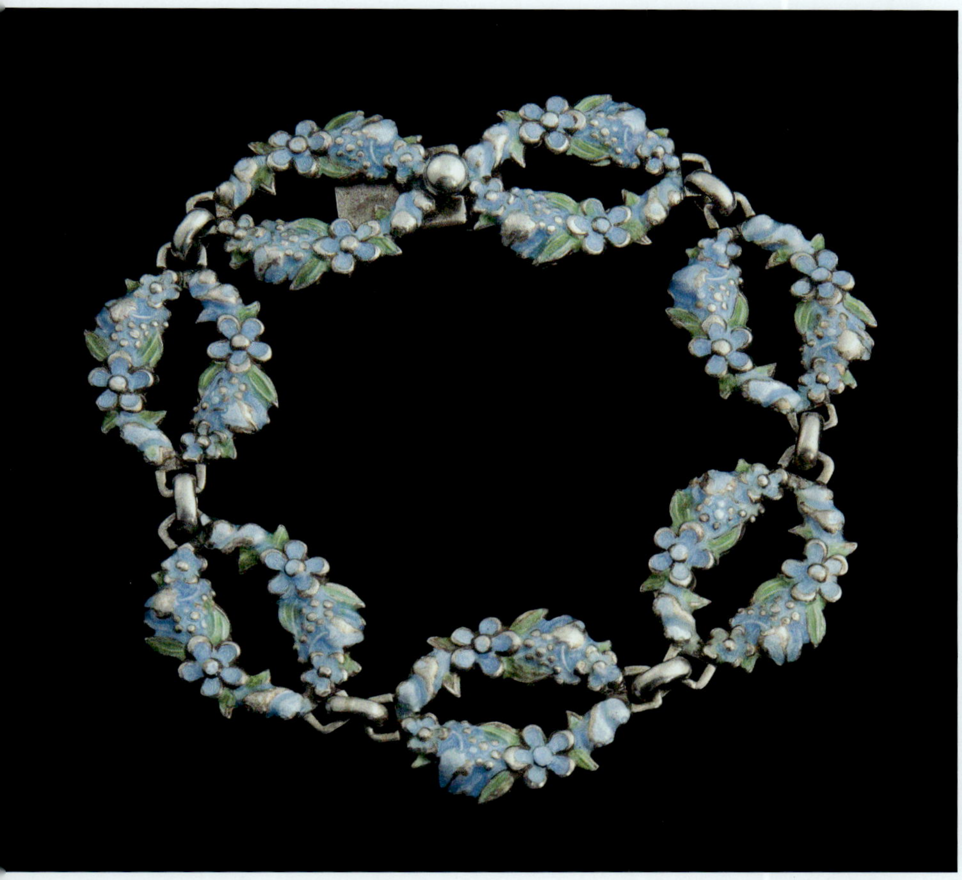

Plate 31 Margot de Taxco 5600. *Ovals in Tiny Blue Flowers* bracelet, 8 x 0.75 in. Elizabeth Finch Collection

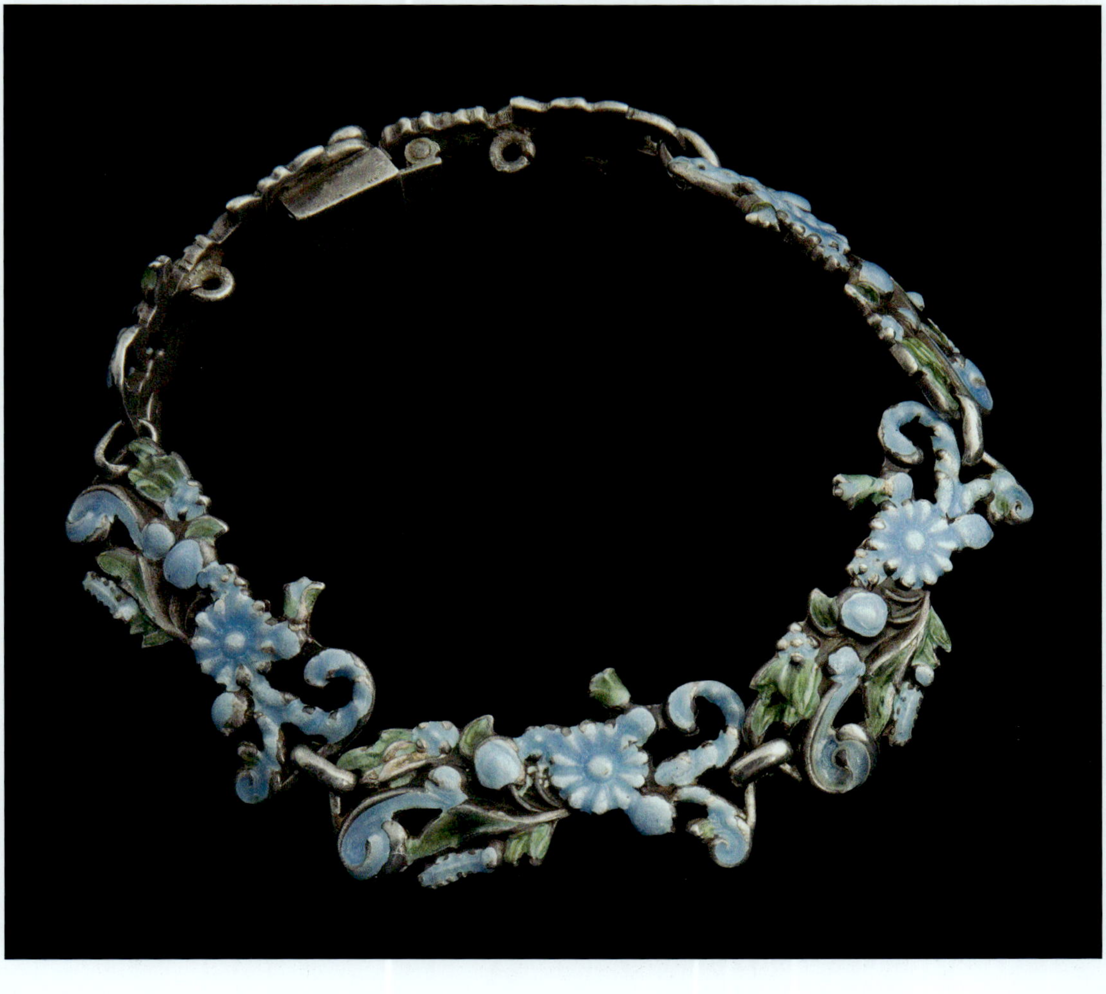

Plate 32 Margot de Taxco 5607. *Tiny Blue Flowers with Green Leaves* bracelet, 7.75 x 0.63 in. Elizabeth Finch Collection

Drawing 5546 *Twirled Lines*

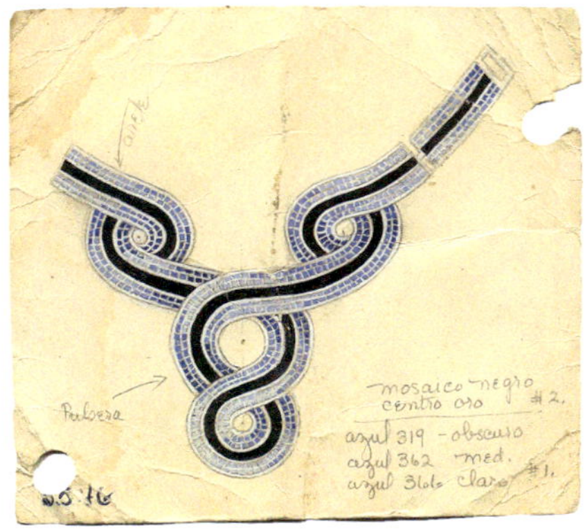

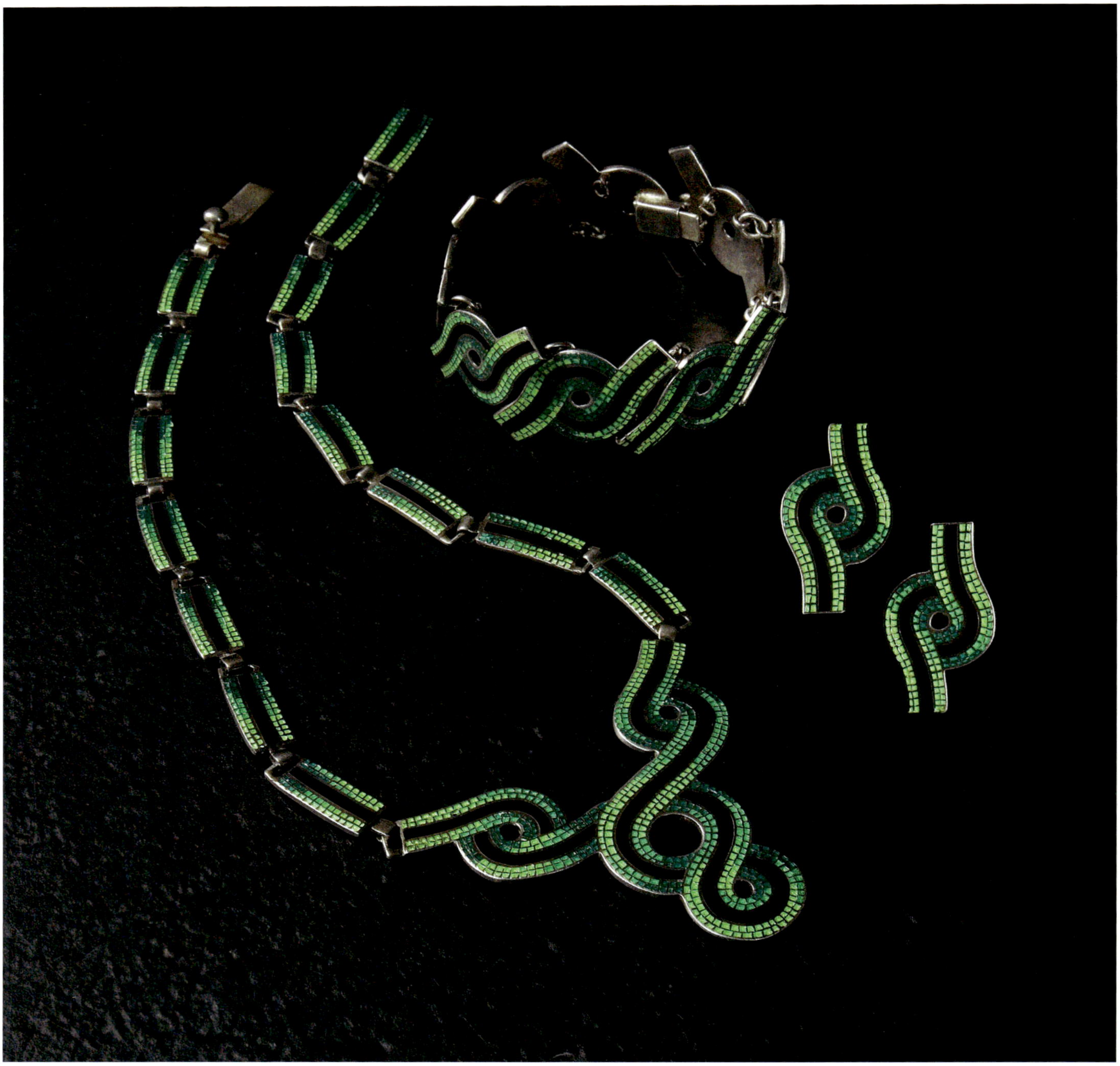

Plate 33 Margot de Taxco 5546. *Twirled Lines Set* in green and black. necklace 16 x 2 in.; bracelet 1.5 x 6.5 in.; earrings 1.5 x .75 in. Carole Berk Collection

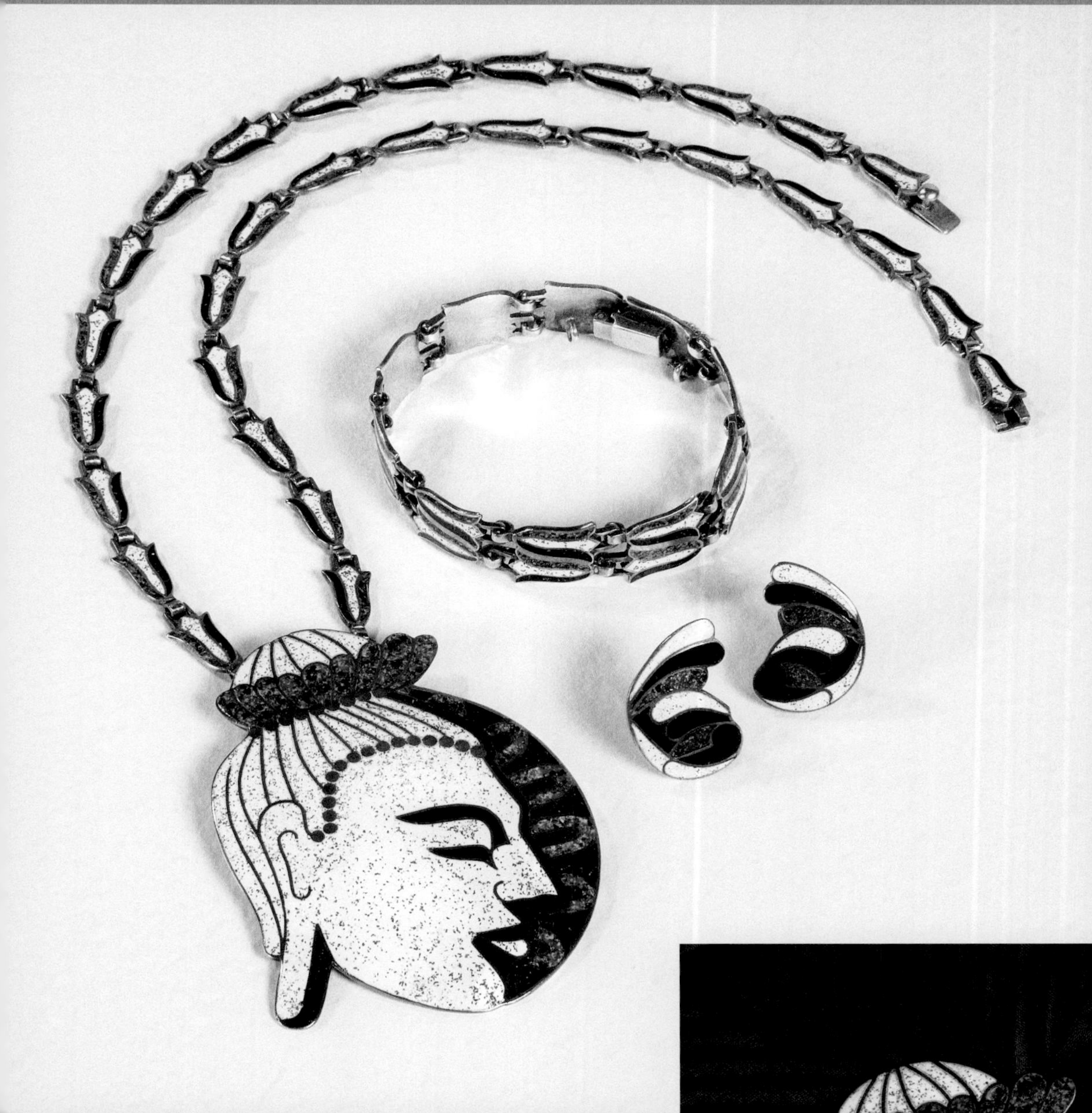

Plate 34 Margot de Taxco 5806. *Thai Figure Set*. necklace 22.25 in.; brooch/pendant 3.13 x 2.5 in.; bracelet 7 x 0.75 in.; earrings 1.25 x 1 in. Things Finer in Santa Fe Collection

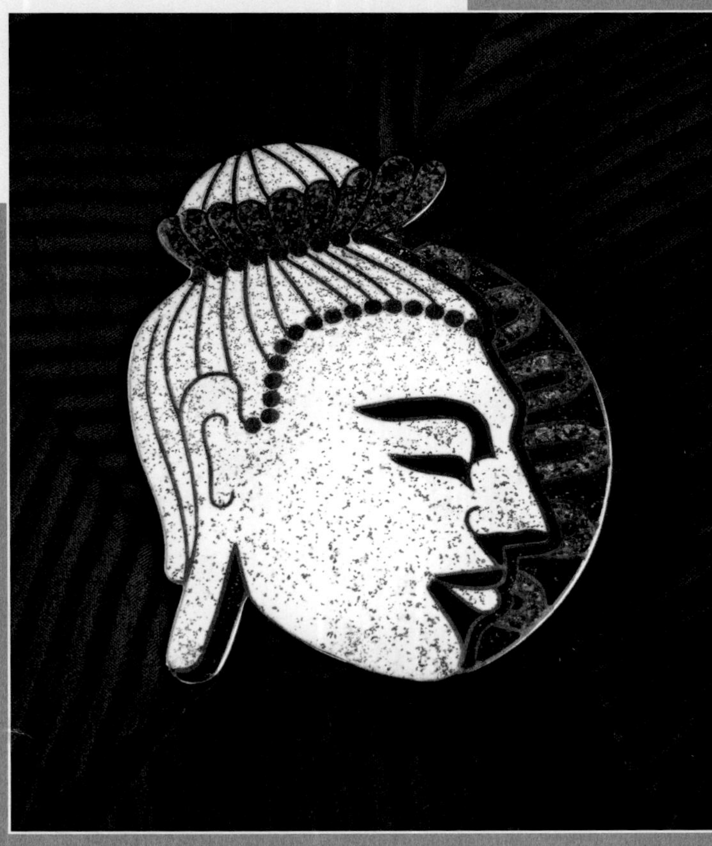

Detail of 5806 *Thai Figure*

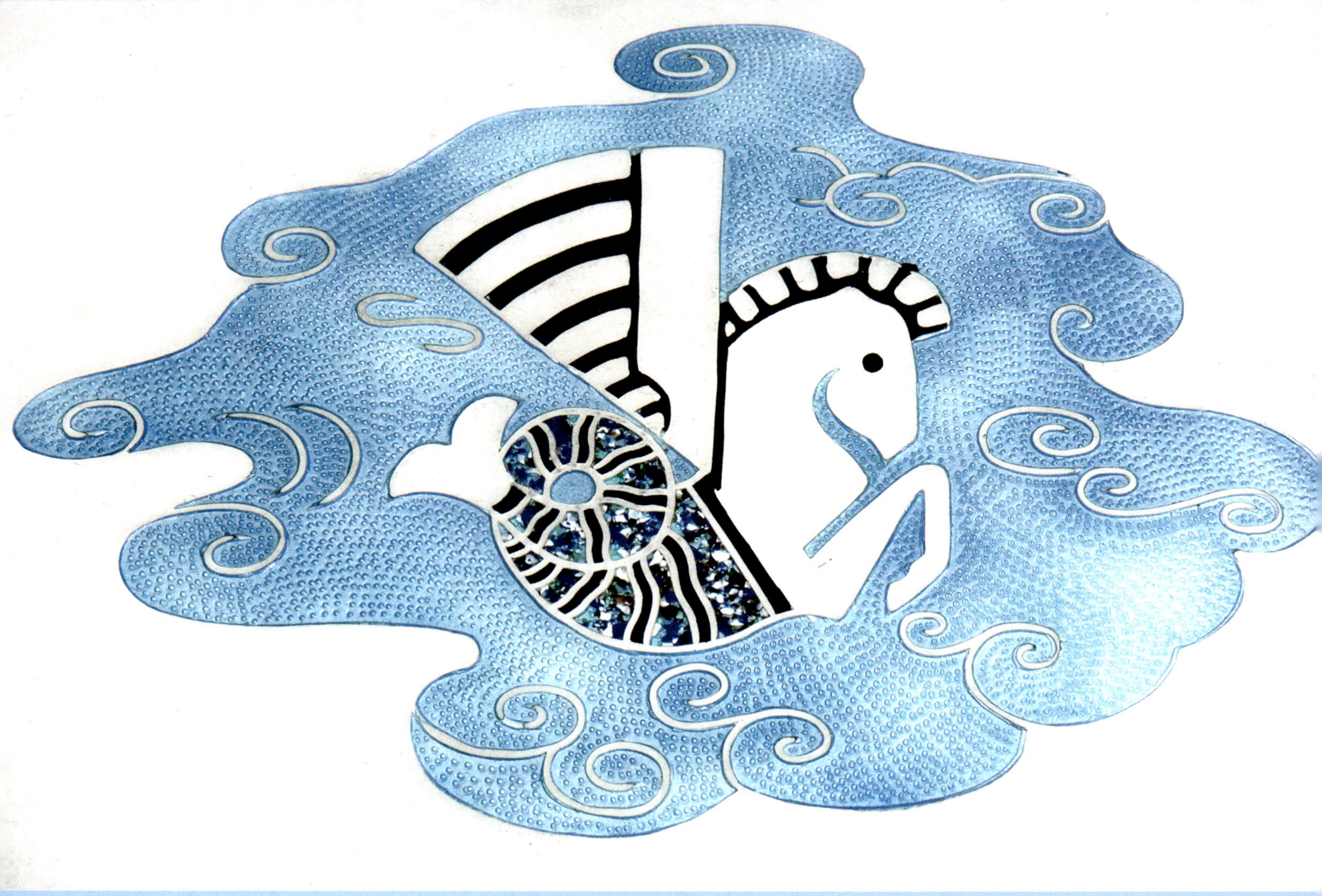

Detail of 5360 *Hippocampus*

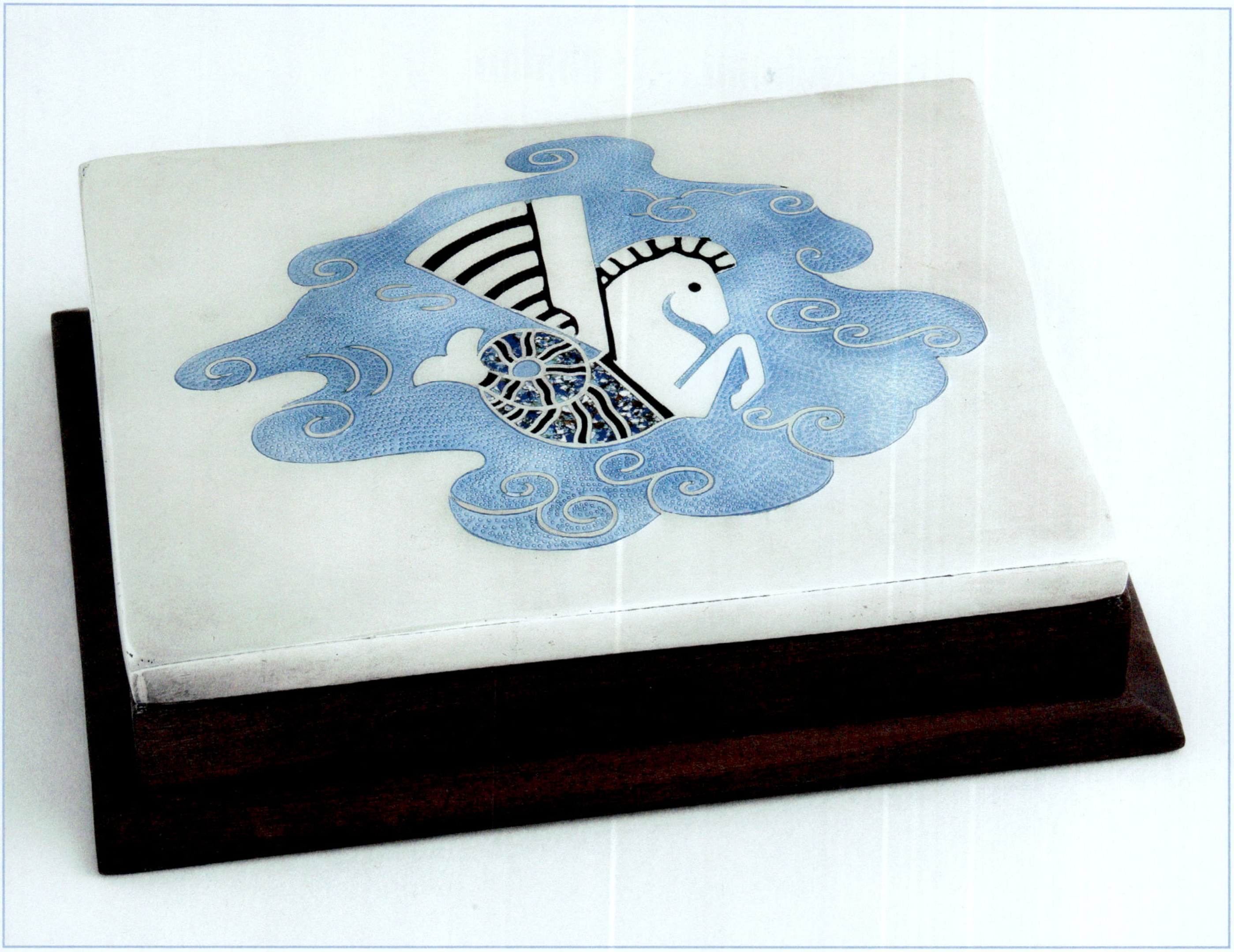

Plate 35 Margot de Taxco 5360. *Hippocampus Cigarette Box* 1.5 x 5.75 x 4.25 in. Private Collection

Part One

Life Before Taxco

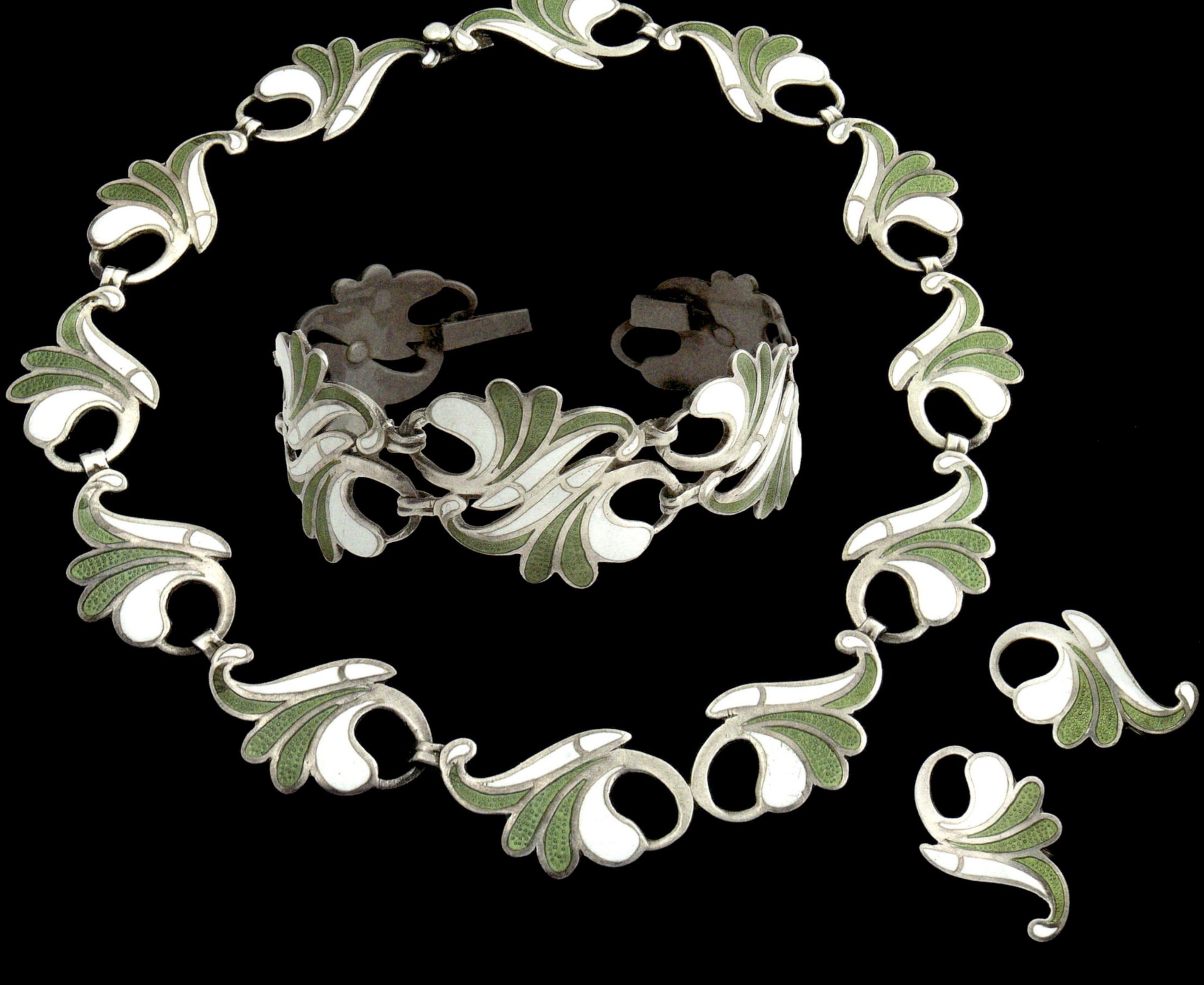

Plate I-1 Margot de Taxco 5626. Green and White *Curved Leaf Set*. necklace 15 x 0.75 in.; bracelet 6.25 x 1.38 in.; earrings 1.88 x 0.88 in. Betsy Vought Collection

Chapter I
Early Childhood

Mystery surrounds many aspects of Margot's early years, symptomatic of the anonymity that can accompany poverty, especially among women. In the late nineteenth century, records of immigration were incomplete and fragmentary, making it difficult to pinpoint the date and place of Margot's mother's arrival into the United States. Margot and her mother, Albina, lived through the San Francisco earthquake and fire, which resulted in the destruction of documents related to births, marriages, and deaths for almost all citizens of that city. Albina spent most of her life as a single woman or widow, and then did not leave a will at time of death, all of which increased the possibility that she (and her daughter) would be overlooked. As a result, Margot's life story has emerged as a patchwork of arcane archival material, brief mentions in newspaper clippings, and the occasional appearance in city directories and census listings. Because much of her life experience before her arrival in Taxco can only be imagined, the narrative of those years must exist within the framework of the few facts that can be extracted from these various documents. Thus with this introductory caveat, we launch the search for Margot's past.

According to Margot's sales manager, Jaime Quiróz, on the long trips he and Margot took to the U.S.-Mexican border, they had hours to spend in conversation and opportunities to share their pasts.[1] Margot informed Quiróz that her birthplace was California and, with this bit of information, the search for the story of her early years began. Because Margot was not always consistent about her age, the discovery of her date of birth required more than a cursory investigation. In the 1930 San Francisco census, Margot listed herself as thirty-one and, in a 1940 border crossing document, she gave her age as thirty-six.[2] The 1900 San Francisco census documents Margot's birth in August 1896 in San Francisco, which means that she was forty-four in 1940.[3] However, as we will discover, this particular entry for Margot is filled with inaccuracies that intentionally obfuscate and have complicated efforts to search out the truth.

The 1900 census lists Margot's parents, Albert and Albina Durand, and their ages, thirty-six and thirty-seven respectively.[4] Albert Durand gave a birth date, 1864 without a month, and his birthplace as Missouri. His parents are listed as immigrants, his father born in France, his mother, in Italy. Albina claimed to have been born in October 1862 and that she had arrived in the United States in 1880. Albert and Albina told the census-taker that they had been married for four years, thus, a marriage that would have taken place in 1896.[5]

Margot's mother, Marie Albine Chabot, was baptized in the Roman Catholic parish church of St. Simon-de-Bagot, Quebec, on October 9, 1861.[6] She was the ninth of twelve children, born to Cecile Jarred dite Beauregard and Jean Baptiste Joseph Chabotte. J. B. Joseph worked a farm in Quebec that he must have inherited from his father, Joseph.[7]

In 1880, at the age of nineteen, Albina emigrated from Canada to the United States[8] to find work in the flourishing textile towns along the Merrimack River in Massachusetts. What may have drawn her to this region was the presence in Massachusetts of an aunt and uncle, her father's sister and the sister's husband.[9]

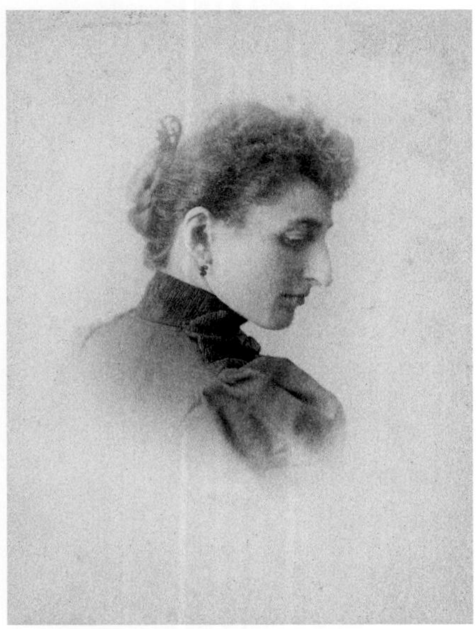

Plate I-2 Albina Chabot. Photograph by Marshall M. Hayden, Lowell, Massachusetts, c. 1880, Elizabeth Freeman Collection

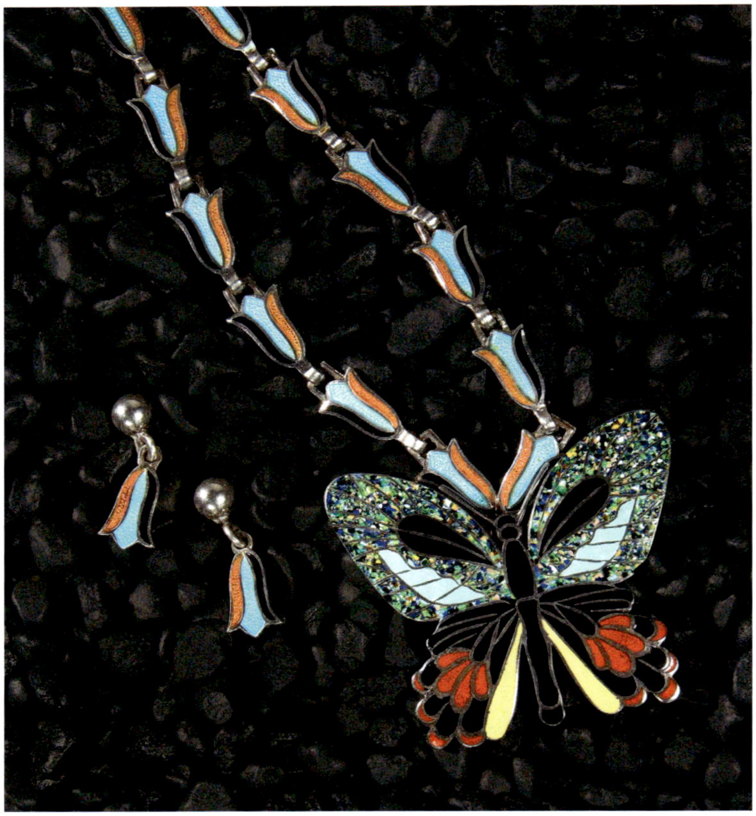

Plate I-3 Margot de Taxco 5628. *Butterfly of Many Colors* brooch 2 in. x 2.13 in. James Black and James Tracy Collection

Albina and her family would have been welcomed into the substantial French Canadian enclave in Lowell, where French was spoken and cultural practices were maintained. According to one author, "Very large numbers of people who were born in the French provinces of Canada immigrated to Lowell and found work in the textile mills. Often relatives were waiting for them with advice, jobs, and housing."[10]

The photograph of Albina as a young woman places her in Lowell, Massachusetts sometime after 1885 (Plate I-2),[11] where Francis C. Lowell had developed a model textile manufacturing industry that employed young women from rural communities. An 1846 description of factory life in Lowell provides a glimpse of what might have been Albina's regimented daily routine: "The operatives work thirteen hours a day in the summer time, and from daylight to darkness in the winter. At half past four in the morning the factory bell rings, and at five, girls must be in the mills. . . . At seven the girls are allowed thirty minutes for breakfast, and at noon thirty minutes for dinner."[12]

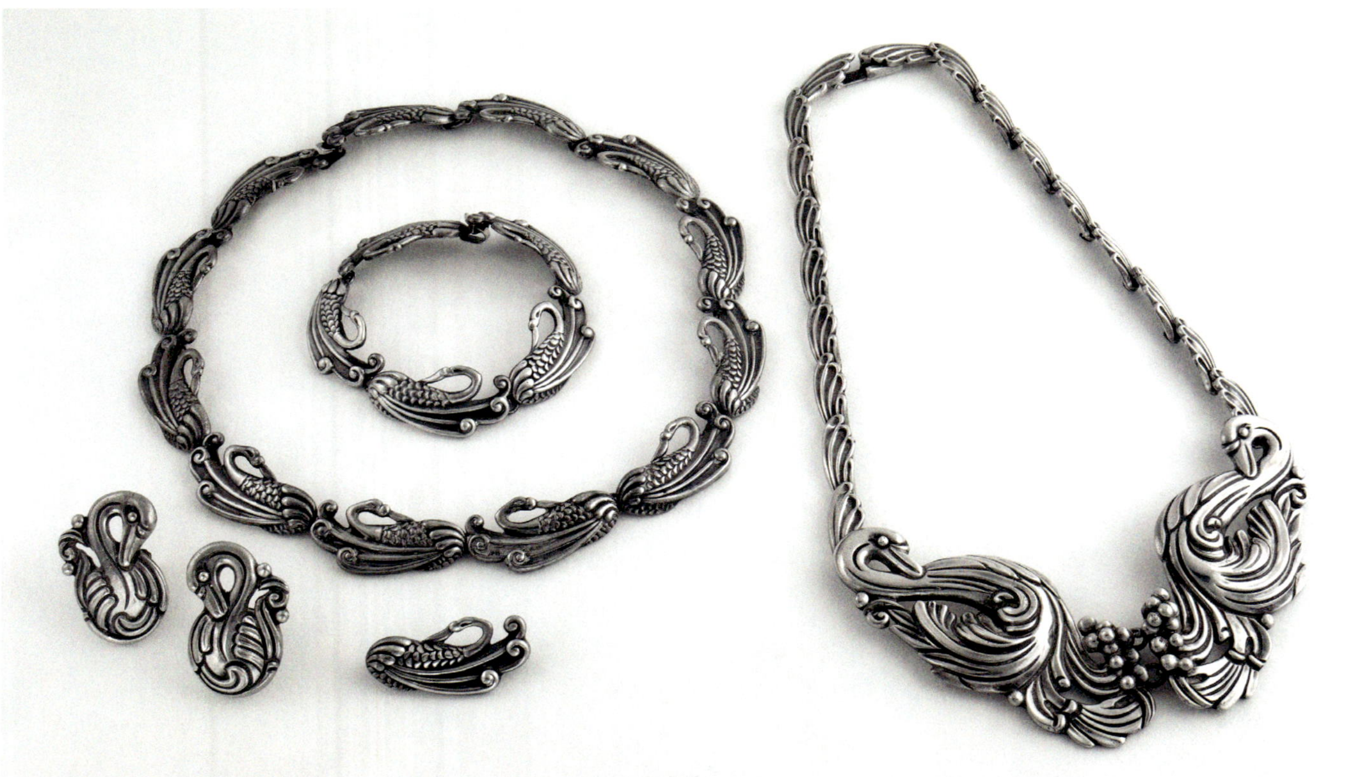

Plate I-4 Margot de Taxco 5320. *Swan Set* in silver. small necklace 14 x 0.75 in.; bracelet 7 x 0.75 in.; brooch 0.75 x 1.25 in.; earrings 1.25 x 0.75 in.; large necklace 15.5 in., central motif 3 x 4.5 in. Private Collection

Albina was an employee of Lawrence Hosiery in 1891-1892 and a boarder at 113 Salem. In 1893, Albina was working as a dressmaker and renting rooms at 59 Read, a house owned by Aimé Gauthier.[13] An officer of *Le Cercle Canadien* for a number of years, Gauthier was a founder of *L'Etoile*, the Franco-American newspaper published in Lowell for over seventy years (1886-1957).[14]

Albina's brother Robert (his given name was Remi Chabot) had come to the United States in 1877, and had traveled to the West.[15] Both siblings seemed to have sought out aunts or uncles on their father's side who were living in the United States and who might provide them with initial support. Anthony Chabot (Antoine Chabot, his given name), one of Albina and Robert's uncles, was a successful entrepreneur in Oakland, California, who had developed water and gas companies in the San Francisco Bay area. Beyond these major enterprises, Anthony Chabot's biographer adds, "Despite his lack of formal education, this self-made, restless, French-Canadian immigrant had widespread interests. He invested heavily in industry and agriculture. Many prominent West Coast industries owe their origins to Chabot's shrewd investments." Chabot was a philanthropist on a grand scale, providing funds for community organizations in Oakland, including the Women's Sheltering and Protection Home, a hospital, and the Chabot Observatory.[16]

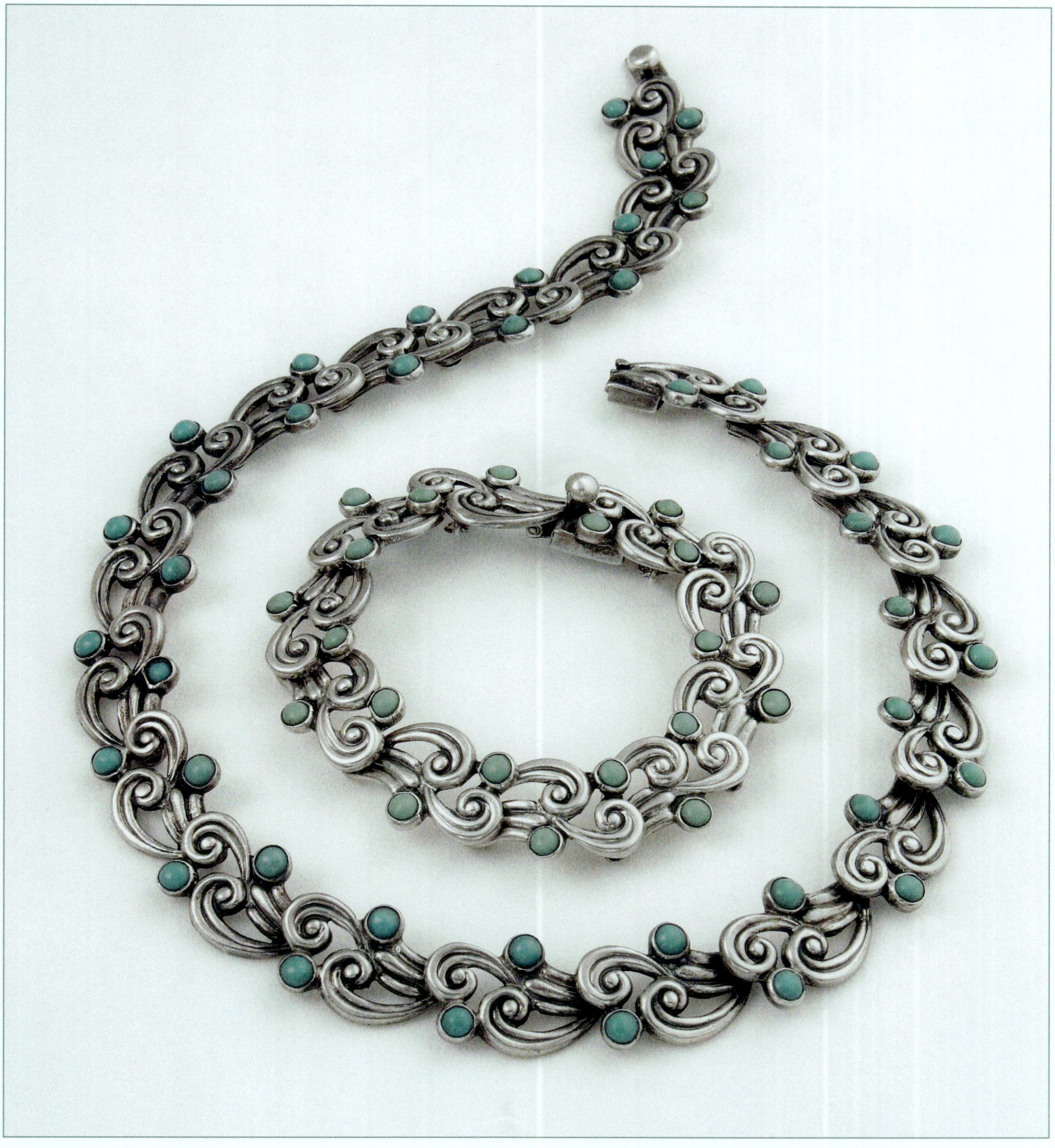

Plate I-5 Margot de Taxco 5312. *Silver Curls with Small Turquoise Cabochons Set*. necklace 14.5 in.; bracelet 7 in. Private Collection

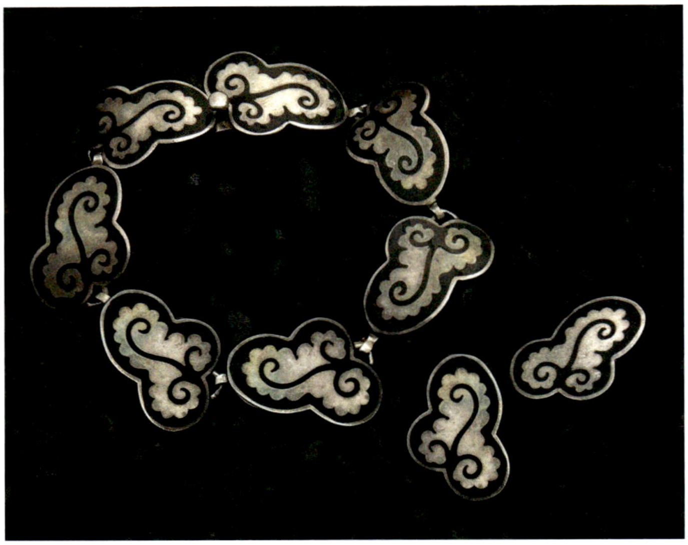

Plate I-6 Margot de Taxco 5570. *Tripartite Curl Set* in black and silver. bracelet 7.75 x 0.75 in.; earrings 1 x 0.75 in. Elizabeth Finch Collection

Upon his arrival, Robert lived in Oakland with his uncle Remi, Anthony's younger brother and business partner, and Remi's wife Emelie. Anthony Chabot had initiated the development of the first cranberry bogs in the Northwest. This remarkable attempt was instigated by a visit from his brother-in-law, who had come from Massachusetts and taken note of the marshy soil. With his assistance, Anthony acquired the plants and found someone with expertise in cranberry cultivation. Anthony's sister refused to leave Massachusetts and Anthony and Remi's nephew, Albina's brother Robert, developed the cranberry industry.

In 1883, Robert departed Oakland for Ilwaco in Washington to take charge of his uncle's cranberry bogs. After eleven years of successful harvests, in 1892, Robert left Ilwaco to establish his own cranberry culture farther up the Washington coastline. By that time, both of his uncles had died, Anthony in 1889, and Remi in 1890. Emelie owned controlling interest in the cranberry company after her husband's death, but she seemed to have lost interest and sold her shares. Emelie's actions probably instigated Robert's decision to leave Ilwaco.[17]

Considering her brother Robert's move to Washington in 1883 and the deaths of her uncles, what prospects could have motivated Albina to leave Massachusetts for California? Did she travel to the West with a relative? The train ride was lengthy and arduous, and dangerous for a young woman, especially if she were traveling alone. The first transcontinental track had only been completed in 1869, so Albina would have passed through vast expanses of land, uninhabited except for the occasional farmer or a group of Indians on horseback.[18]

Albina must have arrived in California between 1894 and 1896, the year she married Albert Durand and gave birth to their daughter Margot. After her marriage, evidence bears out the possibility that Albina remained in contact intermittently with Ellen Chabot, Anthony's daughter, or Emelie, Remi's widow.[19] Family support may have been extended to Albina as early as 1897, soon after Margot's birth in San Francisco.

Albert Durand makes his first appearance in San Francisco in 1897. The city directory provides an address, 2207 ½ Pine between Fillmore and Webster Streets, where he must have been living with Albina and Margot. The reason for this assumption is that, according to the 1897 directory, Albert was working

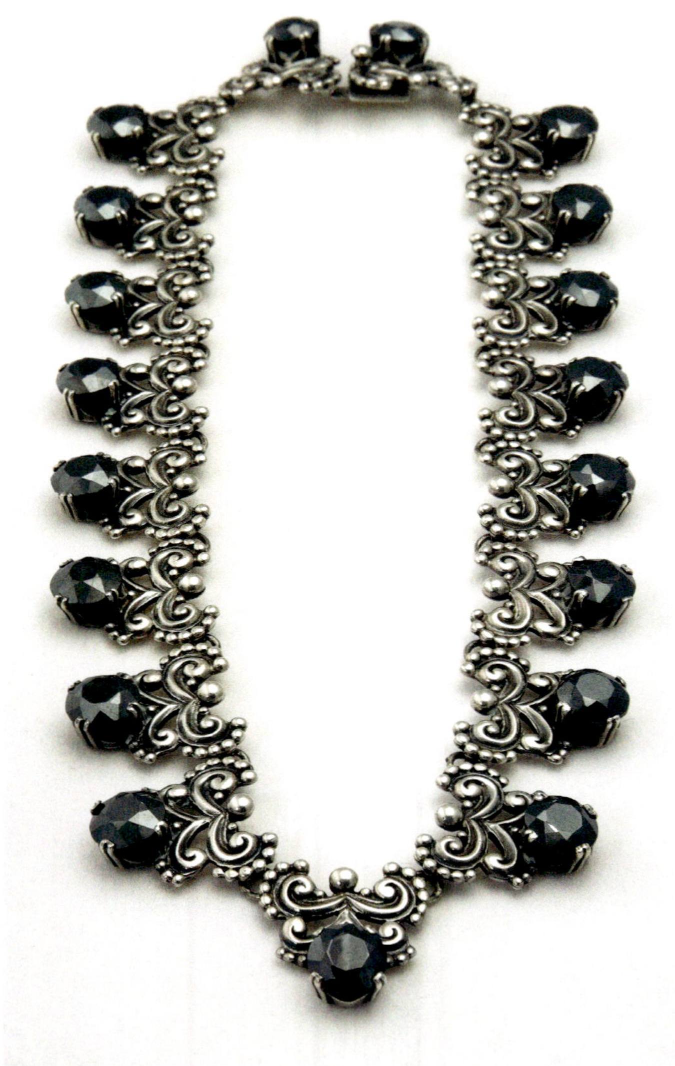

Plate I-7 Margot de Taxco 5302. *Faceted Hematite* necklace 16 x 1 in. J. P. Mather Collection

as a collector for the Gas Consumers' Association. As with most of Albert and Albina's story, coincidences direct assumptions about who these people were as a family. The fact that Anthony Chabot had been heavily involved in the business of gas and water supply in the Bay area points to the possibility that Albert's position with the Gas Consumers' Association was obtained through the influence of a Chabot family member. The only person who could have made this happen for Albert was Ellen Chabot.[20] She was executor of her father's estate and a major stockholder in several of his corporations.[21]

In 1898, Albina and Albert had a second child, Maurice George Durand, who died after only one year of life. During this period, the family was still at the same address, but Albert had become an adjuster for the Gas Consumers' Association.[22] Albert (also listed as Allie) moved with his family to 1109 ½ Leavenworth Street between Sacramento and California Streets in 1899, perhaps after Maurice's death, where they lived for four years.[23]

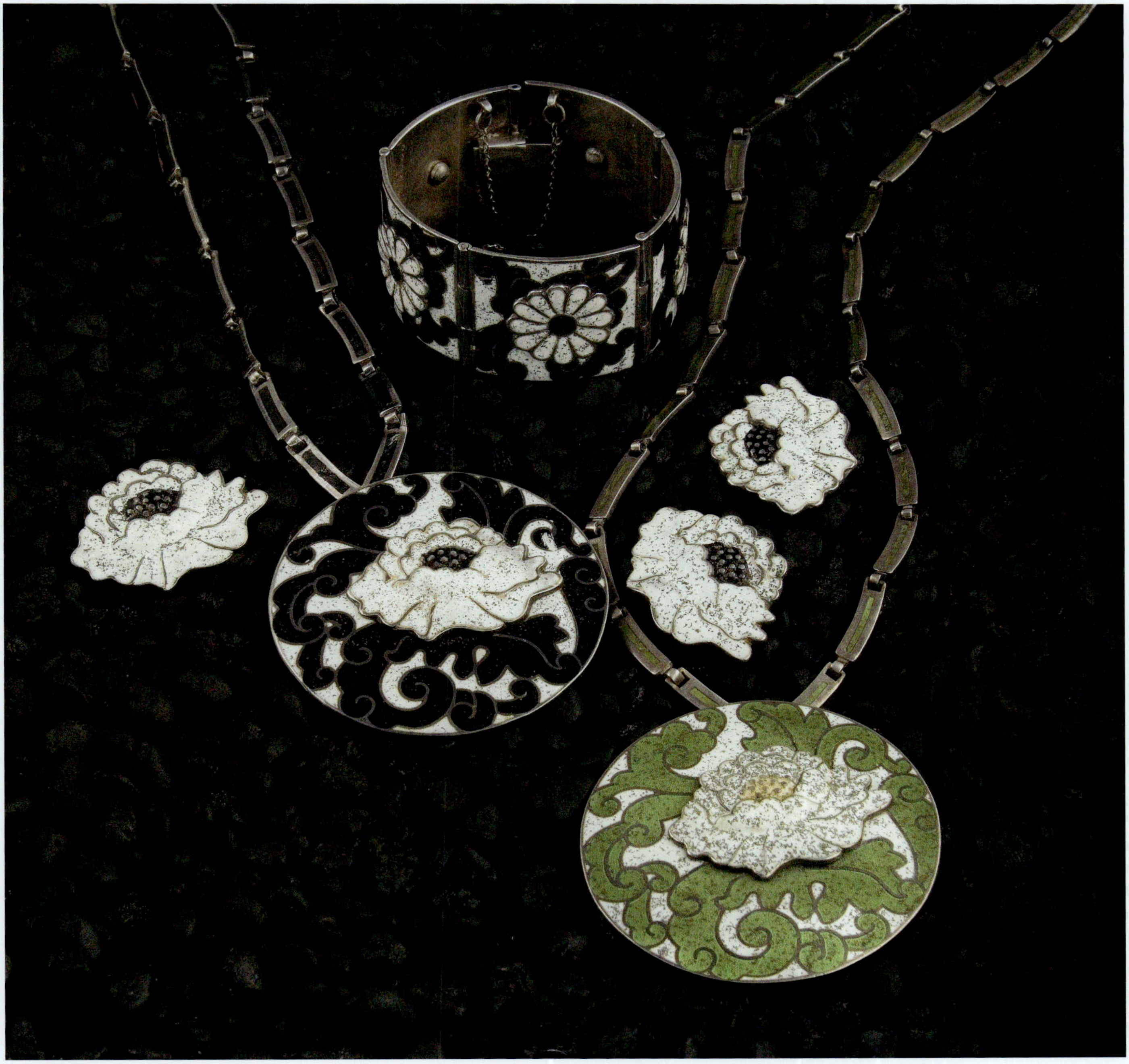

Plate I-8 Margot de Taxco 5773. *Florid Chrysanthemum* with 5744A. black necklace 19 in., green necklace 21 in., pendant 2 in. d.; cuff bracelet 1.25 x 7 in.; brooch and earrings 1.5 x 1 in. Private Collection

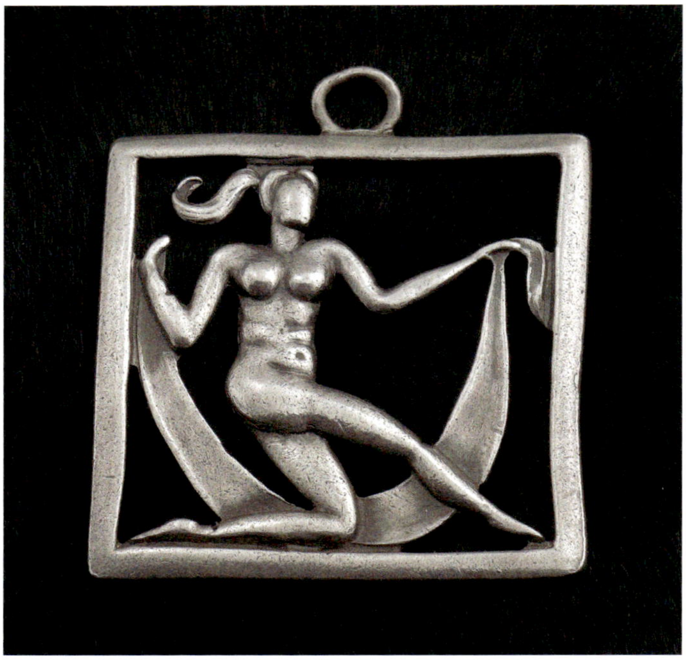

Plate I-9 Margot de Taxco 5446. *Female Nude Square* pendant 1.25 in. sq. Private Collection

Margot's childhood story took a bizarre turn in 1903. The *San Francisco Chronicle* for January 7, 1903 contains the following announcement: "Died in Chair — Allie Durand, an adjuster for the Gas Consumers' Association, while sitting in a chair at his home, 1109 ½ Leavenworth Street, on Monday evening [January 5], fell over dead. Durand was formerly under the care of a physician and suffered from heart disease. He was 36 years of age."[24] Albert Durand was buried at Holy Cross Cemetery in Colma, next to his son Maurice George Durand.[25]

The 1903 city directory gives the same Leavenworth Street address for "Mrs. Allie Durand (widow)."[26] Inexplicably, Albina Durand and her daughter disappear from the San Francisco directory for 1904. It is possible that Albina could have gone with Margot to Washington for a visit with her brother and his family. Robert's daughter Elizabeth Chabot Forrest once mentioned to her daughter that as a child she remembered a visit from Albina and her little cousin Margot sometime before she left for boarding school in 1906.[27]

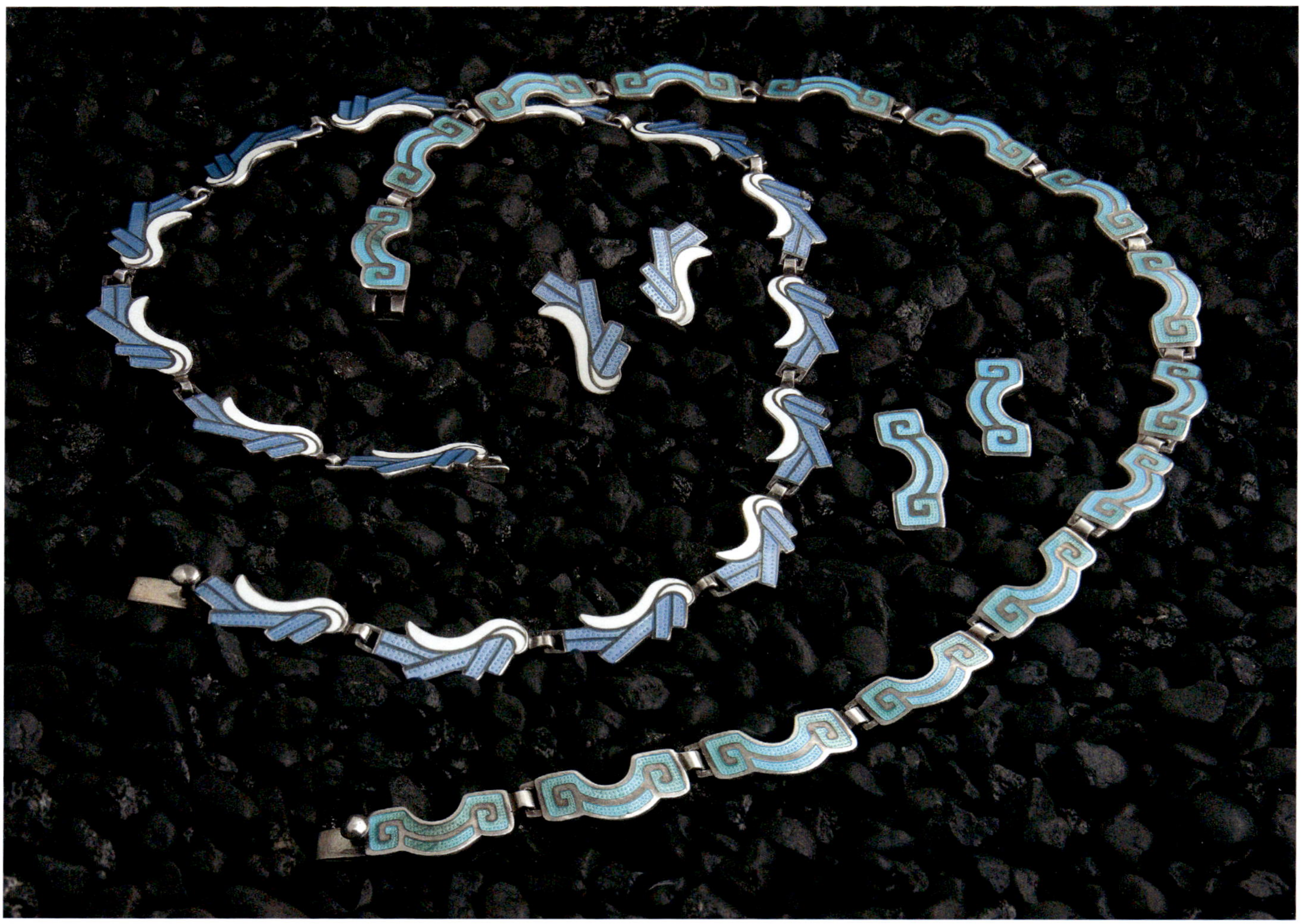

Plate I-10 Margot de Taxco 5692. *Lines from Scroll Set.* necklace 14 in.; earrings 1 x 0.5 in. Margot de Taxco 5796. *Double Scroll Set*. necklace 14 in.; earrings 1 x 0.25 in. Private Collection

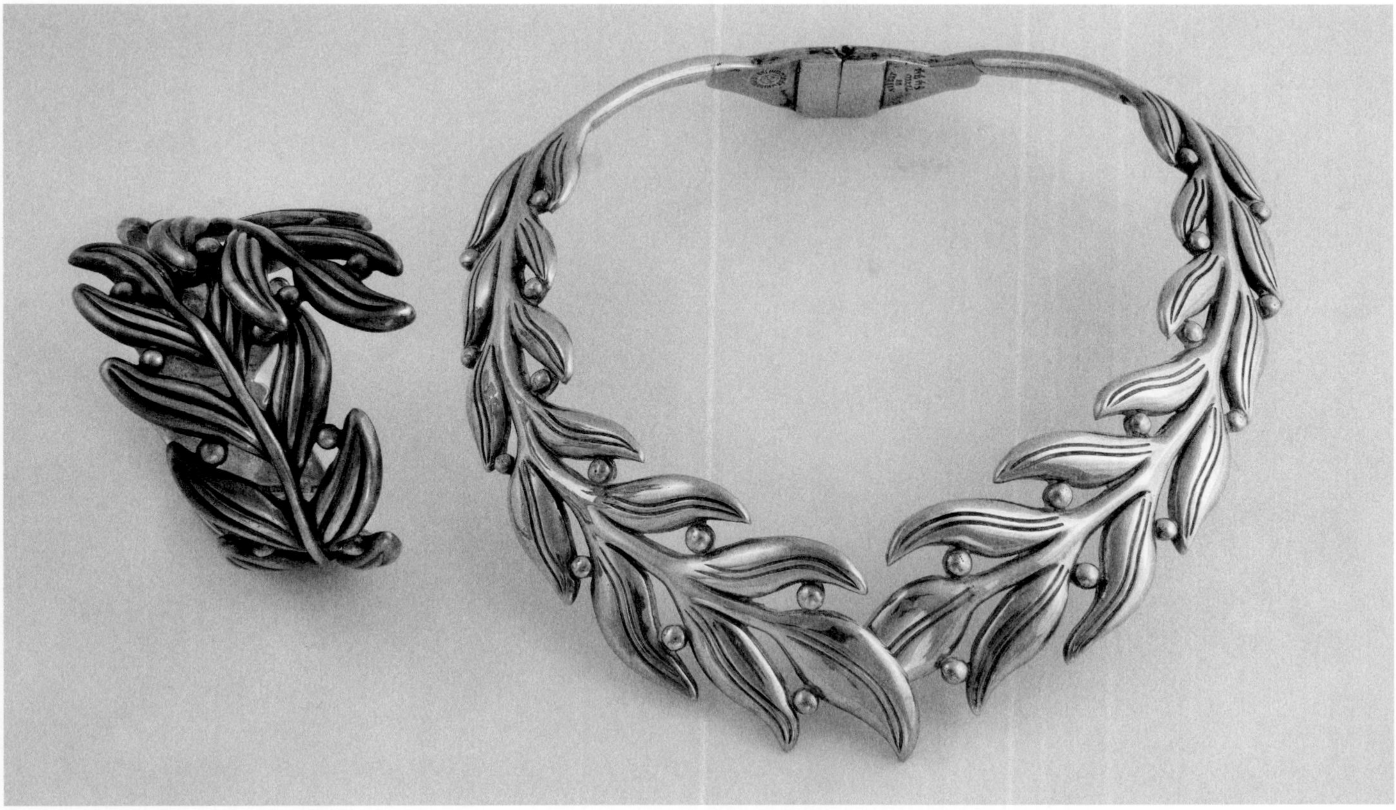

Plate I-11 Margot de Taxco 5474. *Silver Leaves Set*. necklace 15 x 1.5-5.5 in.; bracelet 7 x 1.25 in. Private Collection

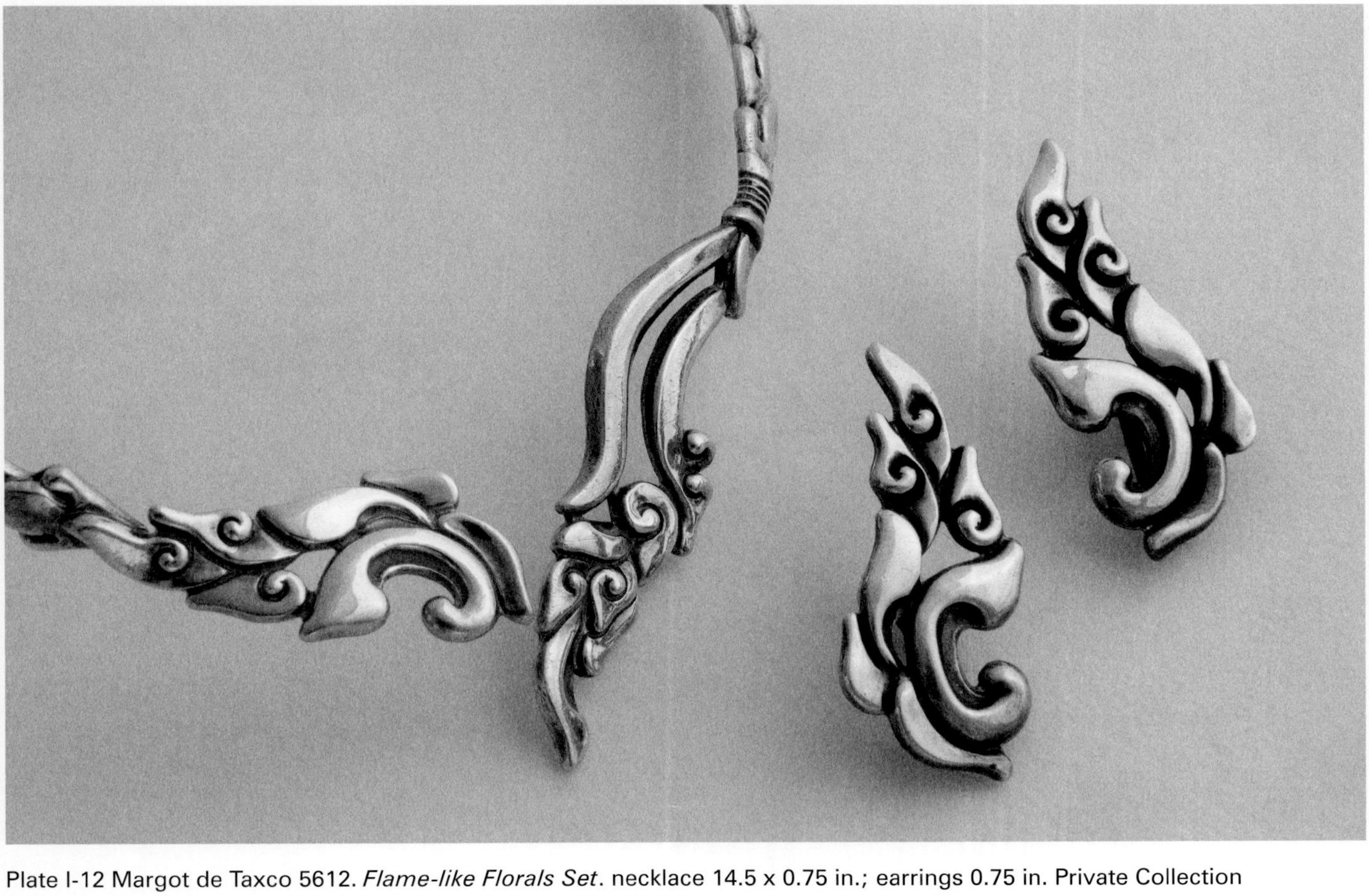

Plate I-12 Margot de Taxco 5612. *Flame-like Florals Set*. necklace 14.5 x 0.75 in.; earrings 0.75 in. Private Collection

It is at this point that almost unanswerable questions arise concerning Margot's childhood and family history. Her mother reappeared in 1905, but she is listed in the directory not as Mrs. Allie Durand, but as Mrs. A. Durrand Voorhies, hairdresser, at the address, 2132 Fillmore.[28] One document exists that has shed light, although indirectly, on the name change from Durand to Voorhies for Margot and her mother. In Margot's 1940 border-crossing permit, the immigration officer who interviewed her made notes on the document. This interview was required since Margot had lost her birth certificate and other family papers during the 1906 San Francisco earthquake and fire. What follows are the immigration officer's notes:

"Claims father, Gr[and]fath[er] born usa. Presented affidavit of one Alice Goyette of San Fr[ancis]co, Cal, stating she is US cit[izen] 3-1-40. Presented letter from US consul MX DF [Mexico City] 10-11-40 showing application for W.S. reg. cert. had been made. Also newspaper clipping from New O[rleans], La. showing death of her Gr[and]father who was former Lieutenant Gov[ernor] of L[ouisian]a."[29]

Thus, in 1940, Margot claimed that her grandfather had been lieutenant governor of Louisiana. This man was, in all likelihood, Judge Albert Voorhies. According to the January 21, 1913 obituary in the New Orleans *Daily Picayune* (which was probably the clipping in Margot's possession), Albert Voorhies was Lieutenant Governor of Louisiana in 1864-1867 and had been a member of the Louisiana state legislature. He had served on the Louisiana State Supreme Court and practiced law in New Orleans. Judge Voorhies was also on the Board of Health for the city of New Orleans.[30]

Albert Voorhies's marriage to Marie Leontine Durand brought together two prominent Louisiana Creole families. Their wedding at the Durand family's Pine Alley plantation near St. Martinville, Louisiana has taken on mythic status: "[Pine] alley, planted in 1829 by slave labor, originally extended to Bayou Teche and was 3 miles long. At the end of the grove is a large sugar cane plantation owned prior to the War Between the States by Charles J. Durand, who has contributed to one of the section's most fantastic legends. In 1850, when the marriage of his daughter,

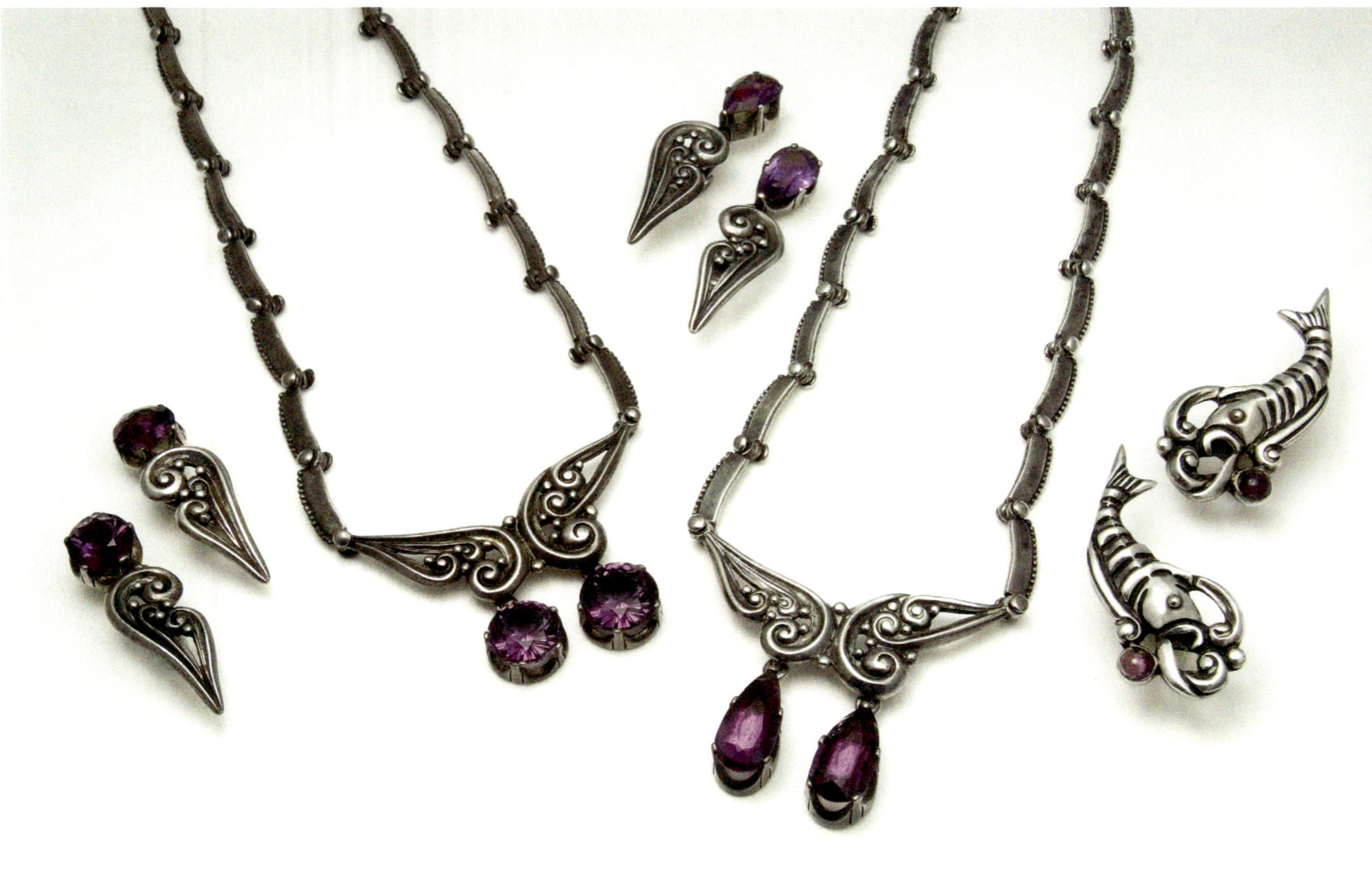

Plate I-13 Margot de Taxco 5344. *Curled Forms Sets*. necklaces 16 in.; earrings 1.75 x 0.5 in. *Fish* earrings 1.5 x 1 in. Private Collection

Leontine, drew near, it is said that Monsieur Durand imported from China a quantity of giant spiders, which spun enormous webs between the trees in the grove; then he sprinkled the webs with gold and silver dust to form a glittering arch for his daughter's wedding procession."[31]

Albert and Leontine Voorhies were presumably Margot's grandparents. They had eleven children, the seventh, Albert A., born on November 23, 1860.[32] His death date, January 4, 1903, was recorded in a Voorhies family history, but no gravesite has been discovered in Louisiana.[33] Thus, it is more than coincidence that the date of Allie Durand's death was January 5, 1903.

The 1900 census now holds more questions than answers in the search for information concerning Margot's father Albert (Allie) Durand. What was the reason for the name change? Judge Albert Voorhies was a prominent man in New Orleans. According to

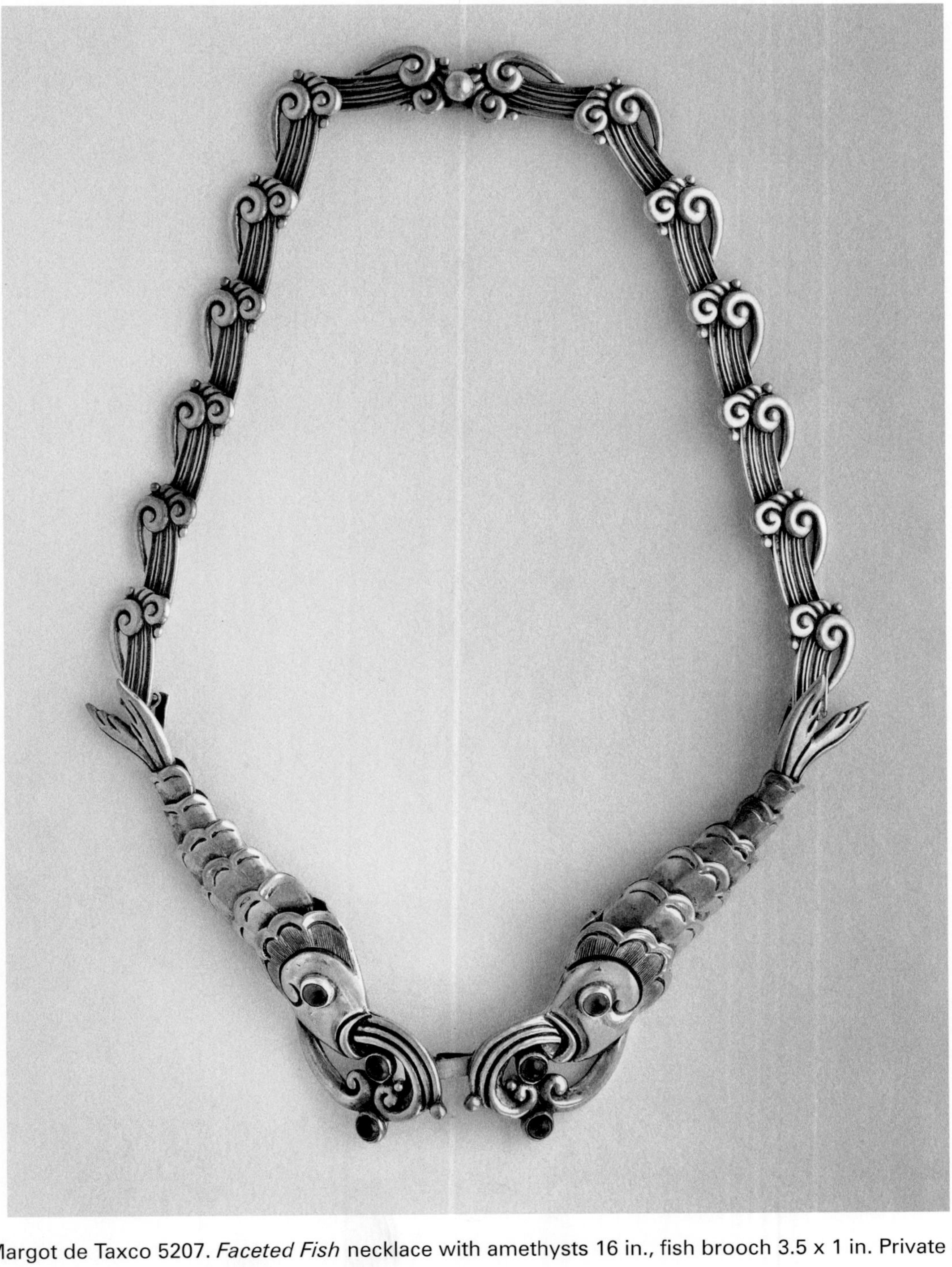

Plate I-14 Margot de Taxco 5207. *Faceted Fish* necklace with amethysts 16 in., fish brooch 3.5 x 1 in. Private Collection

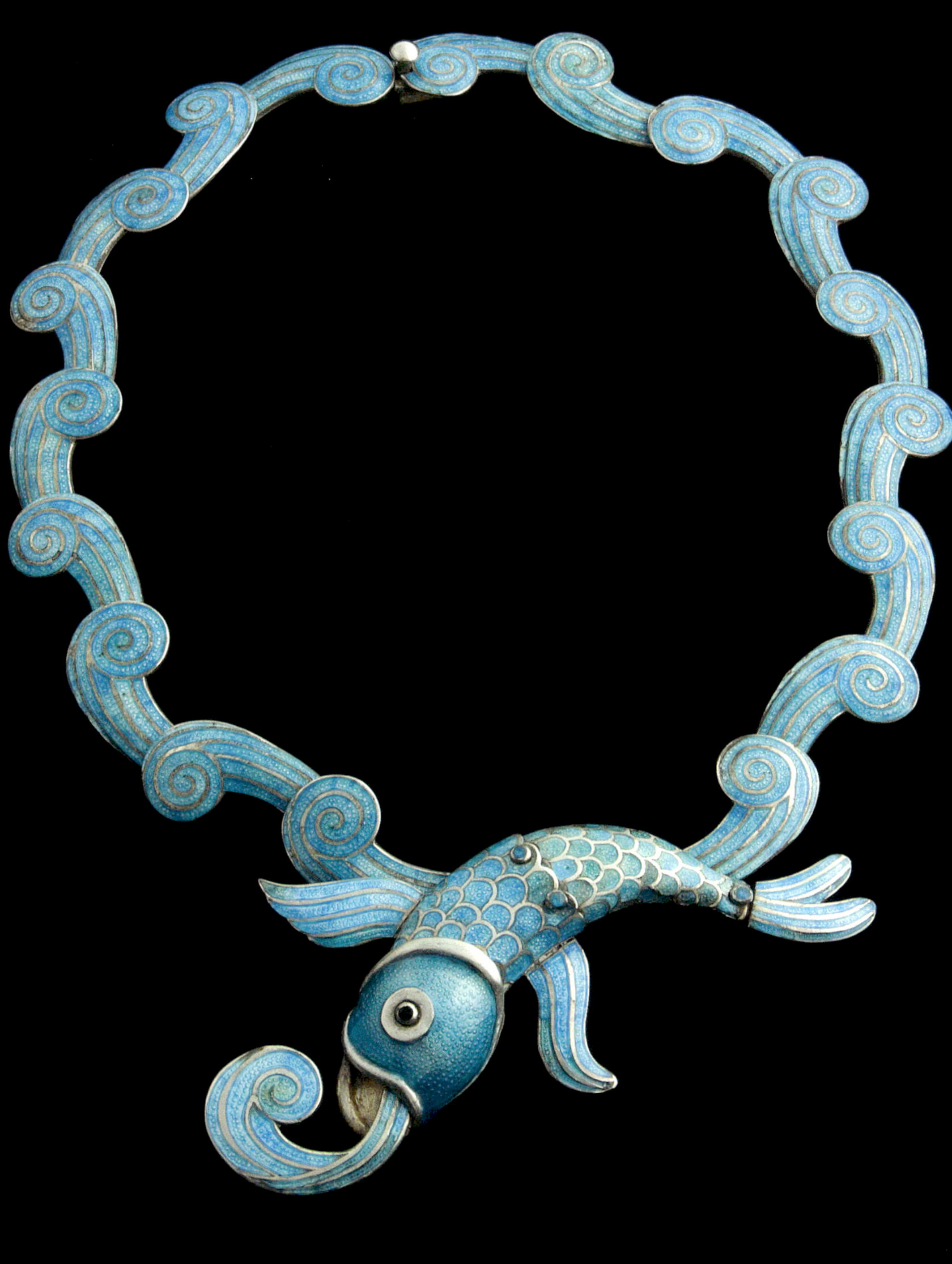

Plate I-15 Margot de Taxco 5644. *Fish in the Waves* in blue, necklace 14 in., central motif 2 x 3.5 in. Betsy Vought Collection

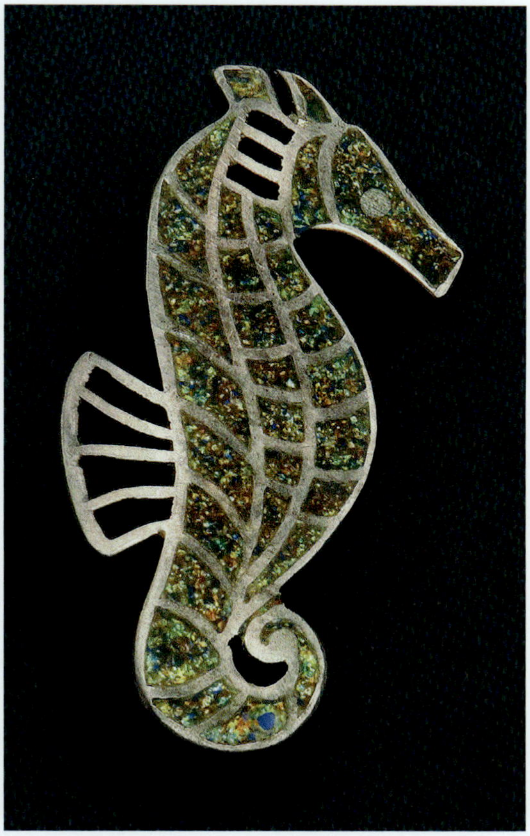

Plate I-16 Margot de Taxco 5424. *Mélange Enamel Seahorse* brooch 1.25 x 0.5 in. Elizabeth Finch Collection

the 1893 city directory for New Orleans, Albert the son was living with his parents at 74 Orleans and working as a clerk at the Board of Health. Since Judge Voorhies was a member of the Board of Health, it is most probable that he helped his son acquire the position.[34]

Upon leaving home, the younger Albert worked for a print shop and had an address change every year between 1894 and 1897.[35] These minor clues from the past indicate that Albert the son may have been a ne'er-do-well who left New Orleans under rather murky circumstances, or he may have had physical or mental health problems that ultimately led to his death from heart disease at an early age.

In spite of the modicum of evidence that has survived, the assumption must be that the younger Albert decided to take on a new identity as he made his way to the West. It is more than a coincidence that he adopted his mother's maiden name, Durand, as his own. He further abandoned his past when he claimed in the 1900 census that he was from Missouri and that his parents were immigrants. While Albert's birth year is known to have been 1860, Allie Durand gave his age as 36 in 1900 and was said to be 36 at the time of his death in 1903. However, even though the birth year occasionally changed, Albert continued to designate November as his birth month whether as a Durand or a Voorhies.

The interval between Margot's birth in August 1896 and the final listing of the younger Albert Voorhies in the 1897 New Orleans city directory raises the ultimate question of parentage. One possible explanation for this discrepancy is that the city directories for New Orleans reflected the addresses of the former year. Another is that Albina was pregnant or had Margot by another

Plate I-17 Margot de Taxco 5644. *Fish and Waves* necklace 14 x 0.63 in., central motif 3.38 in. Margot de Taxco 5693. *Small Fish Set*. brooches 1.5 x 0.88 in.; bracelet 6 x 1.5 in.; ring 1.25 in. Margot de Taxco 5844. *Plump Fish* brooch 1.25 x 2 in. Private Collection

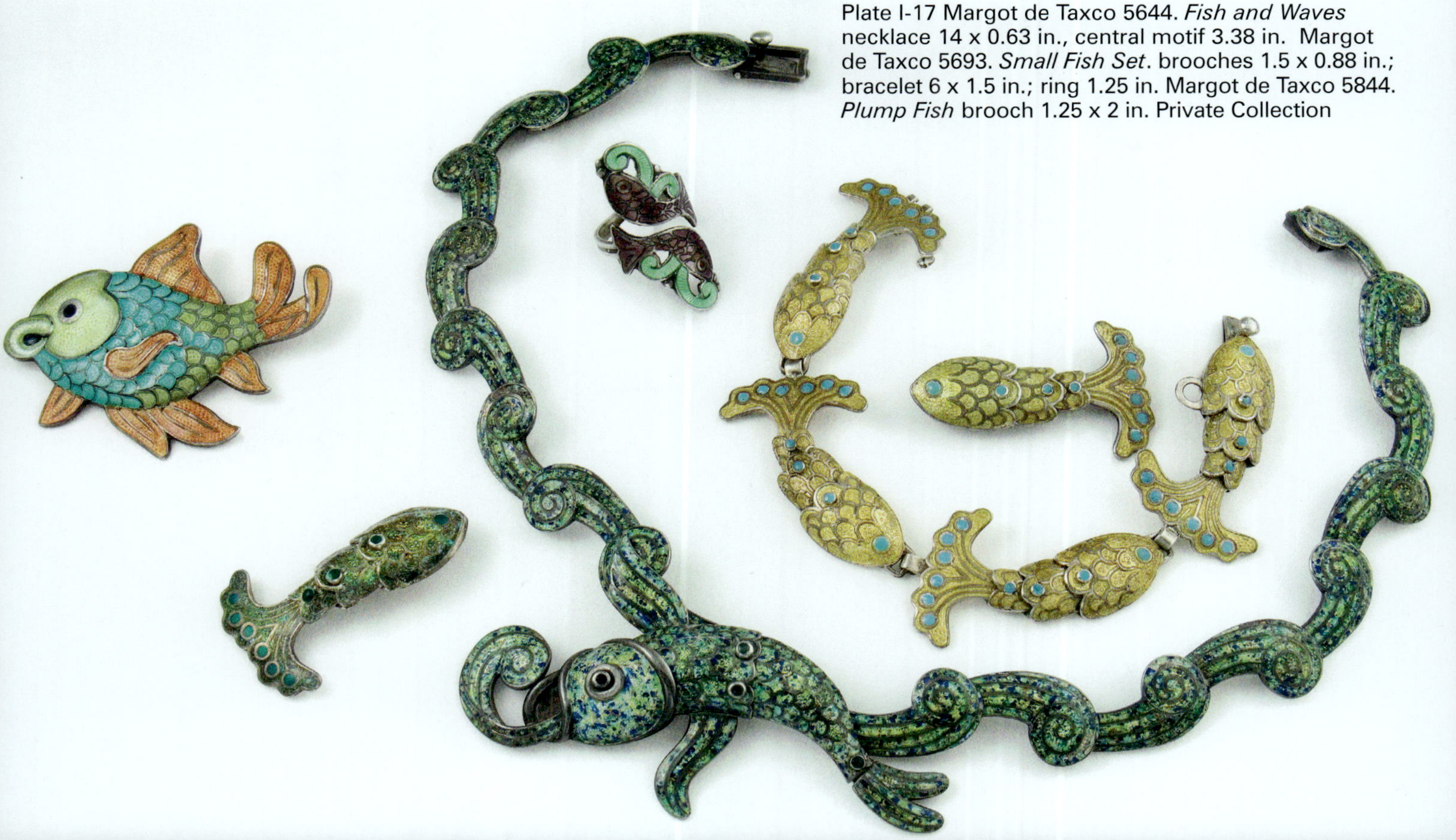

man before meeting and marrying Albert Durand in the latter part of 1896. Albert Durand does appear in the San Francisco directory for 1897, and it is certain that Albert and Albina were married when they had a son, Maurice George Durand, who died in 1899 at the age of one. Albert's burial place next to Maurice, who received the name Durand, seems to be verification of this father-son relationship.

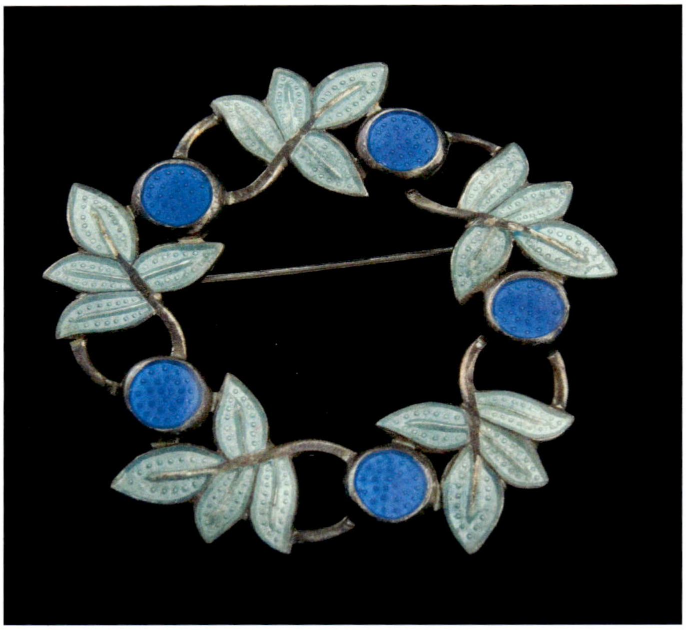

Plate I-18 Margot de Taxco 5614. *Leaf and Berry* brooch 1.75 in. circum. Private Collection

Even though the evidence that exists is tenuous, the scribbled notes on the 1940 border-crossing document continue to point to Albert A. Voorhies as Margot's father. Additionally, the 1930 census states that Margot's father was French Canadian and her mother from Louisiana, origins that were reversed by the census taker.[36] Perhaps the most convincing evidence is Albert's obituary, which reads, "Allie, beloved husband of Albina Durand, and father of Margot Durand."[37] Finally, Robert's daughter Elizabeth Chabot Forrest inscribed the name "Albina Chabot Voorhies" on Albina's portrait photograph taken in Massachusetts, thus indicating the Chabot family's acceptance of Albina's marriage to Albert A. Voorhies.[38]

It must have been a shocking revelation for Albina, to discover that the man she had married seven years earlier and with a name she had taken as her own, was actually Albert A. Voorhies. It can be surmised that Albina discovered Albert Voorhies's papers and thus his identity after his death. During the one year between the listing of "Albina Durand (widow)" in 1903 and of "Albina Durand Voorhies" in 1905, a Voorhies family member most likely corresponded or met with Albina and Margot in San Francisco or in New Orleans. Although no record exists of communication between the Voorhies family and Albert's widow and daughter, the appearance of Albert's death date in the family record substantiates an encounter of some kind.[39]

There is another possible storyline. According to Jaime Quiróz, Margot told him that her mother was French and that her parents had moved from New Or-

Plate I-19 Margot de Taxco 5700. *Dots on a Knot* bracelet 7.25 x 0.5 in. Private Collection

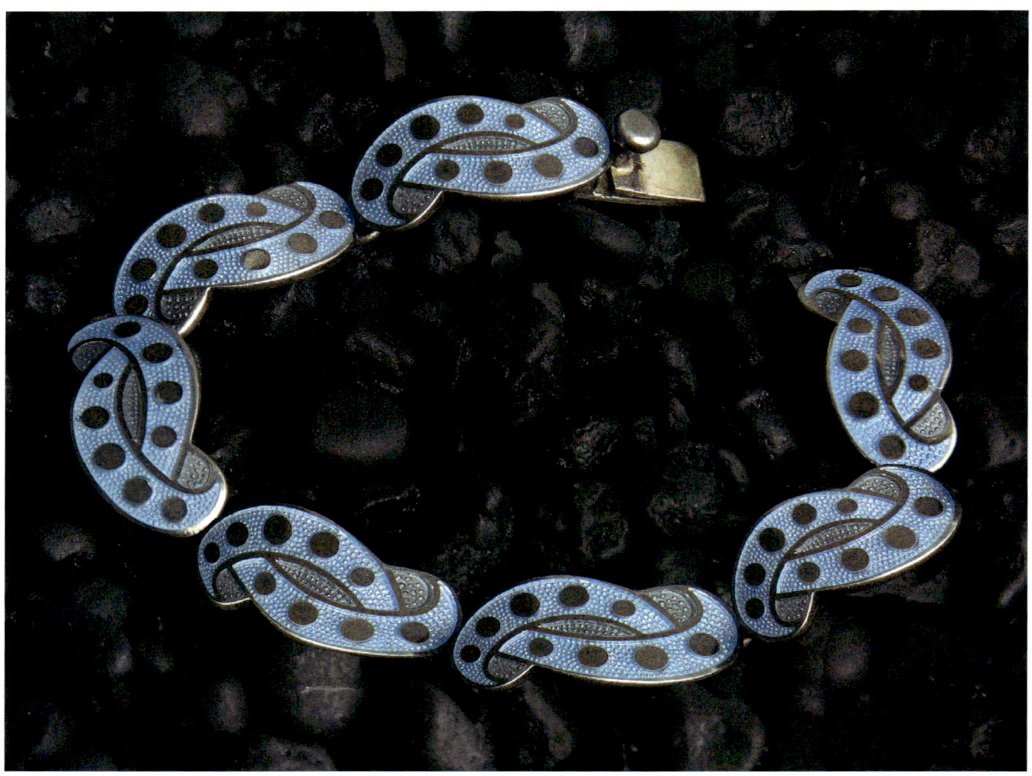

leans to San Francisco. Thus, Margot could have been conceived in 1896 in New Orleans, perhaps precipitating the move West. The Voorhies family then would possibly have been aware of young Albert's name change and of his life in San Francisco with his new family. In truth, Albina does not appear in the New Orleans directories at all and only surfaces in San Francisco in the 1900 census. Albert Durand is listed in the 1897 San Francisco directory, and we can only presume that he was living with his wife and daughter before 1900. If we accept this second scenario, Albina would have known of Albert's name change from the beginning, a concept that is most difficult to imagine. As mentioned earlier, in the obituary in the *San Francisco Chronicle*, Allie Durand is referred to as the "beloved husband of Albina Durand, and father of Margot Durand."

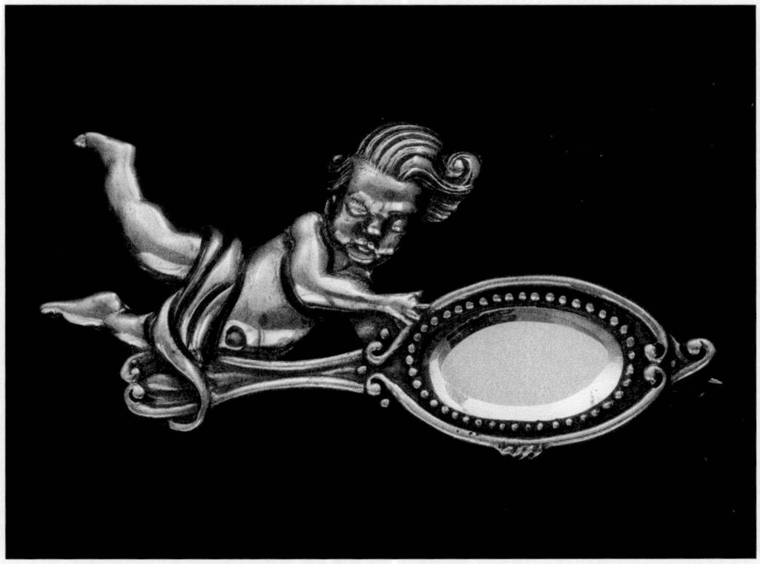

Plate I-20 Margot de Taxco 5441. *Cherub with Mirror* brooch 1 x 2.25 in. Private Collection

After Albert's death, Albina was left to her own devices. She was now a widow with a five-year-old child; most of her family was in Quebec; and her brother was in Washington. Under these circumstances, would it be outside the realm of possibility for Albina to have contacted the Voorhies in order to legitimize her daughter's relationship to a prominent New Orleans family? Months separate the death of the younger Albert and that of his mother Leontine on August 7, 1903.[40] Leontine and Albert Voorhies would have received the news that their son had died across the continent under an assumed name, leaving behind a widow and child. One wonders if Margot was able to meet her grandparents in 1903 or her grandfather in 1904. Did Judge Voorhies provide support, or was the rift between father and son so great that he could not even acknowledge the existence of a grandchild? Whatever money the elder Albert may have offered was not extensive, for Albina was employed as a hairdresser in 1905. That initial support may have been the reason why Margot held onto the notice of her grandfather's death, using the obituary in 1940 to prove her United States citizenship. Ultimately, it would seem that Albert Durand or Albert A. Voorhies was the only father Margot knew.

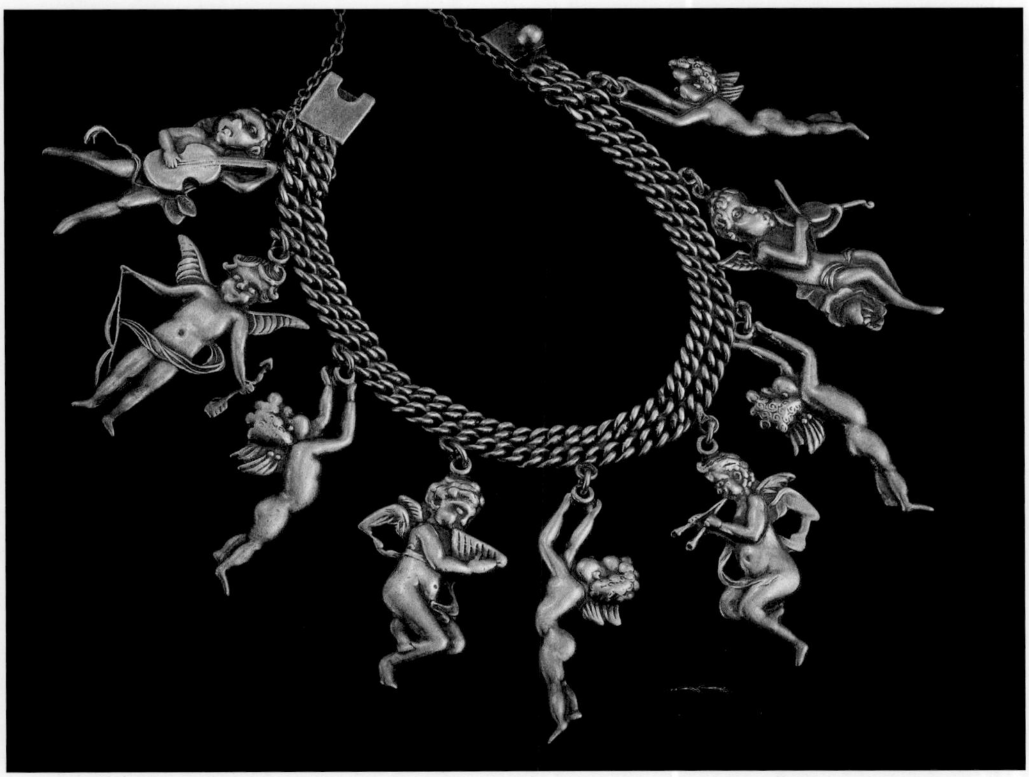

Plate I-21 Margot de Taxco 5441. *Cherubs* charm bracelet, cherub 1.5 in. ht. Private Collection

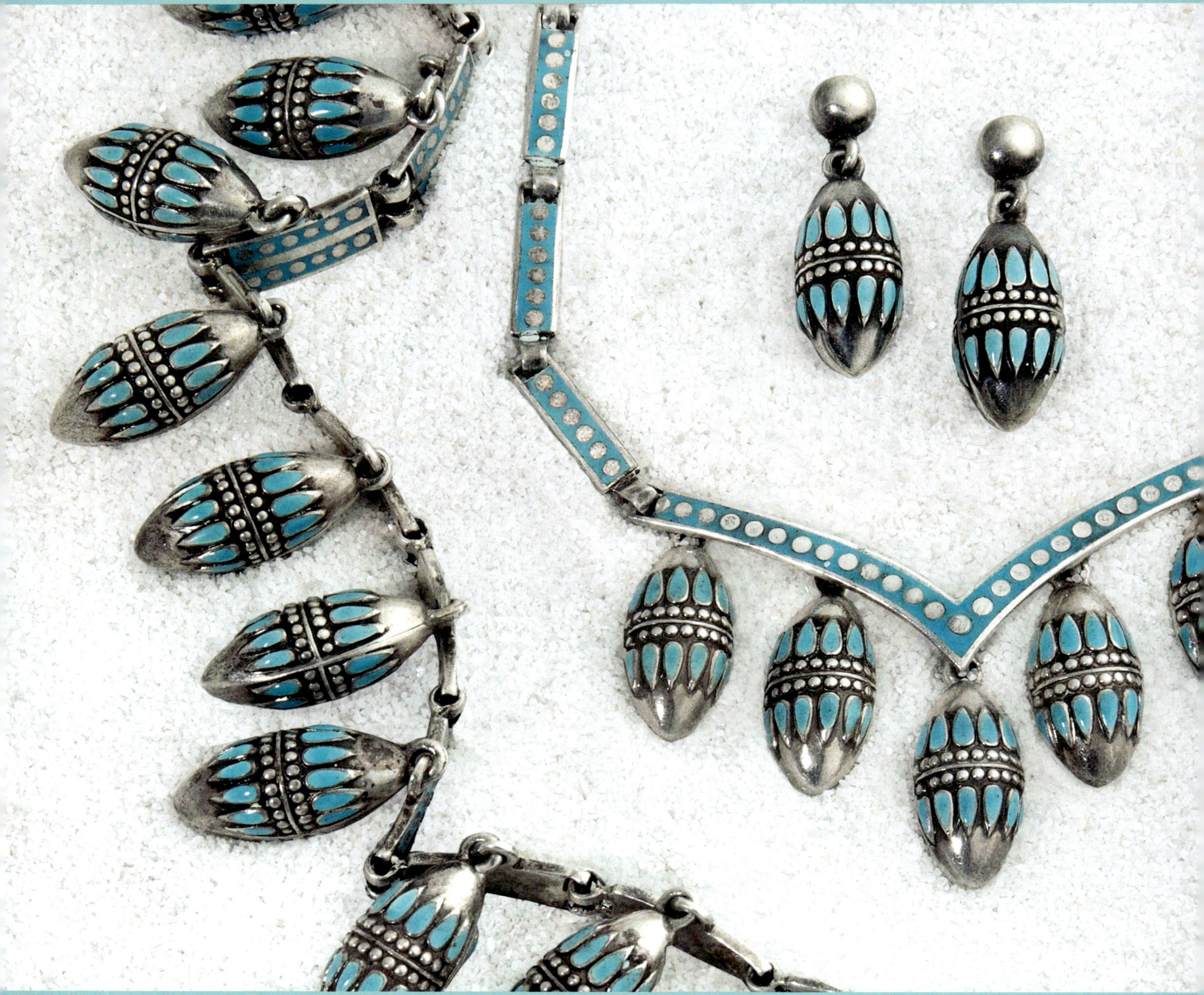

Plate II-1 Margot de Taxco 5701. *Fanciful Pendants Set.* necklace 16.5 in., central motif 2.25 x 2.5 in.; bracelet 8 in., pendant 1 in.; earrings 1.5 in. J. P. Mather Collection

Chapter II
The San Francisco Earthquake and Fire

Margot recalled, in one of her conversations with Jaime Quiróz, that she was a child at the time of the San Francisco earthquake. Indeed, Margot was but nine years of age when she experienced a natural and man-made disaster that reached catastrophic proportions. Author Dennis Smith provides a brief summary of the devastation: "The four-day event took more that 3,000 lives, burned through 28,188 buildings, flattened 522 blocks, destroyed tens of churches, 9 libraries, 37 national banks, the Pacific Stock Exchange, 3 major newspaper buildings, [and] 2 opera houses. More than 200,000 people were burned out of their homes — men, women, and children who found themselves wandering smoke-filled streets with

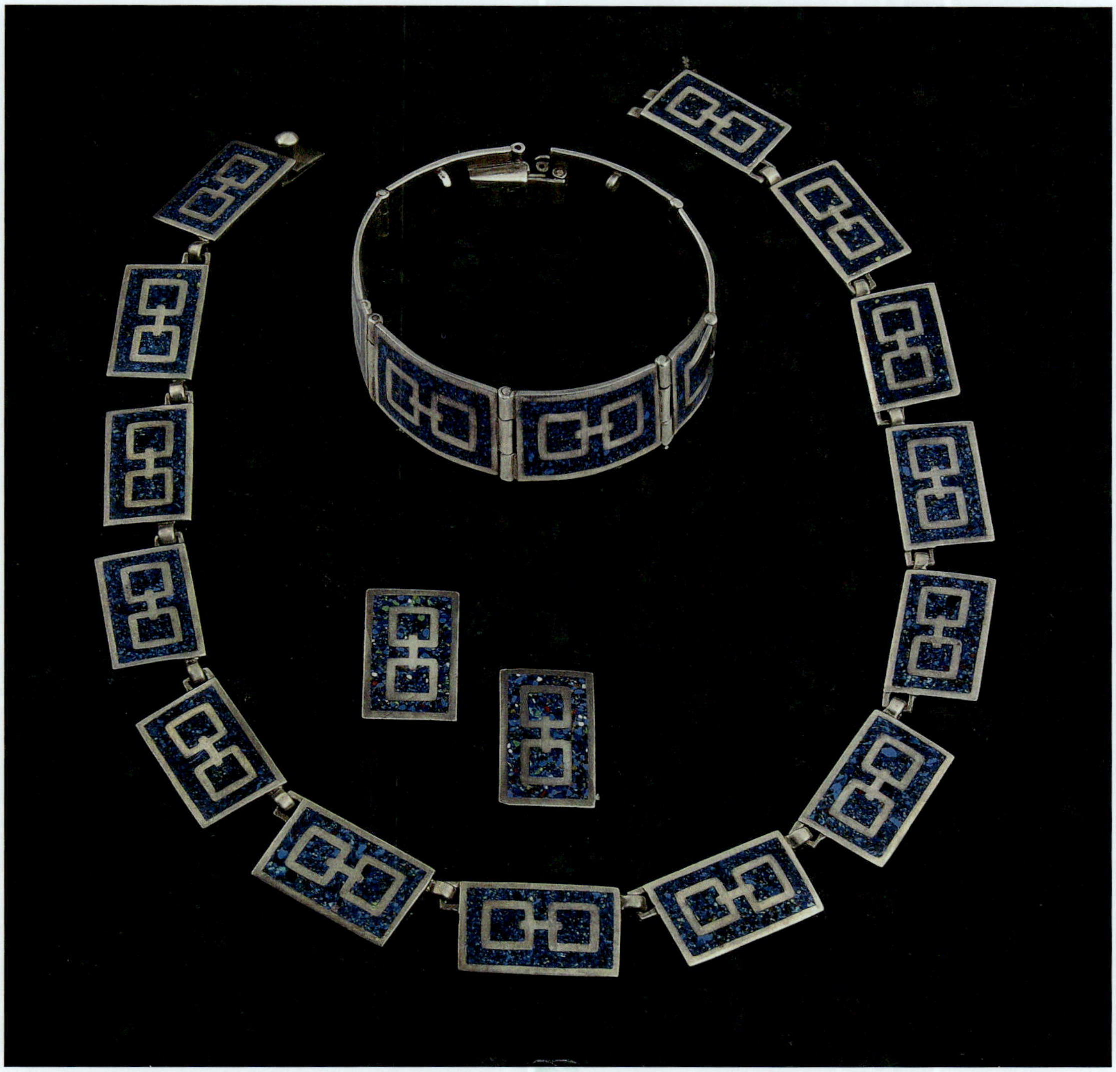

Plate II-2 Margot de Taxco 5399. *Two Squares Set.* necklace 16.5 x 0.63 in.; bracelet 7.5 x 0.63 in.; earrings 1 x 0.63 in. J. P. Mather Collection

no claim to a future except that they were alive and they owned the clothes on their backs."[41]

What would this young child and her single mother have experienced during those four days? The quake began on Wednesday April 18, 1906 around 5:00 in the morning. According to eyewitnesses, the sound was like thunder coming from the center of the earth. The earth's surface heaved in waves, shaking buildings, reducing many to rubble. The jolting of the quake and the aftershocks were terrifying in their violence. One man spoke of the earth moving up and down, and from side to side, so that there was no recognizable center of gravity. The second quake occurred three hours later and was also accompanied by aftershocks, although not as powerful as the first. For miles in every direction, the quake and its aftermath brought massive destruction: "The shaking came from the west, eight miles out to sea from San Francisco, and caused destruction as far as fifty miles to the east and two hundred miles north and south from Monterey to Humboldt County."[42]

Natural forces caused incalculable damage and loss of life, but the tragedy was only beginning to unfold. Major fires, brought on almost immediately by ruptured gas lines and other causes, were unquenchable because of a lack of available water in many parts of the city. Even more terrible was the use of dynamite to create firebreaks, a spectacularly bad decision that led to engulfing firestorms. At 5:00 p.m. on Thursday, an eyewitness wrote, "At that time, San Francisco was hidden in a pall of smoke. The sun shone brightly upon

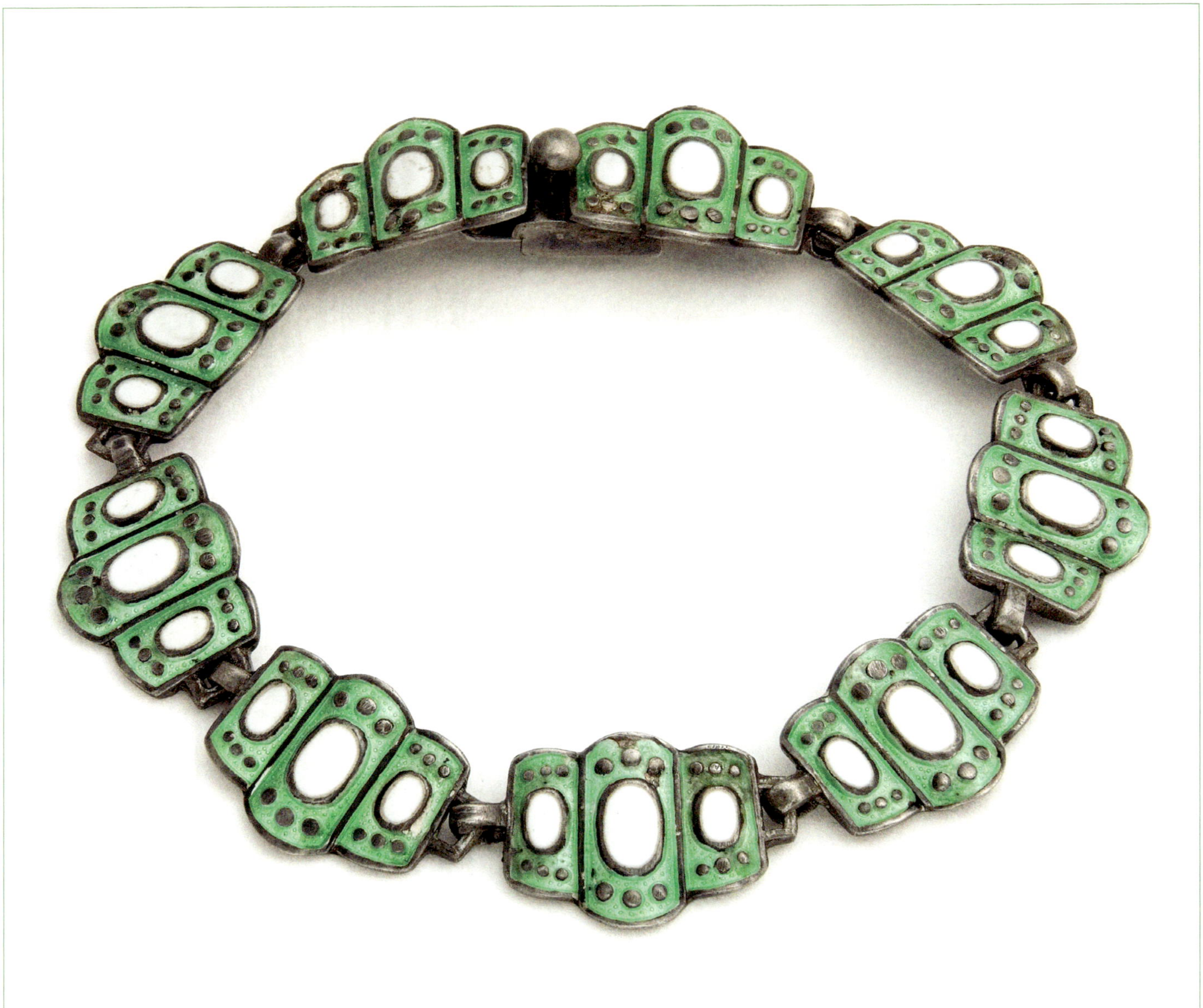

Plate II-3 Margot de Taxco 5835. *Semi-Precious* necklace in green 14 x 1.25 in. Private Collection

it without any seeming penetration. Flames at times cleft the darkness. This cloud was five miles in height and at its top changed into a milk white." [43]

Margot and her mother's home on Fillmore between Sacramento and California Streets was in a sector of San Francisco known as the Western Addition, west of Van Ness Street. In the aftermath of the tragedy, one eyewitness, Mrs. James Watkins, wrote, "Nothing is left except a small residence section on Pacific Heights, a miserable little second-class business street (Fillmore St.) and small residences in the outlying districts."[44]

While most of the devastating fires occurred in the eastern segment of the city, a major fire at Laguna and Sutter Streets was five blocks from Margot and Albina.[45] It is not entirely certain that the fire reached them: "Fires were raging in the western addition, the vast area west of Van Ness, but because water was available there, for the most part the firemen got the upper hand." However, there were areas in the west that continued to burn because the water stream turned to a trickle as a result of breaks in the pipes.[46]

Anna Sprague Cox, in a letter to her uncle, wrote, "Houses seemed to melt away before your eyes, to disappear like magic yet with a sort of orderly deliberation. A line of red would run along the eaves, then flames would issue from the roof and upper windows. Another moment and no building was there, just a red mass. . . . If the dynamiting had been confined to the east side of Van Ness where the lots were deep, all our

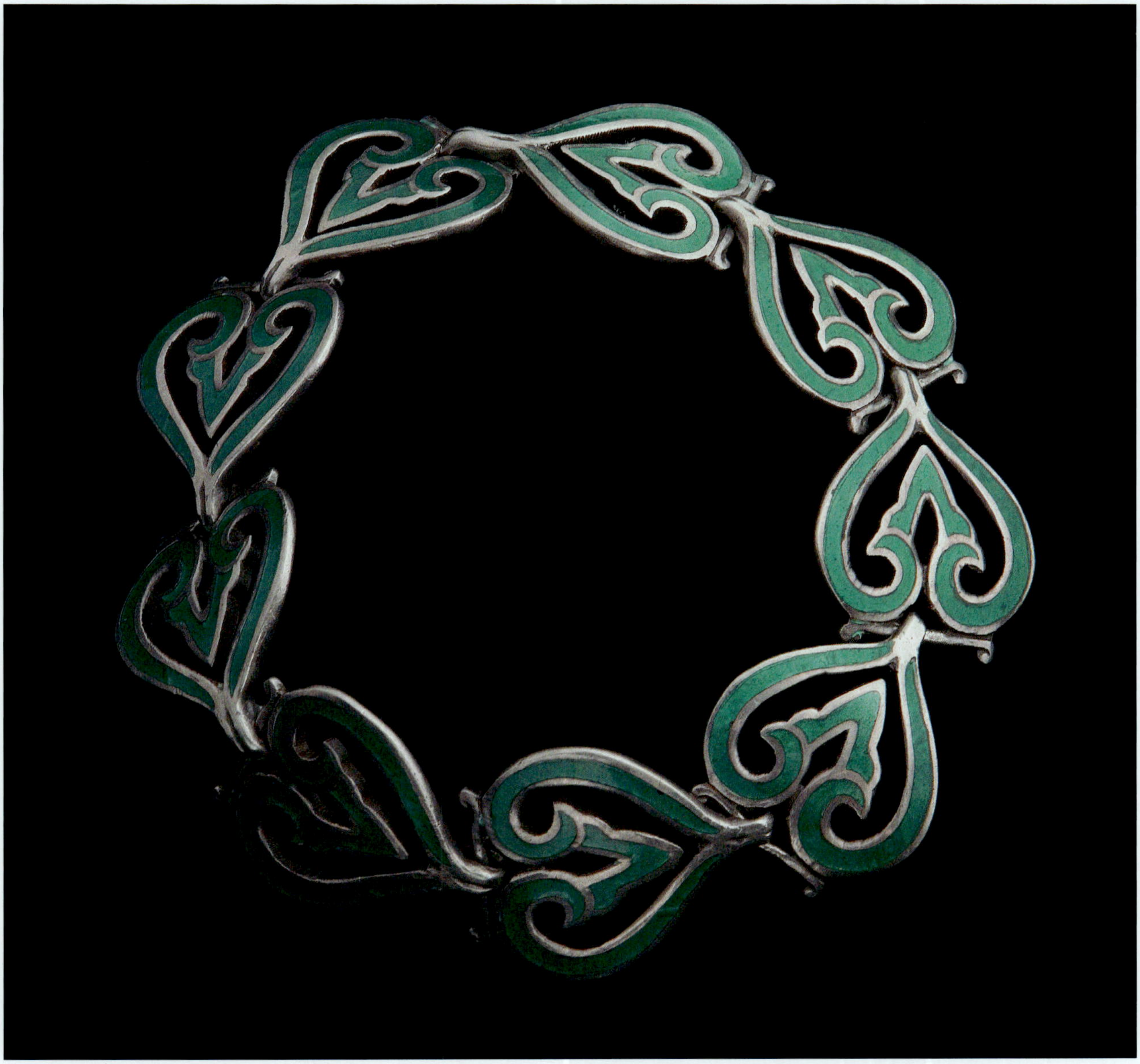

Plate II-4 Margot de Taxco 5381. *Green Hearts* bracelet 8.5 x 0.63 in. J. P. Mather Collection

Plate II-5 Margot de Taxco 5533. *Open Scroll Set*. bracelet 7.25 x 1 in.; earrings 1.5 x 0.75 in. Private Collection

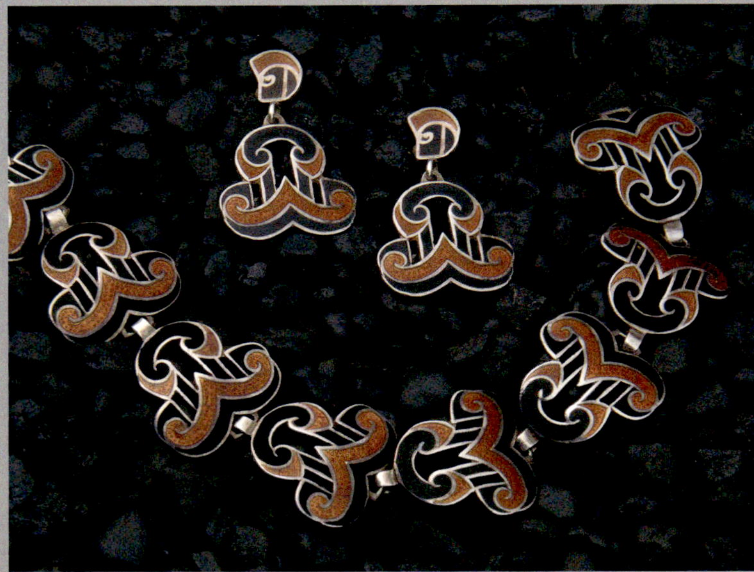

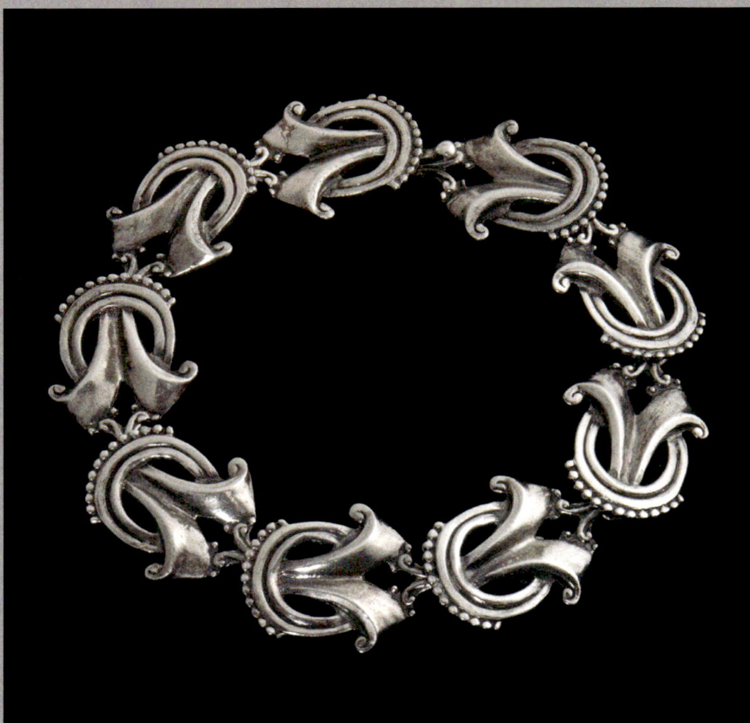

Plate II-6 Margot de Taxco 5306. *Furls and Circle* silver bracelet 7.75 x 0.88 in. J. P. Mather Collection

Plate II-7 Los Castillo 615A. *Circle of Swirls* (eagle and circular Los Castillo mark) necklace 15.5 x 1.25 in. Elizabeth Finch Collection

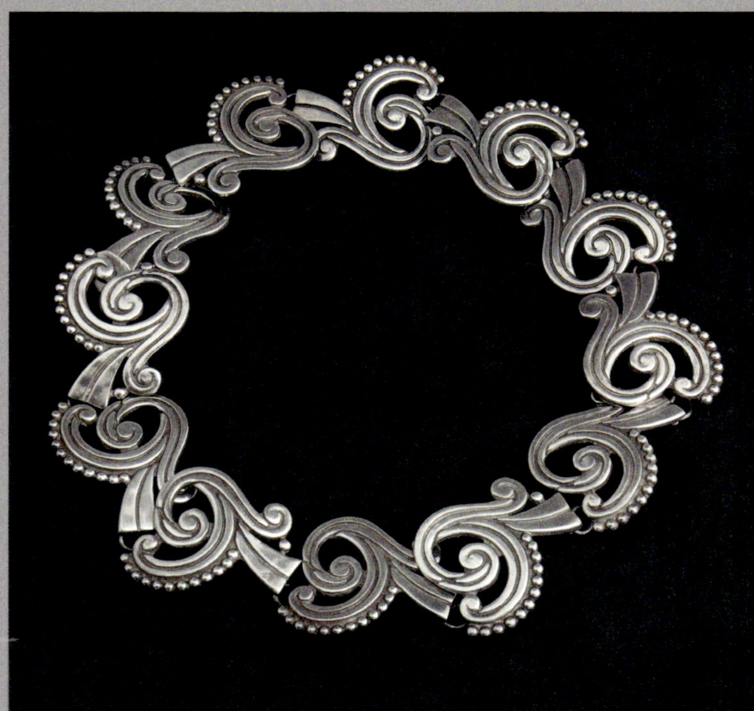

part of town and North Beach might have been spared. Our block burned between eight and nine Thursday night. By that time we were miles away." Anna and her family received assistance from Henry Monroe, husband of Lola (Clayes) Monroe, as they made their way out of San Francisco to the ferry and across to Berkeley. Albina and Margot would meet Lola and her sister Margaret at another point in their lives. As the fire raged in the Western Addition, all evidence indicates that Margot and Albina left their home, either to escape a burning building or the chaos and devastation that surrounded them.[48]

Extraordinary Circumstances

Until the discovery of Margot and Albina's names in the 1915 Berkeley city directory, their whereabouts for a period of fourteen years after the earthquake were unknown. This one mention of their living in Berkeley nine years after the earthquake indicates that the two of them had left the city of San Francisco in 1906. Albina and Margot most likely joined the thousands who fled toward the ferry that would have taken them across the bay to Oakland.[49] Did they seek out Remi's widow Emelie or their cousin Ellen? Emelie was socially prominent, maintaining her membership on the boards of several of Anthony Chabot's philanthropic ventures, for which she may have actually been the original driving force.[50] Albina's brother Robert had lived with Emelie and Remi for years, so it is possible that Albina and Margot could have been invited into Emelie's household right after the earthquake.

In 1906, Ellen Chabot was living alone in the elegant residence that her father had built. Ellen's stepmother, whom Ellen had cared for devotedly, had died in 1904. Ellen appeared often in the social pages of the *Oakland Tribune*, where it was reported that, in February 1906, she was enjoying the social season in Washington, D. C. and had been invited to the White House.[51] Considering the social divide between the

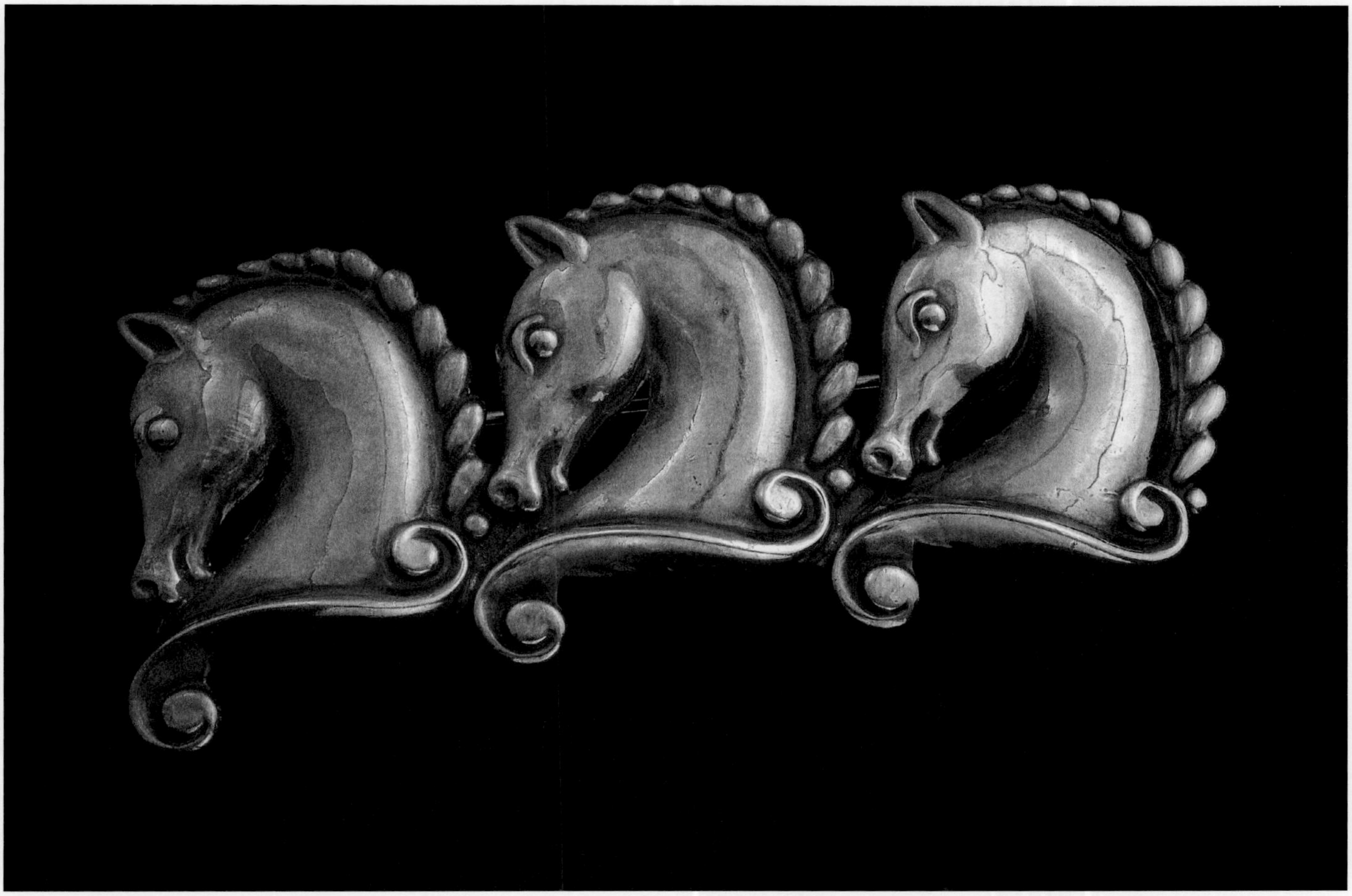

Plate II-8 Margot de Taxco 5160. *Three Horses* brooch 1 x 2 in. Private Collection

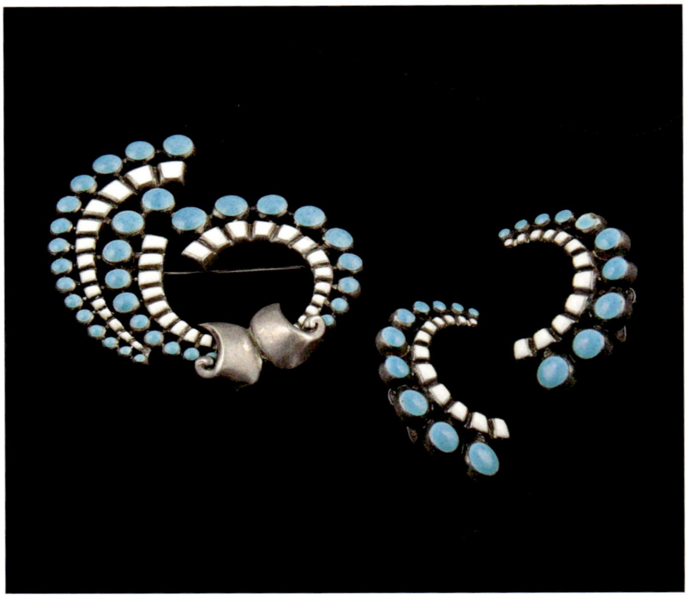

Plate II-9 Margot de Taxco 5601. *Two Color Dot and Square Scroll Set.* brooch 2.25 x 2.25 in.; earrings 1.5 in. Private Collection

cousins, it is hard to imagine that Albina and Margot would have been asked to spend more than a brief time in either Emelie or Ellen's home after the earthquake. They may have required immediate assistance from a family member, but then it seems they moved on to find their own way.

Margot and Albina do not appear in the 1910 census records as members of Emelie Chabot's household and, by 1910, Ellen had married Henry Bothin and moved out of the Oakland area to San Rafael.[52] A strong relationship must have developed between Margot and her mother during those years when they were forced to confront great challenges with little or no support. It is quite likely that Albina left San Francisco with few possessions and only her own resourcefulness in what had to have been a frantic effort to feed, clothe, and house the two of them. Albina had been a hairdresser in the years following her husband's death, and she continued that line of

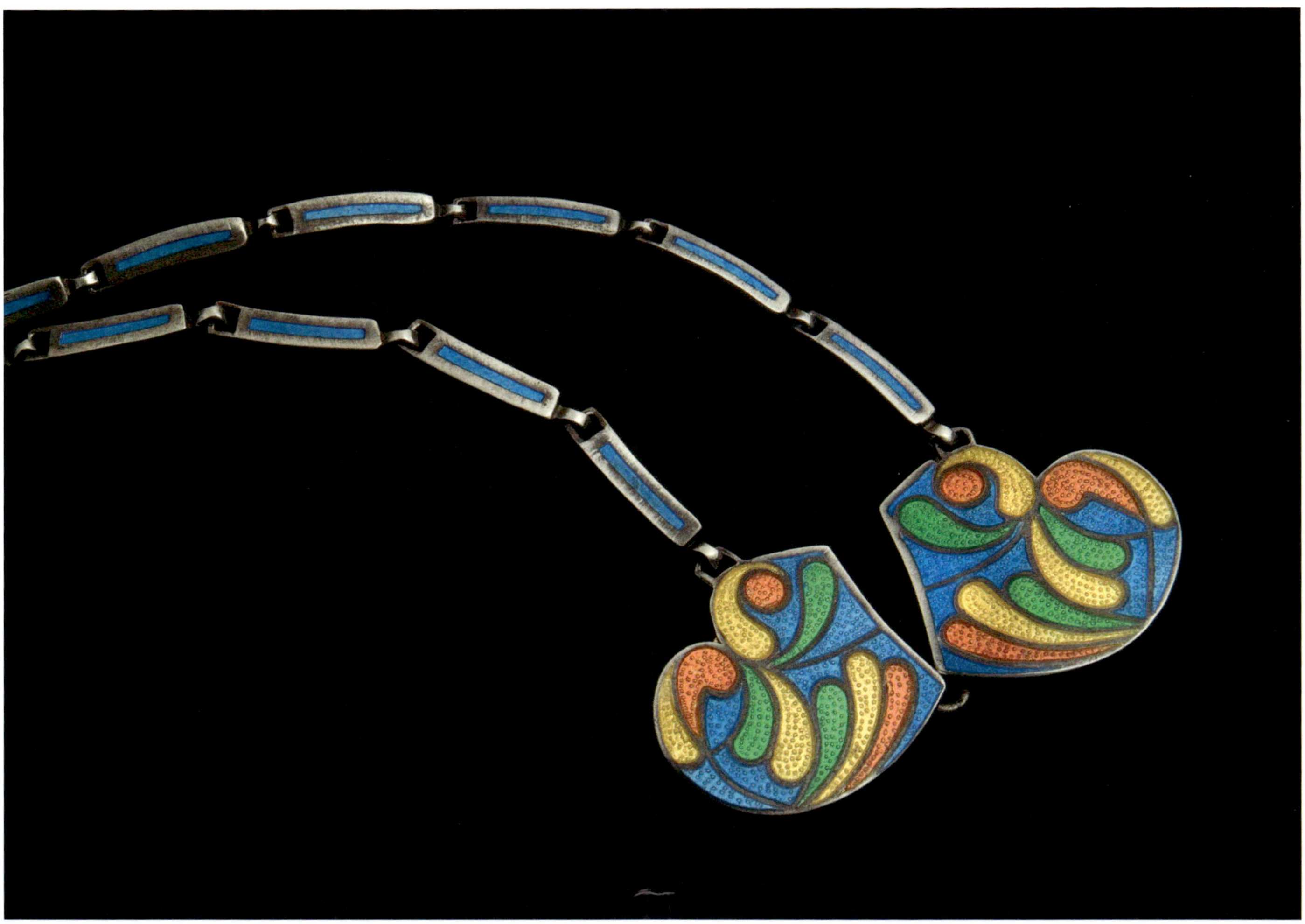

Plate II-10 Margot de Taxco 5662. *Bright Multicolor* necklace 15.5 x 1.13 in. Peregrine Galleries Collection

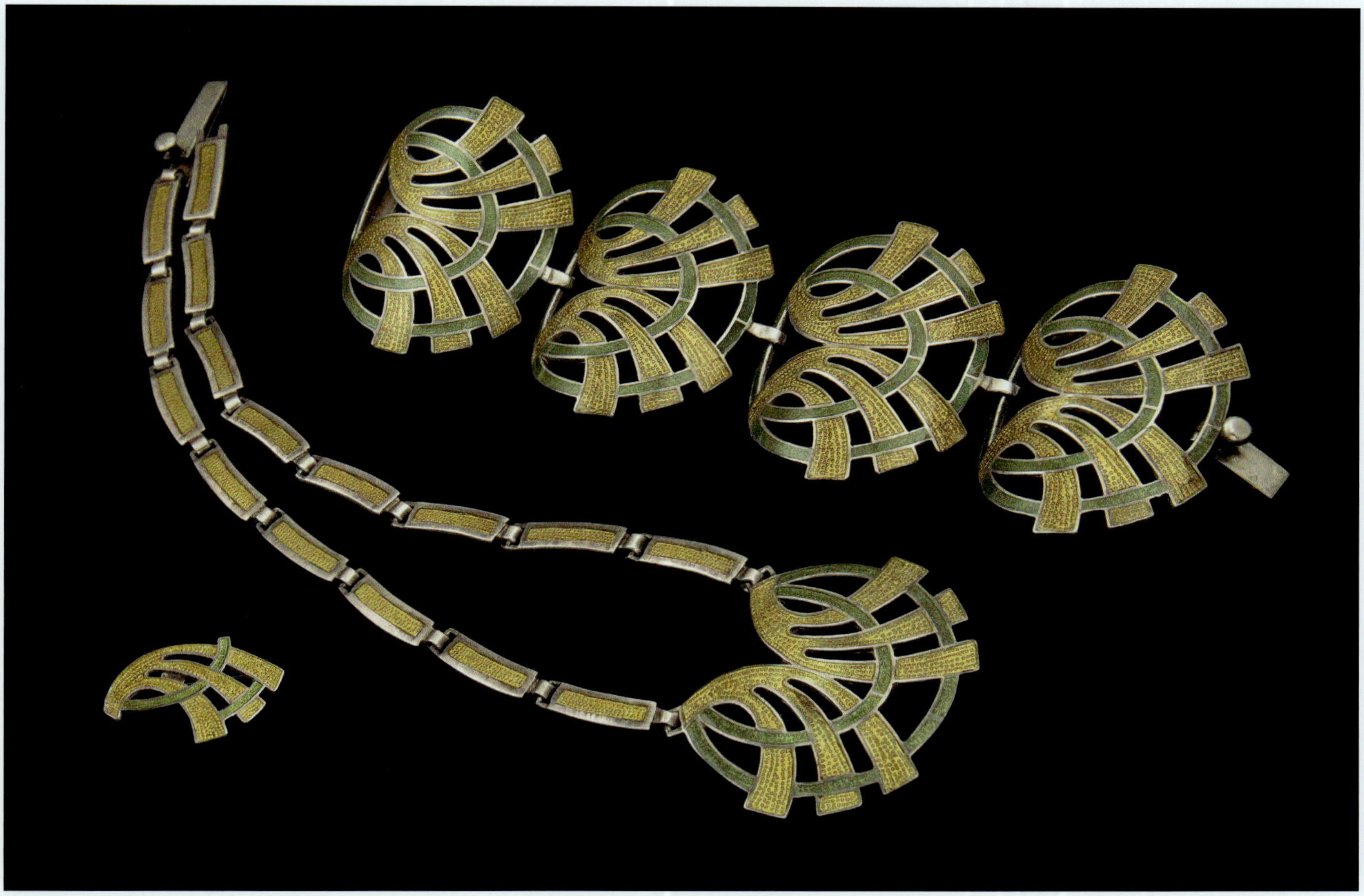

Plate II-11 Margot de Taxco 5652. *Mirrored Flared Lines Set* in yellow and green. necklace 15 in., central motif 1.5 x 1.88; bracelet 6.5 x 1.88 in.; brooch 0.75 x 0.75 in. Private Collection

work after leaving San Francisco. The 1915 listing in Berkeley also indicates that Albina set an example for her daughter, instilling in her the necessity of taking responsibility for her own wellbeing.

The listing for Mrs. A. D. Voorhies, hairdresser, provides a business address, 2135 Center, while the one for Margot Van Voorhies, hairdresser, 2131 Center, was the address for the residence.[53] Albina and Margot must have formed a partnership and, together, opened a beauty salon. Worthy of note is Margot's decision, at the age of nineteen, to change her name to Van Voorhies. She may have wanted to differentiate herself from her mother or, more simply, she may have preferred the sound of the name, a decision that would not have been surprising for a nineteen-year-old to make. Moreover, by changing her name, Margot was following a precedent that had already been set by her father.

Both Albina and Margot went to work as young women and remained unmarried until they were over thirty. Their passage into adulthood seems rather prosaic today in the twenty-first century, but Margot and her mother came of age even before the radical changes to women's lives in the 1920s. Margot followed the quite "modern" example of her mother who was a hairdresser in 1905, at a time when women were only beginning to bob and dye their hair and use cosmetics. Before 1920, "a bobbed head was seen as a sign of dangerous radicalism."[54] Considering the choice of vocation made by Albina and her daughter, it is worth remembering that, in the first decade of the twentieth century, "the successful female entrepreneurs of the era often worked in the business of beauty — one of the first parts of the economy where a woman could be investor, owner, and boss."[55]

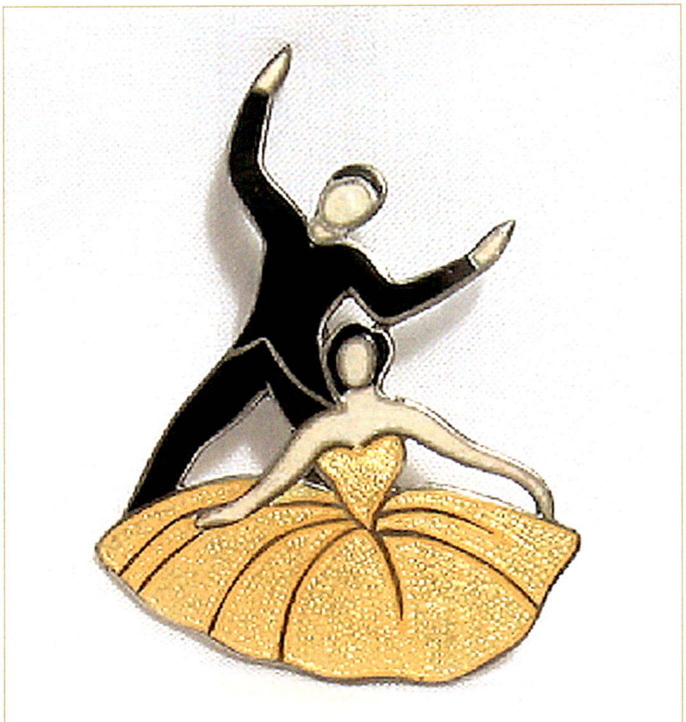

Women of the nineteen twenties received the right to vote in political elections.[56] They wore their hair clipped short in a masculine style. They abandoned the corset to embrace silk stockings and skirts that went only to the knees instead of all the way to the ankle to allow more freedom of movement. While Margot and Albina conjure up the image of the flapper — independent and willing to break with convention — they were hardly self-indulgent and free to choose their lifestyles. The two women's decisions were born out of necessity, but were daring and bold nonetheless. The fact that Albina and Margot chose to establish their own business in 1915 is in contrast with many of their contemporaries who, in this same predicament, would have pursued traditional roles in the wider community, as teachers, nannies, factory workers, or servants in the grand residences of the wealthy.[57]

Plate II-12 Margot de Taxco 5421. *Dancers* brooch 2 x 1.5 in. photograph provided by collector. Sheila Pamfiloff Collection.

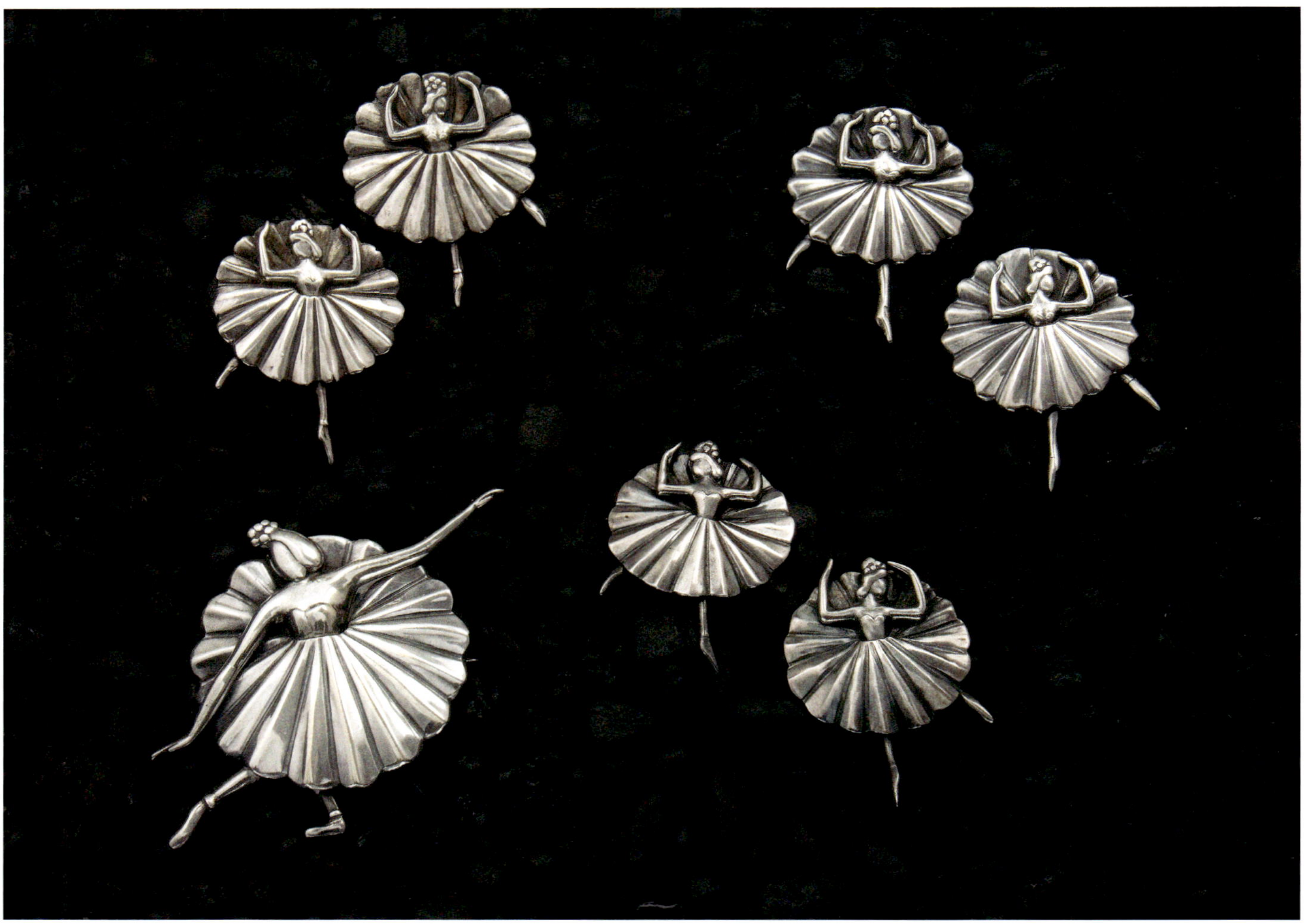

Plate II-13 Margot de Taxco 5200. *Large Ballerina* brooch 2.25 x 1.5 in. *Small Ballerina* brooches 1.5 x 1 in. Private Collection

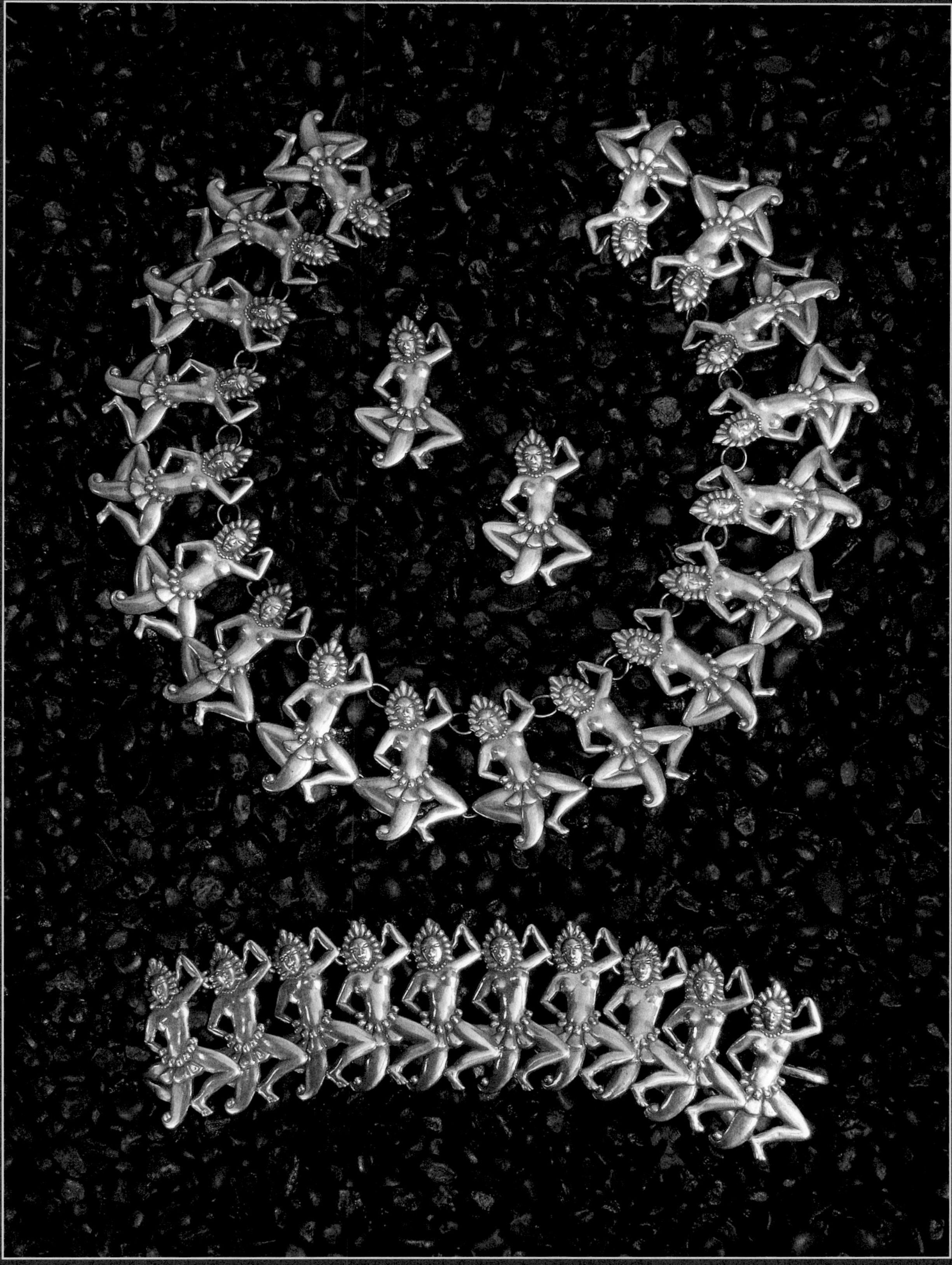

Plate II-14 Los Castillo 519. *Thai Dancers Set* in silver. necklace 14 in. inside x 1.88 in.; bracelet 7 x 1.88 in.; earrings 1.88 x 1.38 in. Silver Huntress Collection

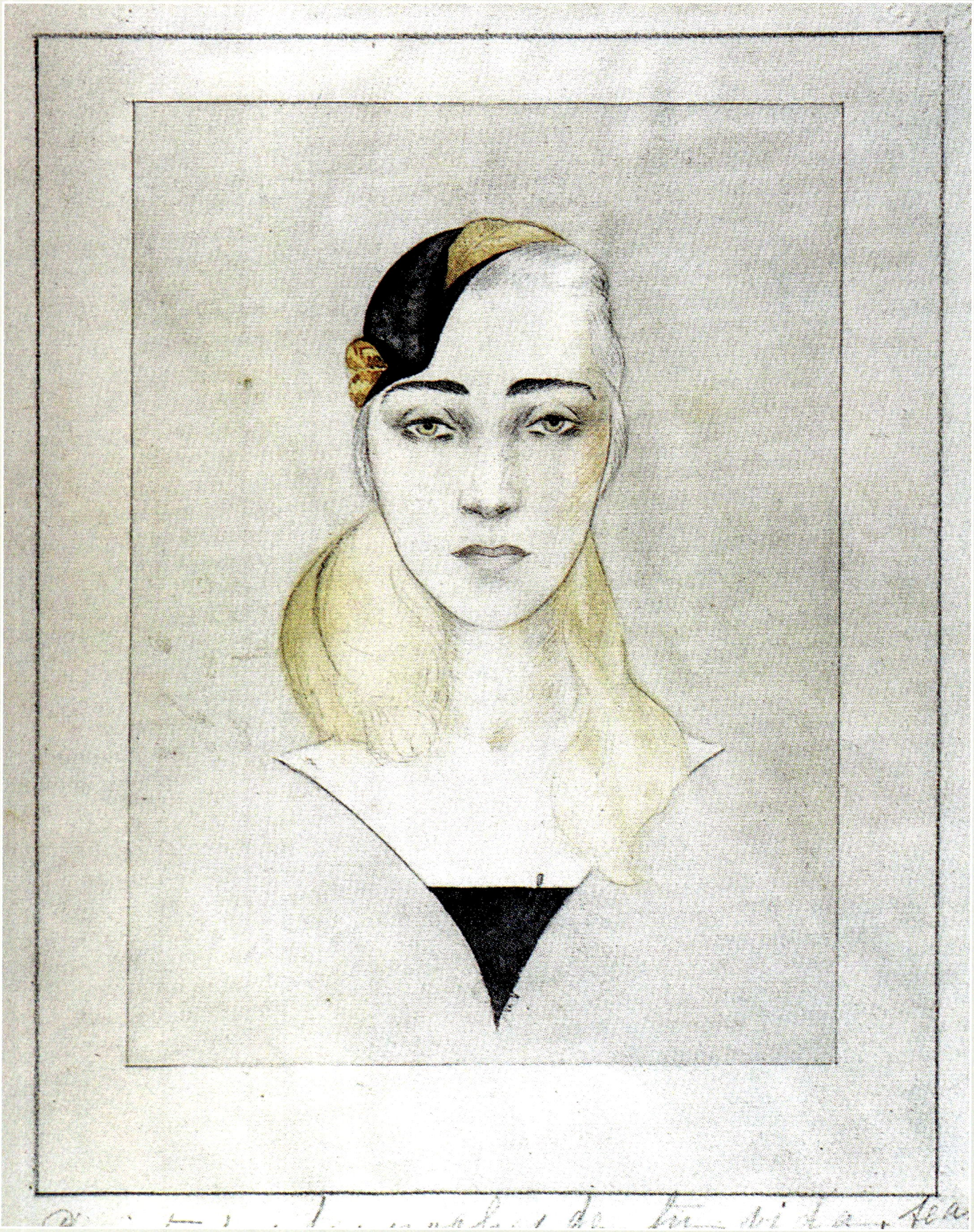

Plate III-1 Margot Van Voorhies. *Self-Portrait*. 1932. pencil/watercolor drawing. 11.88 x 8.5 in.
Jill Crawford Collection

Chapter III
Return To San Francisco

Like many San Franciscans, Margot and her mother returned to San Francisco after the quake. Albina, who continued to refer to herself as "Albert's widow," followed Margot in changing her name to Van Voorhies. In 1919, they moved into rooms at 1677 Sacramento and, during this period, were not operating a business. Margot was working as a clerk with the Travelers' Insurance Company and Albina was apparently unemployed.[58]

In 1920, Margot was a bookkeeper and the following year, a stenographer.[59] Margot's ability to achieve this level of employment raises the question of her education. While Albina probably had little opportunity to attend school during her childhood, Margot was in San Francisco until she was nine and then reappeared in Berkeley as a hairdresser when she was nineteen. It can be assumed that Margot received a primary school education in San Francisco before the earthquake and then might have been in school intermittently until she became a partner in her mother's salon. Her level of formal education aside, Margot was bright enough to continue learning as she moved into different areas in the workplace.

It is almost incontrovertible that Margot received some form of art education. The people who knew her in Taxco have all spoken of Margot's artistic sensibilities, that she was both a painter and photographer.[60] Margot could have attended art school in San Francisco

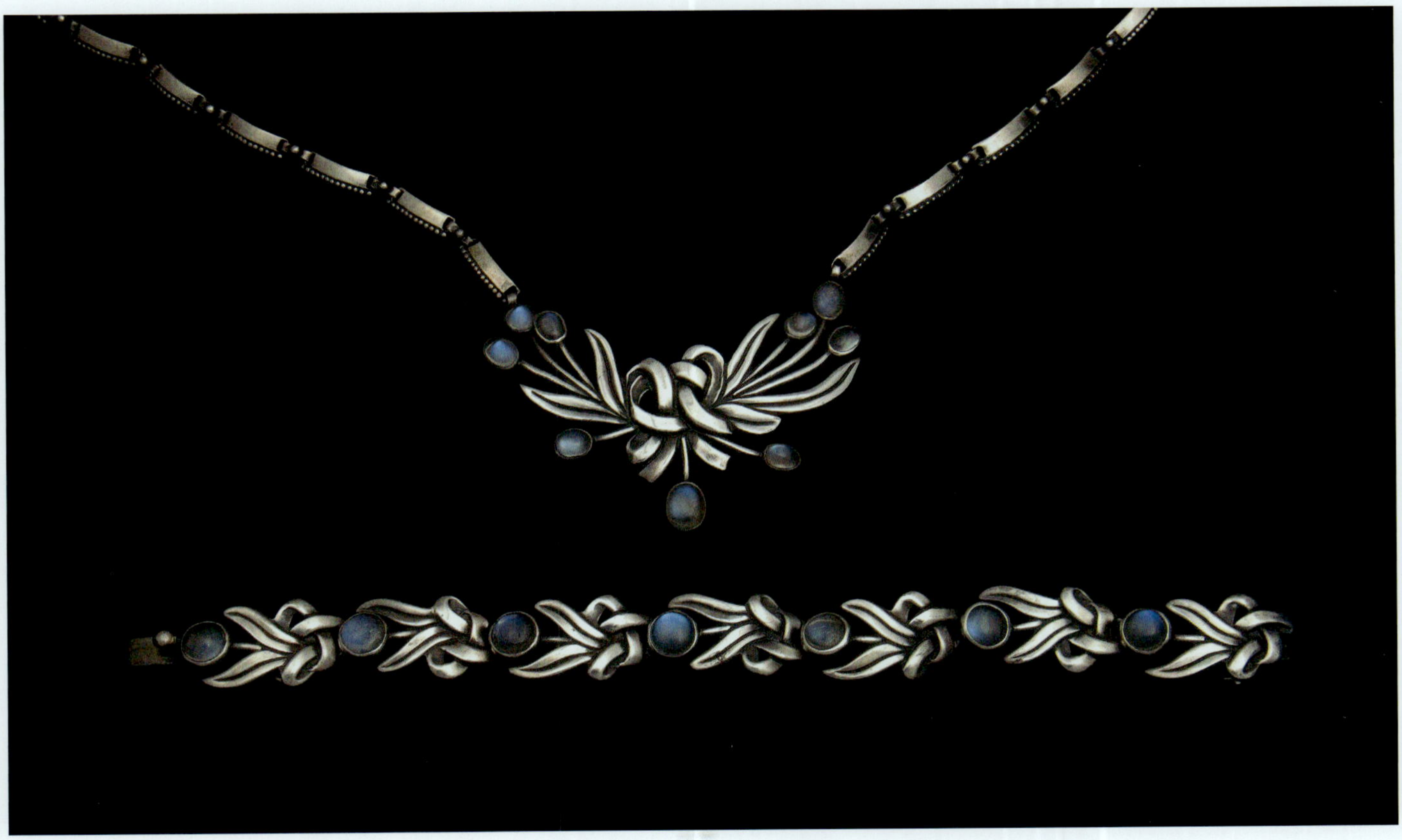

Plate III-2 Margot de Taxco 5379. *Blue Stone Flowers Set.* necklace 15 in., central motif 1.5 x 2.5 in.; bracelet 7 in. Private Collection

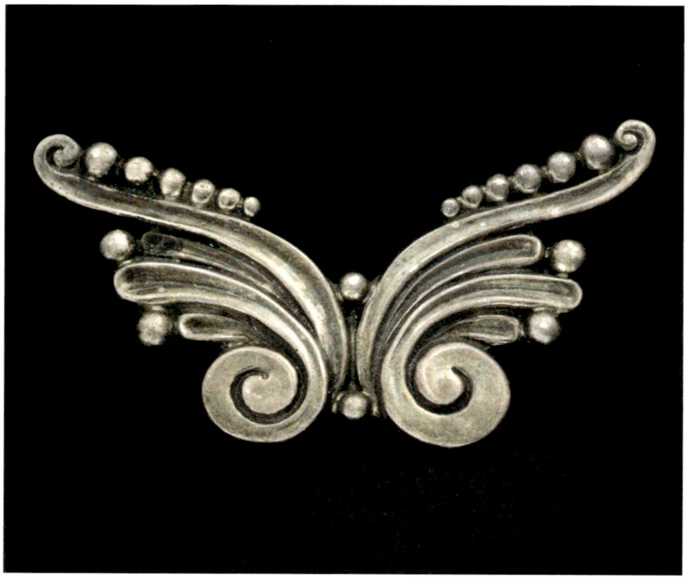

Plate III-3 Margot de Taxco 5337. *Curling Lines* butterfly brooch 1 x 2 in. Private Collection

and her mother could have joined her, for Albina refers to herself as an artist in the 1926-1927 city directories. Opportunity for art education existed, as evidenced in an advertisement from the 1919 city directory: "The San Francisco Art Association (San Francisco Institute of Art, Palace of Fine Arts, Exposition Grounds) — The California School of Design — Maintained by the Association: teaches drawing, painting, modeling in all their branches; decorative and commercial design, handicrafts and teachers' course." [61]

At some point during the years 1920-1921, Margot made the decision to live on her own. Albina was at 1008 Larkin from 1922 to 1927[62] and was once again employed as a hairdresser in 1923. The 1924 city directory listing indicates an interesting turn of events: "Albina D. Voorhies with Clayes, r 1008 Larkin." It would seem that Albina was living with someone named Clayes. According to the 1926 San Francisco city directory, Margaret Clayes was the owner of a company that manufactured and sold art goods inspired by Asian art. Her business listing in the San Francisco directory indicates that the company was not only owned by a woman, but that the top management positions were all held by women, including her sister, Lola Clayes Monroe.[63]

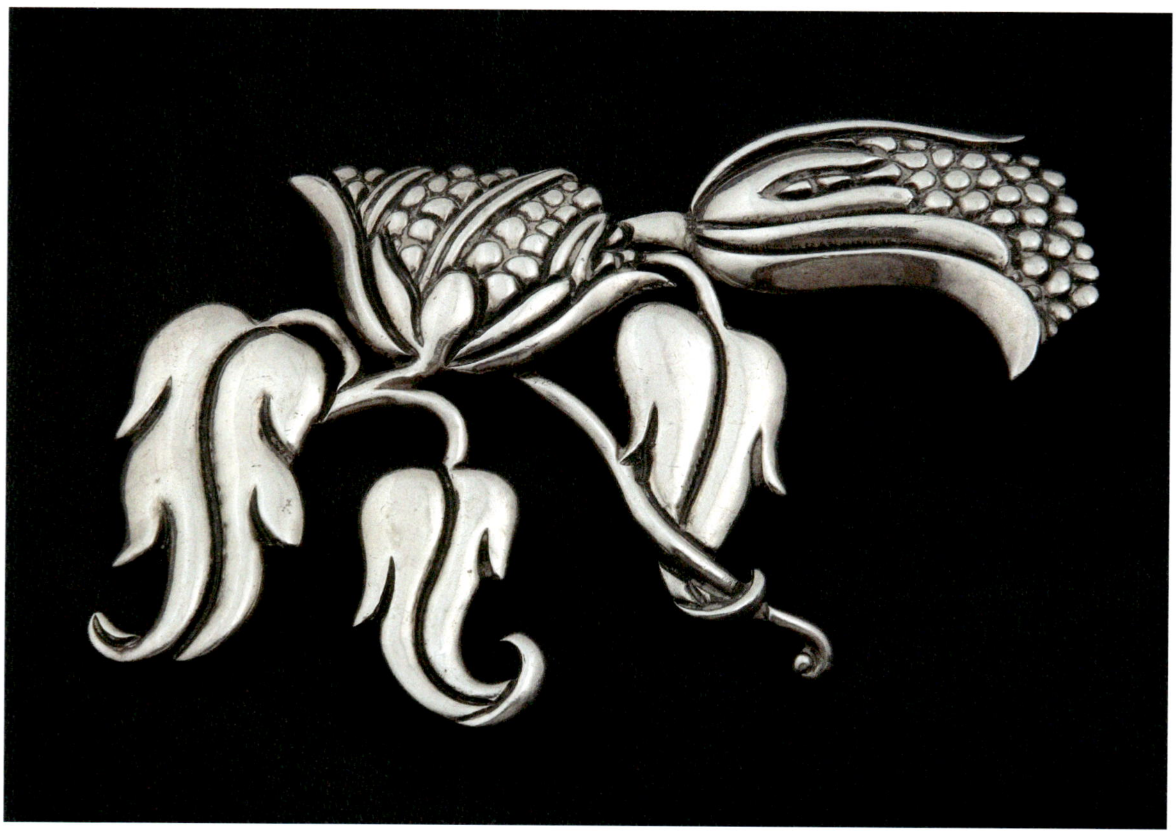

Plate III-4 Margot de Taxco 5271. *Floral Spray* in silver brooch 2 x 2.75 in. Stella and Fred Krieger Collection

Margaret Clayes was a world traveler and entrepreneur. In the *1922 Directory of Manufacturers of San Francisco*, Miss Clayes announced that she was selling candy boxes and art goods at 284 Post and had her factory at 278 Post.[64] As early as 1915, she was already considered a successful businesswoman. In the Official Catalogue of Exhibitors for the 1915 Panama-Pacific Exposition in San Francisco, Miss Clayes was selling "fancy articles" in the Liberal Arts Palace. A second entry in the catalogue is even more intriguing: "Clayes, Miss Margaret, 177 Post Street, San Francisco, California. Models of Interior Decoration for a Chinese Room. Liberal Arts Palace."[65]

During this period, Albina was referring to herself in the directories as an artist.[66] This abrupt change of employment in 1926-1927 has one possible explanation — Albina could have been working for Margaret Clayes. According to Miss Clayes's business card, besides the store in San Francisco below Union Square at 284 Post, she also had shops in Los Angeles, Santa Barbara, and in the Del Monte Hotel in Del Monte, California, where she sold "gifts unusual, and Oriental Applied Art." The number of outlets would have required a large inventory. Artisans in Miss Clayes's factory could have produced many of the craft items that were sold in these shops. One of the artists in Margaret Clayes's factory may have been Albina Voorhies, although it is almost impossible to know if her position required her to work as a designer, supervisor, or as an artisan.

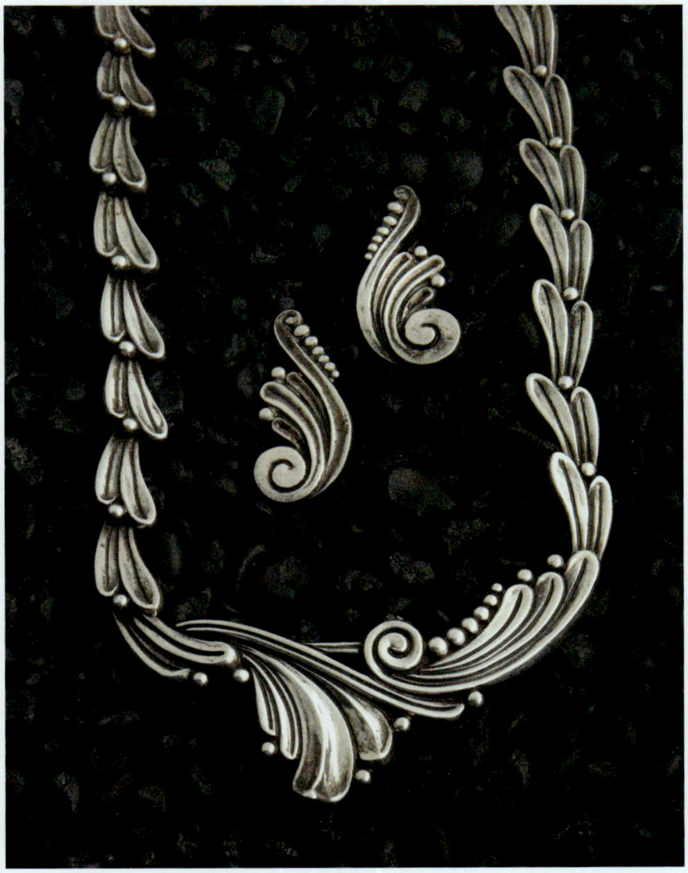

Plate III-5 Margot de Taxco 5337. *Curling Lines Set.* necklace 15 in.; earrings 1 x 0.6 in. J. P. Mather Collection

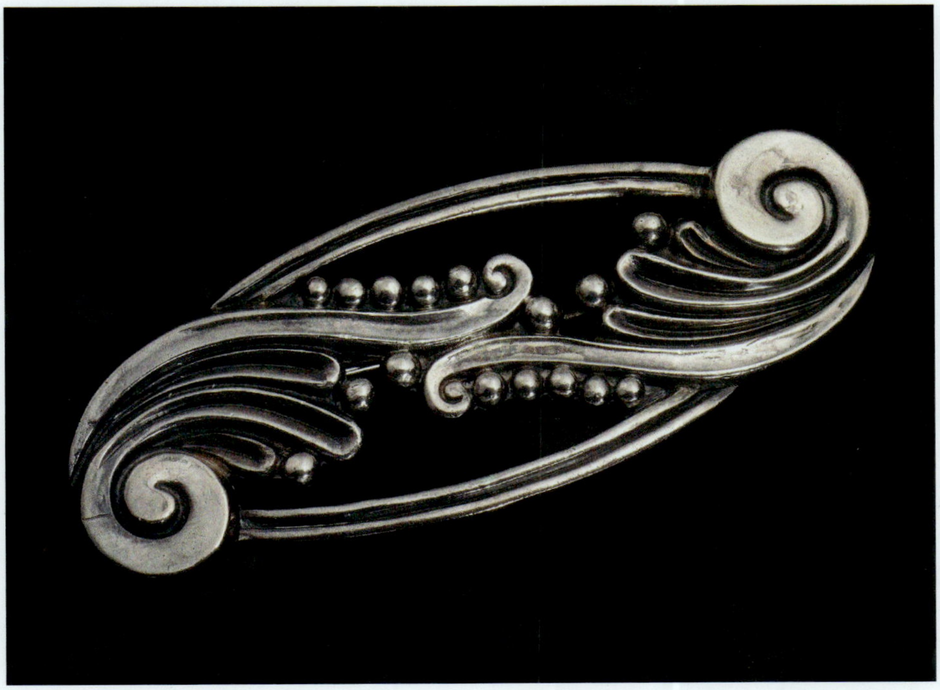

Plate III-6 Margot de Taxco 5337. *Curling Lines* brooch 1 x 2.25 in. Private Collection

The term used by Margaret Clayes, "Oriental Applied Art," on her business card indicates that many of the objects made in the factory reflected the influence of Chinese and Japanese art. In 1920, Ms. Clayes registered a design for a ceramic flower holder she called, "The Chinese Launderer."[67] While the vase was produced by another company, it reveals Ms. Clayes's stylistic tendencies.

Albina's friend and possible employer, Margaret Clayes, could have been a strong influence in Margot's life. Margot had a role model in Margaret Clayes, a woman who created her own designs and then administered their production in her factory. Margot also had before her a prototype for an arts and crafts manufacturing company. Beyond

Plate III-7 Margot de Taxco 5807. *Shell* in white necklace 15.63 in. Elizabeth Finch Collection

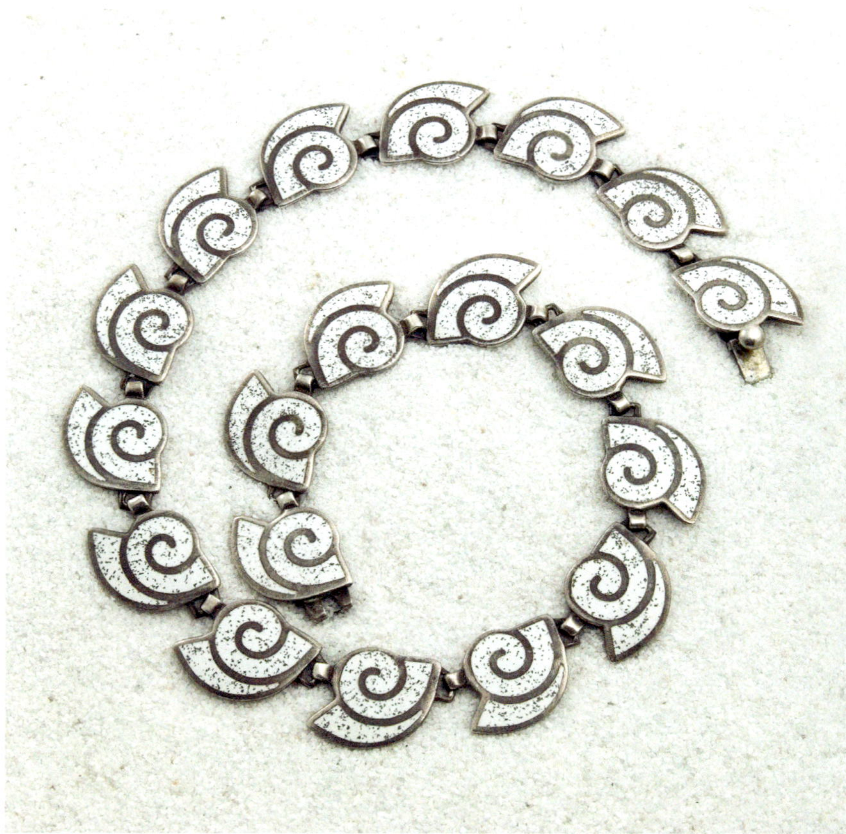

Plate III-8 Margot de Taxco 5807. *Shell Set* in mélange enamel. bracelet 7.5 x 0.63 in.; earrings 0.88 x 0.63 in. Elizabeth Finch Collection

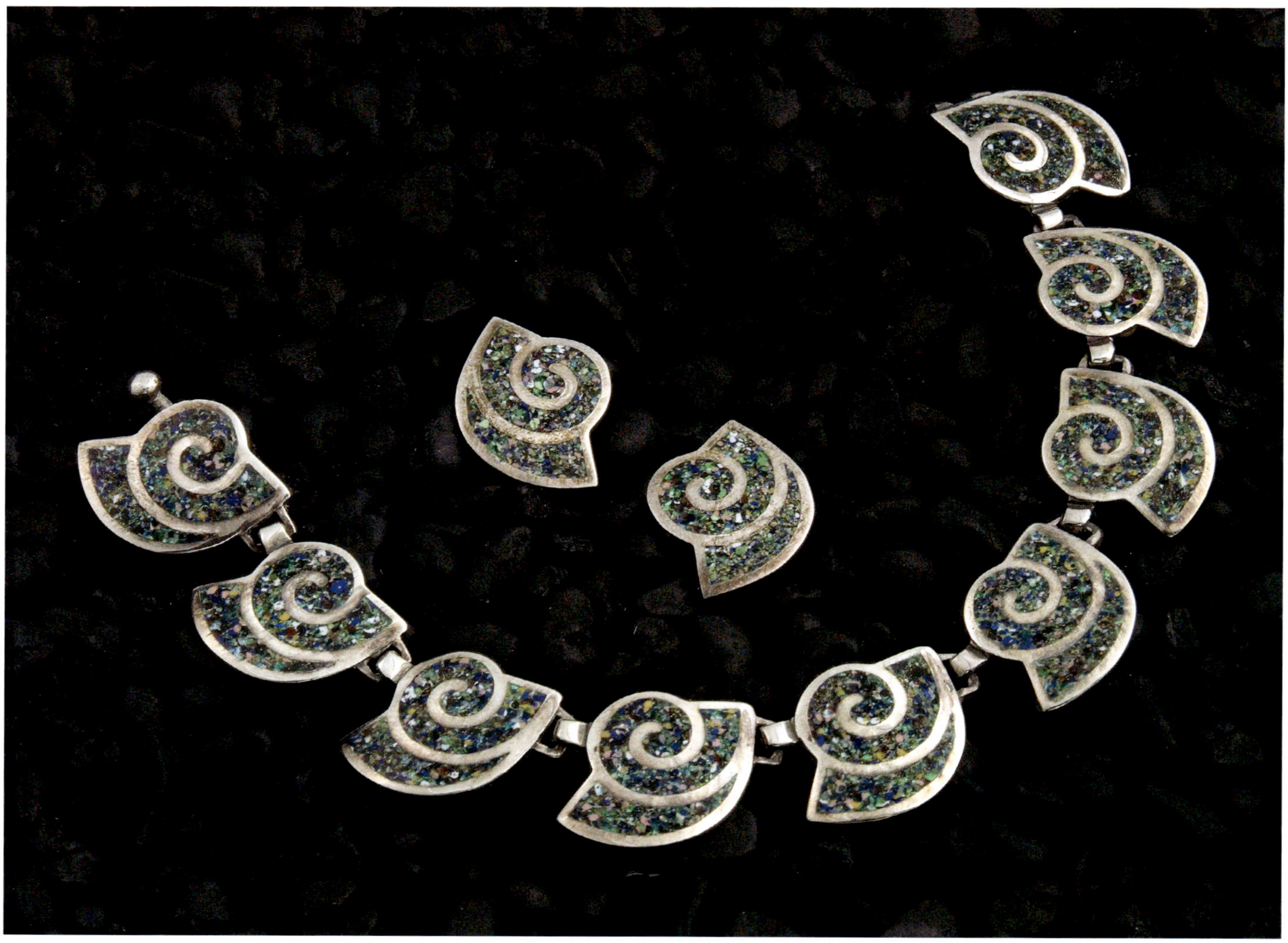

these significant paradigms, Margaret Clayes's admiration for Asian art could have made a lasting impression on Margot, considering the fact that, in her own designs years later, Margot often returned to the Asian aesthetic.

In 1927, Margot married Francis Evans Carr.[68] Like her mother, who married Albert Durand when she was thirty-four, Margot married late, at the age of thirty. The decision made by Margot and Albina to remain single until they were in their thirties distinguishes them from the population at large. During the first half of the twentieth century in the United States, the median age for a woman's first marriage was twenty-one.[69]

Another relevant set of statistics relates to Margot's decision to work for a living. On average, in the 1920s and '30s, just slightly over twenty percent of the female population in the United States was in the labor force.[70] Thus Margot broke from social norms with her decision to work outside the home and, even more so, by supporting her husband. In the 1930 census, Evans Carr is listed as a student and Margot had returned to hairdressing. Her salary must have been the only source of income during this period while her new husband was finishing his course work.

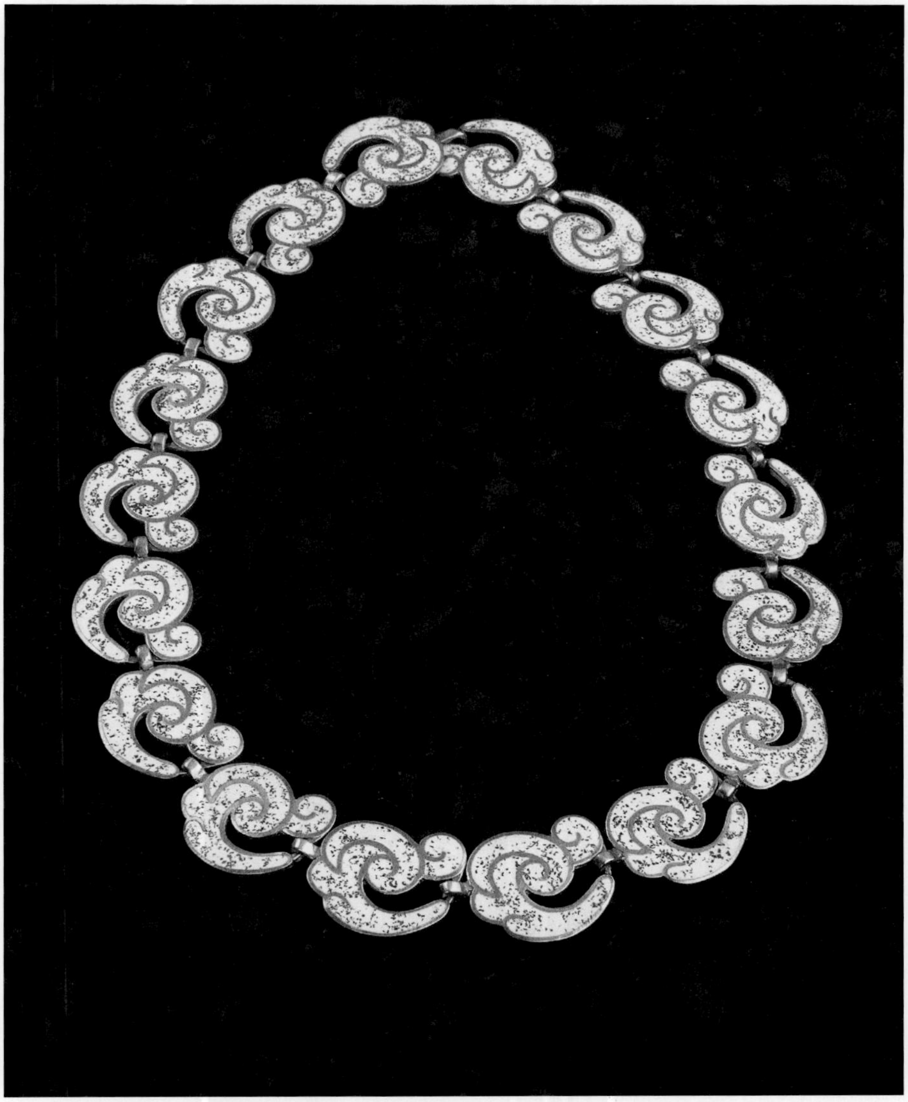

Plate III-9 Margot de Taxco 5855. *Interconnected Scrolls* necklace 15 x 1.25 in. Private Collection

Evans was at the University of California Berkeley studying dentistry and Margot was working as a cosmetologist in a dry goods store.[71] There is every possibility that this store was O'Connor, Moffatt, & Company (OM&C), which later became Macy's West. In 1928, OM&C had moved from Post Street to a new store at 101 Stockton, where Margot would have been employed.[72] Years later, in 1938, two friends, Dorothy Dunnicliff and Consuelo Barreto, came as tourists to Taxco with Margot. Dorothy and Consuelo were employees of O'Connor, Moffatt, & Company at the time, so the three women could have met and developed friendships as co-workers at OM&C.[73]

Albina became the owner in 1929 of a small neighborhood grocery store. This change in the direction of her life was an unusual one, but not inconsistent with her desire to remain financially independent. Albina found and rented space from a friend for the little grocery market at 1513 48th Avenue and lived behind the store at 1515 48th.[74]

Evans and Margot Carr moved to 1355 Willard the same year. In the first of two city directory listings for 1929, Evans and Margot are at the same address as a couple. In the second, Margot distinguishes herself with a listing of her own: "Mrs. Margot Carr, hairdresser. 1355 Willard."[75] Margot had already been living alone for several years, so it is not surprising to find that she had taken the initiative to present herself apart from her husband. Margot was supporting herself and Evans, and so must have chosen to publicly maintain her professional status. Clearly, even after marriage, Margot's independent nature, cultivated in her by her mother, continued to be a defining aspect in her approach to life.

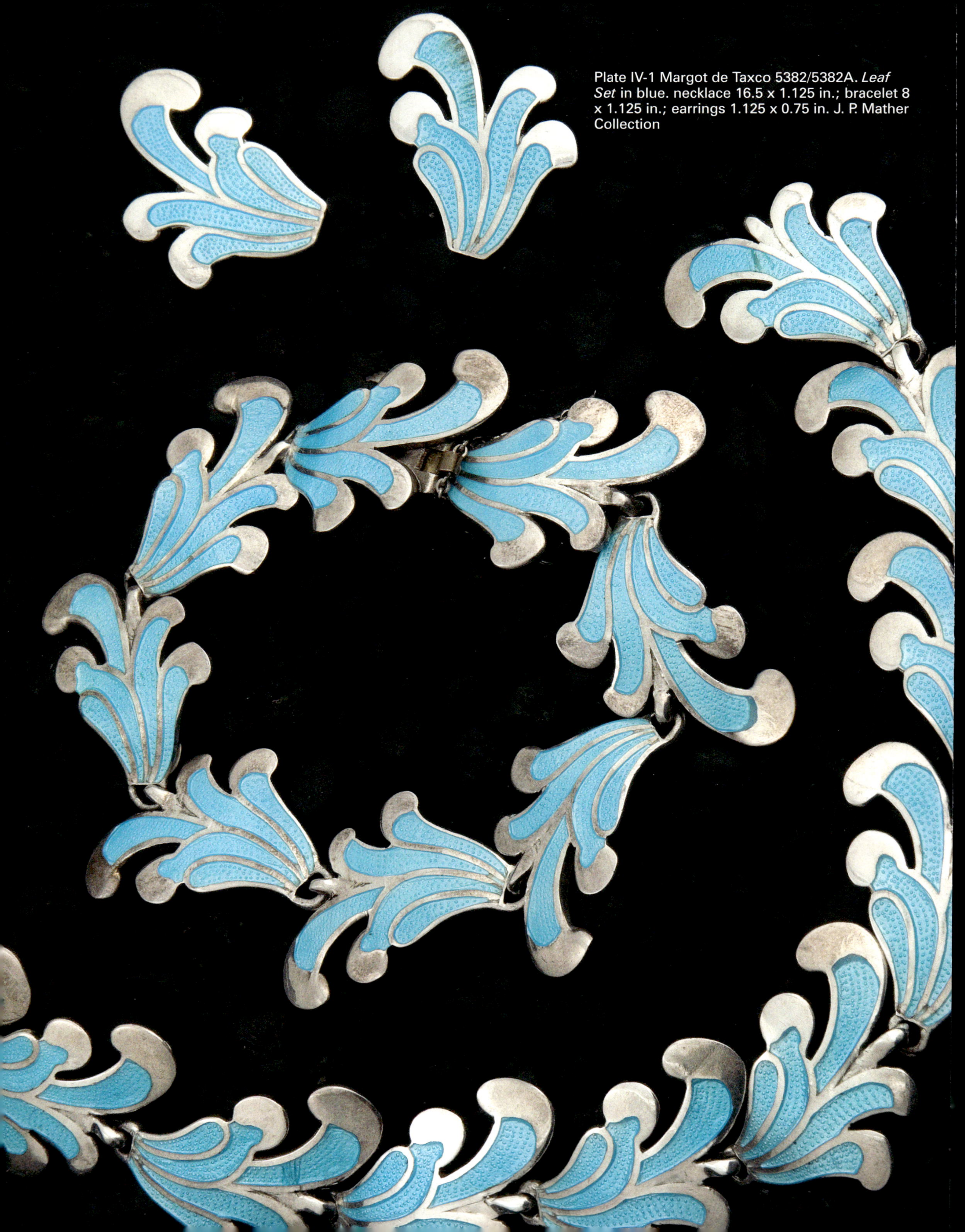

Plate IV-1 Margot de Taxco 5382/5382A. *Leaf Set* in blue. necklace 16.5 x 1.125 in.; bracelet 8 x 1.125 in.; earrings 1.125 x 0.75 in. J. P. Mather Collection

Chapter IV
Albina's Death

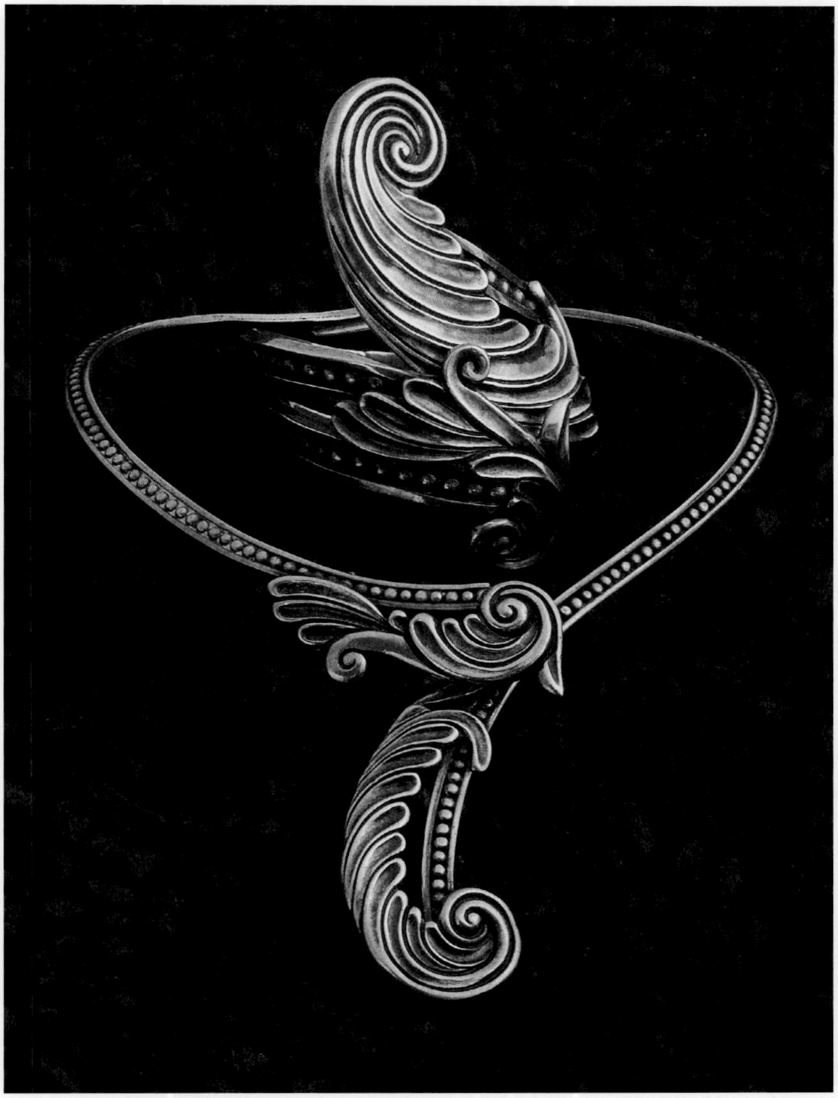

Plate IV-2 Margot de Taxco 5338/5339. *Silver Swirls Set*. spring necklace (5338) 14 in., central motif 2.5 x 1 in.; spring cuff bracelet (5339) 8 x 2.88 (at widest point) in. Elizabeth Finch Collection

Considered "one of the most fiendishly cruel murders in modern times,"[76] Albina's death is quite terrible to contemplate today, much as it was when headlines screamed the story on the front page of newspapers across the country. The first mention of the crime appeared in the February 18, 1931 edition of the *San Francisco Chronicle*: "A little old lady, who preferred independence to a 'safe' life with relatives, perished terribly yesterday in the grocery where she proudly made her living. . . . The victim was Albina Voorhies, 83, widow and relative of the prominent East Bay Chabot family."[77]

There is no question that the crime was premeditated. Charles Simpson, the son of Albina's landlord, admitted to the robbery and killing and described it in great detail for the press. He stated that he had planned the robbery because he had heard and believed a rumor that, over the years, Albina had stockpiled a large sum of money in her store or apartment.[78]

On the evening of February 16, 1931, at just after 8:00 p.m., Charles Simpson lurked outside the store. When Albina approached the door to lock up for the night, Charles came out of the shadows and asked to make a quick purchase. Albina knew Charles well, and so she invited him to enter. [79]

Charles was carrying a club under his coat and, when Albina turned and bent over a box to get cookies for him, Charles pulled out the club and struck her with such force that the lid of the box severed one of Albina's fingers. Albina was able to pull herself up and began running towards her apartment. At the doorway, Charles hit her again and she fell unconscious. He dragged her into the apartment and tied her to a chair that he placed in the middle of the room.

He began searching the apartment and store for the money and came up with only about $3.00, which he removed from the cash register. At this point, Charles began to panic and decided to burn down the building. He took a can of coal oil and poured it over Albina and struck a match. As the flames envel-

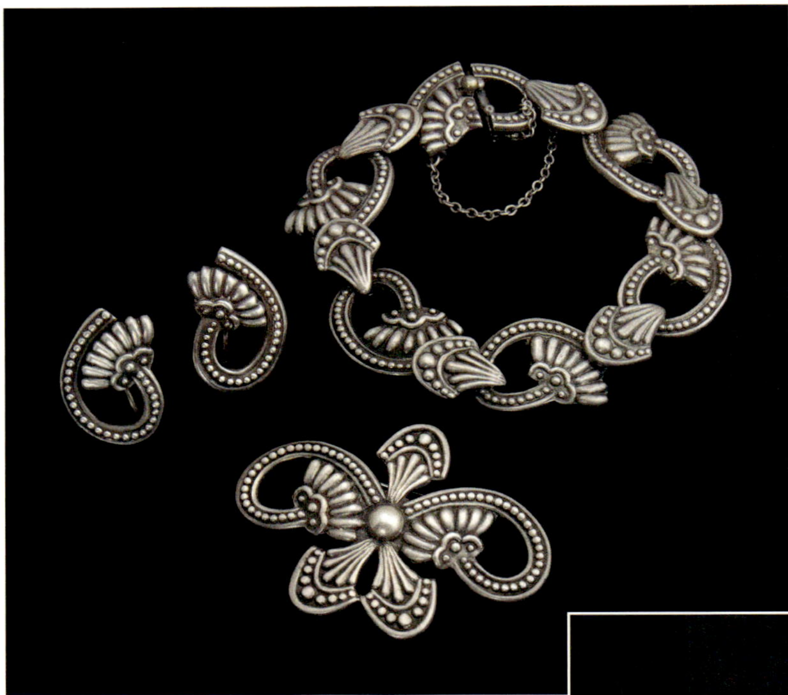

oped the poor woman, who at this point had regained consciousness, she began to scream. To silence her, Charles hit her several more times with the club. She fell out of the chair and he dragged her to the bathroom, where he hit her again and left her to die. He ran from the store, leaving his cap behind. He buried his blood-soaked overcoat and the club in a nearby park on his way home.

Albina's neighbor, Elizabeth Neil, reported billowing smoke coming from the rear of the store. When police arrived, they entered through the open door and discovered the store ransacked and in disarray. In the living quarters, the fire was still burning in the bedroom and flames were coming from the bathroom. It was here that the police found

Plate IV-3 Margot de Taxco 5256. *Stylized Flower and Dots Set.* bracelet 7 x 0.5 in.; brooch 1 x 1.25 in.; earrings 1 x 0.5 in. Private Collection

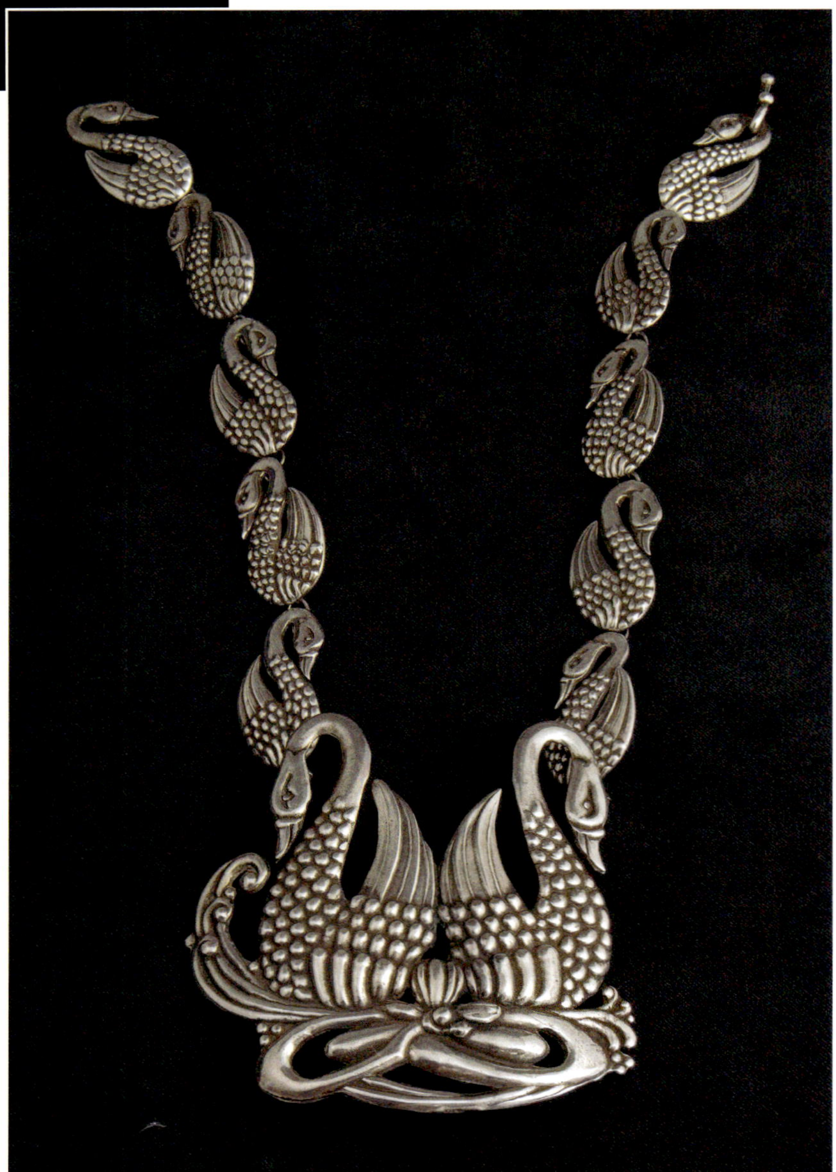

Plate IV-4 Los Castillo 516. *Swan necklace* in silver 15 x 0.63 in., central motif 2.88 x 3 in. Elizabeth Finch Collection

Albina's battered body, "with its remnants of her clothing still aflame." [80]

When confronted with the cap that he had left behind, Charles admitted to the slaying. He stated that he had committed the crime because he had promised to take his girlfriend to the movies that night and needed the money. What horrified the police, the press, and their audience was that Charles had bludgeoned and burned this poor woman to death for a mere $3.00, and then had revealed no remorse. One of the more grisly press reports was the reenactment that appeared in the *San Francisco Examiner* as a series of on-site photographs, complete with Charles pointing towards white chalk drawings of Albina, indicating where he had tied her up and where she had collapsed and died.[81]

On February 18, 1931, Mr. and Mrs. Evans Carr of 1355 Willard Street provided information about Albina for the Department of Public Health, giving her birth date as October 8, 1863 and birthplace, Quebec, Canada. They verified that she had been in the United States for fifty-one years and that she was a widow, whose husband's name was C. Voorhies [sic]. This same document indicates that Albina's body was cremated and sent for burial to the Mt. View Crematory in Oakland, California.[82] Albina's obituary stated without equivocation that the funeral service was to be strictly private.[83] Anthony, Remi, and Emelie were buried at Mountain View Cemetery. Robert, Albina's brother or, more likely, her cousin Ellen could have provided a place for her ashes and Ellen may even have attended the funeral.

The press hounded Margot for interviews, and her husband apparently intervened and took some of the questions: "Evans Carr, dental student and son-in-law of the dead woman, told police that she [Albina] did a daily business of about $20 and never banked her money. Her daughter, Mrs. Margot Carr, was overcome by the tragedy. She revealed that she had frequently urged her mother to abandon the store and make her home with the Carrs — but that Mrs. Voorhies had preferred to 'make her own way.'" [84]

In the beginning, Charles Simpson pleaded not guilty by reason of insanity, but he eventually withdrew the plea and admitted his guilt. He was convicted of murder on April 14, 1931. At the age of eighteen, on July 17, 1931, Charles Simpson was hanged at the San Quentin penitentiary.[85]

Albina apparently did not leave a will. Her estate was not large enough to incur inheritance tax, with a net value of less than $4,000.00. The few next of kin listed included her daughter Margot and her brother Robert Chabot, who was living in Culver City, California. Margot was made administratrix of the estate and received a statutory fee of $499.39. Albina's inventory included a promissory note or claim against Robert Chabot for $1,069.35.[86] This loan to her brother is most curious since, at the time of her death, Albina was living in very modest circumstances, and her brother was a successful businessman and real estate salesman.

The mention of Albina's connection to the "prominent East Bay Chabot family" in the *San Francisco Examiner*, the burial at Mountain View, and the loan to her brother Robert indicate that she remained in contact with her extended family. With this in mind, Albina's insistence on her financial independence, in hindsight, seems heroic. She was self-sufficient most of her life and, in her last years, instead of living with her daughter and new son-in-law, Albina opened a

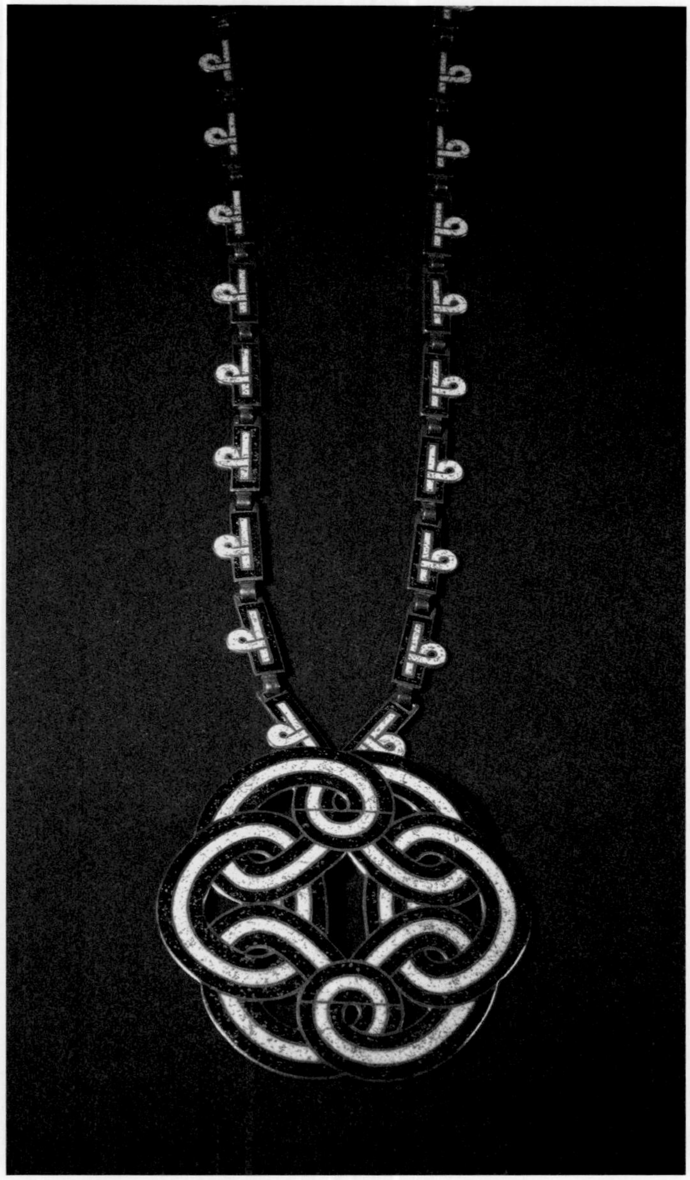

Plate IV-5 Margot de Taxco 5752. *Spiraling Circle* necklace 23 in., pendant 2.5 in. d. Private Collection

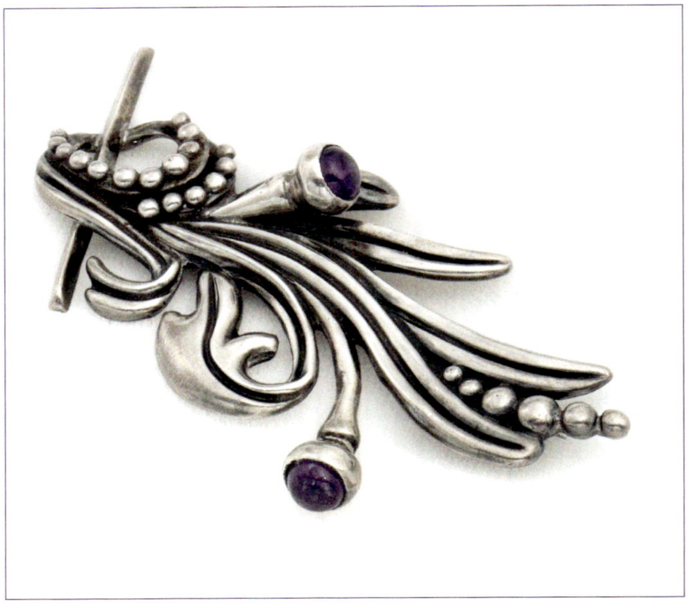

Plate IV-6 Margot de Taxco 5819. *Floral with Amethysts* brooch 2.75 x 1.5 in. Private Collection

corner grocery. When she was widowed in 1903 and left homeless with a child in 1906, she might have looked for assistance and, in all likelihood, this support was extended to her by the Chabots and possibly by the Voorhies. At the same time, Albina seemed to resist dependence on family members. Considering what little is known of Albina Chabot Voorhies, it would appear that this resistance had less to do with an awareness of her place in the social hierarchy than with a profound strength of character and self-determination.

In 1932, Margot and Evans were living at 1150 Union. Evans had become a dentist and Margot does not list employment, a choice that could have been the result of the trauma of her mother's death. Margot's self-portrait (Plate III-1) is dated April 23, 1932 and signed "Margot." Below the portrait is an inscription in Spanish: "May all the nights of your life be good ones, and may all your days be filled with happiness."[87]

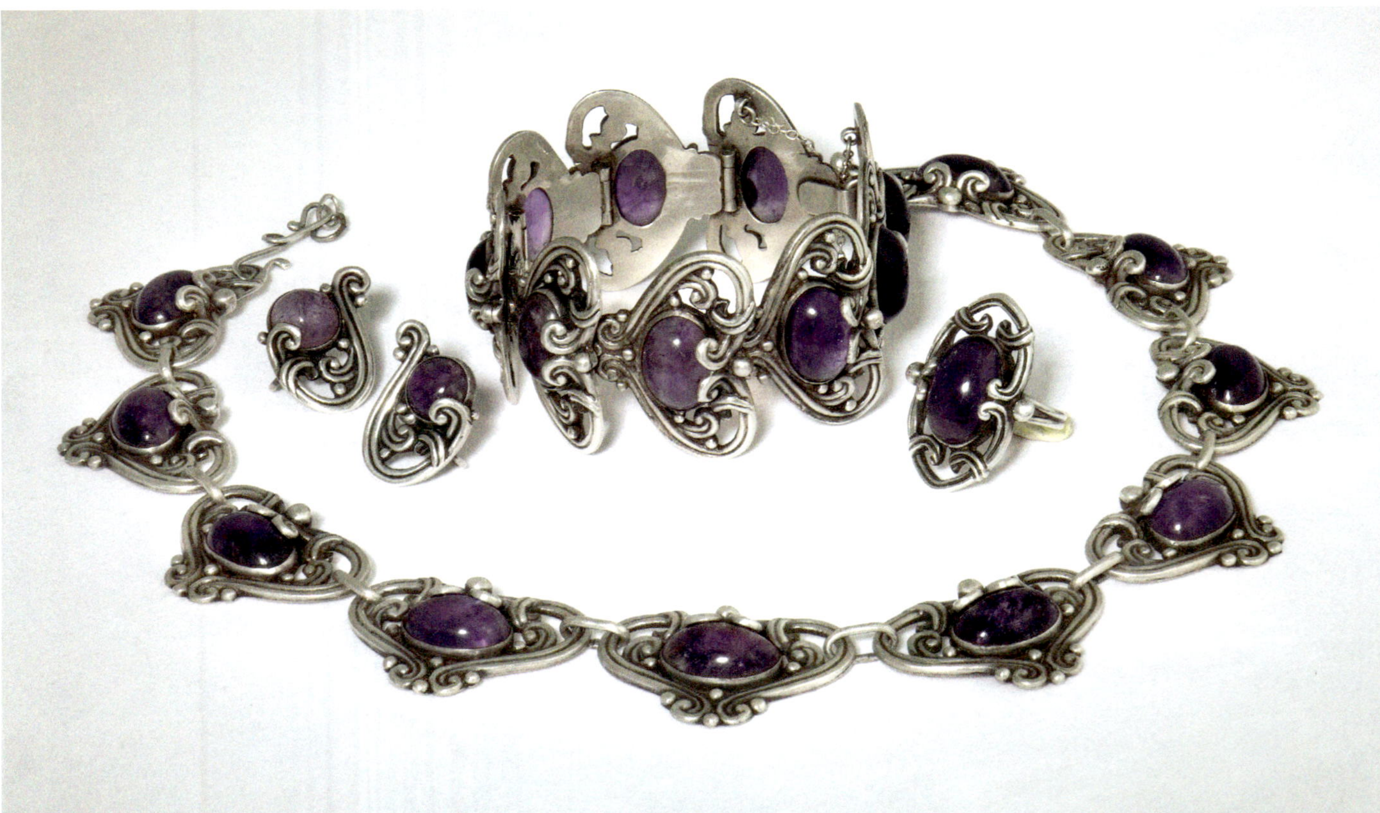

Plate IV-7 Margot de Taxco 5213. *Silver Hearts with Amethysts Set*. necklace 16 x 1 in.; bracelet 8.75 x 1 in.; earrings 1 x 0.75; ring size 8. J. P. Mather Collection

Given the date and the romantic sense of the inscription, this drawing may have been intended as a gift for Evans Carr. This pencil and watercolor sketch suggests that in the year after her mother's death, Margot may have attended art school. The portrait was purchased in Mexico City, so it seems that either Margot gave the sketch to someone in Mexico or kept it until her death.

In 1933, Margot resumed her separate listing as a beauty operator and, in the 1934-1935 San Francisco city directory, stipulated that she was employed by the United Artists Beauty Salon located at 133 Geary. Margot continued to work for the same company in 1936 and Evans had his office at 450 Sutter. On July 1, 1936, Margot and Evans filed for divorce. The following year, Margot remained at 770 California and was working for the beauty shop at 133 Geary. The only address provided in the 1937 directory for Evans was 450 Sutter.[88]

Jaime Quiróz remembers discussions with Margot about her mother's death. During those terrible months, she developed an abhorrence for the press resulting from the harassment she had received in the aftermath of her mother's murder. Margot told Quiróz that, at the time, she felt herself slipping into emotional illness. Ultimately, the horror of her mother's death caused stress in the marriage and led to divorce.[89]

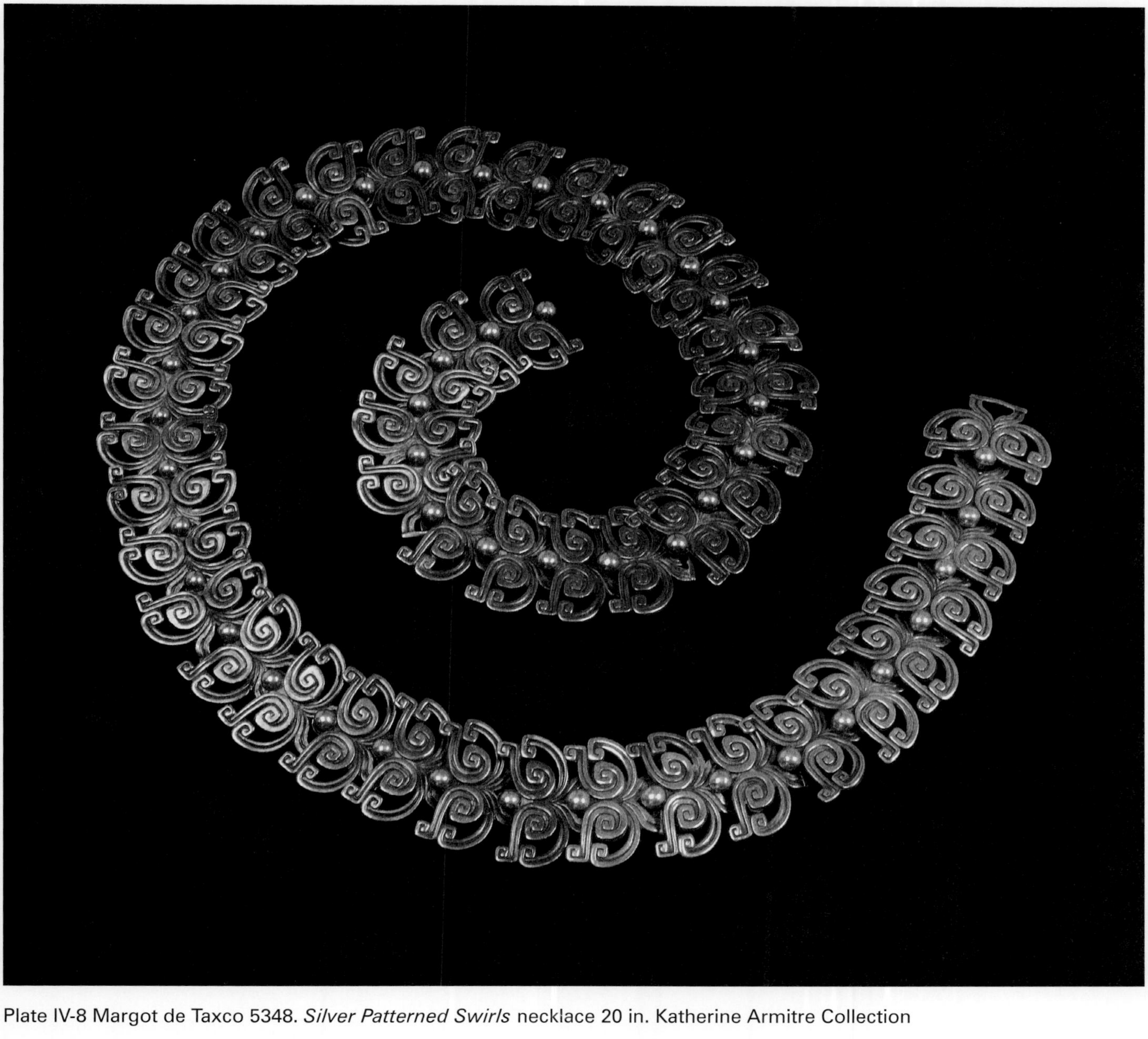

Plate IV-8 Margot de Taxco 5348. *Silver Patterned Swirls* necklace 20 in. Katherine Armitre Collection

Plate V-1 Margot de Taxco 5158. *Symmetrical Curls* in silver and copper. Necklace 18 in. circum.; bracelets 7 x 1 in.; earrings 1.5 x 0.75 in. Private Collection

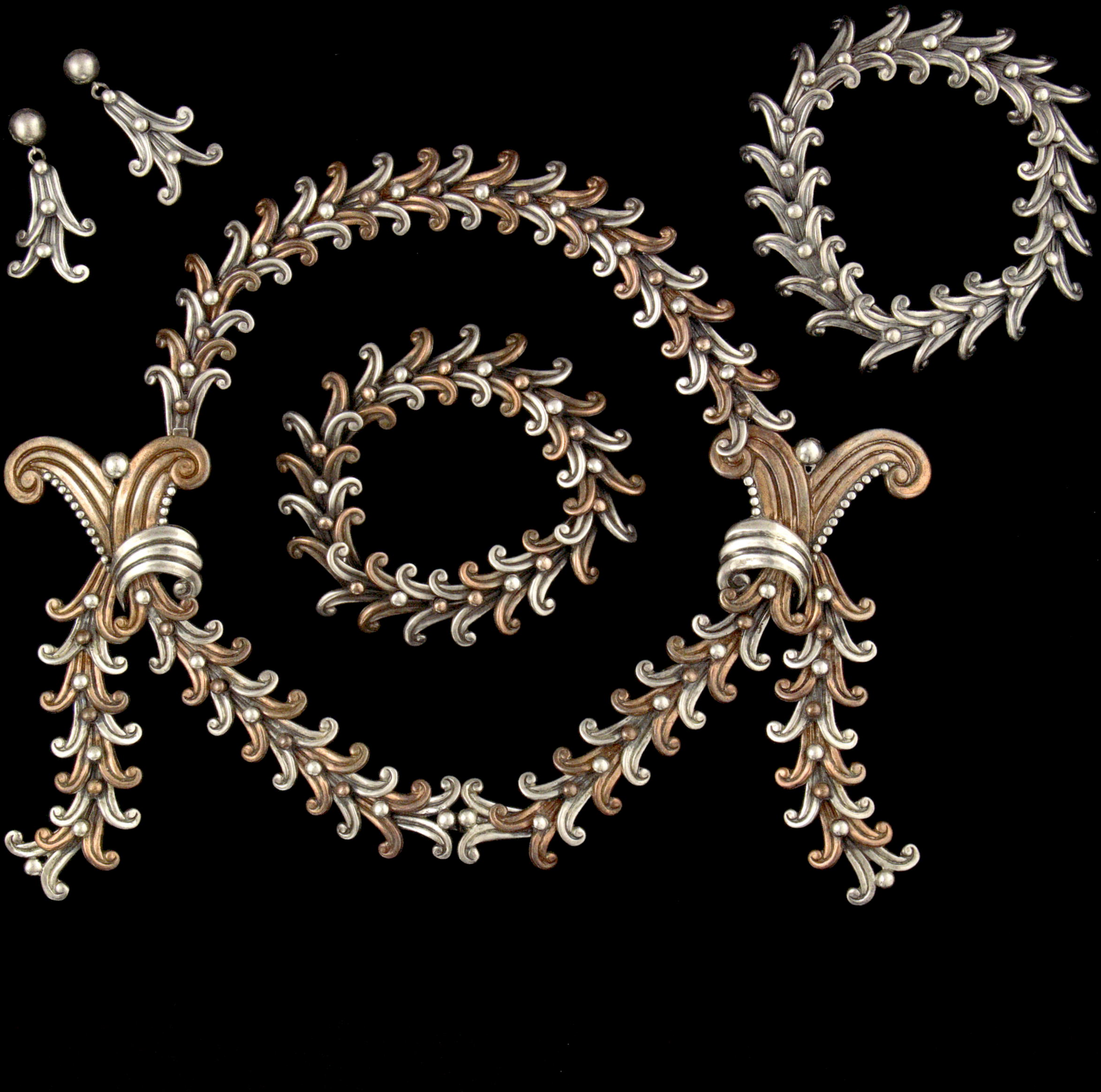

Chapter V
Off To Mexico

Margot made a dramatic change in her life in 1937. She and a group of friends took a trip to Mexico and Margot did not return with them to San Francisco. Defying convention, Margot rented a place in Mexico City at Chopo 139 and, on her own, found employment at a photography shop owned by a German photographer. [90]

Margot took a photograph a few years later entitled "Tarascan type, México" (Plate V-3). Signed "Margot C. de Castillo," the image leads to the identity of the German who owned the photography shop. In Margot's photograph, an old man is seated before a colonial building, wearing a traditional sarape. The rough light-colored stone and the angle of the build-

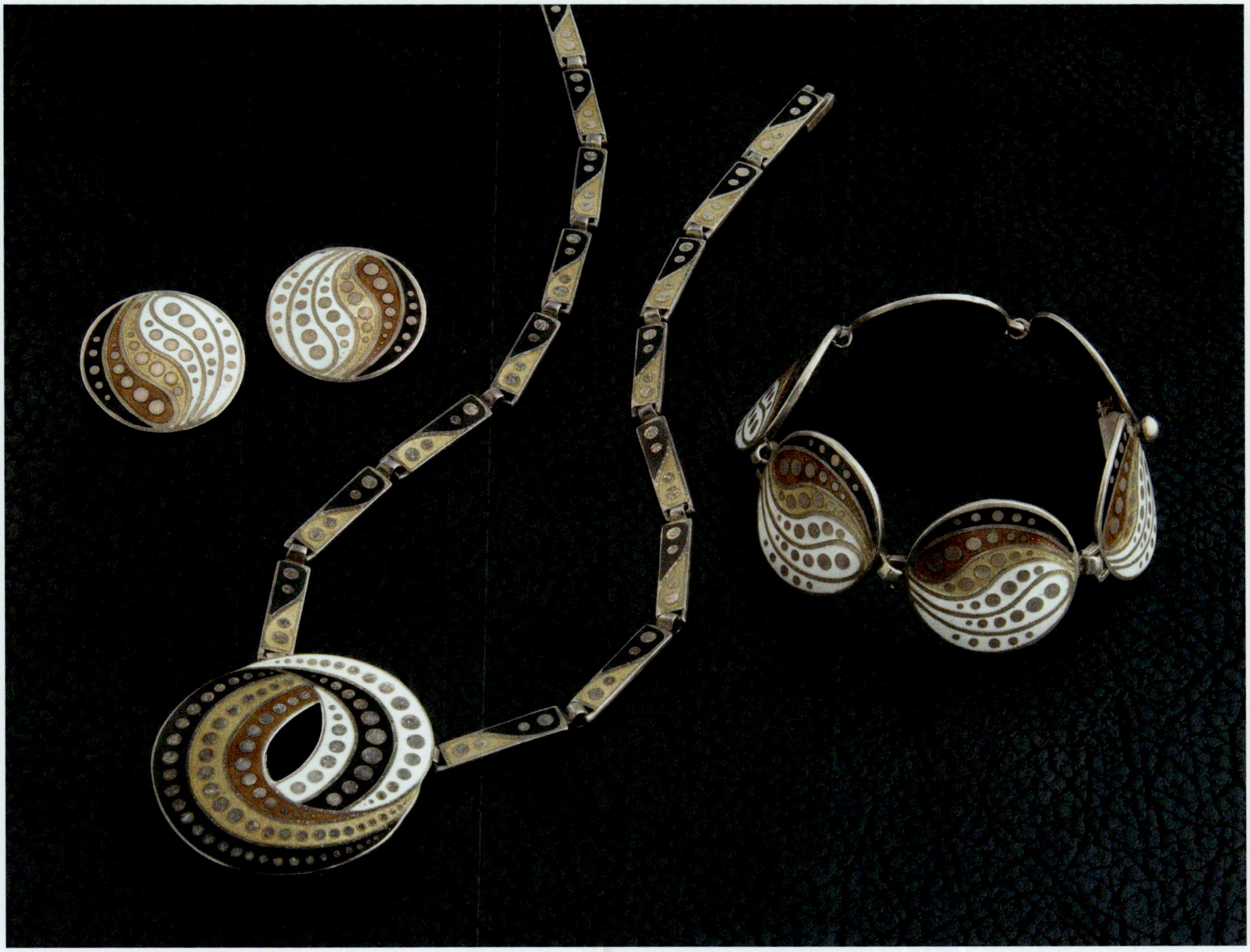

Plate V-2 Margot de Taxco 5710. *Yin and Yang Set* in brown and black. necklace 15 in., central motif 1.75 in.; bracelet 7.5 x 1.13 in.; earrings 1.13 in. Elizabeth Finch Collection

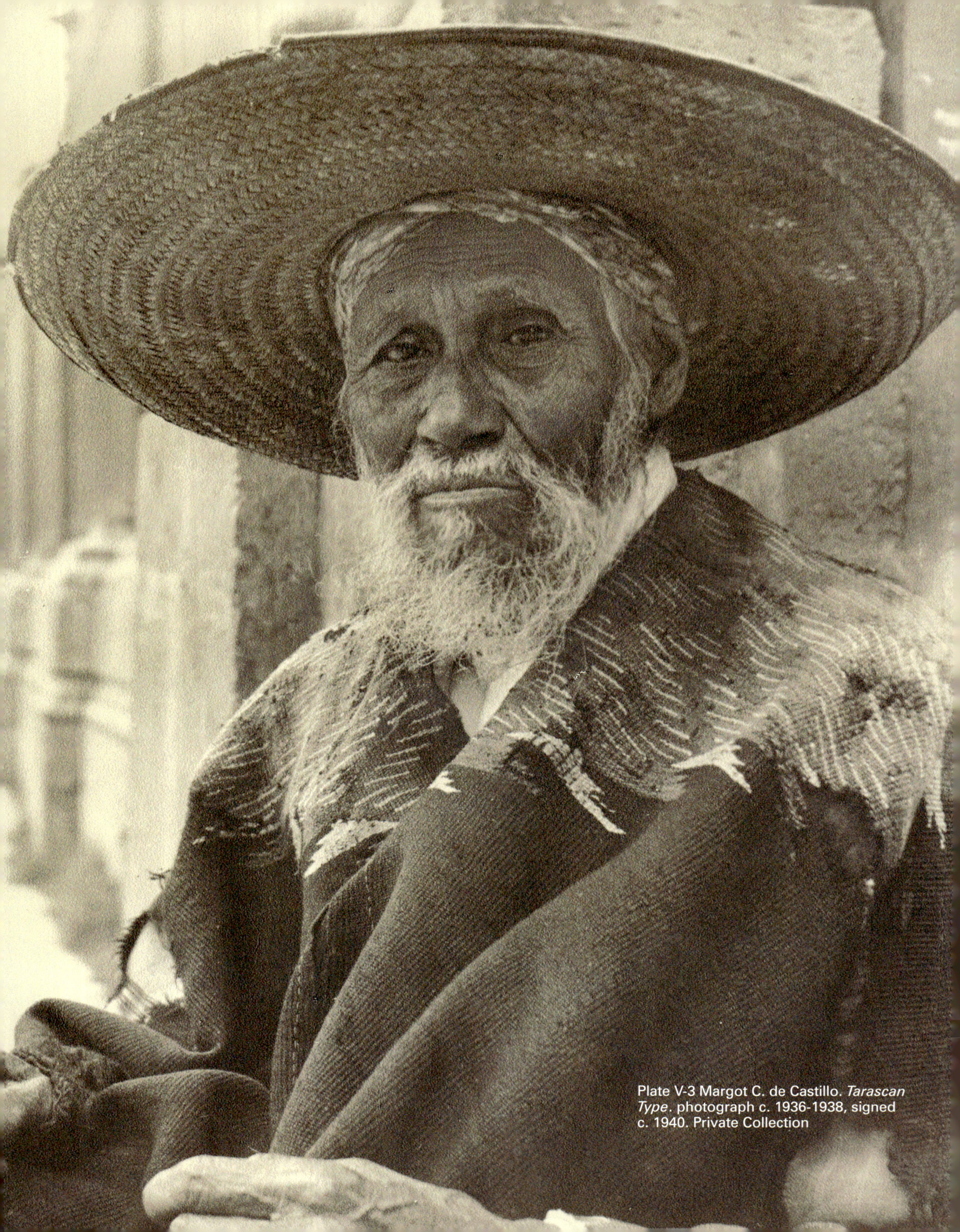

Plate V-3 Margot C. de Castillo. *Tarascan Type*. photograph c. 1936-1938, signed c. 1940. Private Collection

ing extending out of the picture frame in the background thrust the old man towards the viewer. The man's white beard and mustache are in contrast to his skin, which has been darkened and lined by years of working in the sun. He looks directly into the camera, his face framed by his straw hat and the large dark triangle of the sarape. All compositional lines point towards the face. He has a quiet dignity and intelligence about him that must have drawn Margot's attention. She carefully manipulated the lighting to enhance textures, the patterns in the woven straw and sarape, and the shadows that sculpt the face. Hers is a sensitive and carefully composed portrait.

Margot's photograph can be compared to one by Hugo Brehme, which he entitled "El Pastelero de Tacubaya" [The Pastry Vendor of Tacubaya] (Plate V-4). The similarities are striking. Both Margot and Brehme place the subject at an angle to the picture plane. In Brehme's photograph, as in Margot's, the large hat and the shadow it casts frame the man's face. The eyes look directly into the camera, with strength and intelligence. Stylistically, Margot's photograph indicates the influence of Brehme. It can thus be conjectured from her choice of subject and stylistic approach that Margot worked for Brehme.

Hugo Brehme was a German photographer who had arrived in Mexico in 1905. He called his studio "Fotografía Artística Hugo Brehme," indicating that he considered his work in photography as art. He sold photographs and post-

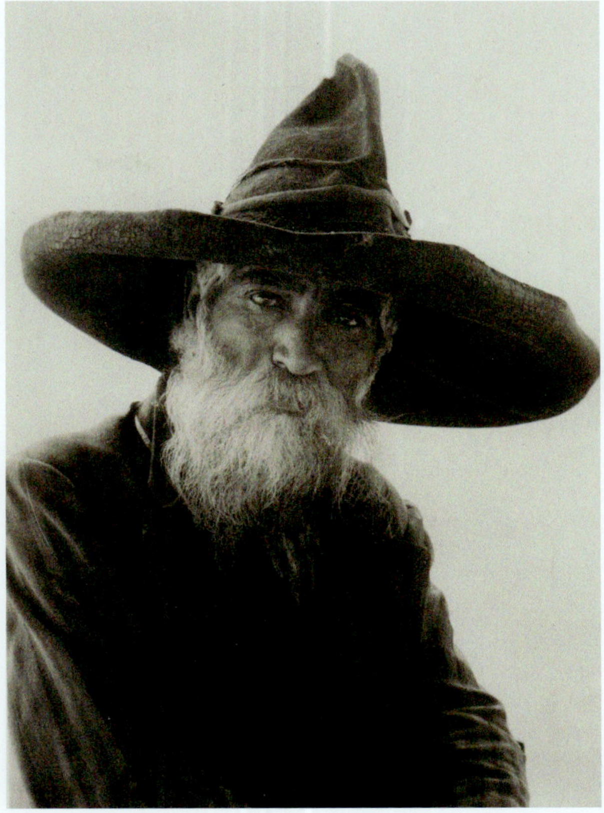

Plate V-4 Hugo Brehme. *El Pastelero*. photograph, c. 1925. Susan Toomey Frost Collection

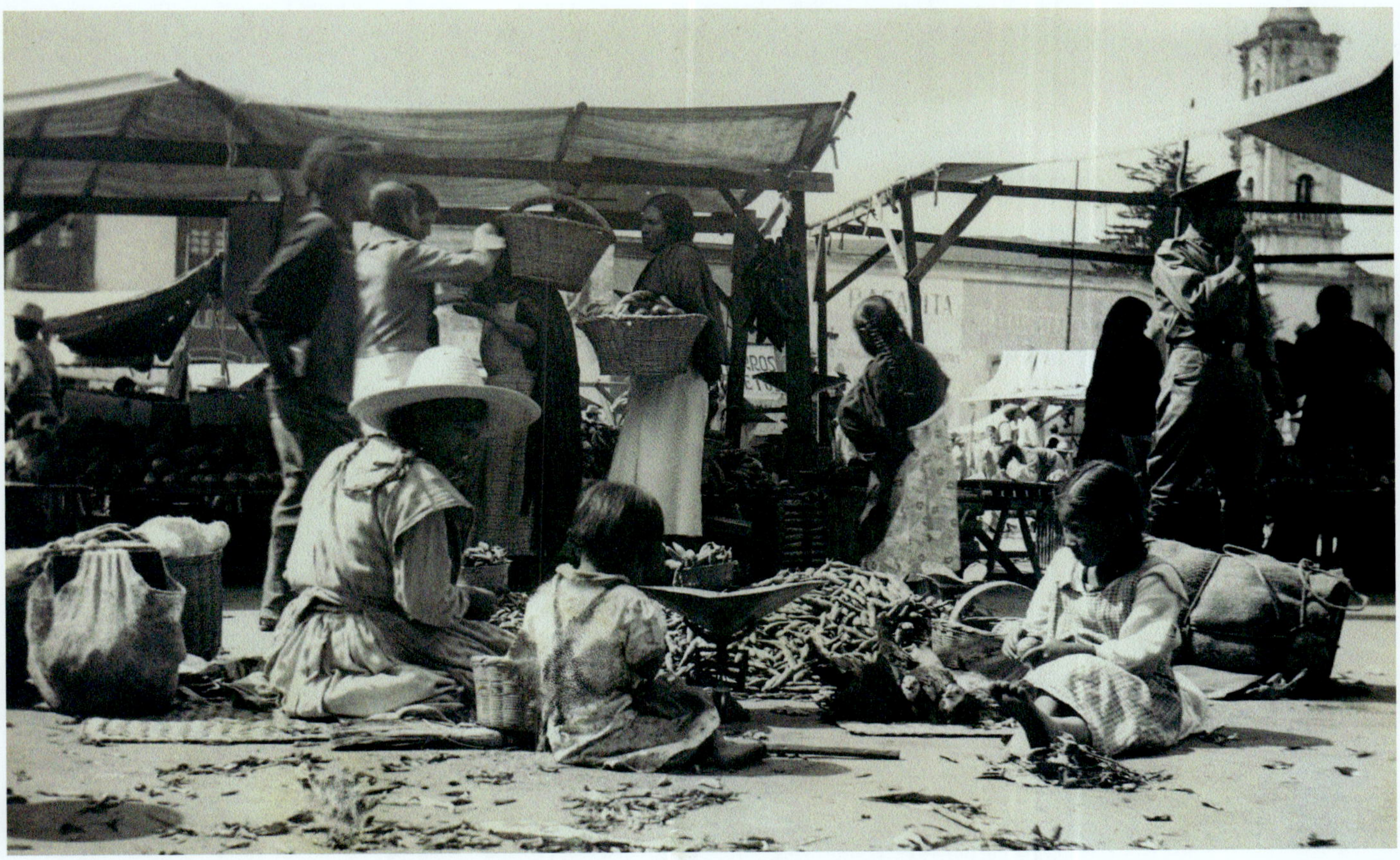

Plate V-5 Margot C. de Castillo.*Toluca Market, Mexico*. photograph c. 1936-1938. Violante and Consuelo Ulrich Collection

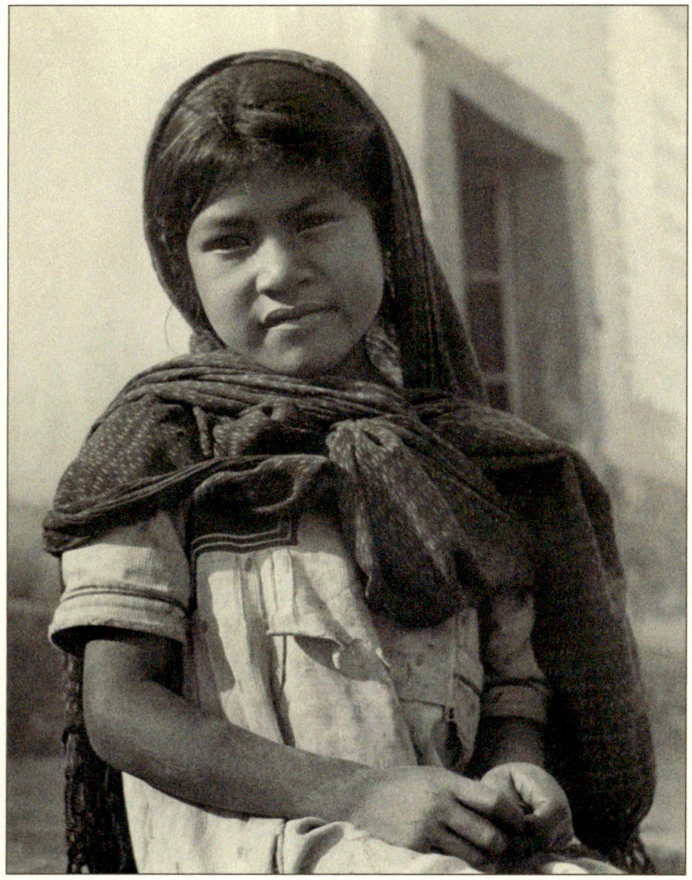

Plate V-6 Margot C. de Castillo. *Manuelita.* photograph c. 1936-1938, signed c. 1940. Private Collection

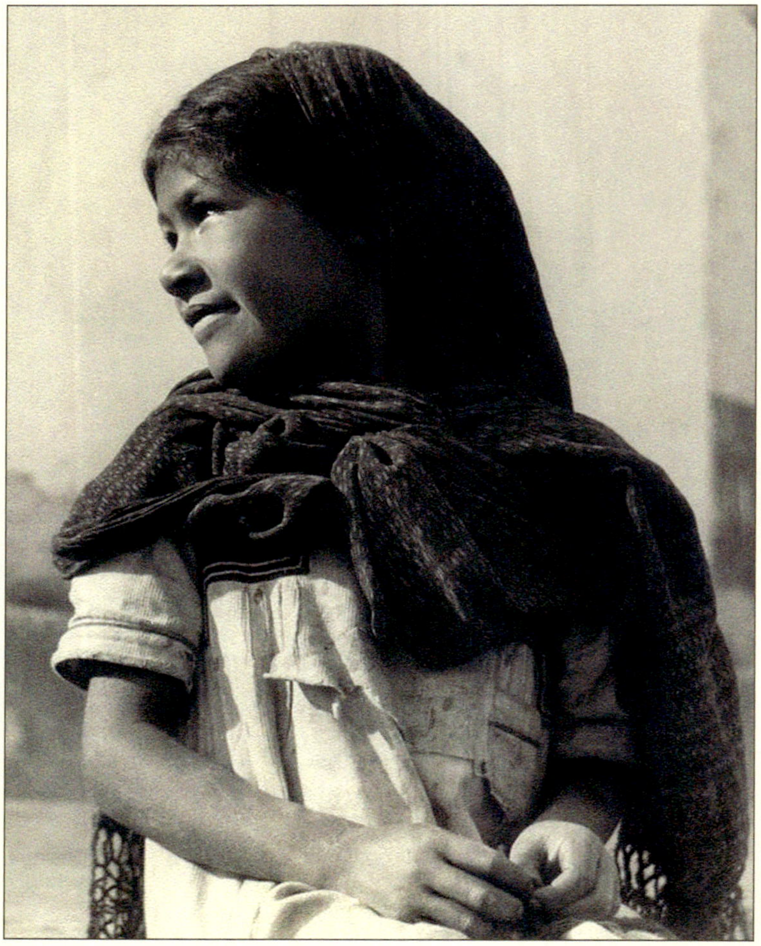

Plate V-7 Margot Van Voorhies. *Muchacha de Tepozotlán*. photograph c. 1936-1938. Rocio Sierra-Honigmann Collection

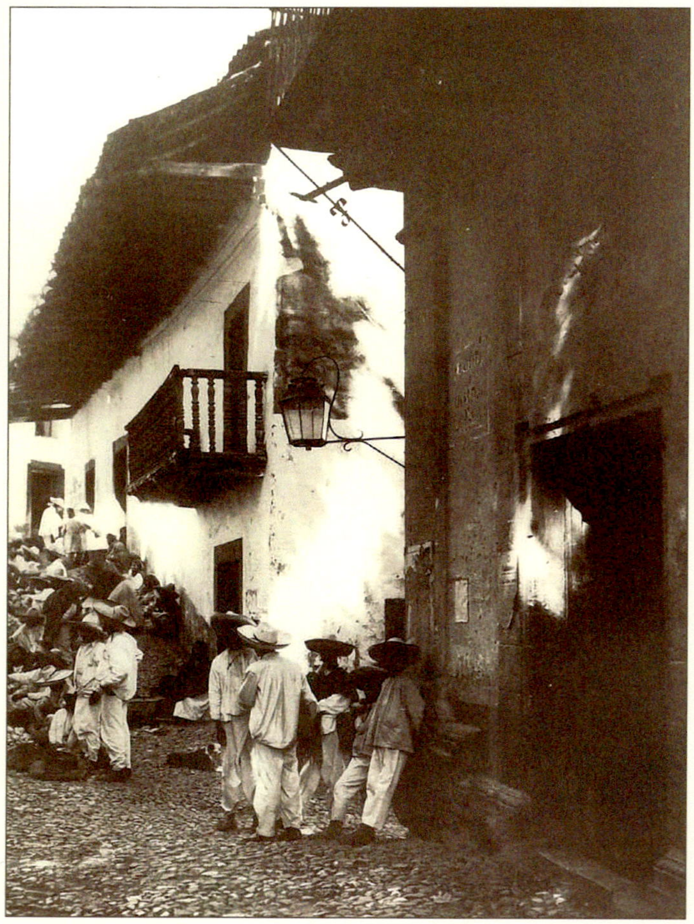

Plate V-8 Hugo Brehme. *The Market in Taxco's Zócalo*. photograph. Susan McCloskey Collection

cards of pre-Columbian pyramids, colonial buildings, the volcanoes, and village life (Plate V-8) and published books of his photographs. In 1937, Hugo Brehme had his studio upstairs at Madero #1. Anita Brenner, in her tourist guidebook, distinguished Brehme's studio from the only other camera shop in Mexico City as a place where the visitor could expect to find outstanding souvenir artistic photographs of Mexico rather than cameras and film.[91]

While Margot's two photographs of Manuelita, also called the" Muchacha de Tepozotlán" [Young Girl of Tepozotlán], vary in pose, it is significant that the signatures differ (Plates V-6-7). Margot could have taken the two photographs of Manuelita, "Tarascan Type, Mexico," and "Toluca Market, Mexico" (Plate V-5) before her marriage to Antonio Castillo, since "Muchacha de Tepozotlán" is signed "Margot Van Voorhies."[92] For these photographs, Margot presumably traveled to the states of Michoacán, Toluca, and Tepozotlán from Mexico City while she was working for Hugo Brehme. Four of the five photographs were probably printed or signed in Taxco, the three with the "Margot C. de Castillo" signature and the view of Santa Prisca (Plate V-13), which she did not sign.[93]

María de los Angeles and Margot met in the jewelry store in Mexico City where María was working, and she and Margot initiated a lifelong friendship. María recalled trips to Taxco and Margot's introduction to Antonio Castillo. On August 25, 1937, Margot and two companions from San Francisco, Leona and Aurora Eckstrom, were staying at the Hotel Taxqueño in Taxco. Signing the register as Margot Carr, she returned to the hotel with Mrs. M. A. Thomson of San Francisco on the fifth of December and with Aileen Thomson (Mrs. Marga-

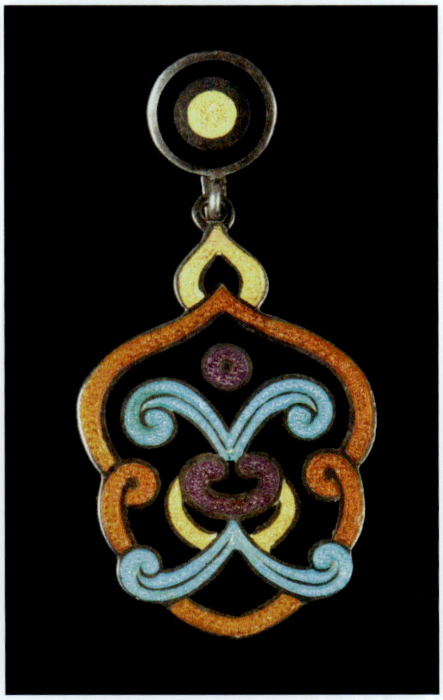

Plate V-9 Margot de Taxco 5531. *Arabesque.* Detail earring. Silver Huntress Collection

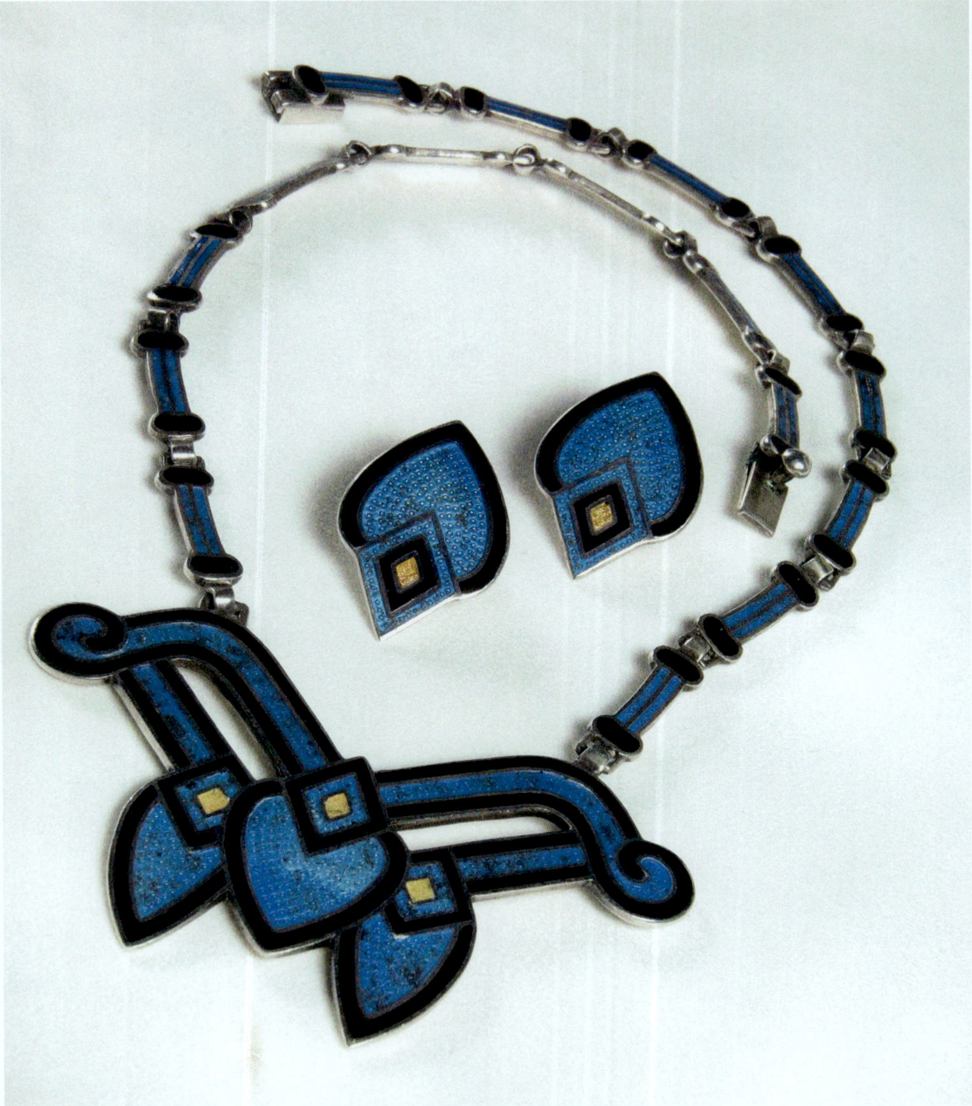

Plate V-10 Margot de Taxco 5717. *Deco Diamonds and Drops Set* in blue and black. necklace 17 in., central motif 2.13 x 3.38 in.; earrings 1.88 x 1 in. Private Collection

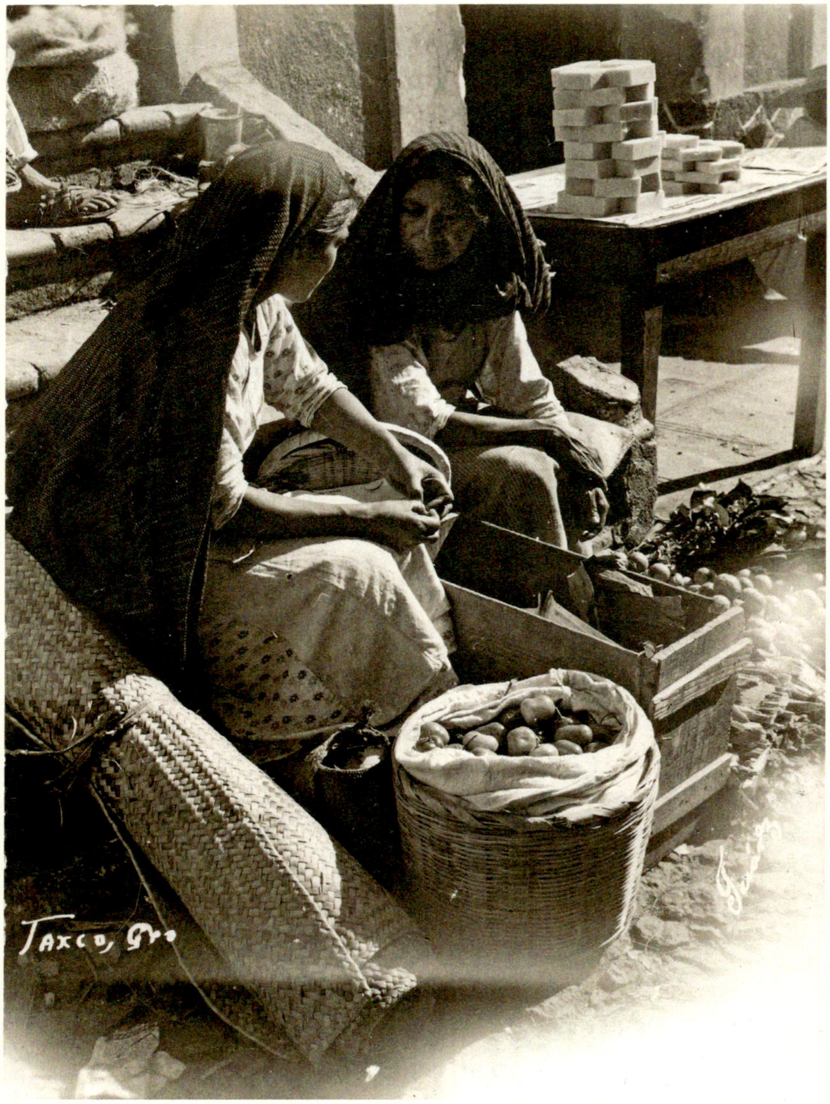

Plate V-11 Anonymous photographer c. 1940. *Two Women in the Market, Taxco*. Private Collection

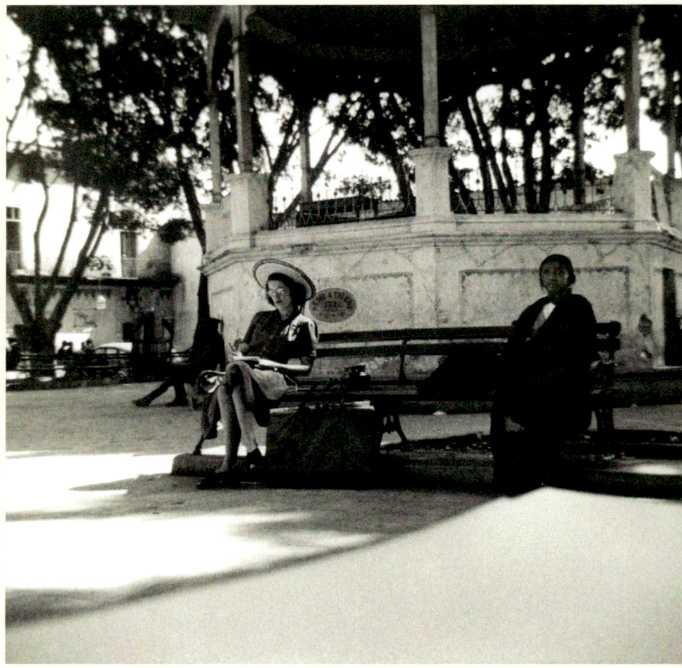

Plate V-12 Anonymous photographer c. 1940. *Tourist and Indigenous Woman on a Bench in Taxco's Zócalo.* Private Collection

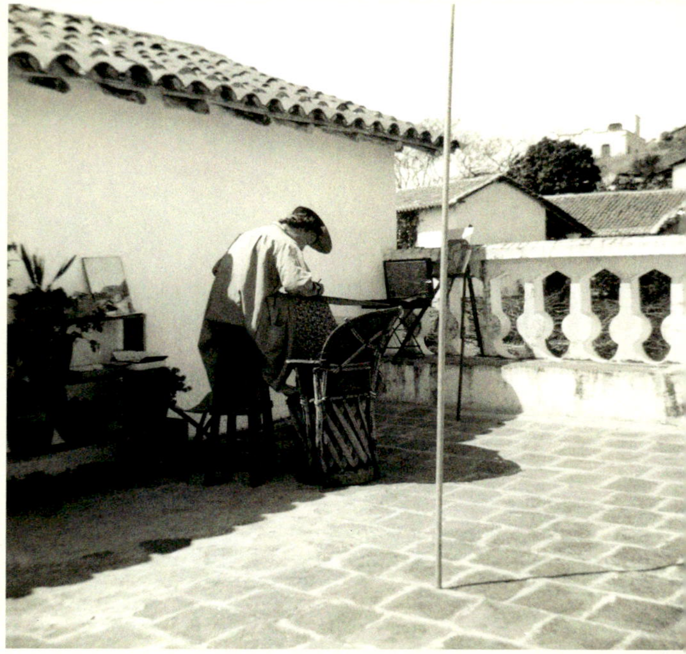

Plate V-13 Anonymous photographer c. 1940. *Painter on the Hotel Taxqueño Balcony*. Private Collection

ret Aileen Thomson) of San Francisco on Christmas Eve. Margot signed the Hotel Taxqueño register on May 10, 1938, accompanied by her friends from San Francisco, Dorothy Dunnicliff and Consuelo Barreto, both of whom were working for O'Connor, Moffatt, & Company where they had met and befriended Margot.[94]

The final entry in the Hotel Taxqueño register for Margot Carr was on July 5, 1938. She arrived with friends Annette and Ruth Bettelheim and Julia E. Smith.[95] At this point, Antonio and Margot had fallen in love and were having a long-distance romance. Margot was twenty years his senior, but the age difference did not seem to be of any consequence for, in 1938, Antonio asked Margot to come to Taxco and marry him.[96]

Margot could not have found a place more distant from her past. Taxco in 1938 was a mountain town surrounded by tiny villages, at once remote and impoverished (Plates V-8, V-11). In many of these communities, the pre-Columbian language Nahuatl was still spoken, and cultural practices harked back to the time before the arrival of the Spaniards. Determined tourists had discovered Taxco and the town now boasted three or four small hotels (Plate V-12). Artists and writers came to live and work, drawn by the beauty of the place (Plate V-13). White houses with red tile roofs lined cobblestone streets that wound their way up the mountainside. Rising above the plaza was the Church of San Sebastián y Santa Prisca, the eighteenth century masterpiece that was at the center of Taxco's spiritual and social life (Plate V-15).

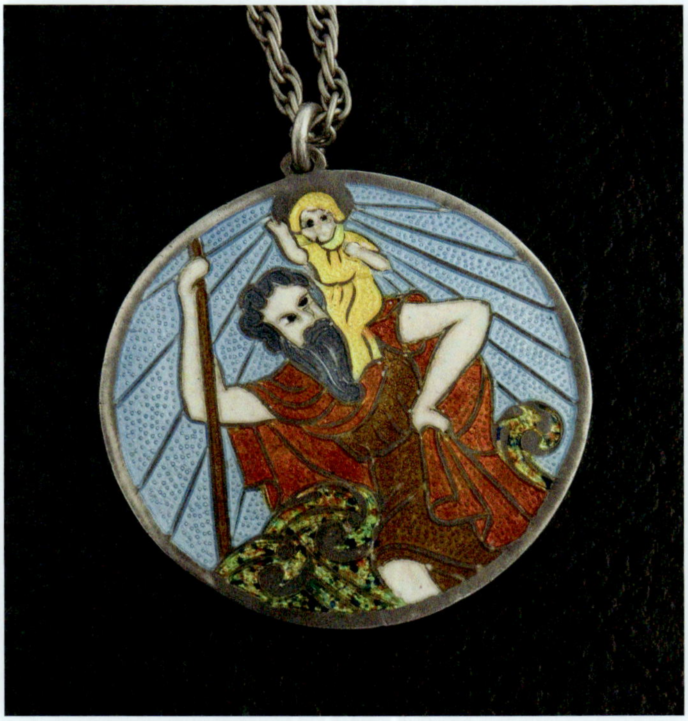

Plate V-14 Margot de Taxco 5537. *St. Christopher* pendant 1.88 in. d. Elizabeth Finch Collection

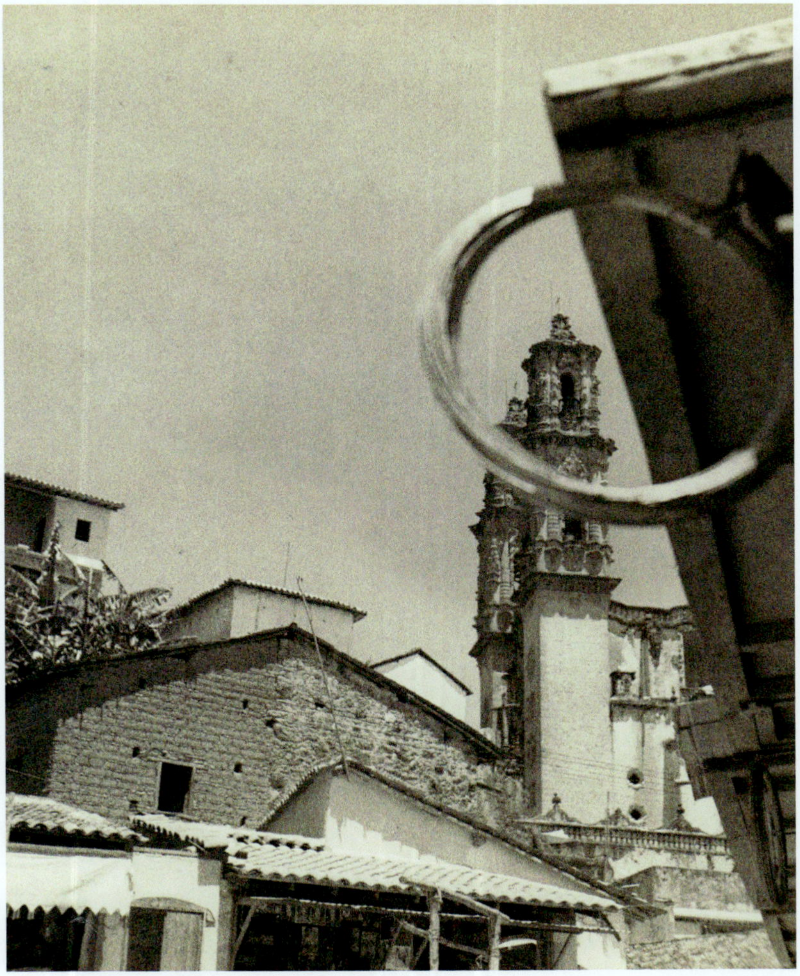

Plate V-15 Attributed to Margot C. de Castillo, c. 1940. *Santa Prisca [Cathedral], Taxco.* Private Collection

Part Two

Life in Taxco

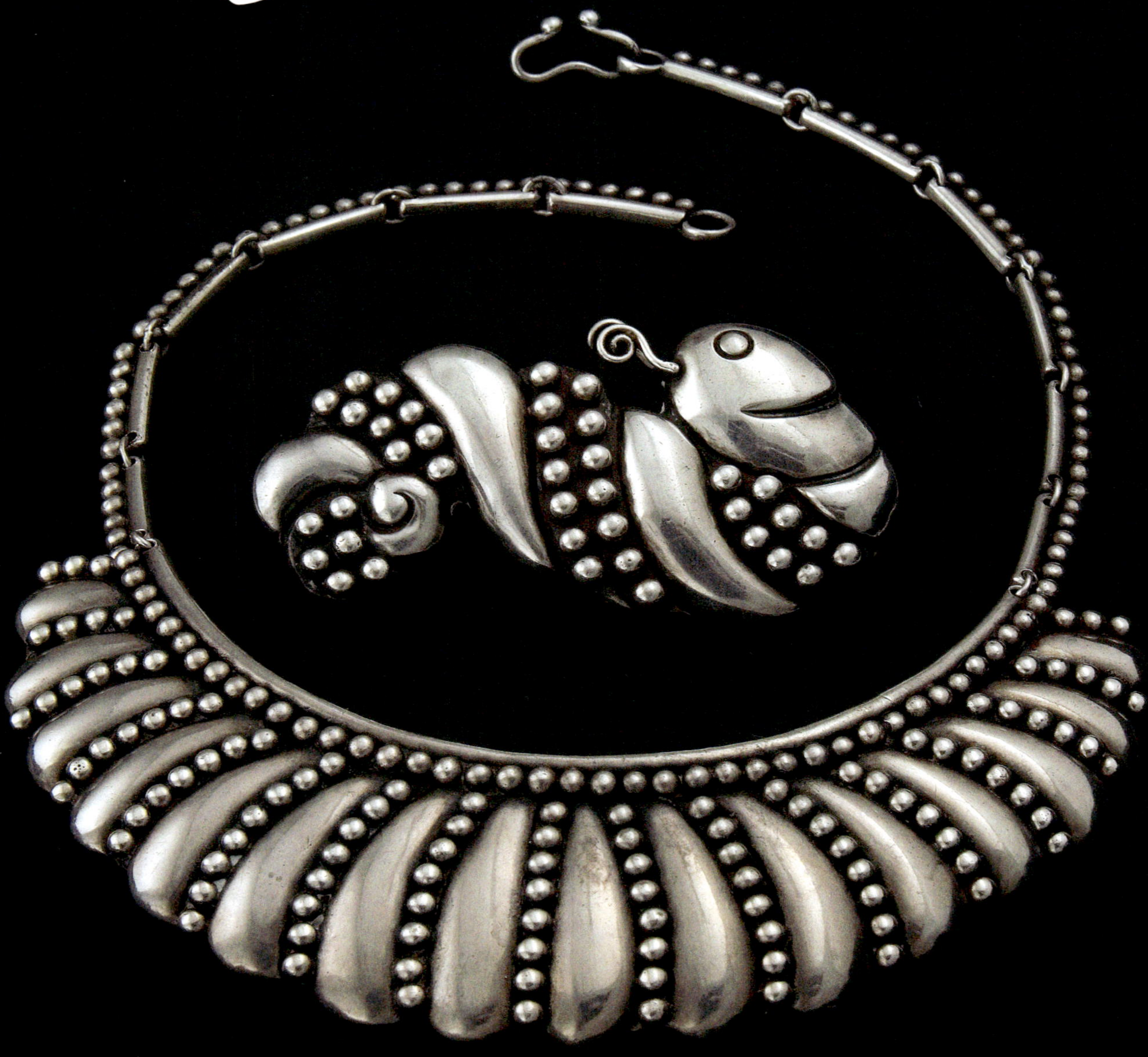

Plate VI-1 Los Castillo 368. *Beaded* necklace, circular mark 16 x 1.38 in. Los Castillo 151. *Caterpillar* brooch 1.75 x 3 in. Private Collection

Chapter VI
Castillos En El Aire

At the time of Margot's marriage to Antonio Castillo, he was working for William Spratling, the founder of Mexico's modern silver industry (Plate VI-3). With the pay Antonio received at Spratling's workshop, Las Delicias, the couple had just enough to cover rent for a small house and the cost of a housekeeper. Thus, it should come as no surprise that, soon after they were married, Margot and Antonio decided to embark on a business venture. They bought a small amount of silver and Margot began to draw designs. In the evenings, Antonio would produce the pieces, which sold quickly. They were so successful that in 1939, Antonio and Margot made the decision to establish their own workshop.[97]

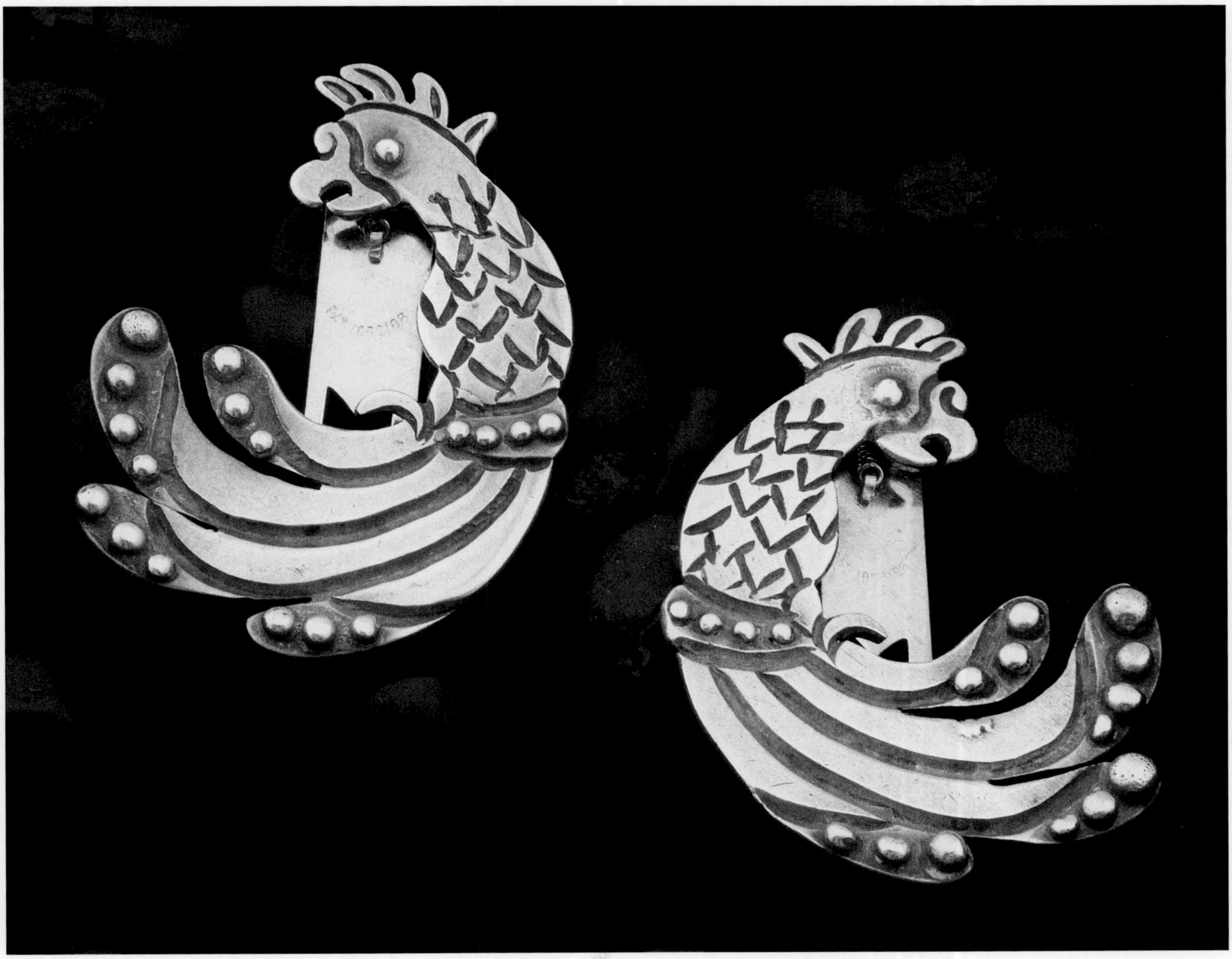

Plate VI-2 Los Castillo 168A. *Rooster* brooches 2 x 1.5 in. Stella and Fred Krieger Collection

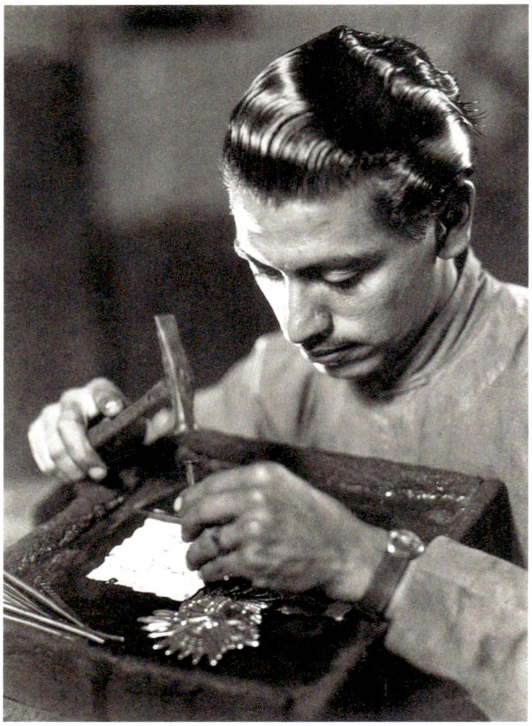

Plate VI-3 *Antonio Castillo as a Young Man, Working at Repoussage*. Photograph in Tulane University's Latin American Library Spratling-Taxco Collection

When Antonio went to Spratling to announce that he was leaving, Spratling received the news with little enthusiasm. Antonio was a valued employee. In later years, Spratling recounted the story of a strike against Las Delicias in which Antonio came to Spratling to admit the fallacy of the workers' actions and that he would no longer participate. At Spratling's request, Antonio went before the strikers to announce his decision, and the strike was ended. Spratling reminisced, "I myself will forever remember his facing the group as an act of courage and integrity, qualities which have added world prestige to Castillo Silver."[98]

There was a reason why Spratling did not particularly welcome the news of Margot and Antonio's business enterprise. From the beginning, small workshops had sprung up around Las Delicias to take advantage of its success. Spratling's new designs were quickly copied by these shops with only minor variations.[99] When Antonio came to him, Spratling wished him well, but with one request, that the Castillos never replicate his designs. Antonio assured him that the concepts for their work would be their own.[100] Spratling's aesthetic was entirely different from Margot's and, while both found inspiration in the art of their adopted country, their interpretations were not going to coincide (Plate VI-6). Thus, Antonio could promise Spratling with confidence that the Castillos would, like Spratling, continue to look forward to new and unique ideas in silver.

In the beginning, Margot and Antonio called their small business, appropriately, Castillos en el Aire [Castles in the Air]. This play-on-words soon became "Los Castillo," a reference to

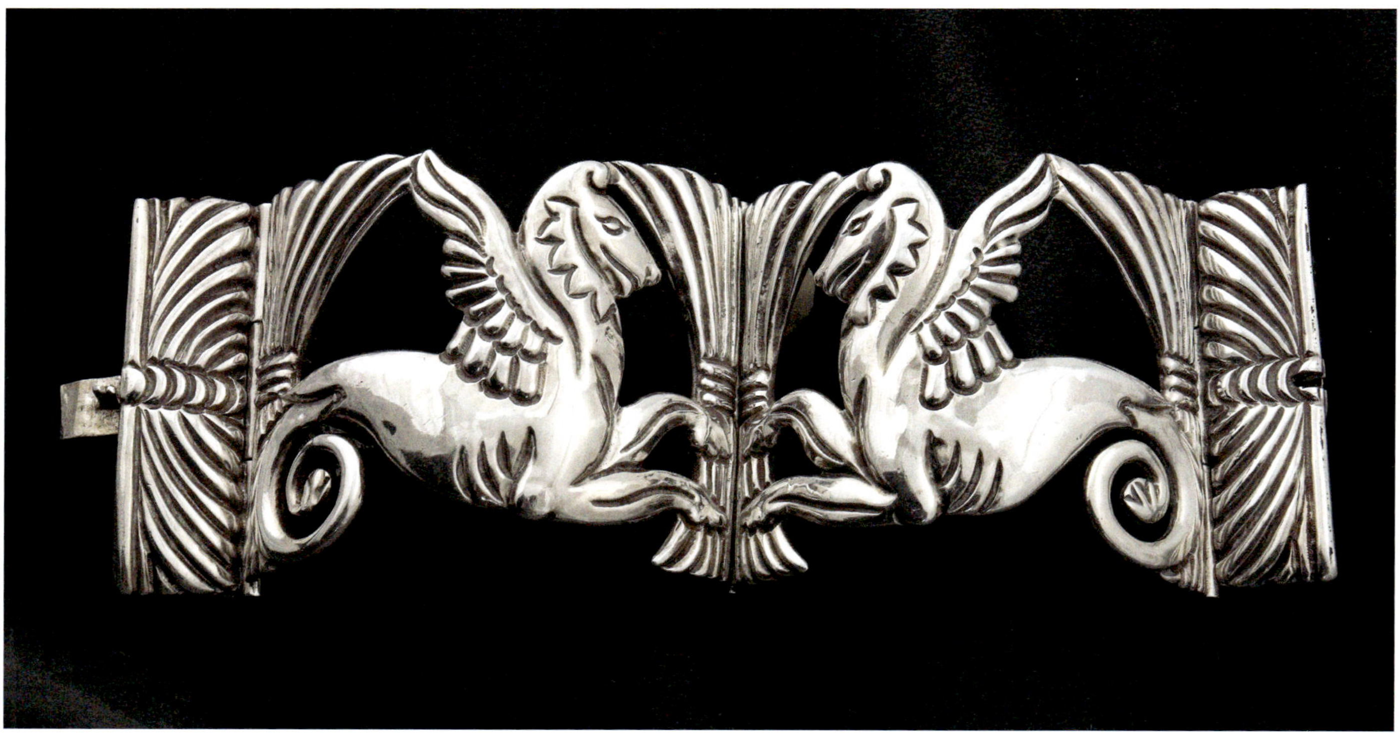

Plate VI-4 Los Castillo 354. *Lion Griffiths* bracelet 2.25 x 6 in. Stella and Fred Krieger Collection

the family's team of collaborators. Antonio organized the workshop with his two brothers Jorge (Chato) and Justo (Coco) and his cousin Salvador Terán, using Spratling's workshop as his model. At Los Castillo, as at Las Delicias, leadership was based on skill and sensitivity to the materials. The *maestros* were given the responsibility of supervising and teaching the young men who showed promise. They also made the final decision on the quality of each piece before applying the Los Castillo mark.

Neither William Spratling nor Margot Castillo was a silversmith — they were artists. Spratling's practice of collaborating with a *maestro* resulted in the successful production of a design. The same was true of the interaction between Margot the designer and the Castillo brothers and their cousin Salvador. Antonio remarked once that Margot's first attempts were two-dimensional patterns drawn on paper.[101] As Spratling wrote of the process, "A ring, when completed in silver, may give an entirely different sensation than it did on paper; it may have to be redesigned over and over and executed six or seven times before it is ready to offer to the public for sale."[102] The designer was able to realize the potential of the material and bring the design to fruition only if he collaborated with a skilled silversmith who held hammer or vertical saw in hand. This collaboration was possible in the workshop setting originally envisioned by William Spratling. In recent years, with the breakup of the workshops and the attrition of accomplished silversmiths, the synergy that can develop between designer and silversmith has become more difficult to achieve.

In studying the early work of Margot and Spratling, what becomes apparent is that each of them took something unique from the pre-conquest aesthetic. They both looked closely to find elements that could be reinvented. Spratling found inspiration in the architectural stone reliefs on the pyramids at Xochicalco and Teotihuacan, and it is probable that Margot did as well. Both archaeological sites are near Mexico City and accessible to the public. At Xochicalco and Teotihuacan, Quetzalcoatl, the feathered serpent, undulates across the stone surface of the pyramid, its entire length covered with feathers, jade symbols, and shells. While these features were deeply carved, they were not meant to appear as realistic, rounded forms. Instead, the light plays across the surface, throwing the carved lines into shadow. The resulting sharp contrast between dark and light creates an overall pattern, filled with movement. These qualities intrigued the artists who were working silver in Taxco, and they were motivated to replicate them (LosC732, LosC263, Plates VI-7, VI-22).

The power of pre-Columbian art defined many of the designs Margot produced during her collaboration with Antonio and the Los Castillo workshop. In some of her most successful jewelry, Margot did not duplicate the pre-Columbian carved stone or polychrome ceramics in silver, but instead she captured the simplicity of line and the formal abstraction of natural form that characterize them. Los Castillo's use of heavy oxidizing created a strong contrast with the reflective surface of the silver, much like the play of light and shadow in pre-conquest carved stone reliefs. While these motifs were produced in a different medium and on a much smaller scale, they retain their strength. These necklaces and brooches transmit the power of the pre-conquest aesthetic, one of the great achievements of the Los Castillo designers and workshop (LosC110M, LosC480, Plates VI-5, VI-8).

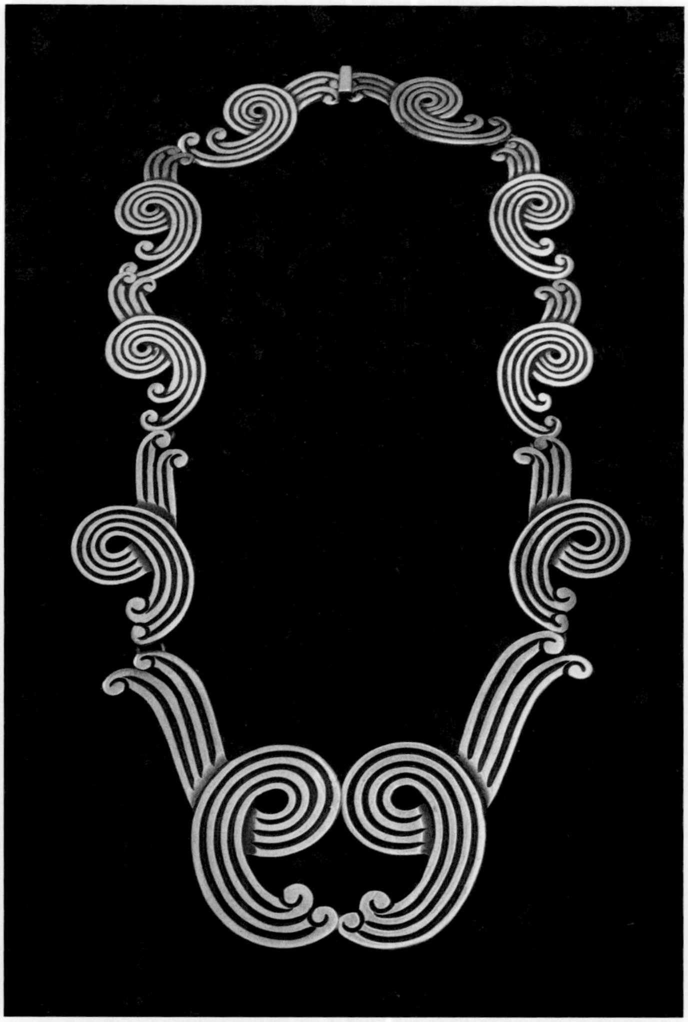

Plate VI-5 Los Castillo 110M. *Pre-Columbian Lines* necklace 14 x 1.5 in. Private Collection

A precedent established by Spratling and adopted throughout the silver industry was his insistence that silverwork should be marked with the name of the producer: "Worthwhile silver requires that it be identified with the name and reputation of its maker."[103]

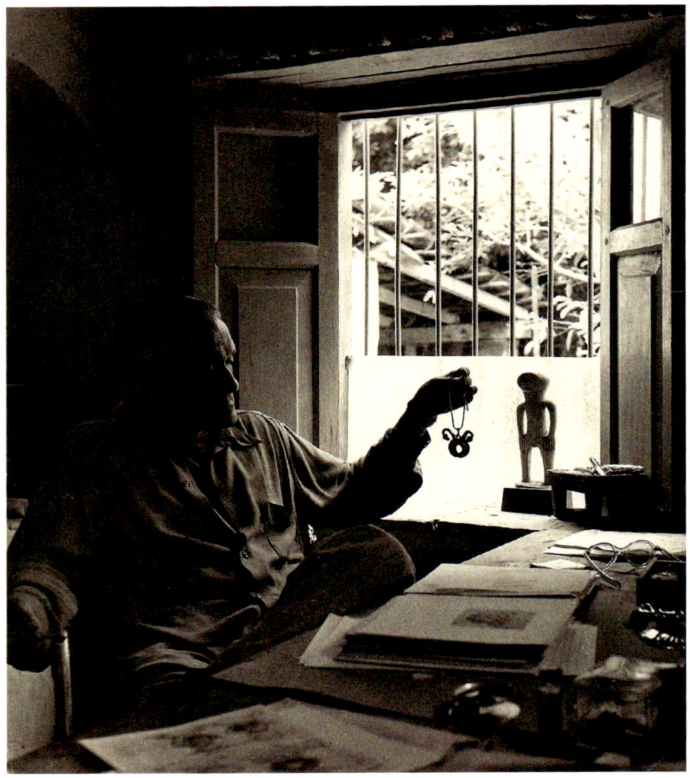

Plate VI-6 *William Spratling*. Photograph c. 1955. Zamora Carbajál Family Collection

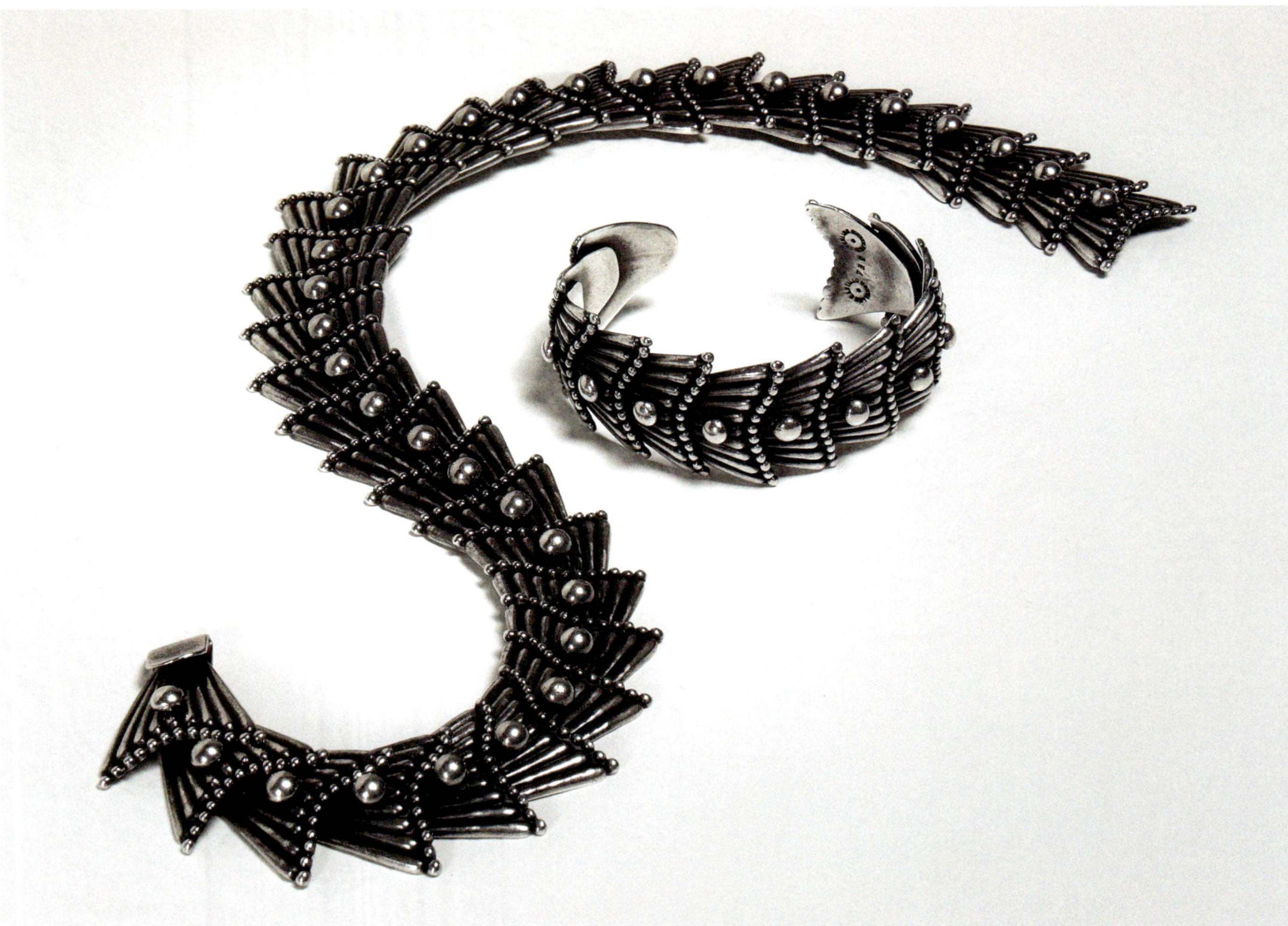

Plate VI-7 Los Castillo 732. *Fish Scales Set*, heavily oxidized. necklace 16 x 1 in.; cuff bracelet 5.5 x 1 in. Private Collection

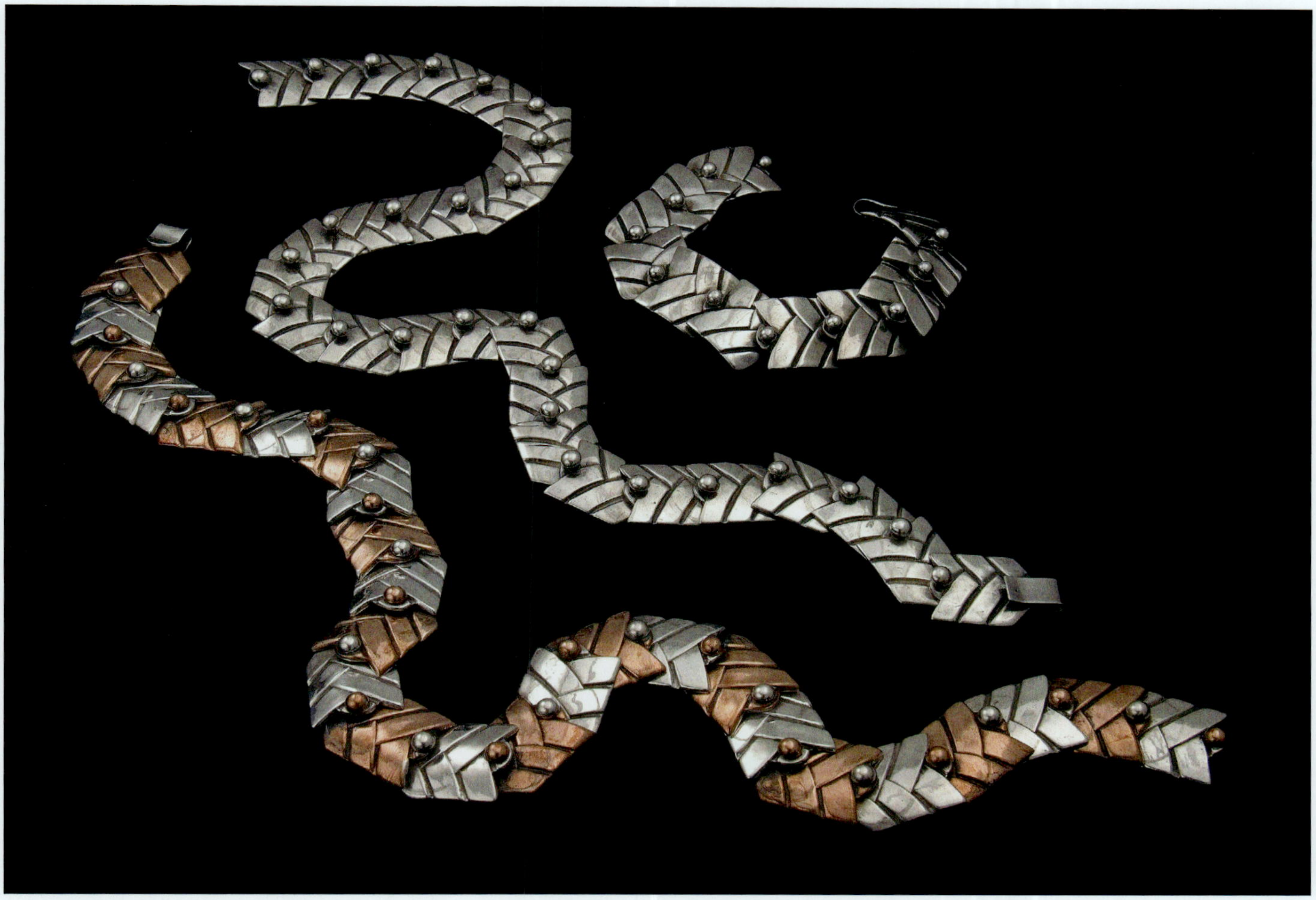

Plate VI-8 Los Castillo 480. *Overlapping V-shapes Set*. Silver necklace 18 x 0.88 in.; bracelet 8 x 1 in. Stella Rubin Antiques Collection. Silver/copper necklace 17 x 0.88 in. Private Collection

Margot and Antonio followed this practice assiduously from the earliest period in their collaboration at Los Castillo. The first series of marks reads, "Los Castillo, Taxco, Sterling, Mexico (or Made in Mexico)."

Contemporaneously, Spratling began to develop a wholesale market in the late 1930s, reaching beyond retail sales at his shop in Taxco.[104] The exportation of silver jewelry and objects to the United States brought about the need for an increase in the number and type of marks. Luckily for historians, Mexican silversmiths were required to indicate the name of the producer, place of origin, and silver (or metal) content. In many cases, this information was die-stamped on the back of the piece in English.

In response to these changes in the marketplace, Margot and Antonio Castillo added another innovation, the inclusion of the inventory number.[105] The numbering system allowed for control over inventory, indicating the full range of designs and providing a sense of what pieces were selling better than others. Production was predicated by the orders that arrived, so that the purchase of silver, copper, stones, and other materials could follow demand. Examples of each design could be made readily available; at the same time, it would not have been necessary to store in the warehouse more than what could be sold quickly. This approach allowed Los Castillo to develop new designs, with the knowledge that there would be capital available for future production. For shops in New York and Chicago, ordering from Los Castillo was facilitated by the ability to indicate a piece by name or number. Whether in Taxco or abroad, the salesmen could make use of either samples of the jewelry or printed catalogues with photographs of each numbered design.

The use of inventory numbers by Los Castillo was an acknowledged improvement in the business of artistic production, and it was adopted by most of the silver workshops in Taxco. Other new ideas emerged as Los Castillo grew. Margot's designs were produced in sets — the necklace came with matching earrings, brooch, and bracelet, the tie bar with cufflinks and belt buckle. Because the Castillos, like Spratling, appreciated the value of technical advances, their application of the die and press facilitated the production of the design element that could be cast in large numbers and then formed into matching necklaces, bracelets, and earrings. Improvements also included rolling mills, belt-driven polishing machines, and blowtorches for soldering. However, none

Plate VI-9 Los Castillo 619. *Circle Swirls* brooch 2.5 x 2.5 in. Stella Rubin Antiques Collection

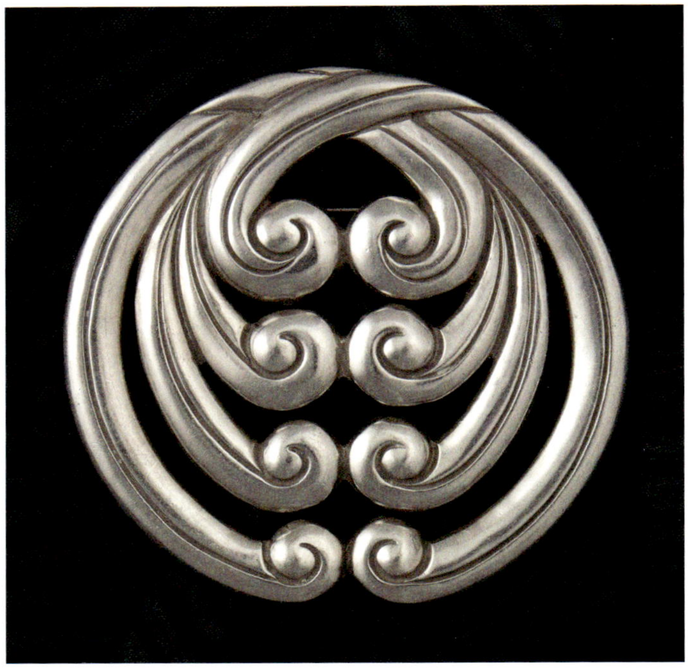

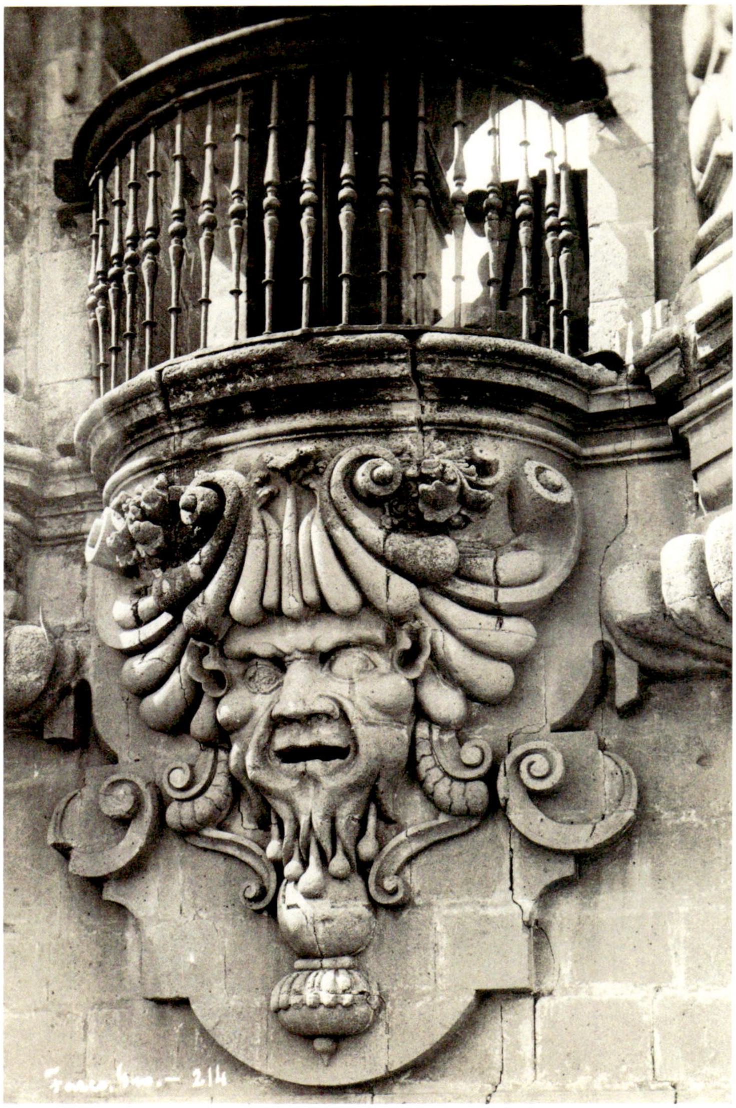

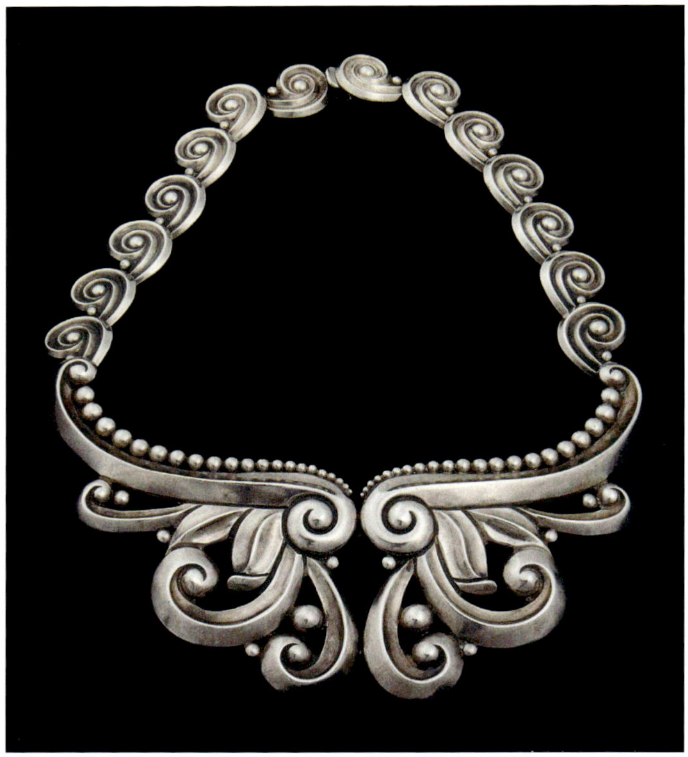

Plate VI-10 Los Castillo 508. *Baroque* necklace 14 x 2 (at widest point) in. Stella and Fred Krieger Collection

Plate VI-11 Decorative Stone Carving, exterior Santa Prisca. postcard. Private Collection

of these advances took away from the many hours of handwork that were required to produce each piece.

Another innovation developed at Los Castillo was the pin on the back of the central motif of a necklace. Male and female hooks on the necklace at either side of the pendant meant that the necklace could be joined and exist without the pendant. When removed, the pendant could be pinned on a jacket to match the necklace, now shorter and more like a choker.

Most Los Castillo necklaces had clasps at the back of the neck that were either rectangular box or flat hook closures or a hook and eye that was fashioned out of a flattened U-shaped wire with beads at either end. The link bracelets were provided with the same types of closures. The cuff bracelets were either made of one piece or were composed of two identical motifs attached by a spring mechanism. In the 1940s, the earrings were given screwbacks.

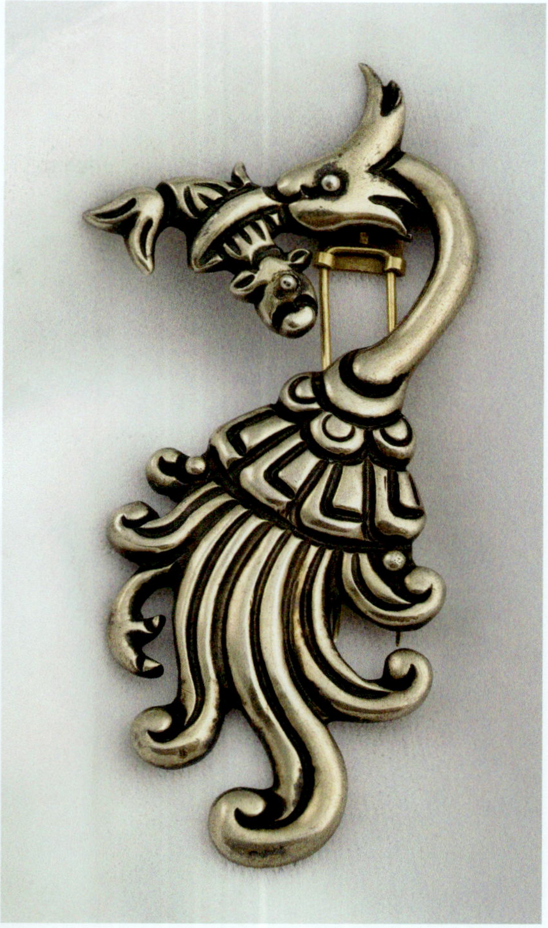

Plate VI-12 Los Castillo 220. *Maya Bird* brooch in silver with gold wash 3.5 x 1. 63 in. Private Collection

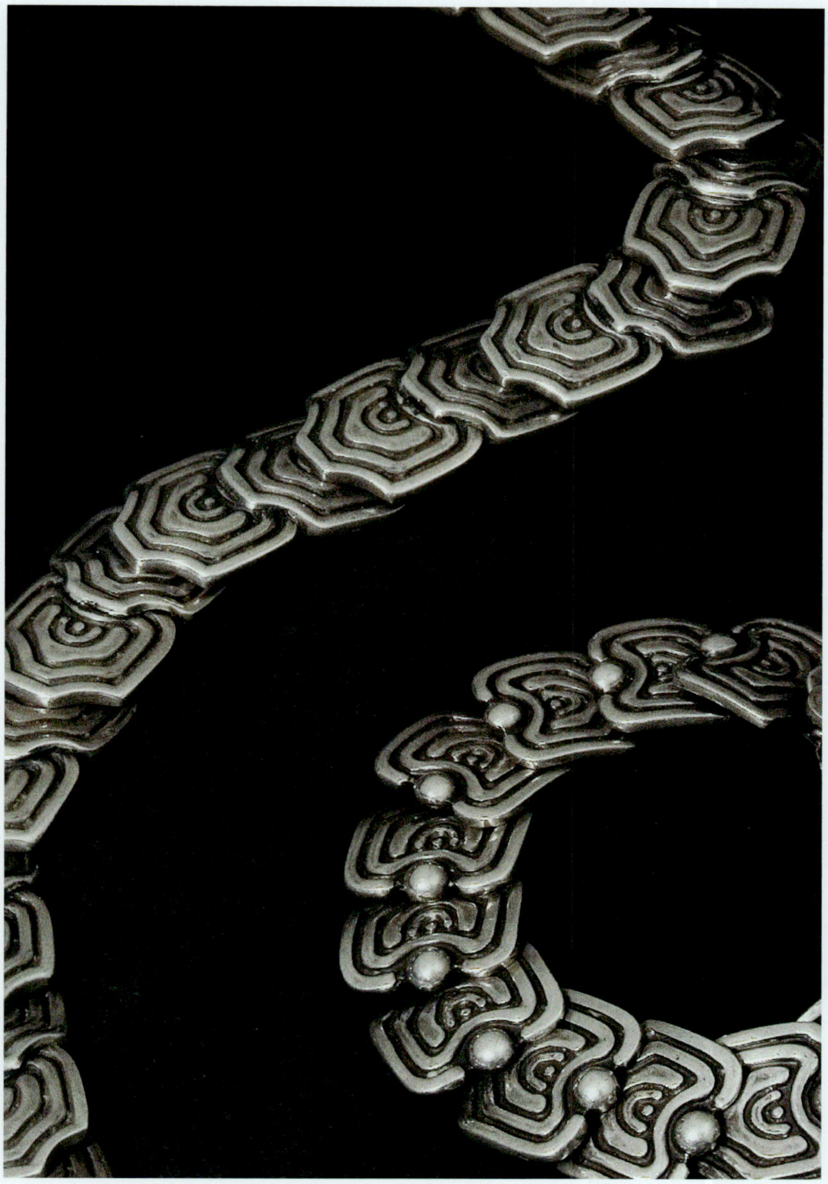

Plate VI-13 Los Castillo 261. *Concentric Lines Set*. necklace 16 x 0.75 in.; bracelet 6.5 x 0.75 in. Stella and Fred Krieger Collection

The interaction among artists at Los Castillo must have been symbiotic, a shared storm of ideas and a creative process that materialized as jewelry. Salvador, Chato, Coco, Margot, and Antonio each contributed an individual artistic vision, so different one from the other. For Los Castillo, the first six years brought experimentation with a variety of techniques, materials, and business practices that was on a grand scale. One outstanding aspect of the jewelry from this period was the beauty of the finish. Rivets allowed for movement and flexibility and were often incorporated into the design (LosC261, LosC536, Pls. VI-13, VI-23). So strong were these joints that the pieces have withstood decades of wear.

Pre-Columbian art inspired motifs that gained their power through repetition, much like the patterns created by cylindrical clay stamps. Margot seemed to have studied the Mayan aesthetic more than other silver designers in Taxco, who were more influenced by central Mexican sculpture and painting. Adapted by Margot, the repetitive sensuous forms fall perfectly into place around the neck and wrist. The pin depicting a water bird with a fish captured in its beak (LosC220, Plate VI-12) is a fine example of sinuous lines, counter-balanced in a dynamic S-curve. Margot borrowed the motif from sculptural reliefs at Palenque, a Maya site in Chiapas.

Margot and the other designers at Los Castillo studied the Baroque and Rococo stone ornamentation on the exterior of Santa Prisca and the carved *retablos* [retables, altars] inside the church. These flowing forms are so exuberant and full of movement, that they seem barely contained (LosC508, Plate VI-10). Often when the Castillos interpreted one of these baroque motifs as a pendant, it was then miniaturized and repeated to form the necklace's links. Baroque form also lent itself to *repoussage*, a difficult technique developed to full potential by the *maestros* at Los Castillo.

The jewelry from the early years occasionally achieves a powerful impact because of its size. *Repoussage* allowed for grand scale without great weight. This is particularly true of the dramatic necklace that at one time had a massive pendant (LosC500, Plate VI-14). As with all of these early repoussé pieces, the shaping of the necklace's elements from the reverse contributes a texture and surface that directly connects the wearer and observer to the maestro's act of hammering. The ability to experience the contribution of the artisan, as it evolves into sculpted form, has made these pieces greatly valued.

Another quality the Castillos incorporated into their jewelry is the importance they gave to whimsy. One can appreciate the jaunty good humor of the caterpillar and fish brooches while admiring their

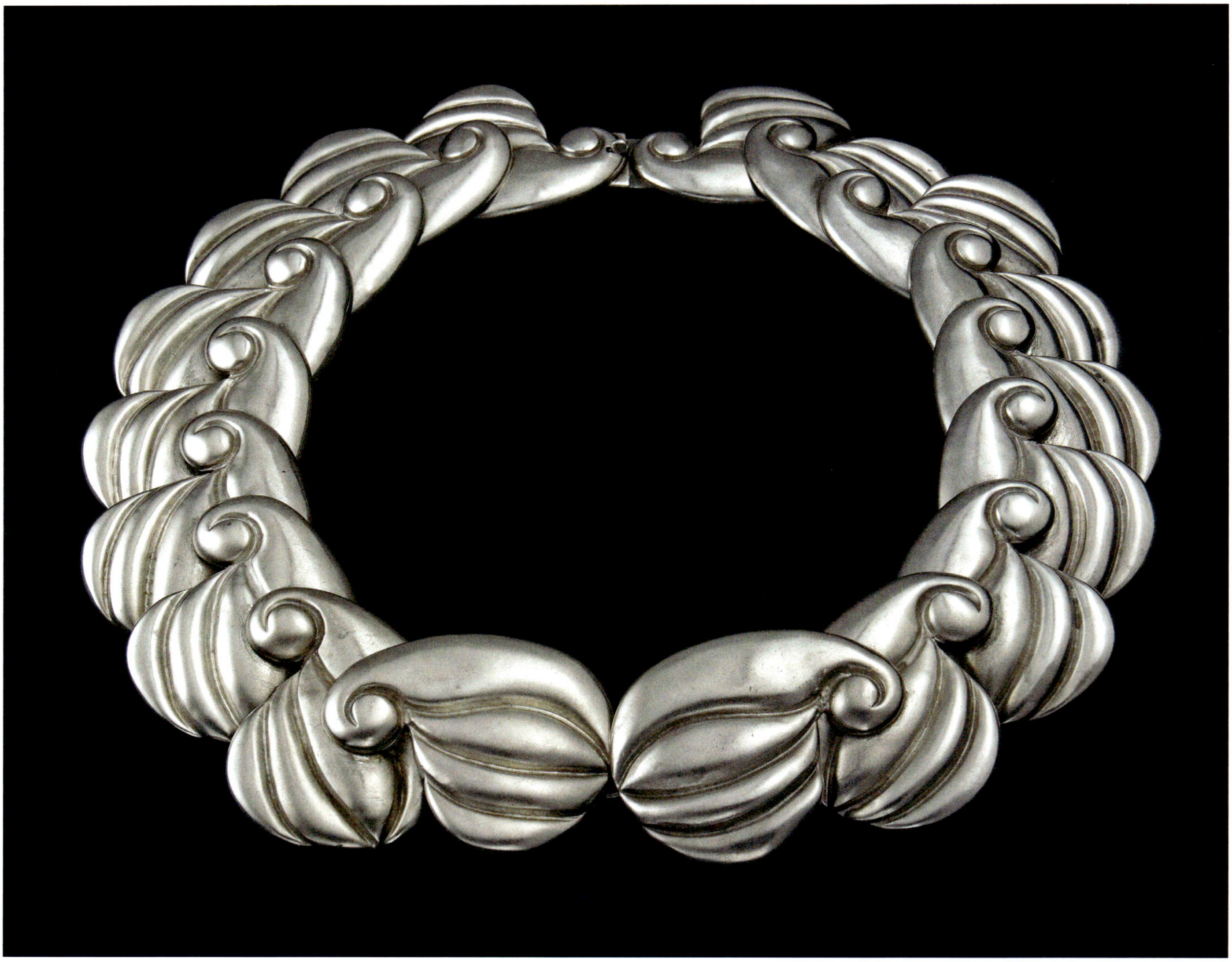

Plate VI-14 Los Castillo 500. *Large Shell* necklace 17.5 x 1.5 in. Private Collection

beautiful finish (LosC151, Plate VI-1). And who would wear a pin representing a fly unless crafted in silver and amethyst quartz (Plate VI-25)?

The designers of Los Castillo found inspiration in the flora surrounding Taxco. These designs are not mere reproductions and they are rarely diminutive and delicate. A flower brooch might be formed of curling leaf shapes around a large amethyst cabochon, making for a large-scale and powerful interpretation. Amethyst quartz was a preferred material because its translucence and color contrasted with the reflective quality of silver. Other materials that were combined with silver by Los Castillo during the early years were agate, onyx, obsidian, abalone, malachite, copper, and cabochons [globes] of glass. The chalcedony necklace (LosC298, Plate VI-15) incorporates translucent blue-tinged stone cabochons with oxidized abstract plant forms, a contrast of rounded form and line, stone and metal.

In a majority of the jewelry of this period, the silversmiths at Los Castillo added beading.[106] A time-consuming technique, it made for a striking addition to these designs, thus well worth the effort. Beading was sometimes graduated to give greater importance to the central motif of the piece. Beads made uniform in size added texture and an interesting contrast when surrounded by oxidizing (LosC154, Plate VI-21). Beads could reappear in the necklace's links, providing continuity to the design. Beading also emphasized line, and could draw together curving forms, so that they seemed more controlled (LosC368, LosC151, Plate VI-1). Beads were also used to imitate seeds or the stigma and anther of a flower.

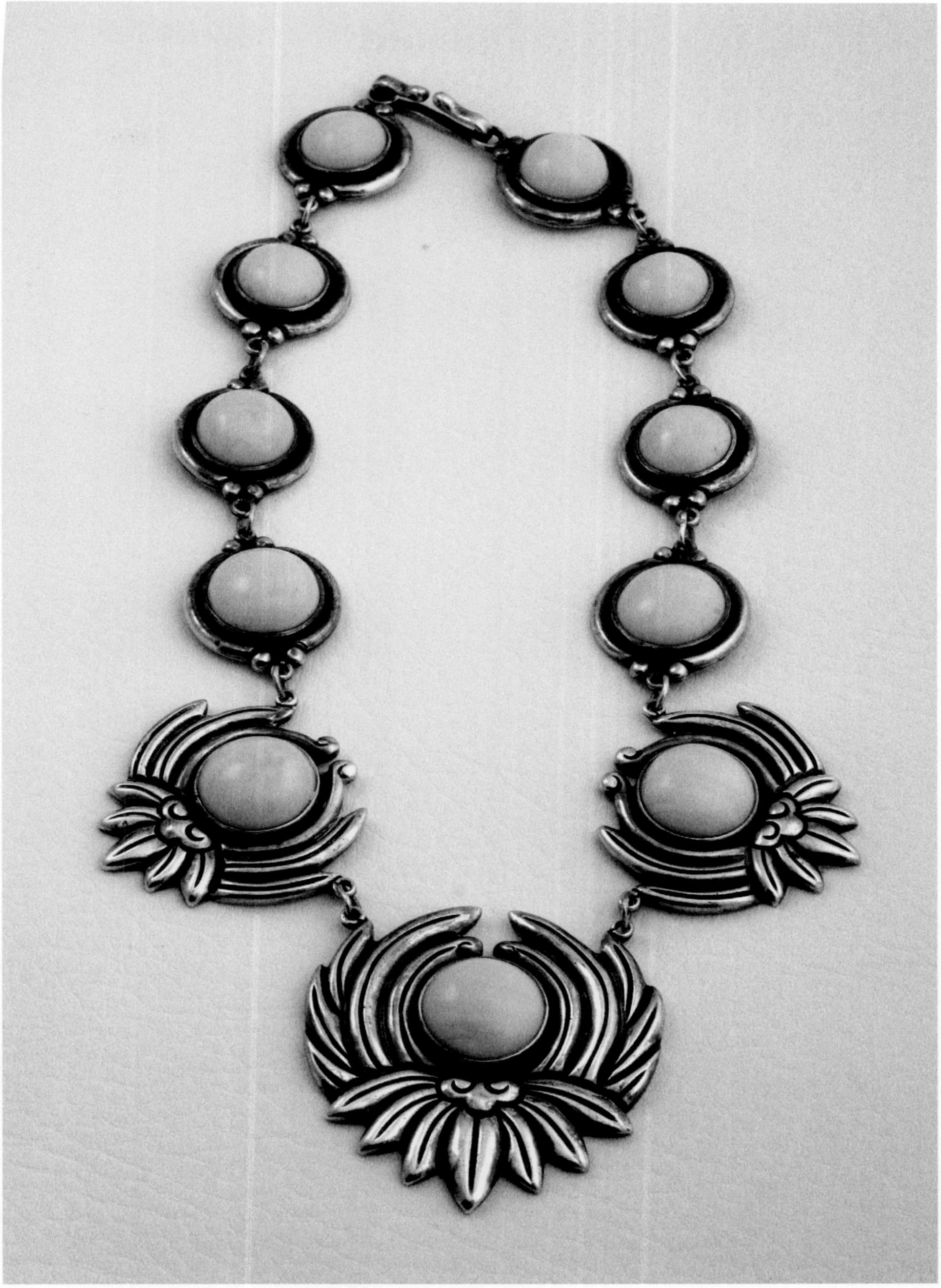

Plate VI-15 Los Castillo 298. *Chalcedony* necklace 15 x 2 (at widest point) in. Private Collection

Margot set small ivory figurines into her jewelry, which she may have acquired when she and Antonio traveled to San Francisco in 1940 and 1943.[107] Was Margaret Clayes the source of these ivories? It is possible, although the San Francisco city directory indicates that Margaret Clayes had retired from her business in 1936.[108] Margot could have contacted her for assistance in purchasing the ivories. She could have bought directly from Margaret or from one of her former suppliers.

In several distinctive brooches from this period, the ivory figures of the Confucian scholar with his hat and rod, the graceful female holding a piece of fruit, and of Buddha (LosC417, Plate VI-16) are possibly Chinese, but more likely Japanese reproductions

Right:
Plate VI-16 Los Castillo 417. *Ivory Figures of Buddha*, Confucius and Woman with Fruit set against Silver Trees brooches, 4 x 3.75 in. Betsy Vought Collection

Below left:
Plate VI-17 Margot de Taxco 5263. *Guanyin* brooch of ivory, silver, and opals 4 x 2.25 in. Private Collection

Below right:
Plate VI-18 Los Castillo 569. *Ivory Netsuke* set in silver, with faceted hematite, 3.25 x 2.5 x 0.75 in. Ron Belkin Collection

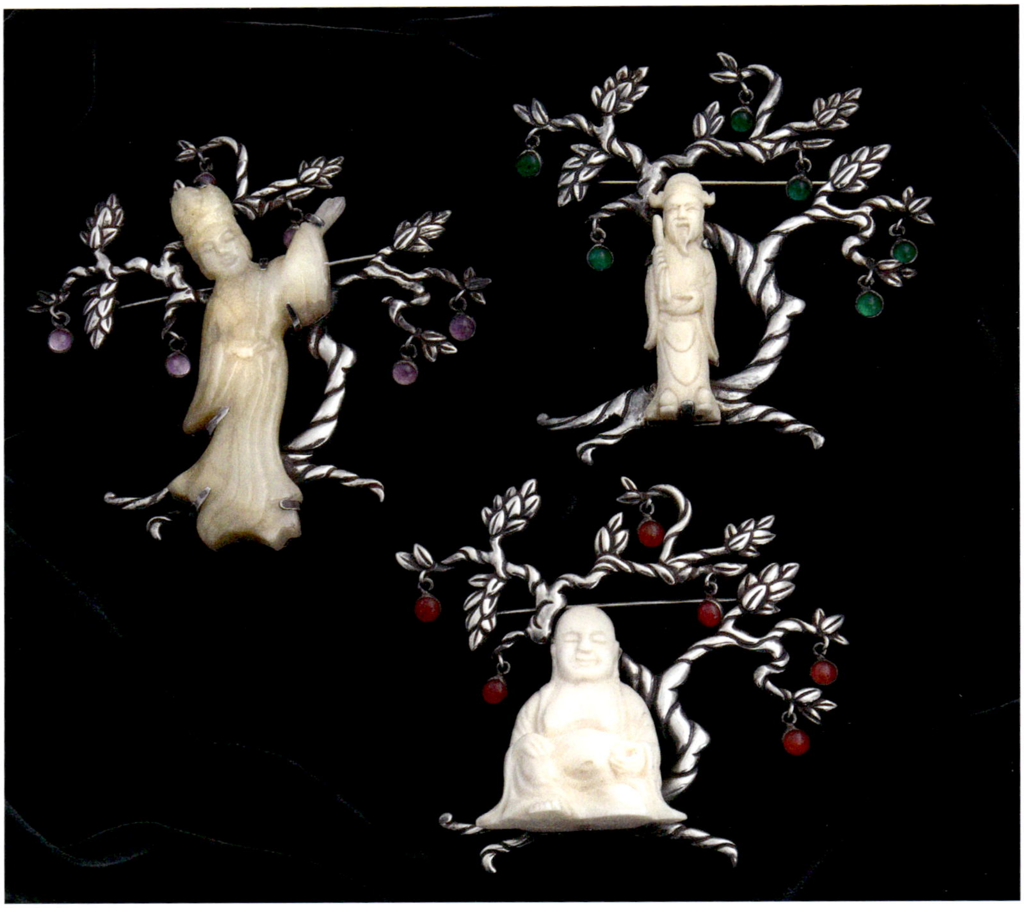

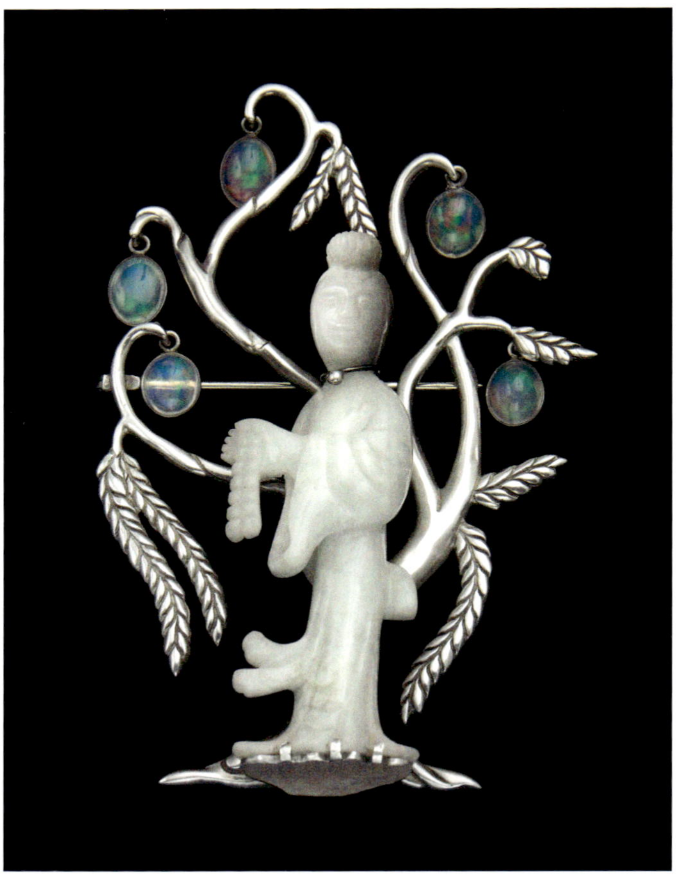

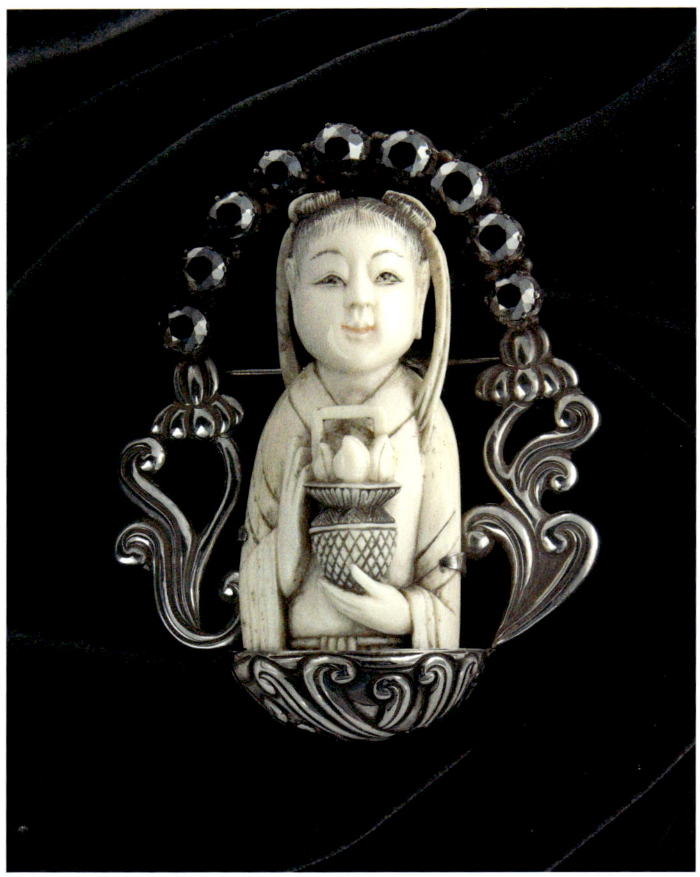

of Chinese figurines. Margaret Clayes could have bought these tiny carved figures in Japan on one of her many trips. According to her passport, Margaret was doing "commercial business" in various countries in Asia.[109]

Margot continued to produce this type of jewelry after she left Los Castillo to form her own company. According to Dr. Robert De Caroli, the Buddhist figure of Guanyin in the Margot de Taxco design "5263" (Plate VI-17) carries prayer beads in her hand and is the Bodhisattva of compassion. She is often associated with medicine and healing and, for this reason, appears beneath a willow tree, known for its medicinal properties, actually, the source of aspirin. This figure would have originally served as an amulet. The trees for the three Los Castillo figures are the same, except for the types of stone cabochons representing the trees' fruit. The silver trees were probably die-cast. The tree for the Margot de Taxco brooch is distinctly a willow.

The *netsuke*, a pendant that hangs from the kimono, is Japanese. This particular *netsuke* (LosC569, Plate VI-18)) is known as the peach girl, for she carries a basket of peaches signifying fertility.[110] The setting designed by Margot for the netsuke is unique and quite stunning. The figure rests on a pedestal with undulant flowers to either side. From the silver base rises a halo of hematites that frames the little ivory figure, referring ever so subtly to the ubiquitous carved saints in niches encountered in Mexican churches and residences.

The *netsuke* appears to have been of a higher level of workmanship than the other ivory figures. Dr. James Ulak provides an analysis of the figurines: "They seem to be of a very generic decorative type that could have been produced just about anywhere in East, Southeast Asia, or even in California. They could have been salvaged from or stripped off of some larger work or may have been made for precisely the kinds of settings described in the images."[111] Thus, it seems possible that Margaret Clayes could have bought these pieces in China or Japan or in Chinatown in San Francisco and then sold or given them to Margot.

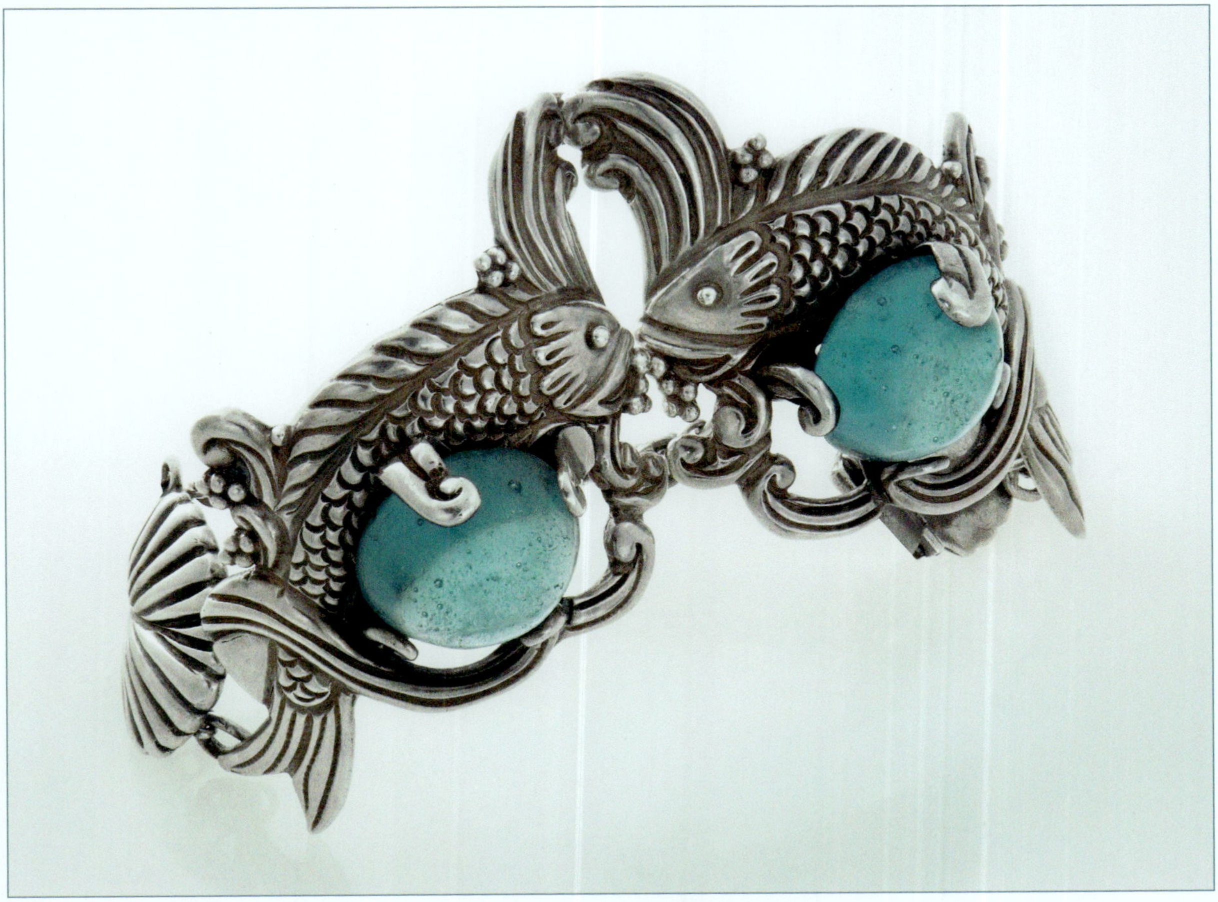

Plate VI-19 Los Castillo 514. *Fish in the Waves* in silver with glass cabochons bracelet 2 x 7.25 in. Stella and Fred Krieger Collection

World War II and Taxco

Ana Brilanti, who founded her silver workshop in Taxco across the Plazuela de Bernal from Los Castillo, recalled Margot's willingness in 1940 to assist her during the first year of building "Taller Victoria." Ana spoke of Margot as the heart of Los Castillo, for it was she who brought passion and artistry to the company.[112]

Margot persuaded María de los Angeles to come work for Los Castillo in 1940, where María tracked inventory before ordering new production from the workshop. According to Siguifrido Pineda (Sigi), María found it amusing to give him orders, in that he was only twelve years of age when he joined the workshop.

As a young man, Sigi knew all the great artists who had acquired and perfected their skills during the early years of Los Castillo. Among the most talented were Enrique Ledesma, Moisés Sotelo, Elpidio Morales, and Angel Rodríguez. Miguel Meléndez was considered the most accomplished *maestro* of *repoussage* among them. Meléndez had learned silversmithing from Antonio Castillo, his *maestro* at Las Delicias, before joining him when he founded Los Castillo. When speaking of the team of talented artisans recruited to work for Los Castillo, Antonio particularly recalled Roberto Aviles, so gifted a jeweler that he was asked to do special projects for Cartier. Antonio also admired the work of Elpidio Morales, Leopoldo Betancourt, and Carlos Mejía.[114]

William Spratling wrote about the profound changes that the war brought to the Mexican silver business: "Buyers from many of the finest stores in America came to Mexico, desperately looking for. . . luxury goods. Europe was suddenly cut off from them. Until the war

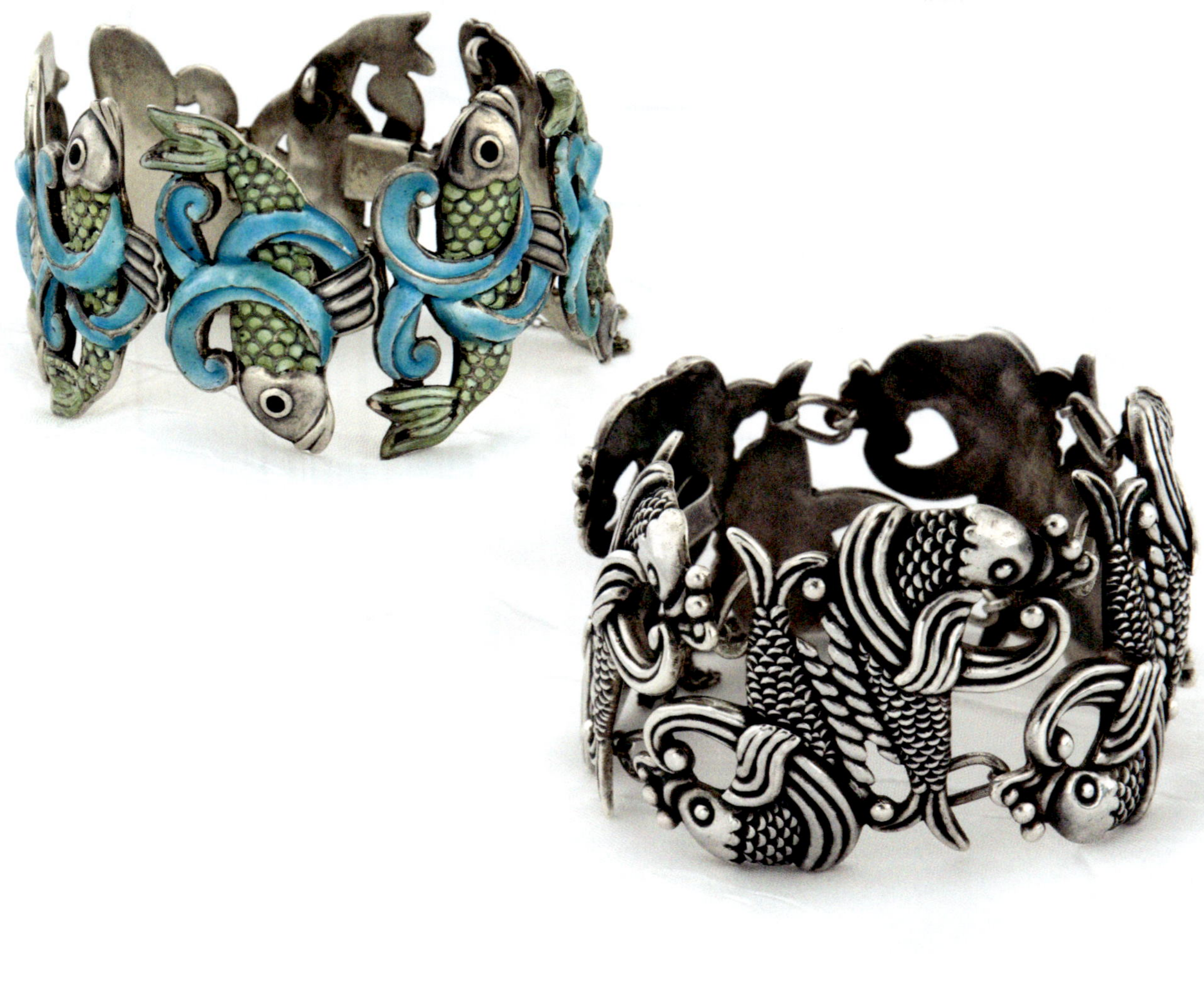

Plate VI-20 Los Castillo 106. *Leaping Fish* bracelet in silver 7.5 x 1.63 in. Margot de Taxco 5847. Enamel Leaping Fish in the Waves bracelet 7.5 x 1.6 in. Silver Huntress Collection

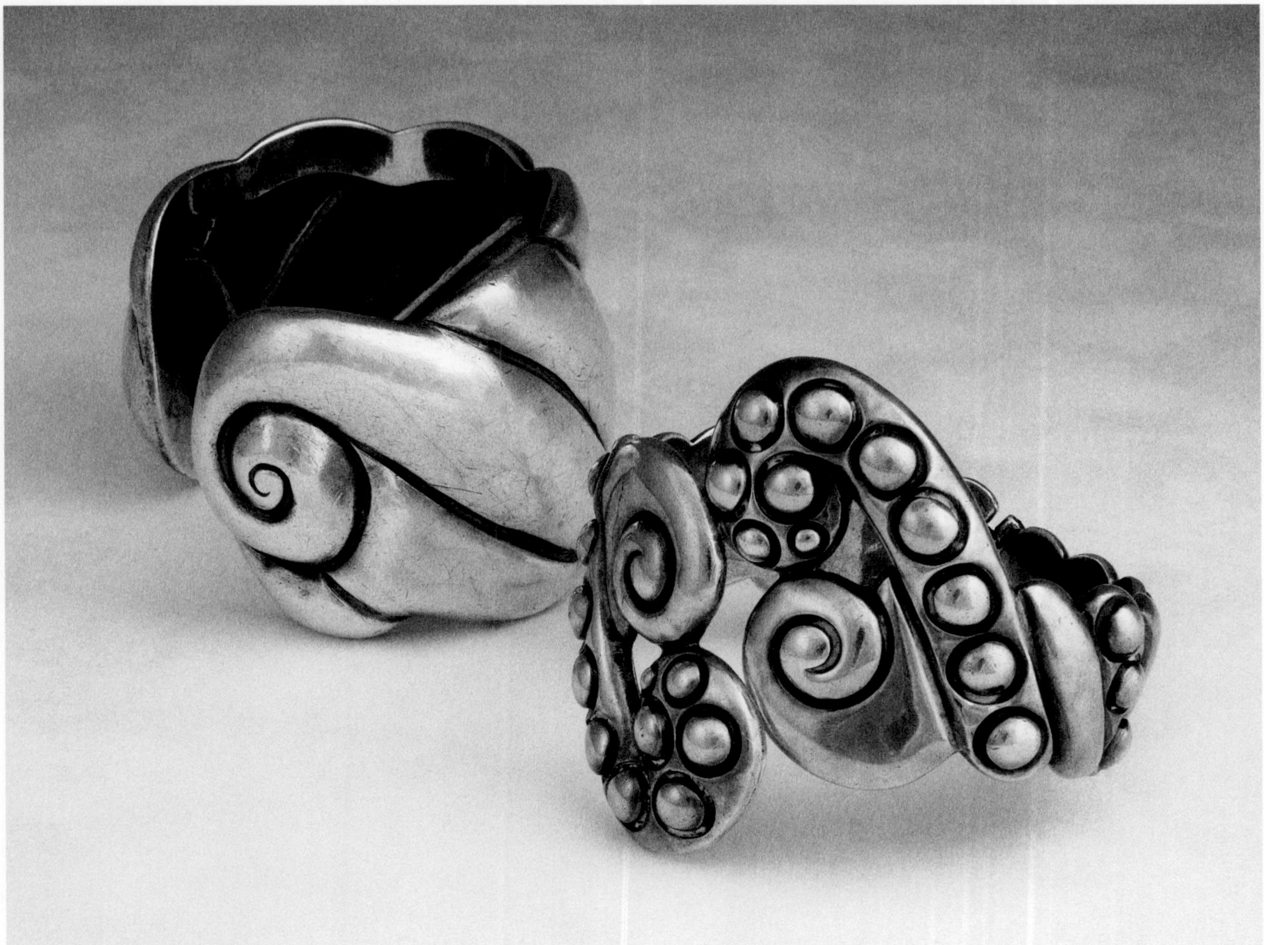

Plate VI-21 Los Castillo 154. *Shells and Dots* spring cuff bracelet 7.75 x 2 in. Margot de Taxco 5192. *Shell* cuff bracelet 7 x 2 in. Elizabeth Finch Collection

occurred, sales of Taxco silver had been mostly retail, and exportation of wholesale orders had been practically nonexistent." Spratling was able to influence U. S. policy through his friendships with Nelson Rockefeller and President Franklin Roosevelt. The levels of American import duties on Mexican silver were drastically reduced, further enhancing the market. [114]

For Los Castillo, success during the war was beyond imagining. At one point, there were over three hundred artisans in the workshop.[115] Innovative creativity remained the cornerstone of Los Castillo's ongoing success and, because the leadership of the company was so vibrant and dynamic, the workshop operated like a community. Antonio's organizational skills contributed to the silversmiths' shared commitment to excellence and the ability of Los Castillo to reach a very high level of production during the war years.

Like most of the workshops in Taxco, Los Castillo had contracts with the U. S. government during the war to produce military identification bracelets and other insignia. Antonio and Margot also worked together to position Los Castillo in the luxury and wholesale marketplace. The company developed an inventory numbering system that became the prototype for all the larger silver workshops in Taxco. Sigi's brother Rafael Pineda was sales manager and, in Mexico City, called on Sanborns and the Lampells, owners of the store India Bonita.[116] At the same time, Antonio and Margot continued to push beyond Mexico to find outlets for their work. They were in San Francisco together in 1940 and, in a letter from Taxco dated December 31, 1943, Freddie Jennison informed Natalie Scott that, "Margot Castillo is off on a visit to the states so she cannot be there [New Year's Eve party]."[117]

In 1944, La Tausca, a jewelry store on Fifth Avenue in New York, presented a collection of Los Castillo silver to its customers (Plate VI-25). The young woman in the advertisement wears a black mantilla, an obvious allusion to the Hispanic heritage of Mexico. The diversity and quality of the work are highlighted in the ad, which describes the Castillos as "Premier Creators and Stylists."[118]

Plate VI-22 Los Castillo 263. *Copper and Silver Set*. necklace 18.5 x 1.25 in.; ear-rings 1.25 x 0.75 in. Stella Rubin Antiques Collection. Silver necklace, 16 x 1.25 in.; bracelet 7.25 x 1.25 in. Private Collection

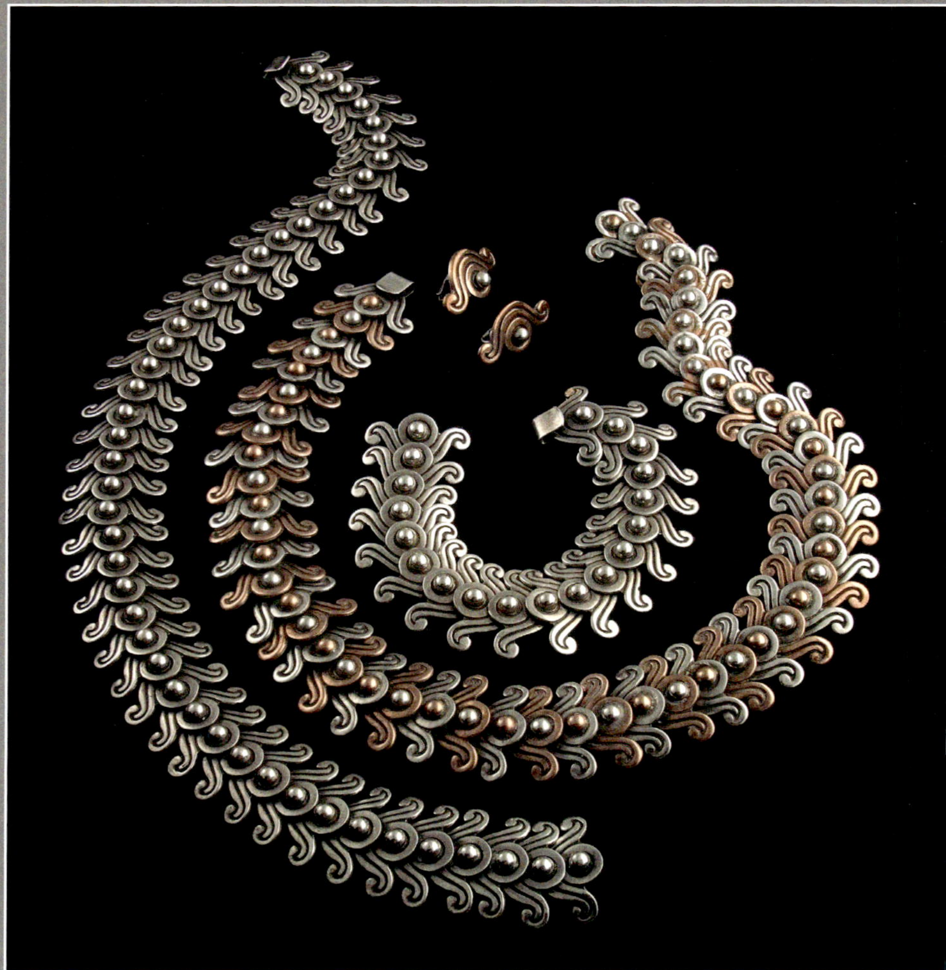

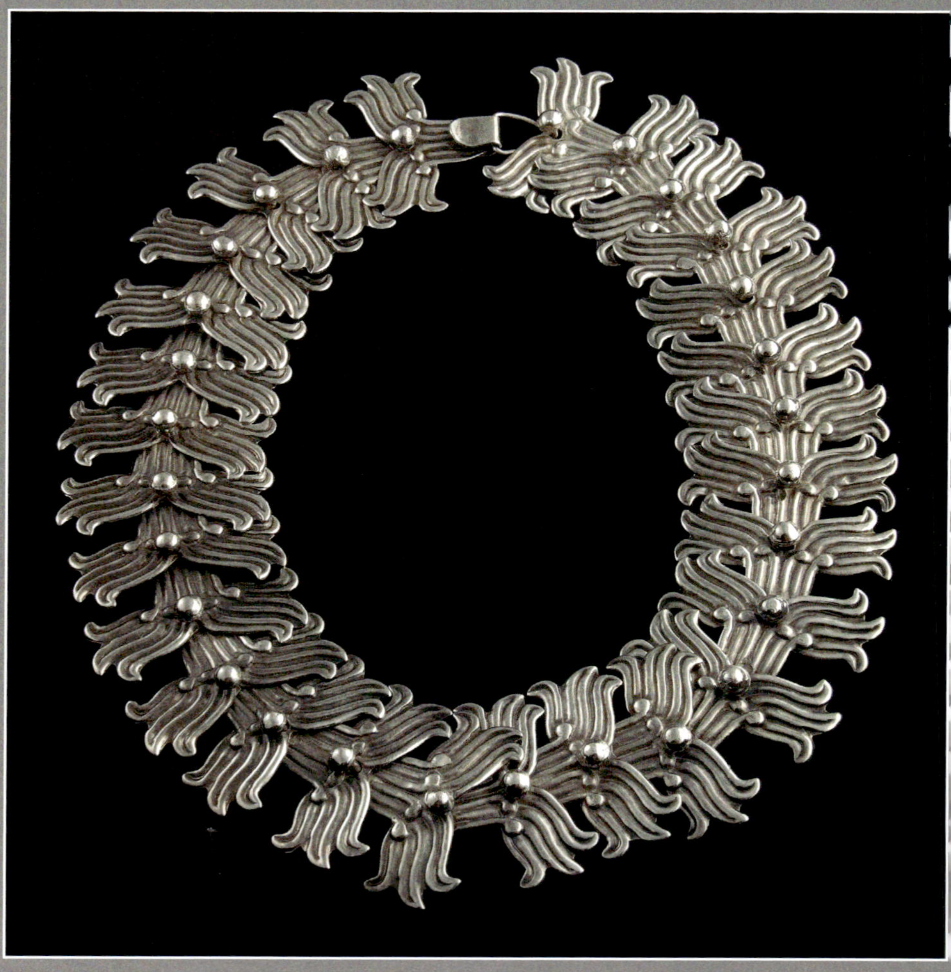

Plate VI-23 Los Castillo 536. *Feather Lines* necklace 19 x 2 in. Stella Rubin Antiques Collection

What should have been a time for celebration was clouded by problems Margot and Antonio were encountering in their marriage. They had lived in a small house across from the ex-convento of San Bernardino de Siena. After establishing Castillos en el Aire, they moved to the apartment over the workshop and store that became Los Castillo, on the Plazuela de Bernal. The workshop expanded exponentially with demand, and Margot and Antonio moved off the property to a house they built together. [119]

Antonio recalled that, while he wanted a partner in life, he also wanted children and was disappointed that Margot was not interested in becoming a mother.[121] This version of the story can only partially explain this aspect of the deterioration of the relationship. Margot was turning fifty in 1946, and would have been beyond her childbearing years. She may not have been willing to admit her inability to have children, just as in the 1940 border crossing document, she had given her age as thirty-six (almost the same age as her husband) when she was actually forty-four.

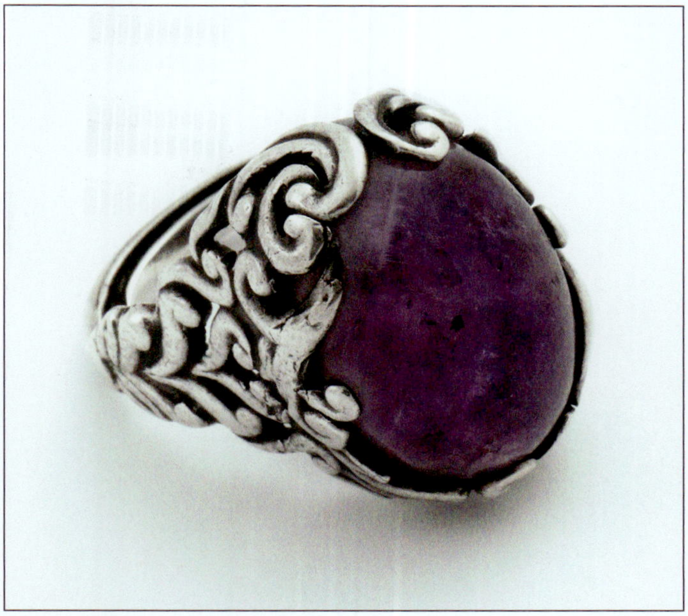

Plate VI-24 Los Castillo. *Amethyst* ring. 1940. Private Collection

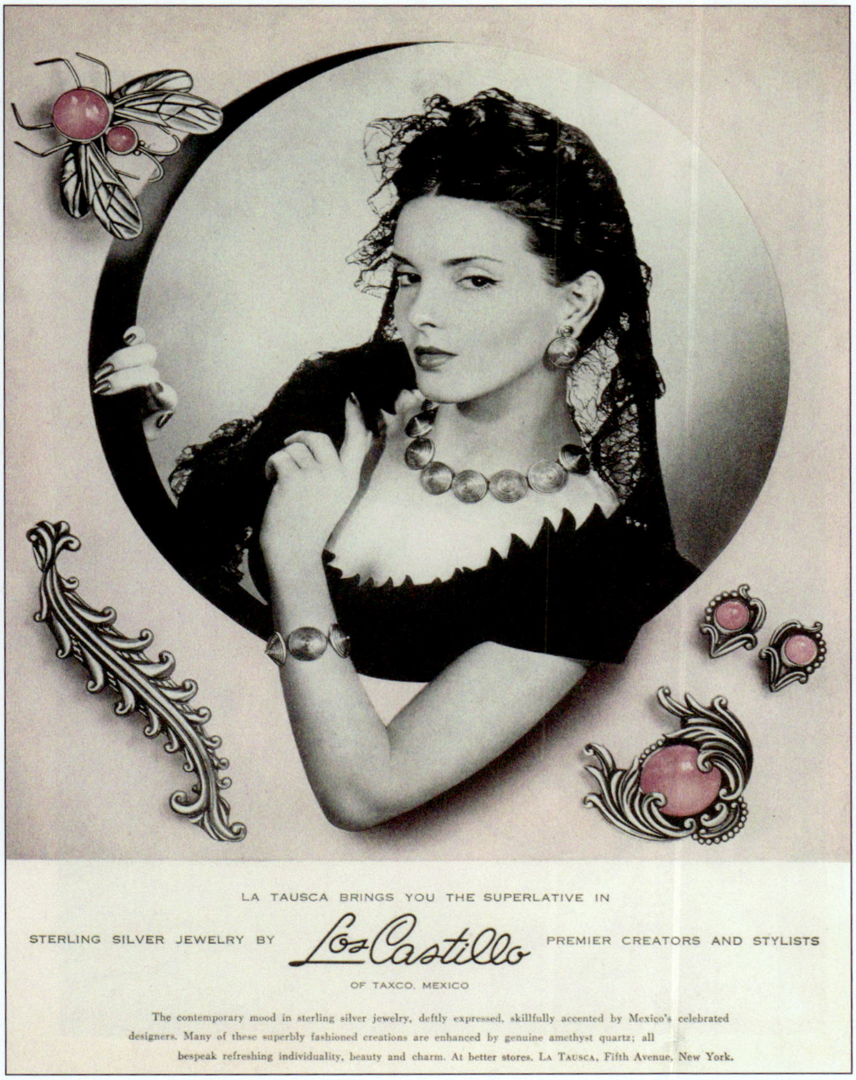

Plate VI-25 *La Tausca,* 1944 advertisement for Los Castillo. Stella Rubin Antiques Collection

In an effort to save the marriage, Margot made the decision to adopt Rafael, the son of a woman so poor she was, in the words of several silversmiths, *sin zapatos* [without shoes].[121] The adoption was initiated in large part because the mother claimed that Rafael was Antonio's son. While this allegation has remained in question as a result of the woman's apparent promiscuity, Antonio was known to have had affairs outside his marriage to Margot. Rafael's adoption gave Margot a son but, unfortunately, did little to bring Margot and Antonio together.

In an instant, Rafael Castillo's life changed dramatically. Margot dressed him and educated him, sending him to the United States to study at a military school. Upon his return, Rafael became a guide and lived in Acapulco, where he later joined the police force. According to María Pineda and Sigi, Rafael learned much from Margot, who had, with remarkable grace, lifted him out of what would have been a life of enervating poverty.

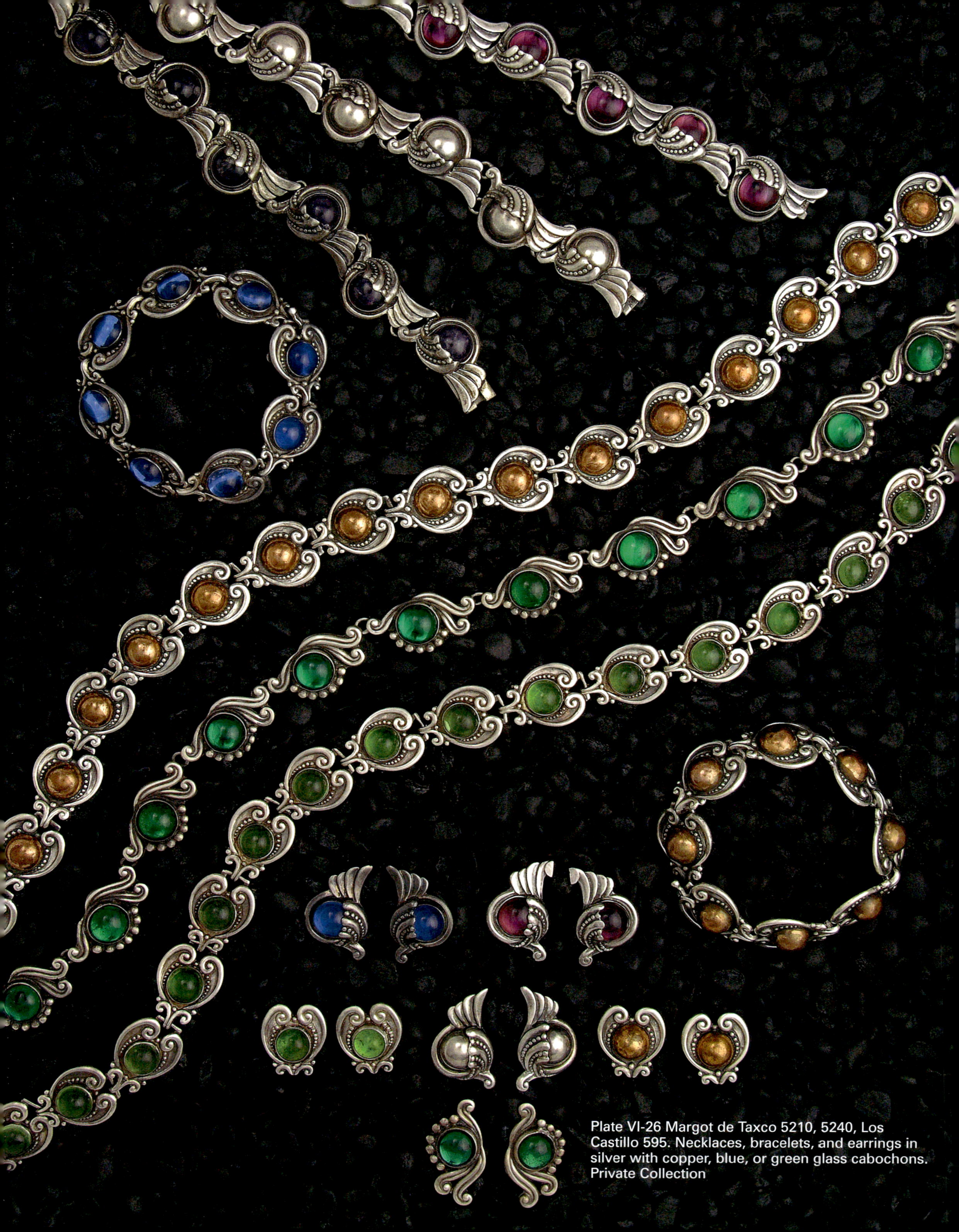

Plate VI-26 Margot de Taxco 5210, 5240, Los Castillo 595. Necklaces, bracelets, and earrings in silver with copper, blue, or green glass cabochons. Private Collection

The marriage suffered rather than improved as Margot and Antonio moved in different directions in their private lives.[122] They agreed to a divorce in 1946. That year, Antonio made an abortive attempt to develop an extension of the business in Puebla. He accompanied Sigi Pineda and Salvador Terán, who were tasked with opening the Puebla store. The store closed after only two years. Salvador returned to Taxco with Antonio, and Sigi went into the military. As part of the divorce settlement, Margot received the house that she and Antonio had built together. This property enabled Margot to put the past behind her and to ensure that her passion for creativity would continue to be at the center of her life.

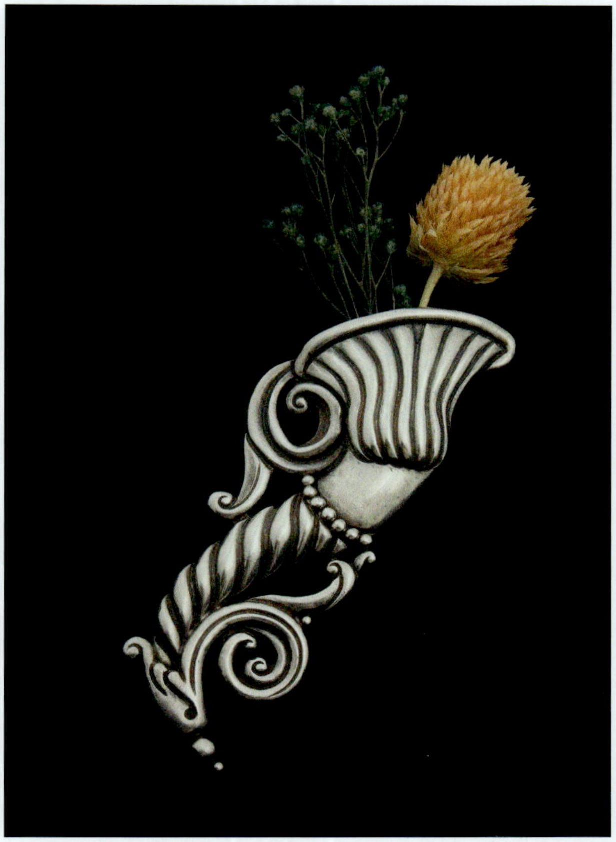

Plate VI-27 Los Castillo 110. *Cornucopia* brooch 3.75 x 1.5 in. Stella and Fred Krieger Collection

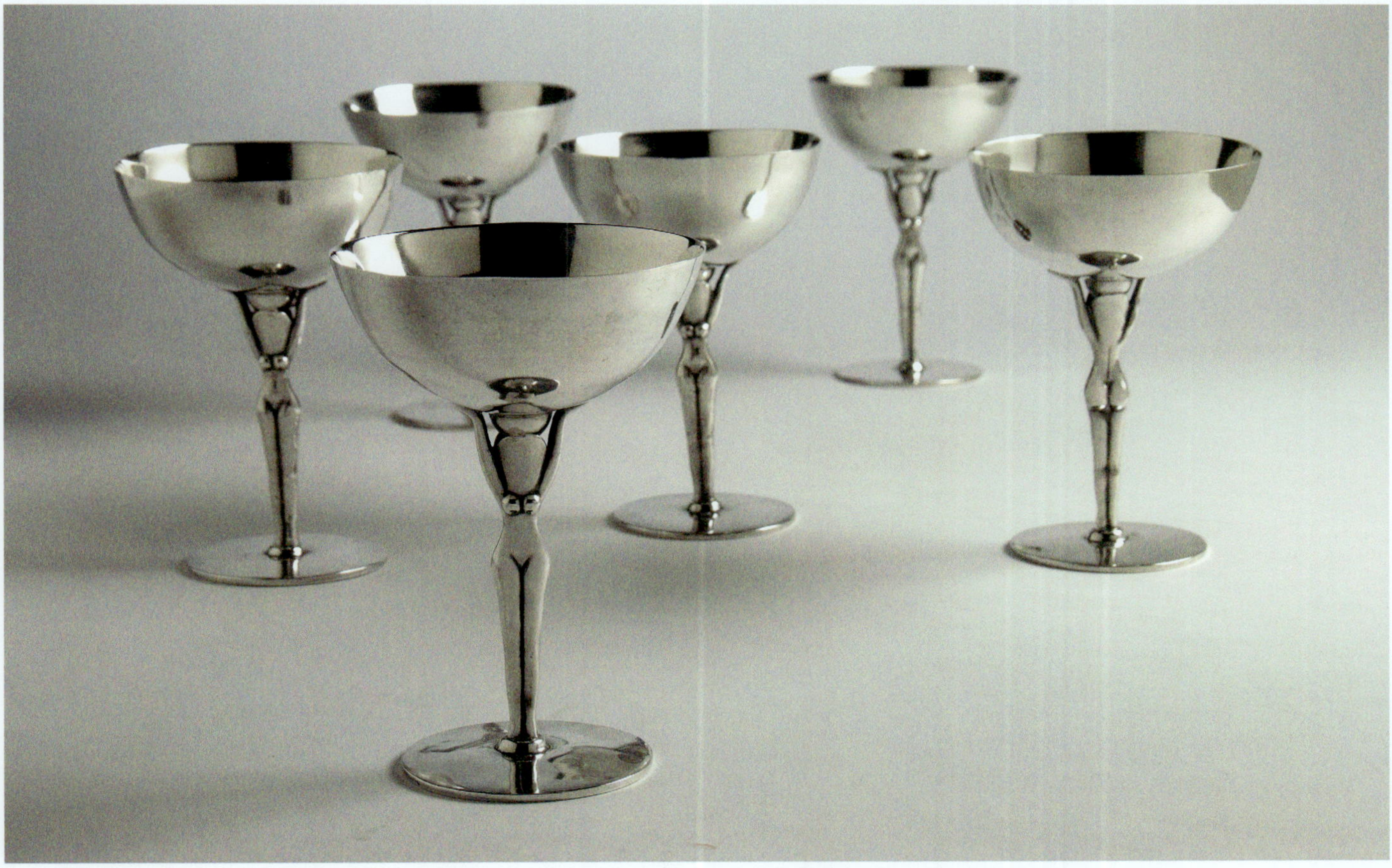

Plate VI-28 Los Castillo. *Faceless Female goblets* 5.25 h. x 11 in. circum. Stella Rubin Antiques Collection

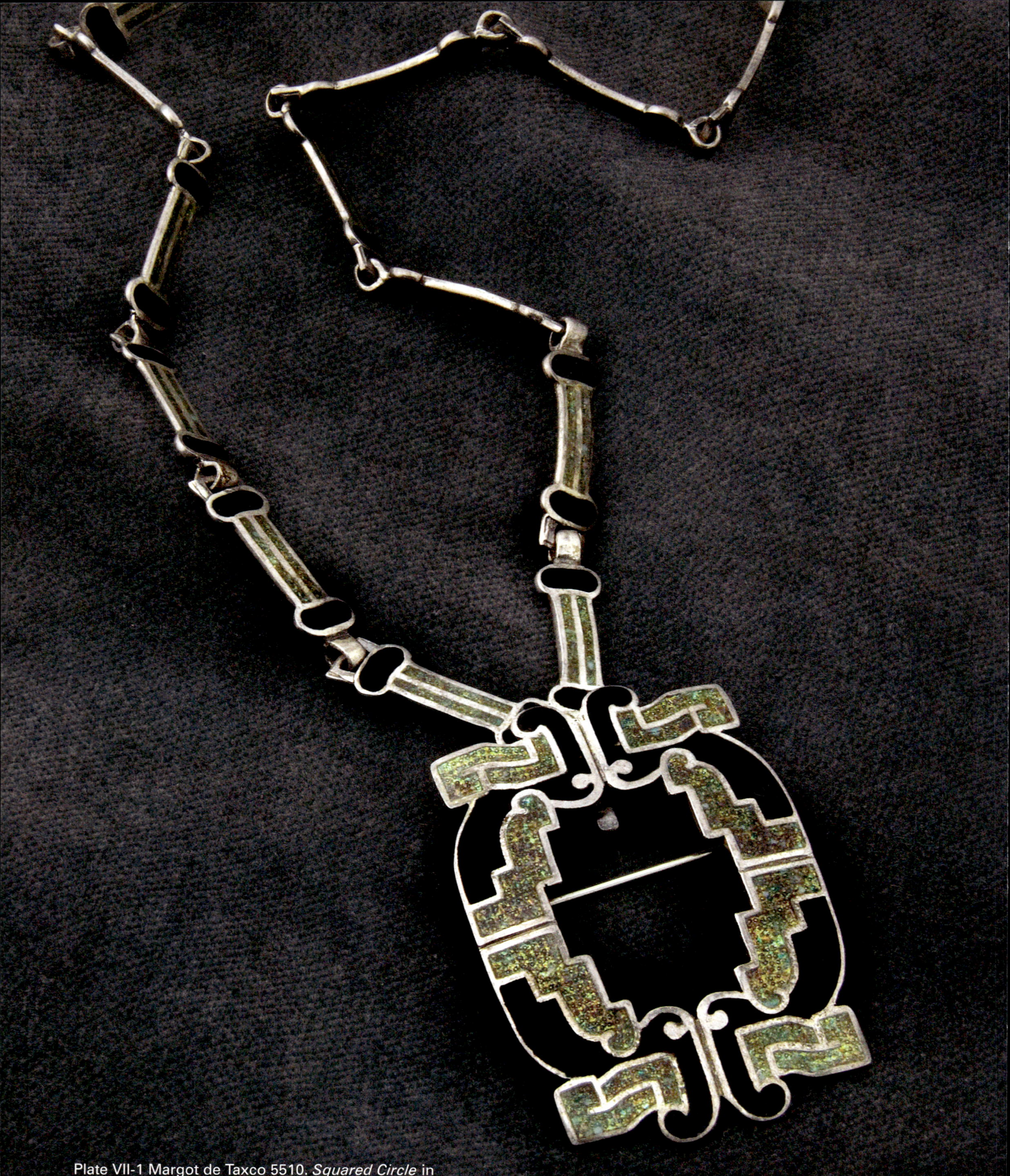

Plate VII-1 Margot de Taxco 5510. *Squared Circle* in black and green necklace 21 in., central motif 2 x 1.5 in. Private Collection

Chapter VII

Margot de Taxco: Silver and Champlevé

After leaving Antonio, Margot went to live with her friend María de los Angeles. This brief interlude allowed Margot to contemplate her options and, in characteristic form, she barely paused before moving forward into her next life adventure. Margot sold the house and, six months after the divorce with the money she had raised, she established a workshop and named the new company, "Margot de Taxco." [123] The fear of competition seemed to have been in the air. Her former collaborators at Los Castillo threatened her with formal expatriation in an effort to keep her from developing her own workshop and business. While the threats came to naught, they were disturbing for their malicious intent, and Margot remained fearful of extradition for the rest of her life. She once told Jaime

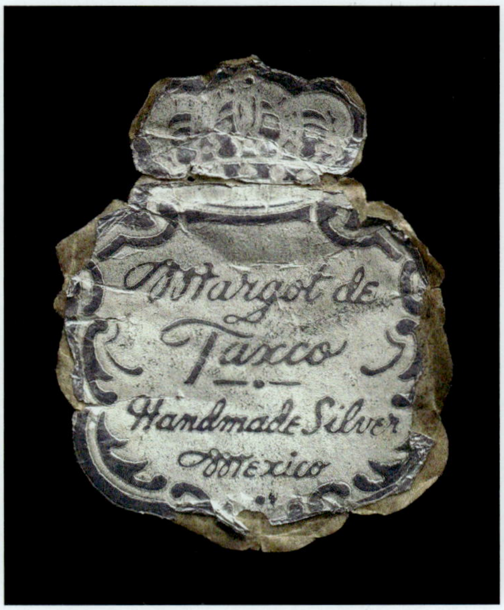

Plate VII-2 Paper Label for Margot de Taxco. Private Collection

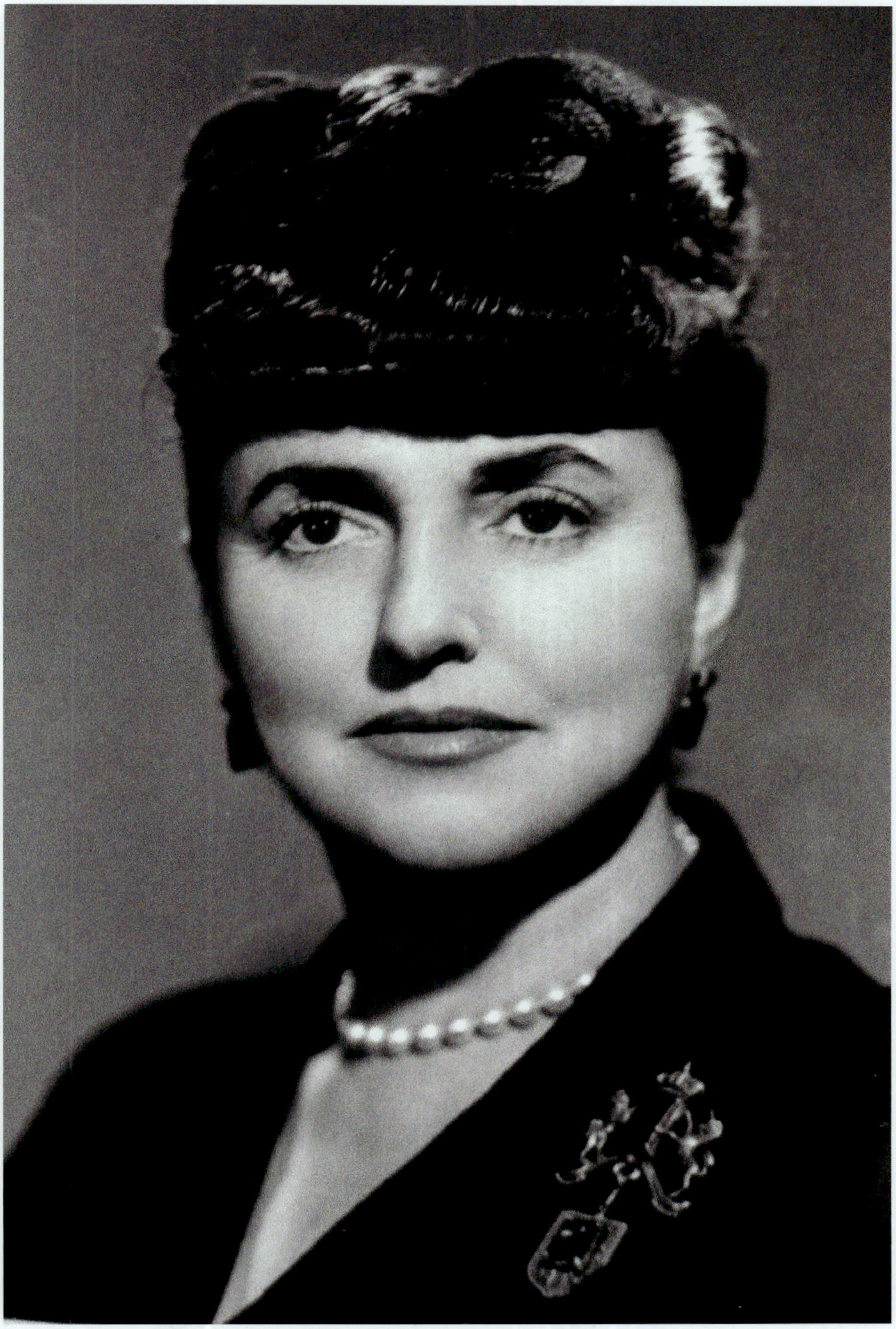

Plate VII-3 Margot Van Voorhies, c. 1950. photograph, Jaime Quiróz Duque Collection

Quiróz, "Unfortunately, when a legal problem arises, I will lose because I have two problems working against me: I am a woman and a foreigner." [124]

Several silversmiths came with Margot from Los Castillo; perhaps among the most important of the group was Miguél Meléndez. A talented *maestro*, Meléndez became the manager of the *taller* [workshop]. Margot had confidence in the strength of her friendship with María de los Angeles, and it was she who Margot asked to run the office for the new company. The silver workshop was located on the second floor of a building one door down from Paco's Bar. It was

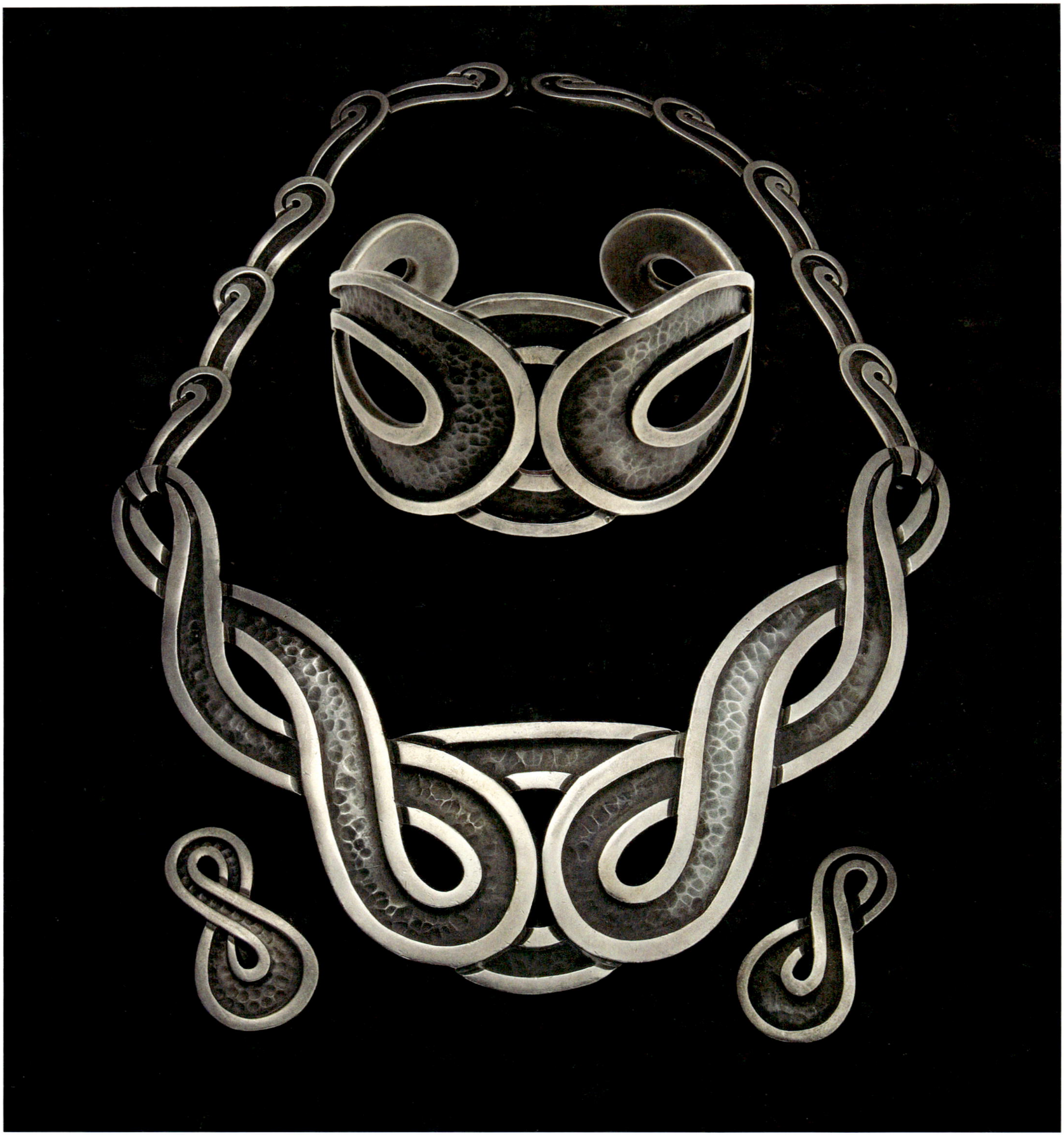

Plate VII-4 Margot de Taxco 5100. *Figure Eight Set*. necklace 16 x 1.63 (at widest point) in.; cuff bracelet 6 x 1.63 in.; earrings 1.63 in. Stella and Fred Krieger Collection

fronted by the store, a large room with floor-to-ceiling windows on three sides, facing out onto the plaza (Pls. VII-13-15).

While Margot carried over very few of her early designs from her years at Los Castillo, a comparison of her Los Castillo necklace, number 615A (Plate II-7) and her Margot de Taxco design 5306 (Plate II-6) provides a sense of a stylistic continuum. The curled forms with flanges that act as a counterbalance, the use of beading to emphasize line, and the repetition of one motif are all aspects of design that Margot brought with her to Margot de Taxco. What is equally important to note is that this particular necklace (LosC615A) has a later Los Castillo mark coupled with the eagle stamp, which indicates that the design remained in production after Margot's leave-taking.

A couple of years after the establishment of the Margot de Taxco silver workshop, a salesman from an American enamel company in Mexico City arrived in Taxco and, fortuitously, Margot found time to meet with him. She became enthralled with the material and its possible applications with silver. As a painter, the idea that she could work in silver and with color was a provocative challenge for her. She may also have recalled the artistry of Chinese *cloisonné*. Margot knew she needed a bright and imaginative artisan to for-

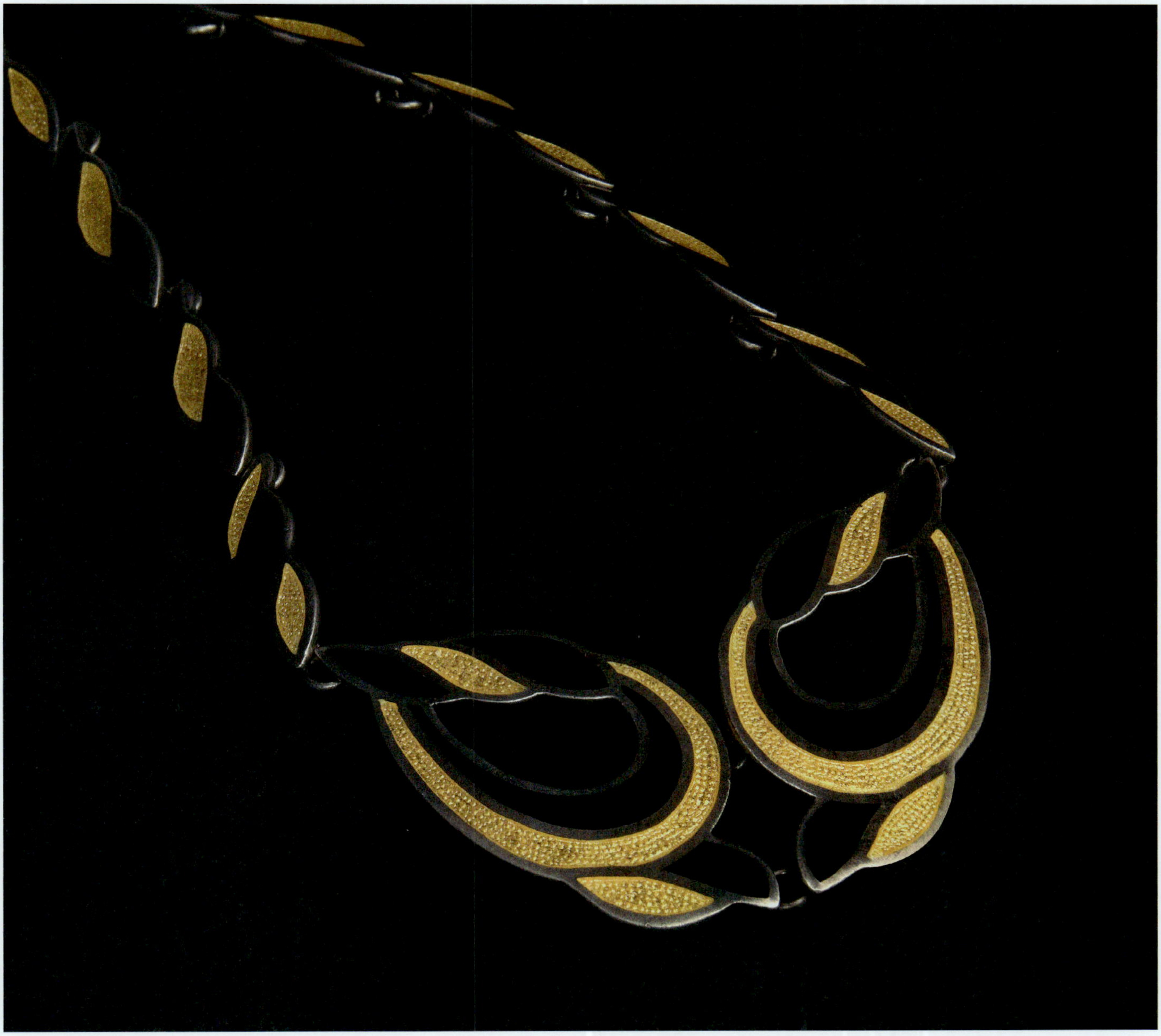

Plate VII-5 Margot de Taxco 5524. Torque in black and gold necklace 15.25 in.; central motif 2.25 x 2.75 in. Private Collection

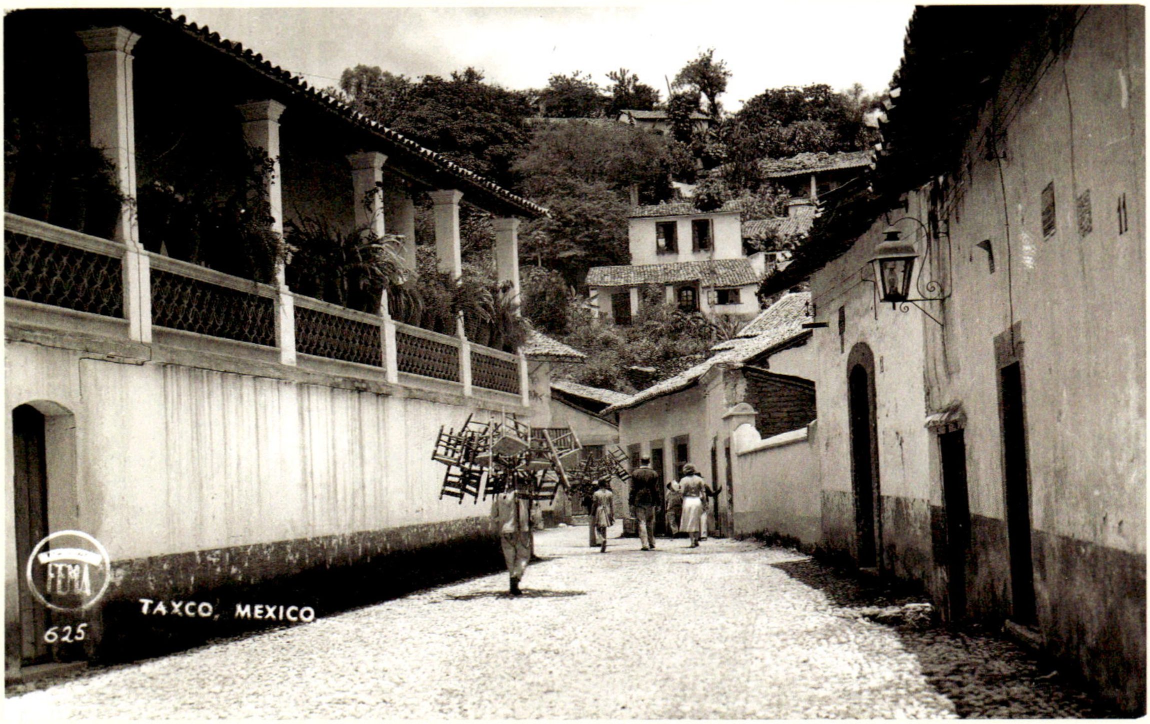

Plate VII-6 Margot's home and the original Margot de Taxco enamel workshop, exterior. Postcard, c. 1945. Private Collection

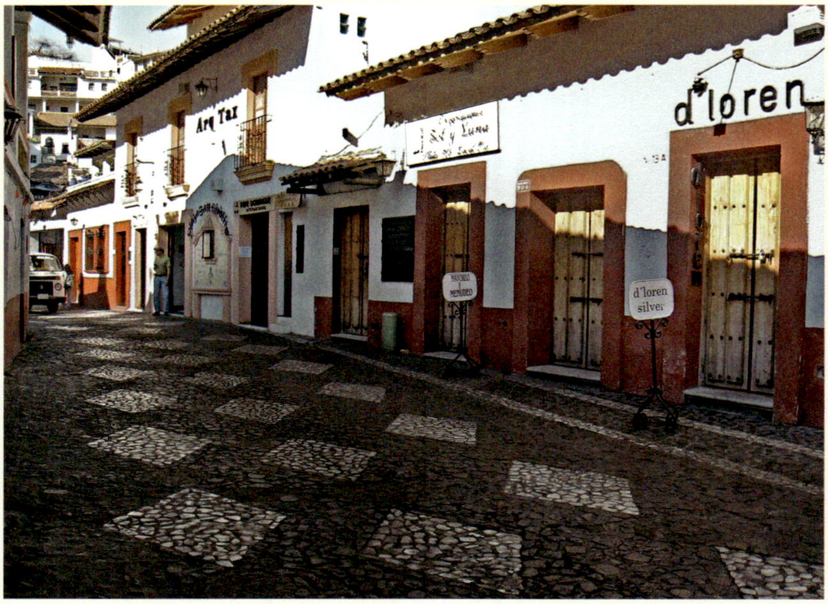

Plate VII-7 The original Margot de Taxco enamel workshop, exterior. Photograph by author, 2008

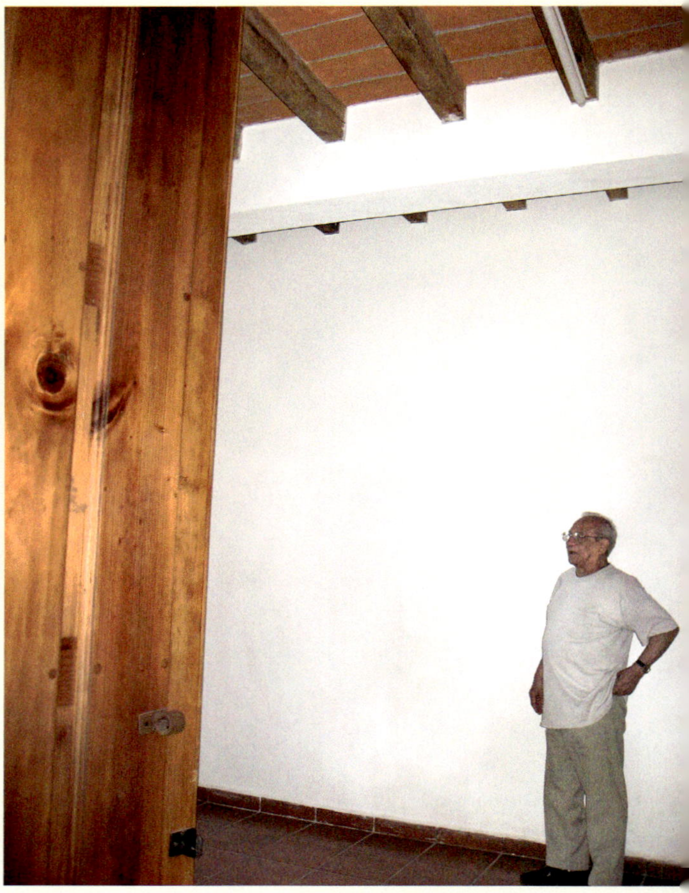

Plate VII-8 Sigi Pineda in what was the Margot de Taxco Enamel Workshop. Photograph by author, 2008

mulate the technical and workshop protocols for the new production. In 1949, she and Miguél Meléndez turned to Sigi Pineda and urged him to join Margot de Taxco.[125]

The enamel workshop was initially in a centuries-old building opposite the Ayuntamiento [City Hall], a five-minute walk from the plaza (Plates VII-6-7). The workshop space was below street level and reached by stairway because the building had been constructed into the side of the mountain. Margot rented the whole building, and she lived above the workshop in an apartment that had ancient ceiling timbers and a fireplace.[126]

For the enamel work, Sigi Pineda created a system that began with Margot's detailed gouache design sketches (Plate VII-8).[127] Arturo Sorio produced the die in steel following specifications in Margot's design drawing. According to Sigi, Sorio had a great sense of humor and an independent streak, one of the few artisans who read extensively.

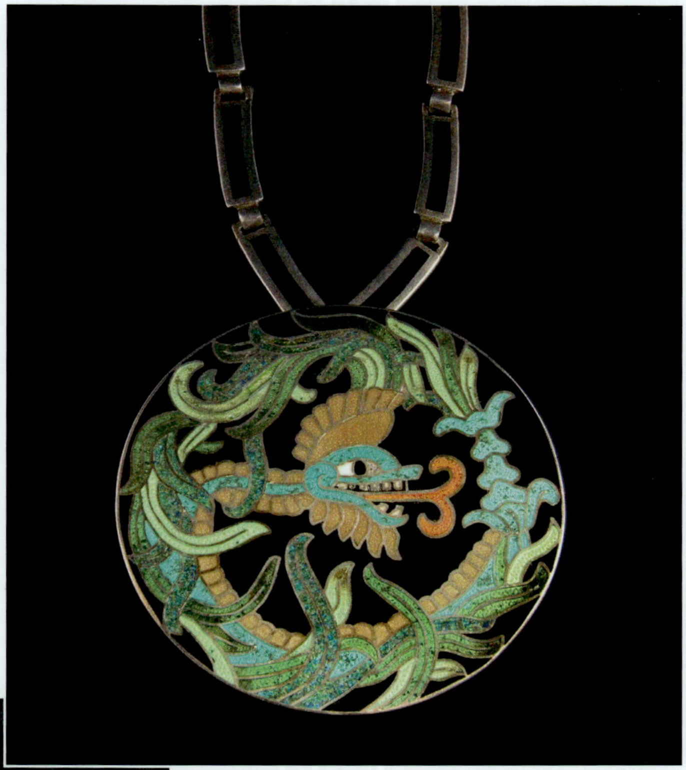

Plate VII-9 Margot de Taxco 5753. *Quetzalcoatl* necklace 22 in., pendant 3.75 in. d. Private Collection

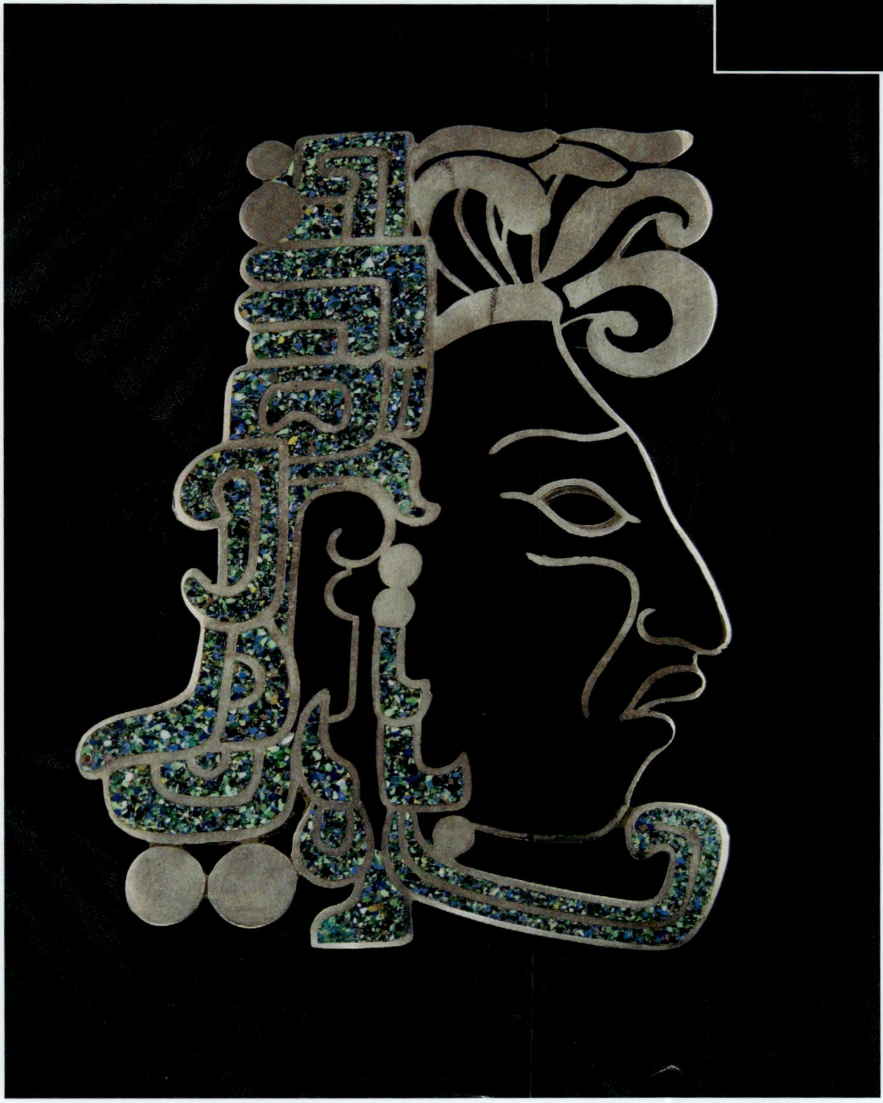

The challenge, in the beginning, was to establish the appropriate depth of the enamel.[128] When the die was pressed into a flat sheet of silver, it created a silver base with raised sides or ridges at the proper height that matched the design drawing's outlines. The raised walls' function was to hold in place a specific color, separated from different and contiguous colors. The various colors could not be painted onto the silver next to each other because they would run together when the piece was heated. The height of the raised walls then became the first challenge to the technical mastery of enameling silver. For the most part, the design lines carved into the steel die were 1-2 millimeters deep, which translated into ridges of that height.

Plate VII-10 Margot de Taxco 5404. *Maya Profile* brooch/pendant 3.5 x 2.63 in. Things Finer in Santa Fe Collection

In the first few months, Melecio Rodríguez would occasionally have to file down the silver because the ridges were too high. He remembers the long hours of experimentation and many failures before Sigi, Arturo Sorio, and Margot decided they had arrived at the proper wall heights, and the dies could go into production.

The creation of a steel die was of the highest difficulty. This hard material had to withstand 45-50 tons of pressure. Thus, in order to carve out the design, Arturo Sorio or Juan González, another die-maker, had to use sharp tools made of silver steel, denser steel than the carbon steel used in the die itself. Before die-casting in silver, the design was cast in plasticine to check for accuracy. Work on the die could take as long as six months before Margot and the die-maker decided it was ready. Jaime Quiróz remembers that the Quetzalcoatl pendant die was especially difficult to produce (5753, Plate VII-9).[129]

The die occasionally produced a texture on the surface of the silver that was visible through transparent enamel. This technique, which Margot referred to as *guilloché* in the instructions for several of the design drawings (5774), may actually have been *basse-taille*. According to Greta Pack, a jeweler, jewelry historian, and friend of William Spratling, "Bassetaille enamel

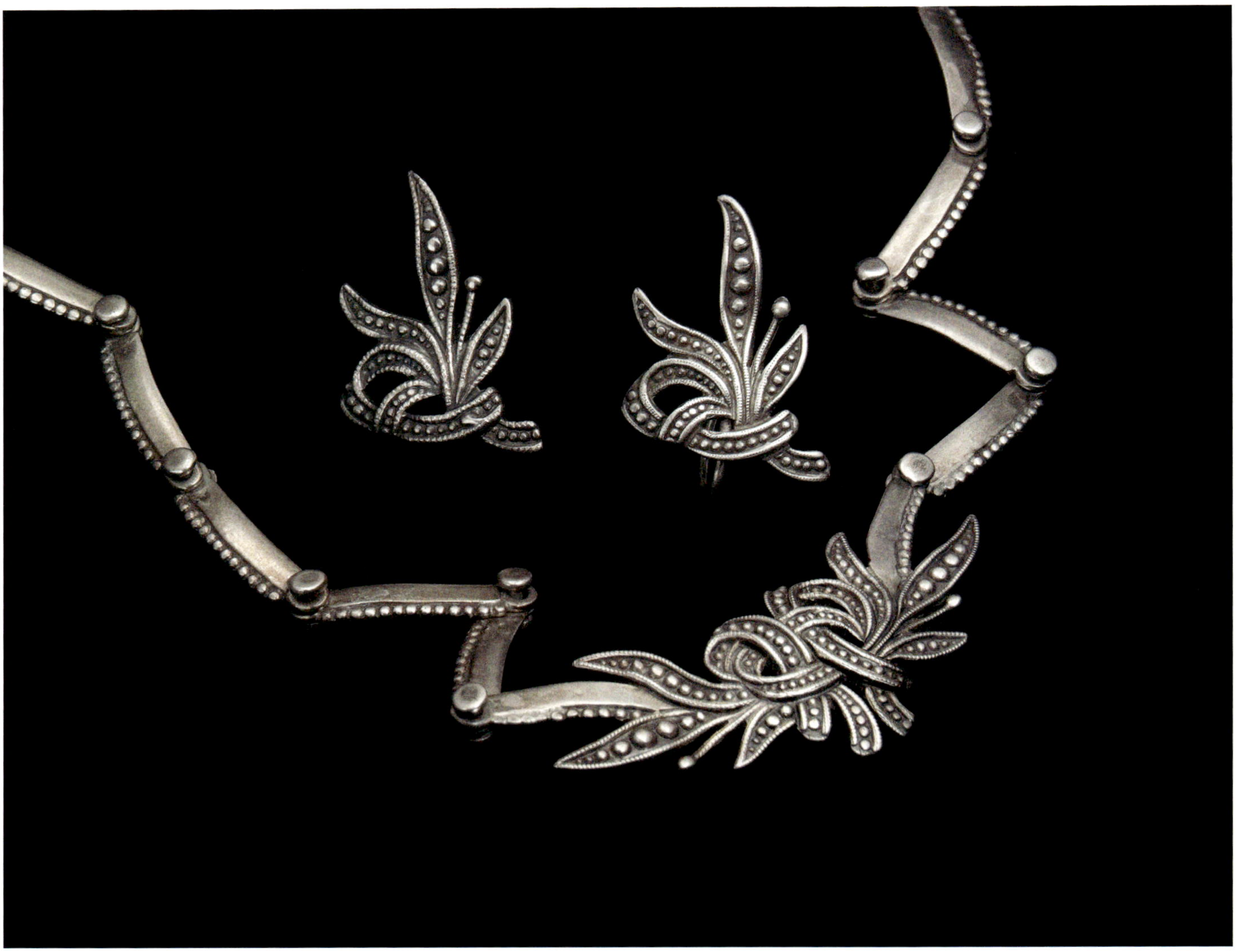

Plate VII-11 Margot de Taxco 5346. *Leaf and Ribbon Spray Set* in silver. necklace 16 in., central motif 1.125 x 2 in.; earrings 1 x 0.6 in. J. P. Mather Collection

is much like *champlevé* enamel. The design is carved or executed in *repoussé* in low relief about 1/30 inch below the surface. Transparent enamel is fused over the design, possibly several times, until a uniform surface is obtained. For this method gold or silver should be used as the reflecting quality is better than copper. The design may be concave or convex and must be executed with great care; interesting shadow effects are obtained as the depth of the enamel varies."[130]

Once the carving in the die was tested and found to be accurate, the die itself was tempered. It was boiled in oil and then allowed to cool for twenty-four hours in the corner of a closed room as protected as possible from temperature fluctuations. If the die cooled too quickly, it would weaken and could not stand up to the pressure exerted by the press.

The first step in production was the die-casting of the silver base for the enamel. Pedro de León was among the first to work in the die-casting division. In the early years, the press was operated by hand at twenty tons or more.[131] The artisans in this section of the workshop were required to die-cast the number of pieces that could be cut out in a day.

The silversmiths in the next section were responsible for using vertical saws to meticulously cut away extraneous silver from the central design. A percentage of the silver bases for popular designs were placed in storage so that the pieces could be enameled and the finished product could arrive in the store more quickly. The silver content for the enamel jewelry was 980 parts silver to 20 parts copper, a much higher level of silver than sterling, with its ratio of 925 parts silver to 75 of copper. The choice of 980 silver content provided a material that was softer and easier to manipulate with a press. The reduction in copper also meant that the jewelry would not tarnish as easily against the skin.

Sigi recruited identical twins, Germán and Cervando López to work in the silversmithing division of the enamel workshop. Germán López later became stock manager. Macedonio Martínez, who had been with Los Castillo, Rafael Serrano, Manuel Flores, Gumecindo Palacios, and Melecio Rodríguez joined the workshop as silversmiths and later worked on this second phase of production. Manuel Quinto was the first stock manager for the enamel workshop.

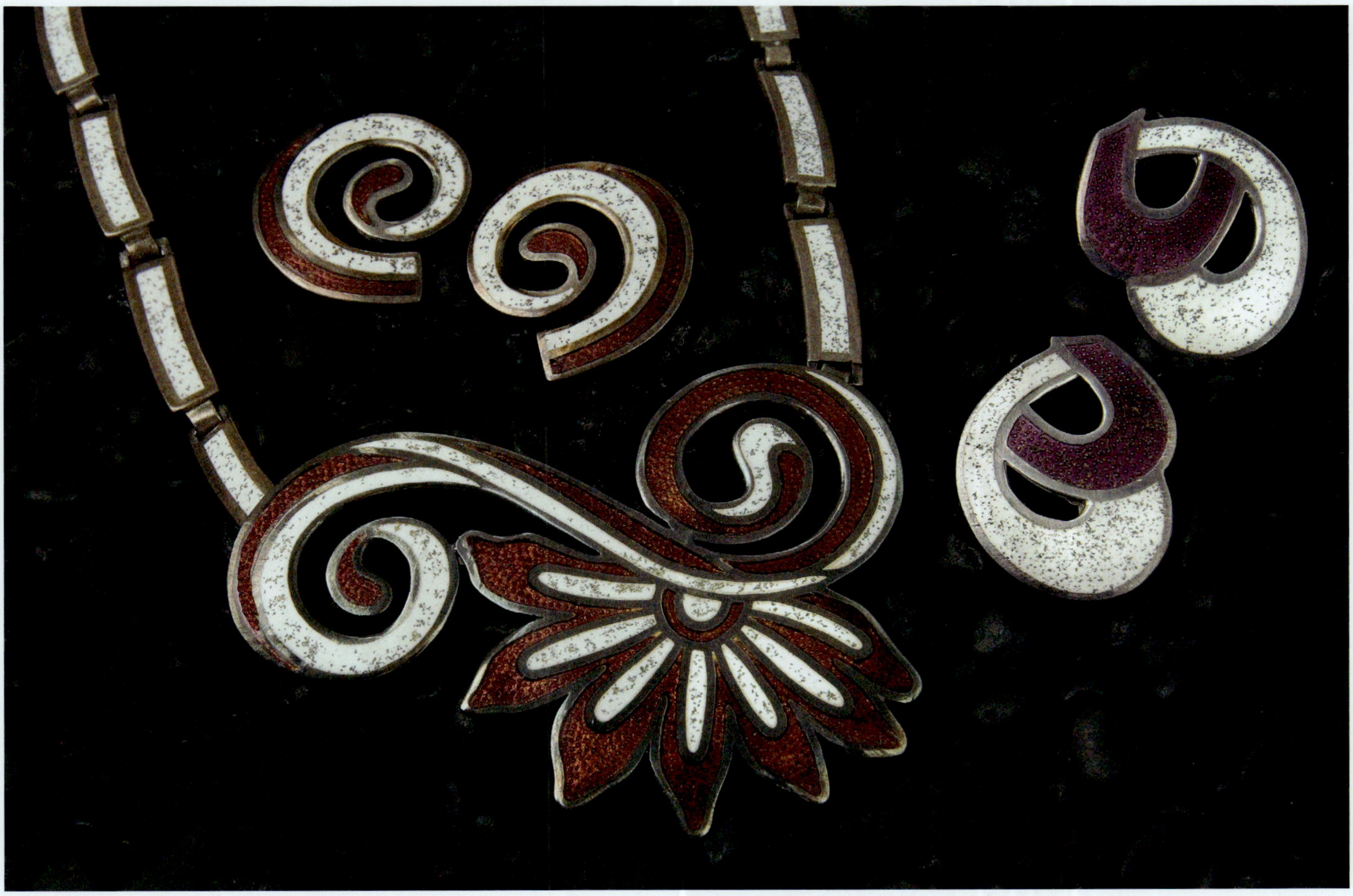

Plate VII-12 Margot de Taxco 5794. *Floral Flanges and Curls Set*. necklace 15 in., central motif 2.25 x 3 in.; earrings 1 x 1 in. Margot de Taxco 5699. earrings 1.25 x 0.75 in. Private Collection

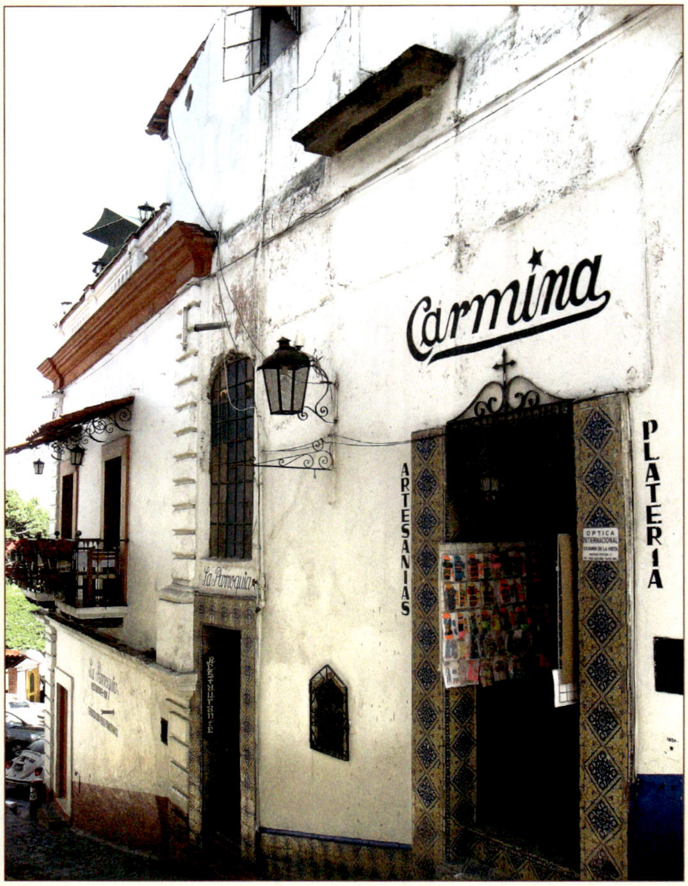

Plate VII-13 Margot's store today, exterior. Photograph by author, 2008

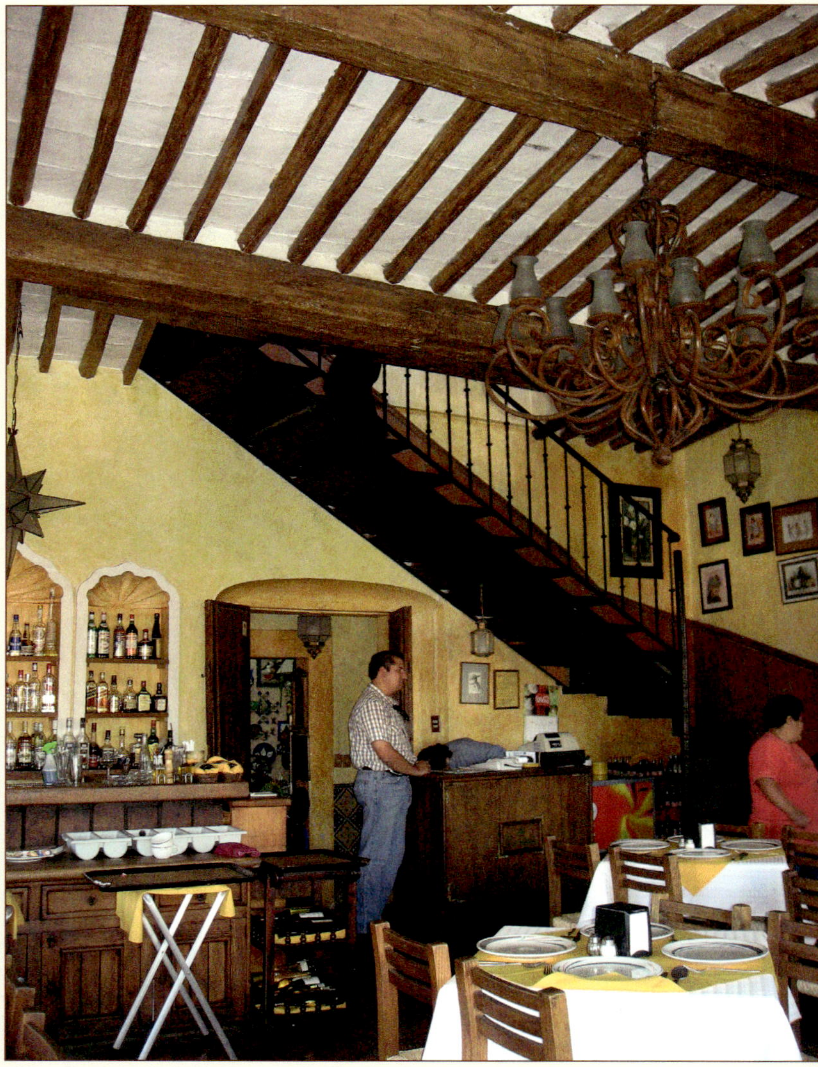

Plate VII-15 Margot's store and silver workshop today, interior. Photograph by author, 2008

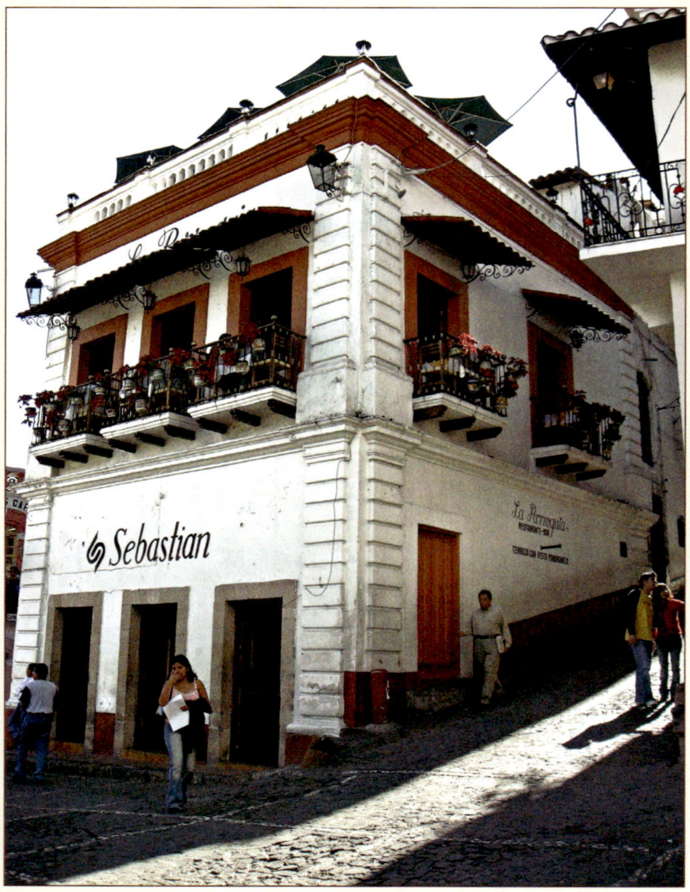

Plate VII-14 Margot's store today on the Zócalo, exterior. Photograph by author, 2008

When Melecio Rodríguez entered Margot de Taxco in 1948, he was one of two hundred workers in the combined workshops. Initially he was placed under the guidance of *maestro* Manuel Hernández in silver jewelry. Because he was so young, Rodríguez was going to be one of the first to leave in the event of a downturn in demand.[132] But Rodríguez had other talents. As early as 1948, Margot de Taxco had organized a baseball team and Rodríguez was one of the star players. At the time, the team members had responsibility for their own expenses incurred when they participated in out-of-town games. Miguél Meléndez found work for Rodríguez during the slow period so that he could be paid by the piece and, in this way, cover his expenses. Rodríguez looks back with amusement at the fact that he got his start in silver because of his ability as a baseball player. Sixty years later, Rodríguez is still working silver in his own workshop where he supervises several sons and grandchildren (Plate VII-18).

The third phase involved finishing the piece. Once the designs were cut out with a vertical saw, the surfaces had to be filed and polished in preparation for the next step in production. Some silver was lost in the polishing process, although workshops attempted to capture the silver filings and use them to produce silver for soldering, which always had a higher copper content than the piece itself. In the fourth phase, the artisans were tasked with attaching the various parts of the necklace or bracelet together, using circular links and jump rings made of solid silver wire.

Sigi supervised the next phase, the enameling of the silver base. At this time, no other workshop in Taxco was employing women. Sigi chose to recruit and hire them for the enamelwork because it required concentration and slight movements of the hand. The areas to be painted were miniscule and the brushes were tiny. Additionally, there was no margin for error with the color, so it took a special talent to paint the enamel at the right level within the ridges on the silver surface of the jewelry. The enamel arrived from the United States as a paste. In numerous pieces of jewelry, a very thin layer of 24-carat gold leaf was placed beneath transparent enamel, again, a technique that required dexterity and precision. Numerous design drawings (5758) indicate that certain colors should be applied "c/oro" [with gold].

Sigi was sympathetic to women and liked working with them. At the same time, Margot was more than willing to provide opportunities for women, so this unique experiment became a grand success. María Elena Flores was the primary artisan and Velarmina Palacios was the *maestra* in the enameling section. Other enamelists were Reina Arrellano, Socorro Ocampo, Oralia Castrejón, and María Pineda. In later years, Vicenta Rodríguez was *maestra* of enameling.

The artisans applied the enamel according to Margot's design drawings and instructions, now in the Spratling-Taxco Collection of Tulane University's Latin

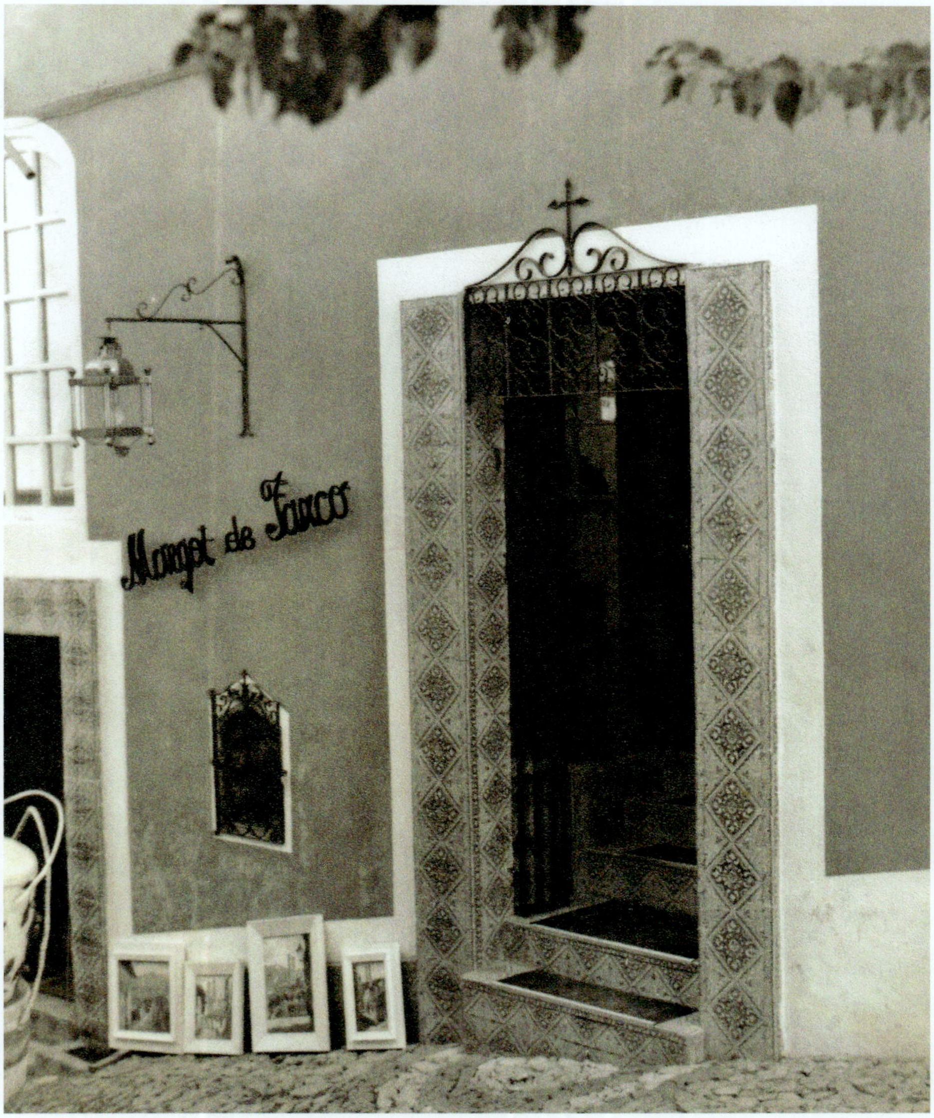

Plate VII-16 Margot's store, exterior. Photograph c. 1955, Zamora Carbajál Family Collection

American Library (Plate VIII-2). Margot preferred the use of a damp paste rather than a powder, which was applied with tiny brushes. After the pieces were painted and the enamel had dried, the women could insert them into one of two kilns at 1400-1600 degrees Fahrenheit. The enamelists used another technique, placing the silver base onto a wire mesh platform. Again, after the enamel had been applied and was dry, the piece was heated from beneath the mesh platform with a torch. Once the metal reached a certain temperature, the enamel fused to the surface. The pieces were then sent back to the men for any soldering that had to be done and for a final polishing.[133]

In the spirit of experimentation that prevailed in Taxco during this period, Margot developed a "secret formula" for the enameling process. For the polychromatic enamels, Margot may have used a larger grain of enamel that would have retained its color when fired, even when touching other grains of equal size. The opaque enamels, which had deeper more brilliant colors, were not ground as finely as the transparent enamels. Each type had its own special beauty, for the transparent enamels allowed for the use of gold leaf and for subtle patterning on the silver surface beneath the enamel and gold leaf.[134]

It should be noted that Margot created a small number of designs for enamel on copper. These were apparently produced for a store on Estero Beach in Ensanada, Baja California, owned by Benjamin Novelo. Novelo specialized in leather belts and commissioned copper buckles from Margot.[135] For these copper pieces, the enameling technique differed somewhat, and the design format was on a larger scale, usually a two- or three-color abstract pattern (5711, Plate VII-21). Besides the buckles, there was some copper jewelry. A bracelet that takes advantage of the color and texture of the medium is number 5715 (Plate VII-20). The piece is made of two opposing interlocking curved forms in which one is in green and the other is given a textured appearance with the polychromatic enamel. These two parts are separated by a thin line of the exposed copper surface, and this coppery color becomes a compliment to both of the enameled segments of the composition.[136]

In 1950, María de los Angeles married Reveriano Castillo and they formed their own workshop "Reveri." They were never in competition with Margot de Taxco because their work was so different in design and technique.[137] In 1952, Margot married Alberto Cuevas (Plate VII-25). He had come to Taxco from Mexico City and opened a meat market with financial assistance from Margot. Alberto played the piano and Margot bought him one, hoping it would become something they could enjoy together at home.[138]

The marriage lasted only a couple of years. This is hardly surprising since, at this time in her life, Margot seemed to relish the fact that she was self-sufficient and free to travel without commitments. She flew on her own via Air France from Mexico City to New York City and on to Paris in April-June 1954, listing herself as Margot van Voorhies Cuevas. On October 2, 1955, on a trip to New York, she appeared in the passenger list as Margot van Voorhies Carr.[139]

A number of years later, Margot took a trip to Japan that inspired a journal she kept until the end of her life. Margot often returned from her travels with small figurines, special stones, and with ideas for jewelry based on what she had seen. Margot also bought art books, which became inspirations for her designs. Sigi recalls that Margot used a mirror to translate the book image into a form that could be more

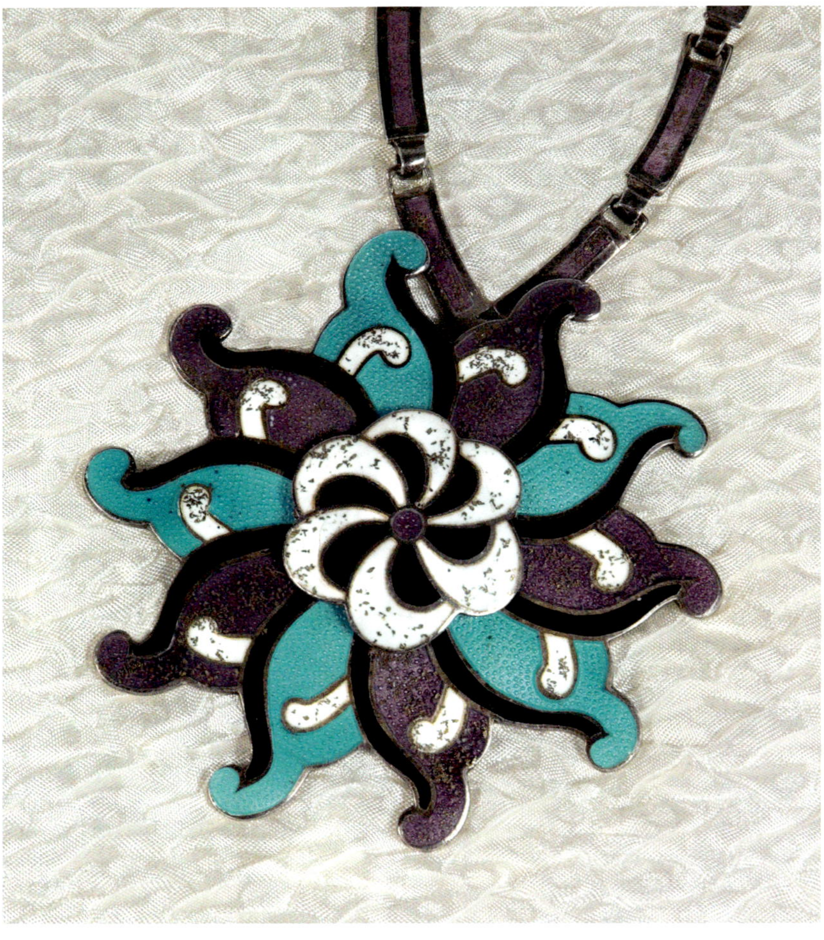

Plate VII-17 Margot de Taxco 5907. *Contra Spirals* necklace 21.5 in., pendant 2.63 in. d. Maestros de Taxco Collection

easily transposed into silver. These books were among her most prized possessions, for they allowed her to venture as an artist beyond Taxco.[140]

According to Jaime Quiróz, Margot had a sense of humor and was a caring person — he remembers that she called everyone "honey." Margot had been a beautician before coming to Mexico and so was always carefully made up and coiffured. She created dress patterns and made all of her clothes; she knitted the jackets and sweaters she wore. Margot probably learned these skills from her mother who, as a young woman, had worked in a textile factory and hired out as a seamstress. Sigi said of Margot, "She was an extraordinary woman, in her manner, her character, her sense of self. As an administrator, she was courageous and tough, and she had high expectations for her employees. Unfortunately, she spoke only rudimentary Spanish and she at times offended by saying the wrong thing or was misunderstood."

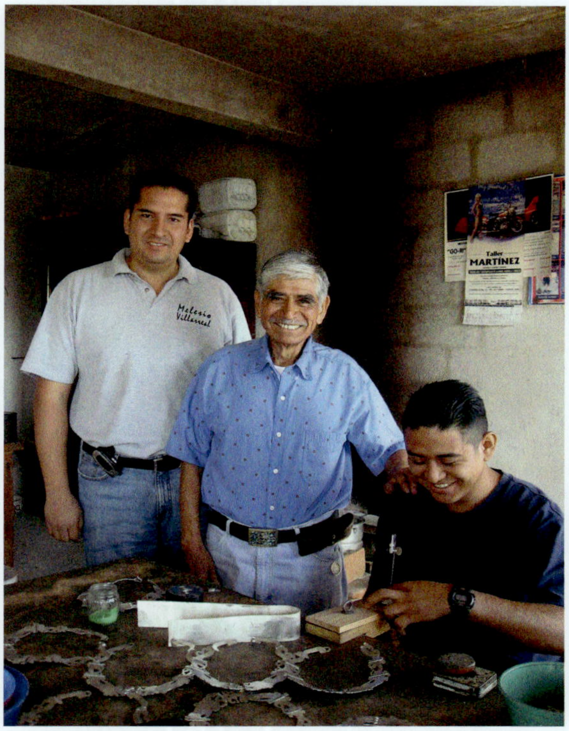

Plate VII-18 Melecio Rodríguez with his son, Melecio, and grandson, Melecio, in the workshop. Photograph by author, 2008

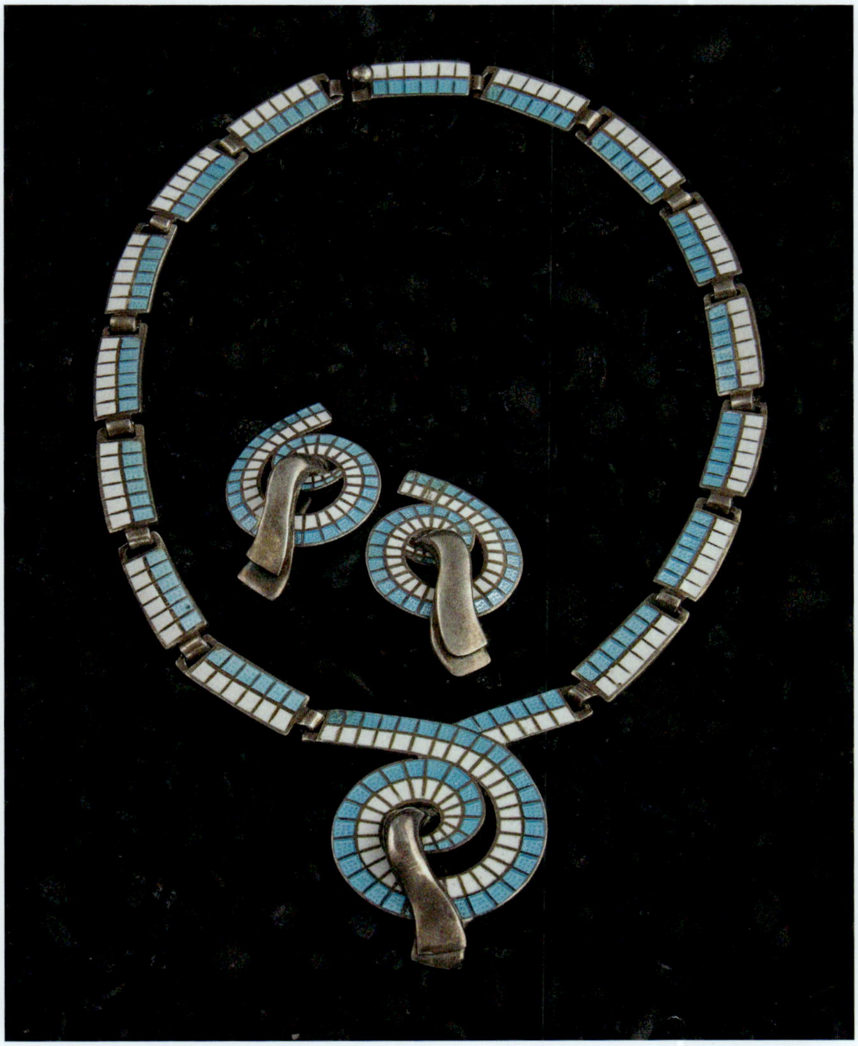

Plate VII-19 Margot de Taxco 5556. *Two Colors Ribbons and Curl Set.* necklace 14.5 in., central motif 2.5 x 2.5 in.; earrings 1.5 x 1.25 in. Private Collection

Trouble brewed less than five years after Sigi helped initiate the enamel workshop. When he joined Margot de Taxco, Miguél Meléndez and Margot gave him an incentive that included the opportunity to make a commission on production, in addition to a base salary. At the time, Meléndez was supervising the silver workshop on the plaza, and left the enamel production to Sigi, then located across from the Ayuntamiento. Within six months, demand grew and the enamel workshop expanded to twenty-eight artisans.

Sigi's advancement began to attract attention among the other silversmiths, who became disgruntled over his success. Their complaints to Margot centered on Sigi's contract that allowed him to manage a workshop and accrue a 26 percent commission on production. Margot's initial reaction was to

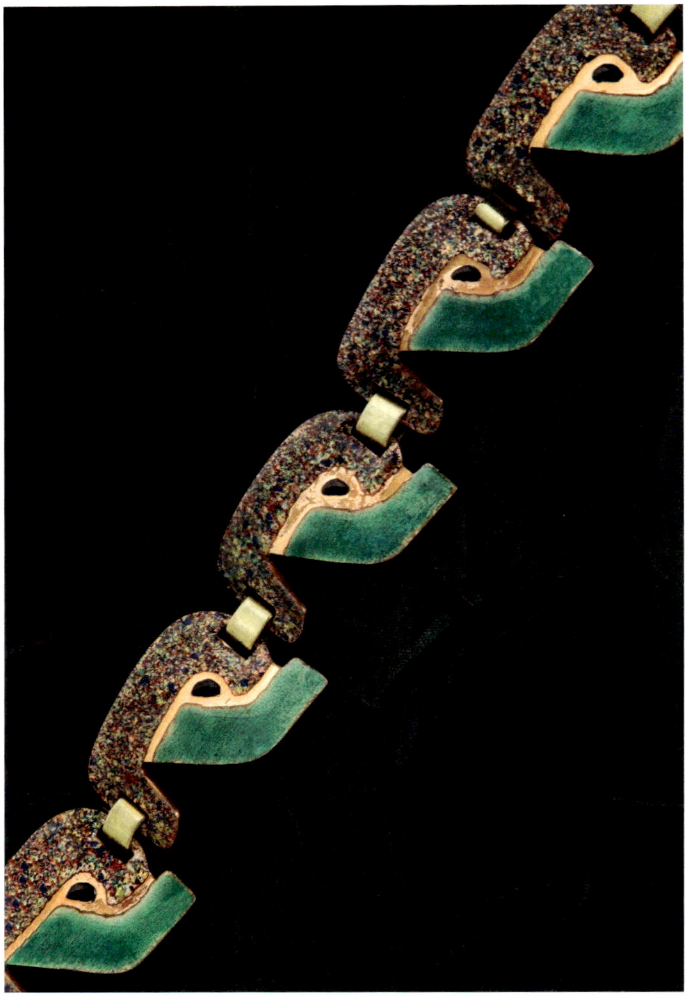

bring calm to the situation by limiting production. This temporary measure solved the problem only in the short term.

The turning point was a suspected robbery of five kilos of silver from the stockroom. Margot became infuriated and accused Carlos Sánchez, the stock manager, of the theft. Even though several took his side, including Sigi, Miguél Meléndez, and Alberto Cuevas, Sánchez suffered from the accusation.[141] Sigi discovered one of the reasons for the seeming disappearance of five kilos of silver. While there probably was some theft that took place, the production of accessories, like pins on brooches, had not been taken into account in the weight of the silver expended. Sigi had kept a record of accessories he had used, but the office had not. Although Sigi showed the discrepancy to Margot, it was too late for Carlos Sánchez, Miguél Meléndez, and Antonio Reina. In 1953, they left the

Plate VII-20 Margot de Taxco 5715.
Enamel on copper bracelet 7 x 0.88 in.
J. P. Mather Collection

Plate VII-21 Margot de Taxco 5711. Enamel on copper belt buckle 2 x 4.13 in. Private Collection

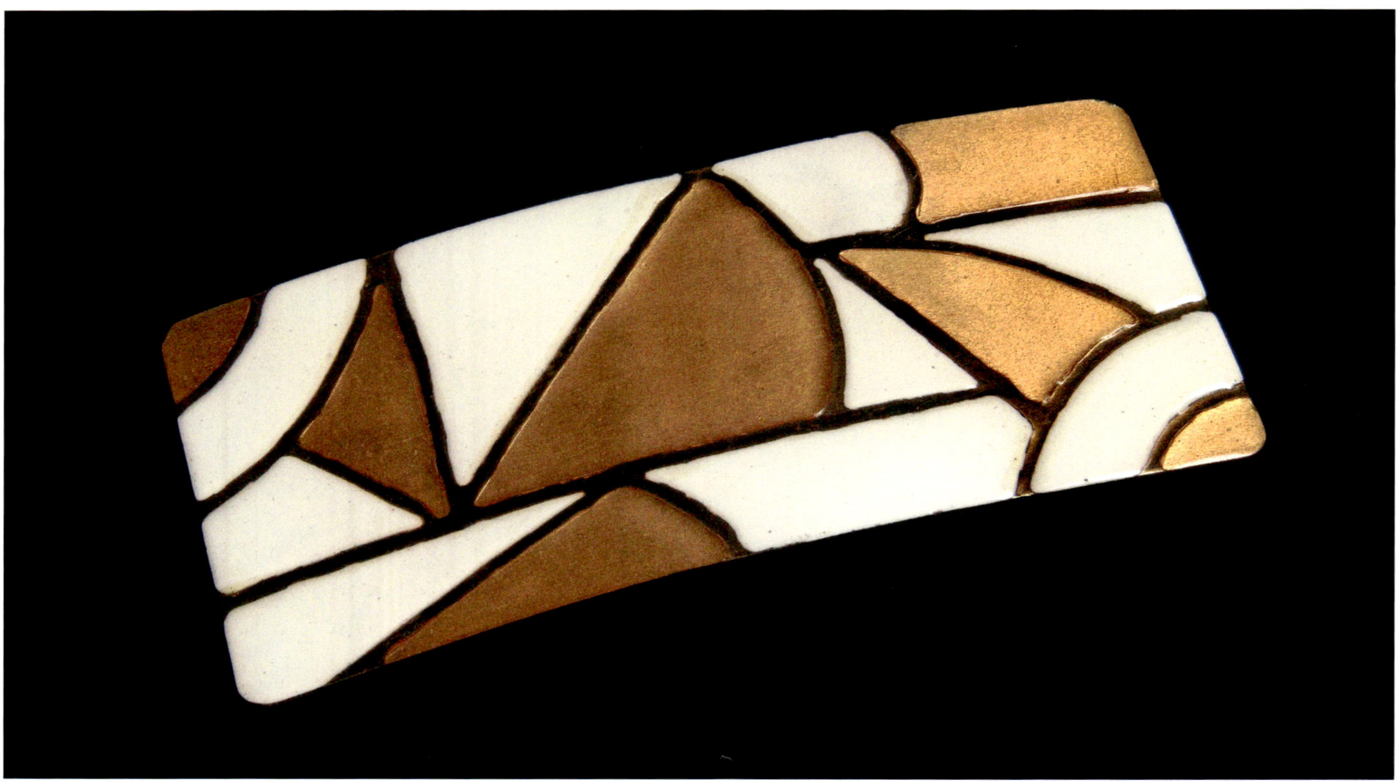

company as a result of a misunderstanding not only with Margot but with her new administrator Felipe Zamora as well. Sigi departed soon after to start his own workshop and store with the money he had earned at Margot de Taxco.

Miguel Meléndez began working out of his home, either on contract with others, including Margot, or for himself. He placed his mark, an "M" with a swirl that resembles a "C" surrounding the "M," when producing his own designs. When he was working for Margot, she would provide the die for her mark, which he would apply to the pieces he was making for her.

Jerónimo Fuentes replaced Sigi as director of the enamel workshop (VII-24). He was assiduous and honest, according to Melecio Rodríguez. At this time, the enamel workshop moved to the plaza. Felipe Zamora had begun working for Margot in 1950 and, in a short time, became director of Margot de Taxco. Zamora

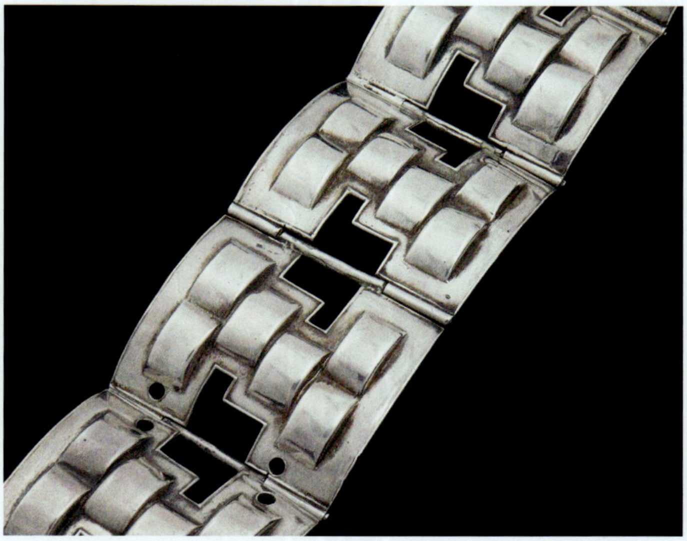

Plate VII-22 Margot de Taxco 5520. *Rectangles* bracelet 1.75 x 7 in. Stella and Fred Krieger Collection

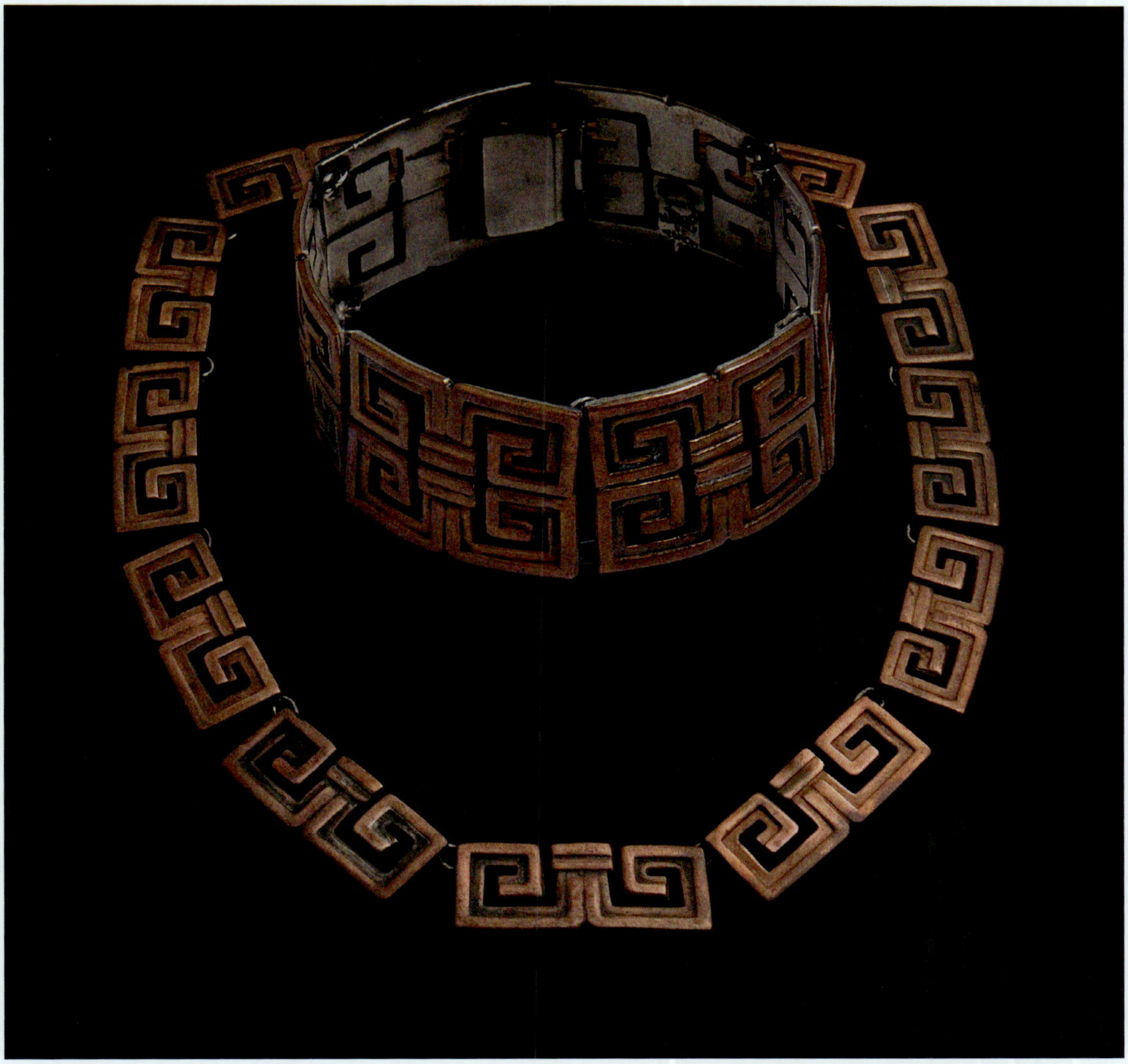

Plate VII-23 Margot de Taxco 5112B. *Pre-Columbian Set* in copper on silver. necklace 18 in.; bracelet 7 in. Private Collection

was handsome and audacious and he swept Margot off her feet (Pls. VII-25, VII-29). He designed houses for himself and for her, and Margot paid for both to be built. Zamora established his own store, fronting on the Plazuela de Gallos, a small square near the plaza across from the building Margot was renting (Plate VII-30). He sold Margot's designs, establishing himself as another outlet for her work.[142] Zamora's shop can still be identified by the large planter he placed out in front, ornamented with applied pieces of obisidian. He also installed a fountain in the *plazuela* and had images of roosters embedded into the cobblestones. A few years later, Zamora left Margot de Taxco to open an interior design practice in Acapulco.[143]

Margot's business card (Plate VII-38) reads, "As the stars are to the night, so are jewels to the woman." In Margot's shop, besides a glittering array of enamel on silver or *champlevé* jewelry, Margot kept one-of-a-kind pieces in a locked Louis XVI-style vitrine. The majority of this special jewelry was produced by Juan González. Adán Alvarado also worked in *repoussage*[144] and, Marcial Chávez, after he left Spratling's workshop in 1962, contracted with Margot to make gold rings.[145] Margot held Juan González in high regard for his artistry in silver, but would comment to him when she commissioned a unique design, "Juan, will this cost me an eye?" The *repoussé* pieces and those incorporating precious or semi-precious gem stones, carved ivory or coral, or tiny carved figurines were the work of this *maestro*.

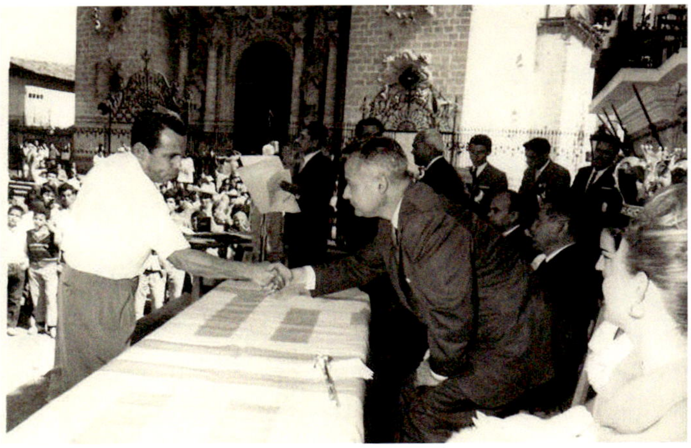

Plate VII-24 Gerónimo Fuentes Receiving an Award for Silver Crowns for the Silver Queens. 1959. Photograph Fuentes Family Collection

Margot also sold hollowware, which was not produced at Margot de Taxco. These Paul Revere-style centerpiece bowls, trays, and melon-shaped tea sets came primarily from the Mexico City workshops of Salvador de la Serna and Pedro López. Margot bought this hollowware outright; she never negotiated sales by commission.[146]

Margot's success did not go unnoticed. According to William Spratling, by the end of World War II, the Mexican government had exacted three different export taxes on silver. This action had an immediate negative impact on the market, as Americans turned

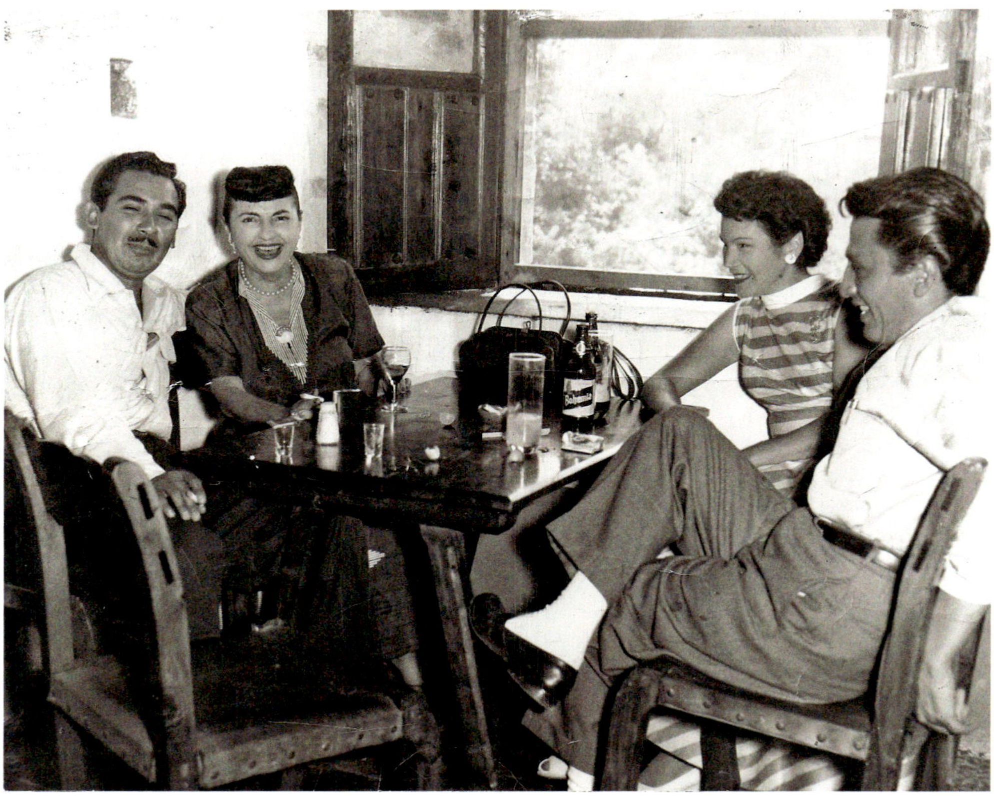

Plate VII-25 Margot with her husband, Alberto Cuevas, Felipe Zamora and his wife, Berta Carbajál, photograph c. 1954. Zamora Carbajál Family Collection

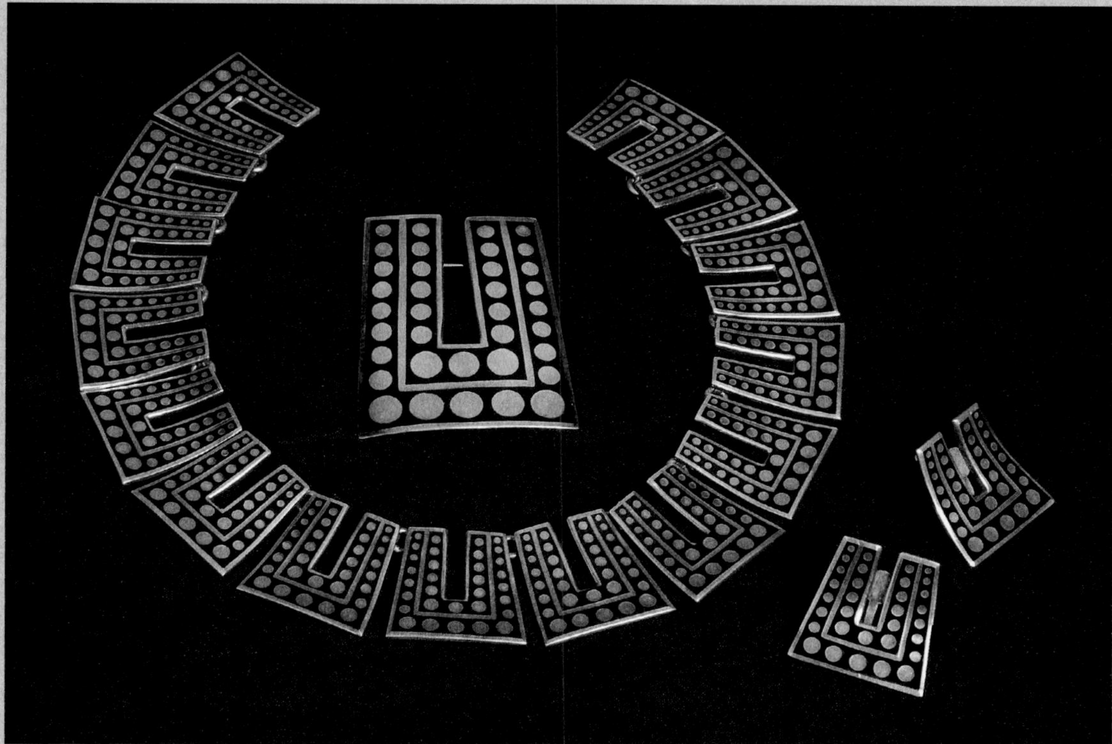

Plate VII-26 Margot de Taxco 5586. *Modern Dots Set* in black. necklace 14 x 1 in.; brooch 2.12 x 1.75 in.; earrings 1 x 1 in. Things Finer in Santa Fe Collection

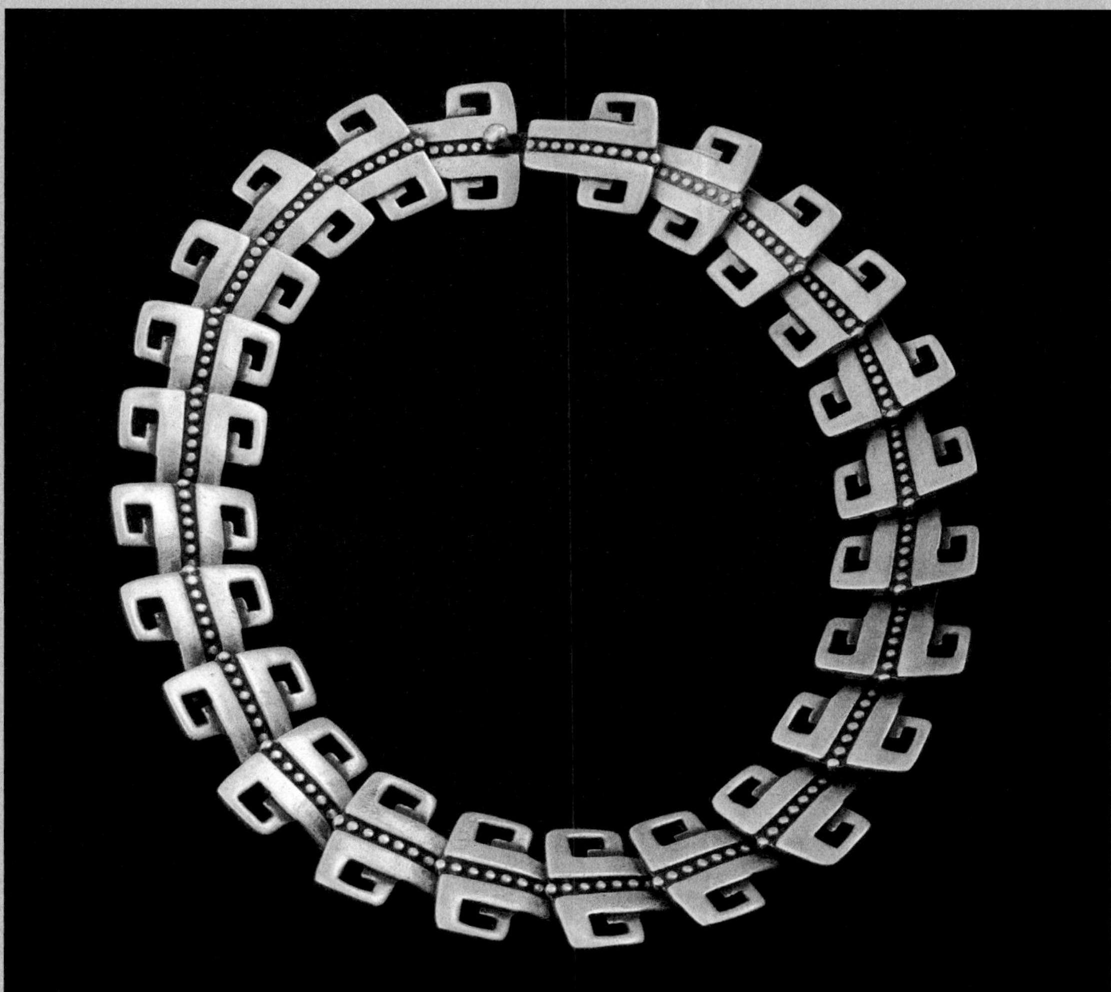

Plate VII-27 Margot de Taxco 5285. *Deco Pre-Columbian* in silver necklace 15 x 1 in. Stella and Fred Krieger Collection

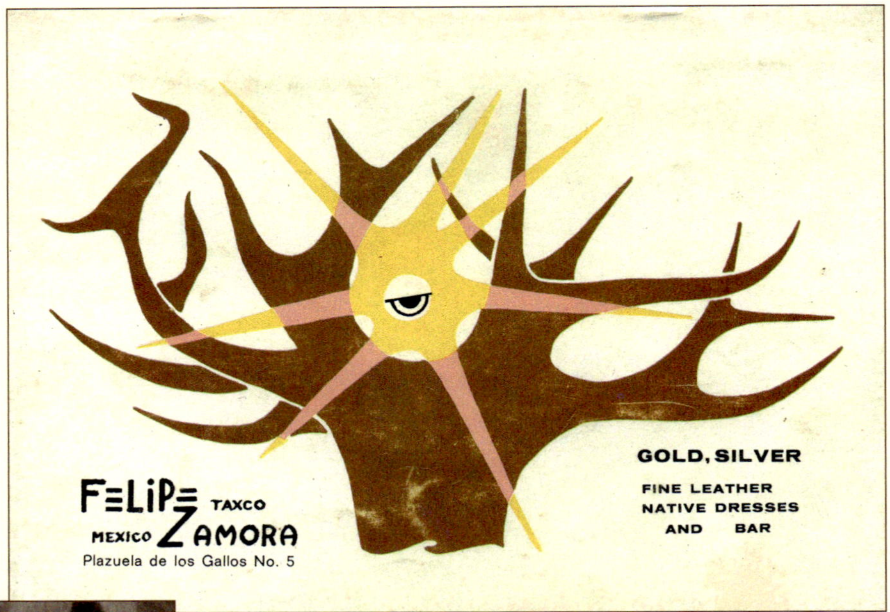

Plate VII-28 Advertisement for Felipe Zamora's store. Private Collection

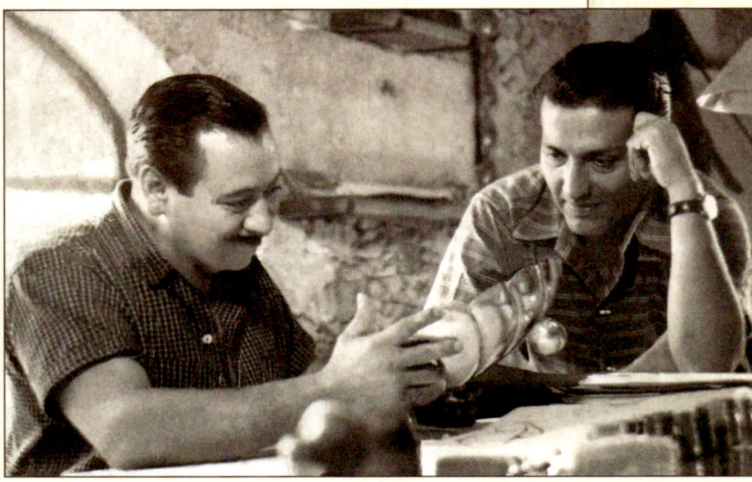

Plate VII-29 Chato Castillo with Felipe Zamora. photograph c. 1950. Zamora Carbajál Family Collection

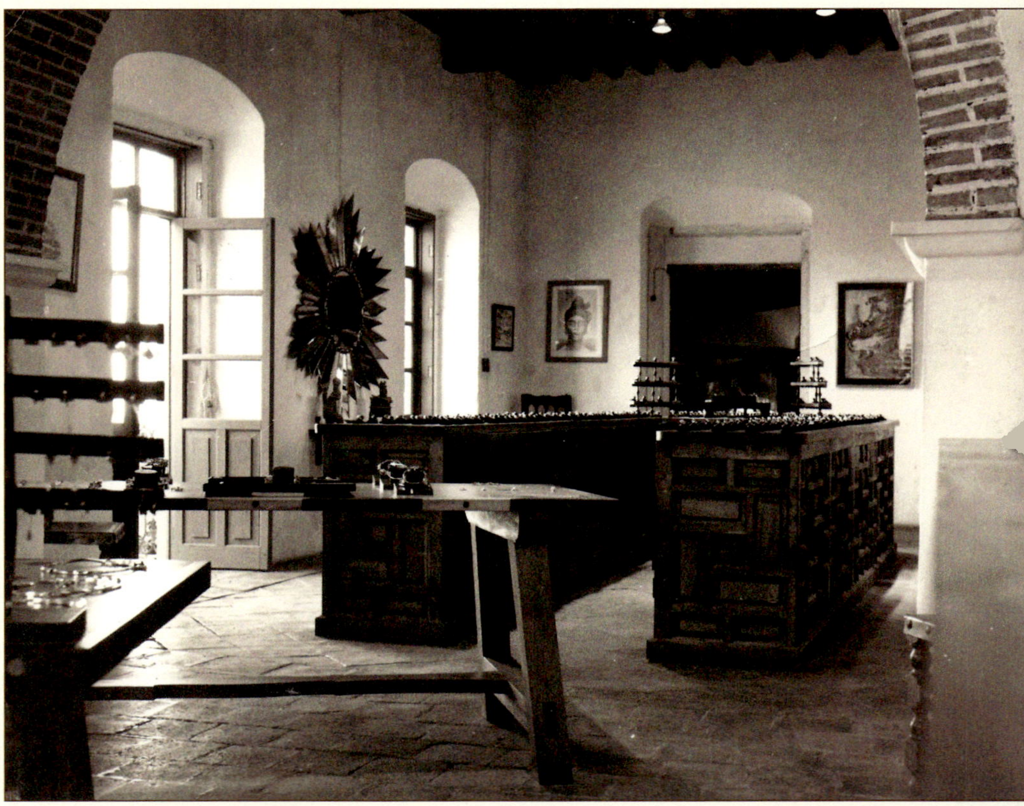

Plate VII-30 Felipe Zamora's Store, interior. Photograph c. 1955. Zamora Carbajál Family Collection

from the Mexican silver designers and returned to their European sources. However, Spratling noted in 1955 that the competition had been formidable but had produced the positive effect of "forcing improvement in design and quality and eventually making possible the greater prestige enjoyed in the world market by Taxco today. Among the workshops that have survived and have achieved their individual characteristics of design as original work recognized not only in Mexico but abroad, are Los Castillo, Taller Borda, Piedra y Plata, Margot de Taxco, Antonio Pineda, Enrique Ledesma, and the writer, who founded Las Delicias in 1931. . . . These are the principal initiators and creative producers who are now well established."[147]

In 1953, a fire broke out in the workshop. One of the silversmiths was using kerosene when the container tipped over and a blowtorch touched off a blaze. Because the space was restrictive and cramped, the fire caused the tragic deaths of two enamelists, María Elena Flores and Alicia. An eyewitness recalls the organization by the community of a bucket brigade that finally put out the fire. [148] The owner of the building refused to renew the lease and Margot moved out of the center of town onto the highway, where she established a new store and workshop. In 1960, a second fire took place

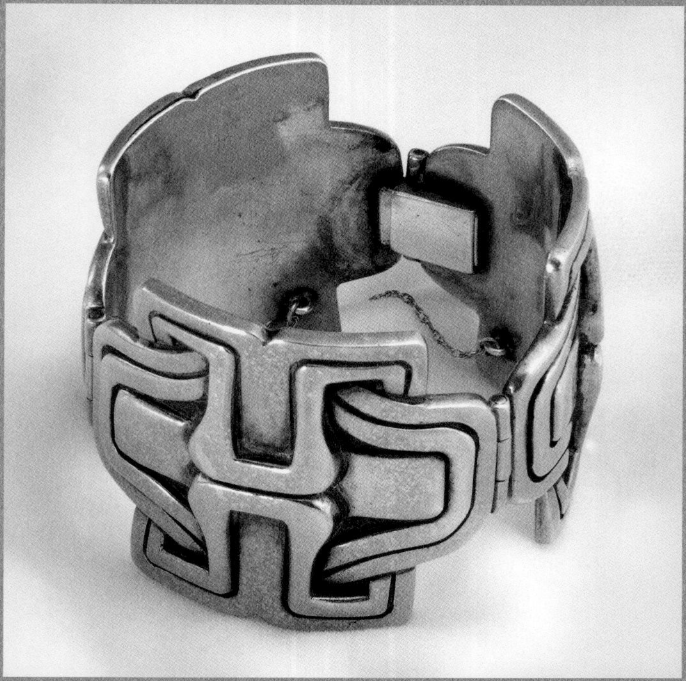

Plate VII-31 Margot de Taxco 5415. *Pre-Columbian Design* in silver cuff bracelet 6.5 x 2 in. Private Collection

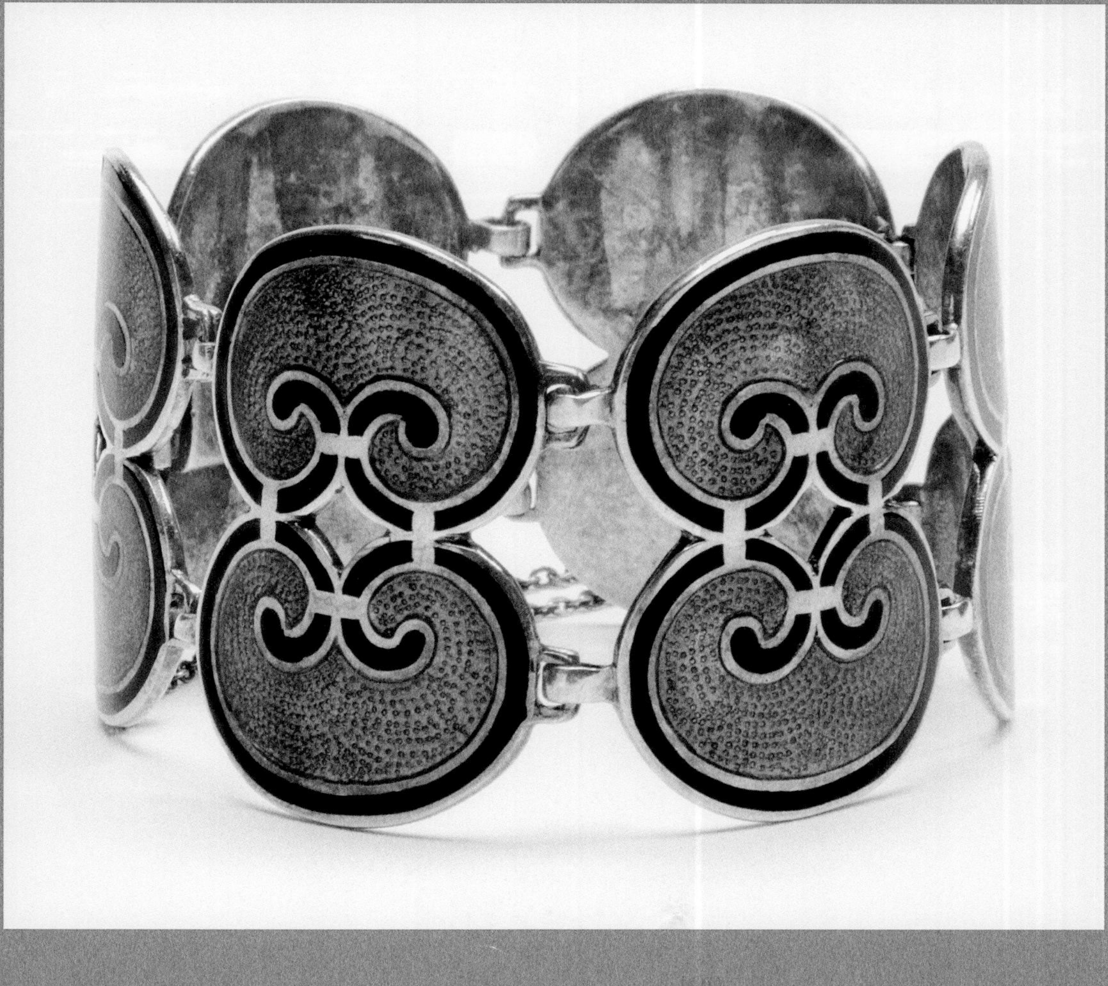

Plate VII-32 Margot de Taxco 5627. *Curled in Oval* bracelet 6.5 x 15 in. J. P. Mather Collection

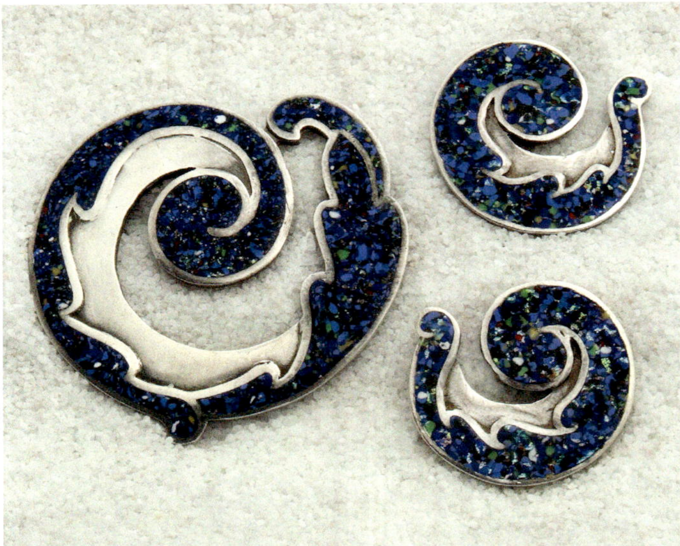

Plate VII-33 Margot de Taxco 5384. *Floral Spiral Set* in blue. brooch 1.5 x 1.38 in.; earrings 0.88 x 0.88 in. J. P. Mather Collection

in Margot's workshop when a still-lighted cigarette was tossed into a trashcan. Added to the damage was the level of theft that took place as the fire burned.

The photograph of Margot sitting in front of a display of her jewelry was taken in the store on the plaza (frontispiece). The tables were circles, so that the salesperson could stand in the center, which is where Margot appears to be standing in the photograph. This same format continued in the large store on the highway. The store was at street level, with the workshops below in the two floors that hugged the mountainside, the enameling on the second, and the silver workshop on the third or bottom floor. When the tourist buses arrived, they stopped in front of Margot de Taxco and the visitors poured out of the buses and into the shop. The owner of the building took notice and raised the rent to the point that Margot was forced to move again. Her last store was across from Rancho Alegre, owned by Pedro Pérez, who paid to have the

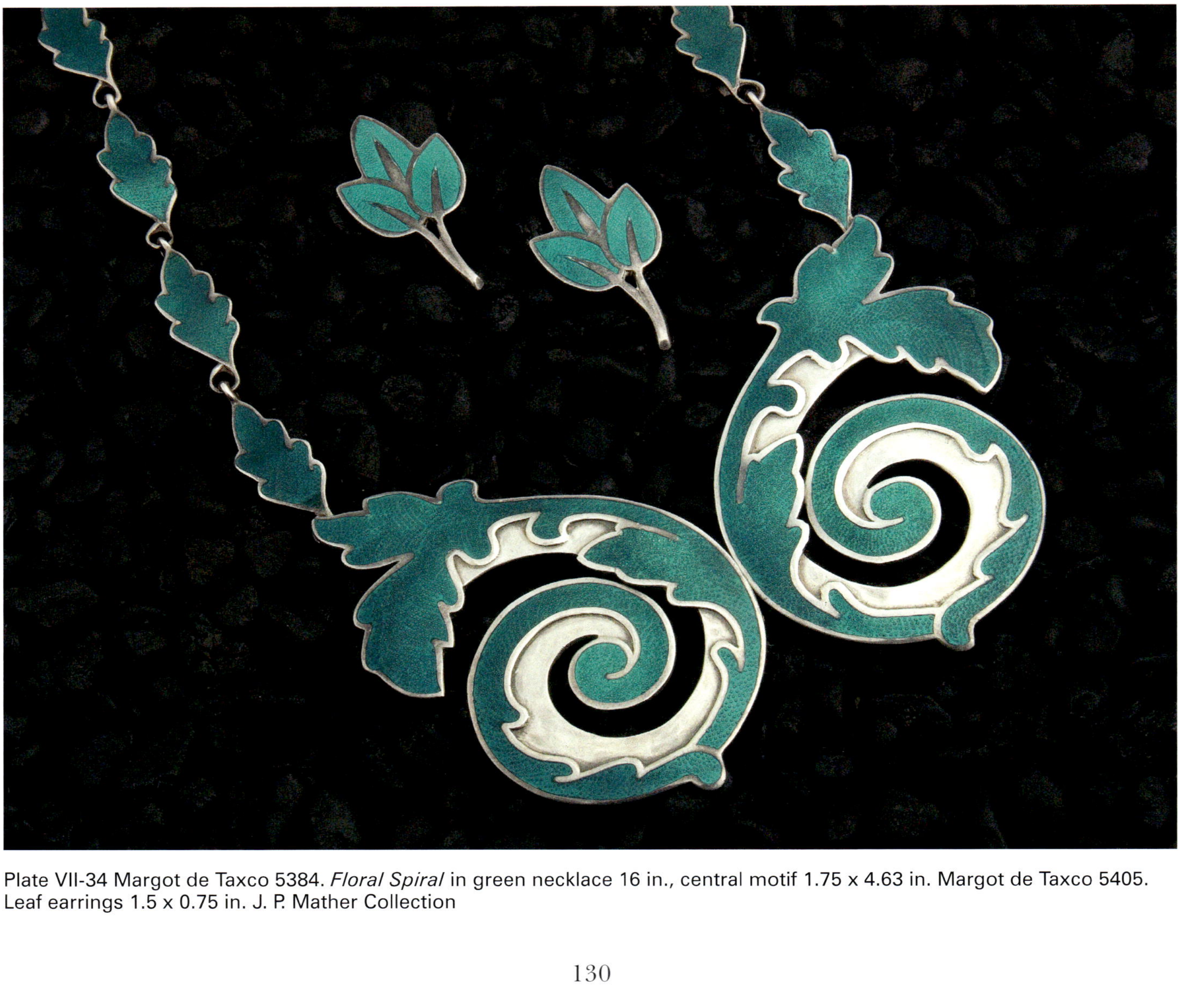

Plate VII-34 Margot de Taxco 5384. *Floral Spiral* in green necklace 16 in., central motif 1.75 x 4.63 in. Margot de Taxco 5405. Leaf earrings 1.5 x 0.75 in. J. P. Mather Collection

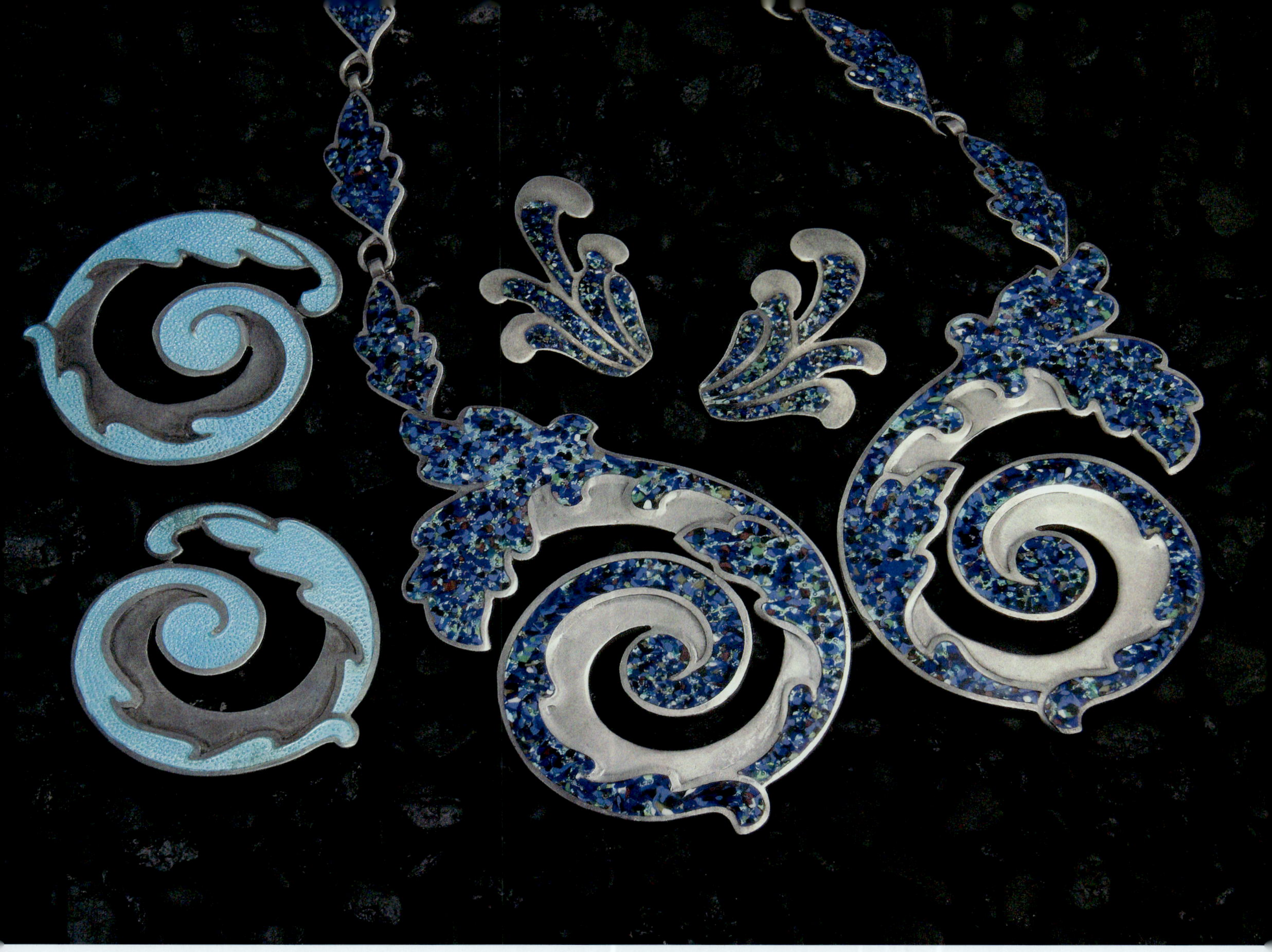

Plate VII-35 Margot de Taxco 5384. *Floral Spiral Set* in blue and mélange enamel. necklace 15 in., central motif 1.75 x 4.63 in.; earrings 1.25 x 0.75 in. Margot de Taxco 5382. earrings 1.25 x 0.75 in. Private Collection

buses stop in front of his store. Nonetheless, the tourists continued to stream into Margot de Taxco. Because the space was much reduced, Margot had the tables cut in half and joined so that each faced the opposite direction from the next (Plate VIII-26).

Betty Hays, owner of the Phoenix in Washington, D.C., recalls a glittering display of enamel jewelry. The store, in each of its manifestations, was filled with color. Margot, she says, was elegant and delicate, a dramatic looking woman, always perfectly coifed. When she reached her seventies, Margot allowed her hair to go grey and sprayed it with a silver color. Jaime Quiróz used the words, "muñequita de pastel," [a confection], to describe her.[149]

Jaime Quiróz joined Margot de Taxco in 1961 as a salesman in the store, and became director of sales in 1964. Sanborns carried Margot de Taxco's enamel-on-silver in Mexico City, as did Cecilia Ríos in her stores in the Hotels Camino Reál, María Isabel, and Fiesta Palace. Sanborns had stores in Acapulco and Monter-

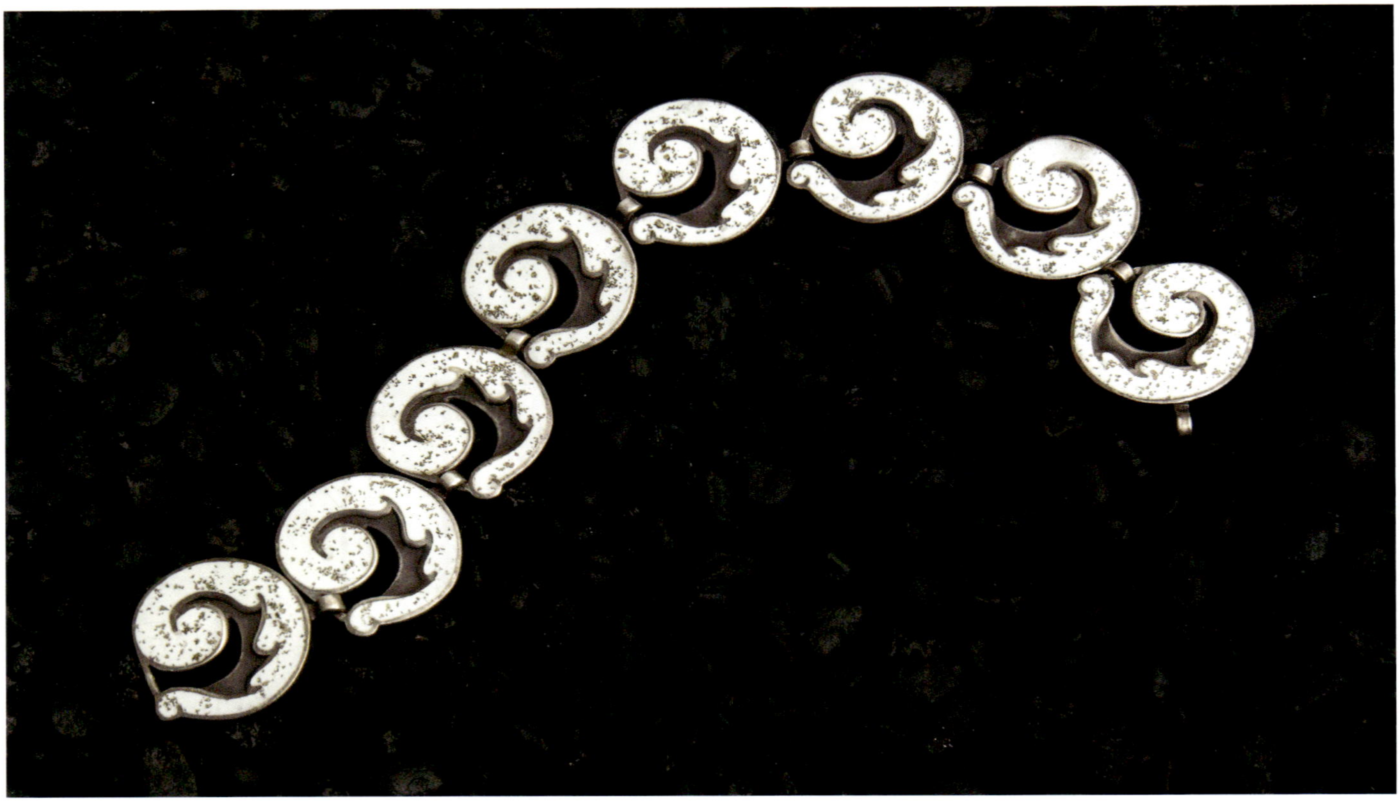

Plate VII-36 Margot de Taxco 5384. *Floral Spiral* in white bracelet 7 x 1.25 in. Private Collection

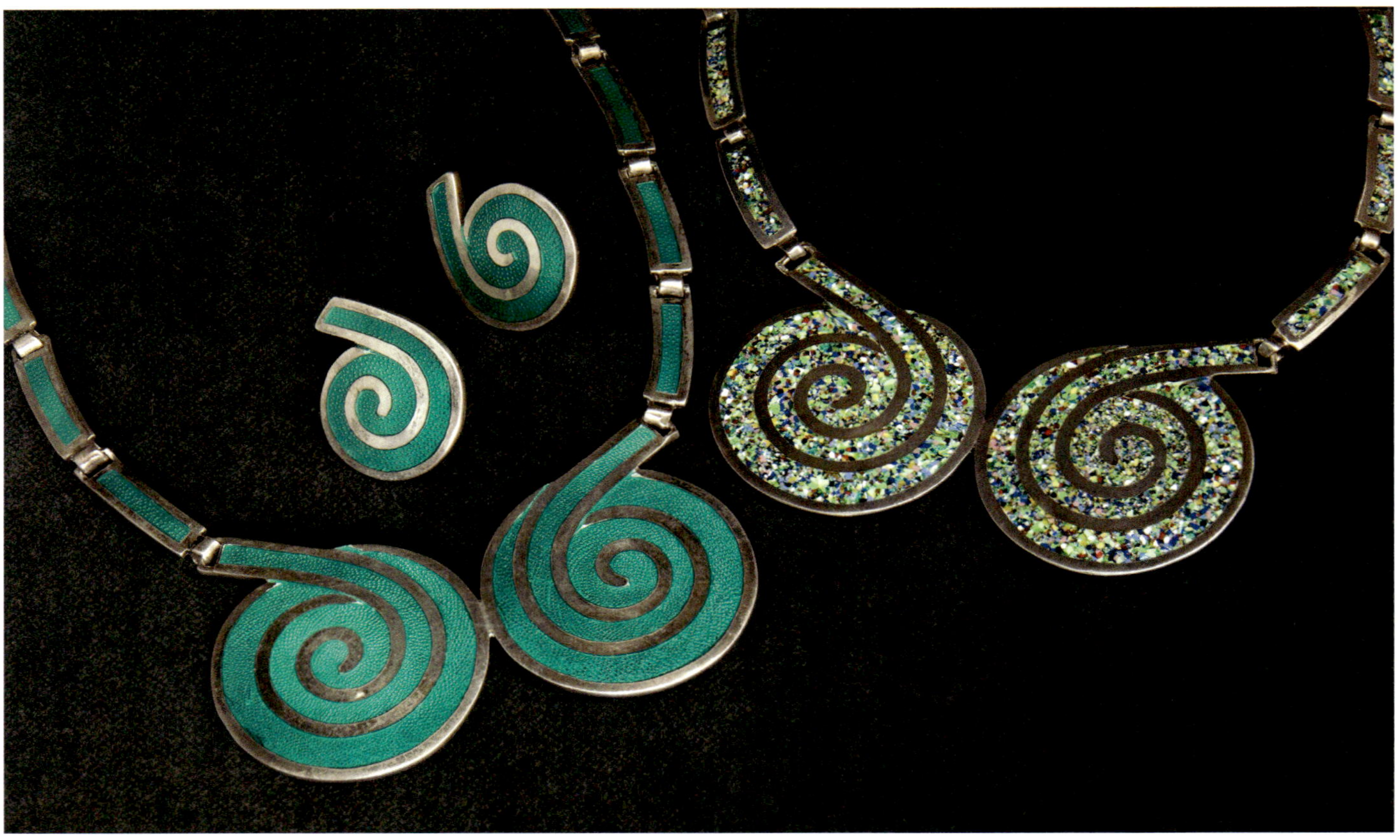

Plate VII-37 Margot de Taxco 5357. *Circular Spiral* in green and mélange. necklaces 15 in., central motif 1.75 x 2.75 in.; earrings 1 x 0.75. J. P. Mather and Private Collections

rey as well as the one in Mexico City, and Quiróz would visit all three stores once a month to check inventory and take orders. Quiróz also traveled in Mexico to San Miguél Allende, Puerto Vallarta, Guadalajara, Hermosillo, Nogales, Tijuana, and Donald Graham's store in Mazatlán.

In the United States, Margot attempted to keep a single outlet in each large city: Fred Leighton in New York City, Ghirardelli Square in San Francisco, the Hays family at the Phoenix in Washington, D.C., Lucy Ledman in Chicago, and Carlos Díaz in Tucson. Quiróz often traveled with Margot after 1964. They would take a sample line with them and return to Taxco with orders. The pieces were then shipped to the outlets. All business was transacted through the Margot de Taxco store.

On these selling trips, Quiróz took with him the two lines, in silver and enamel on silver, to show prospective clients. Each outlet had a particular preference, so inventory was always changing and increasing in diversity. Most of Margot's designs were sold as sets. One specialty that was particularly successful was the array of unique large-scale repoussé cuff bracelets (5192, Plate VI-21). These would have been the work of Juan González. Favored also were the zodiac pendant charms (5220-5232, Pls. VIII-23-24) and the serpent sets (5554, back cover). For Margot, these were years of great success, fueling ever greater variety and innovative design.

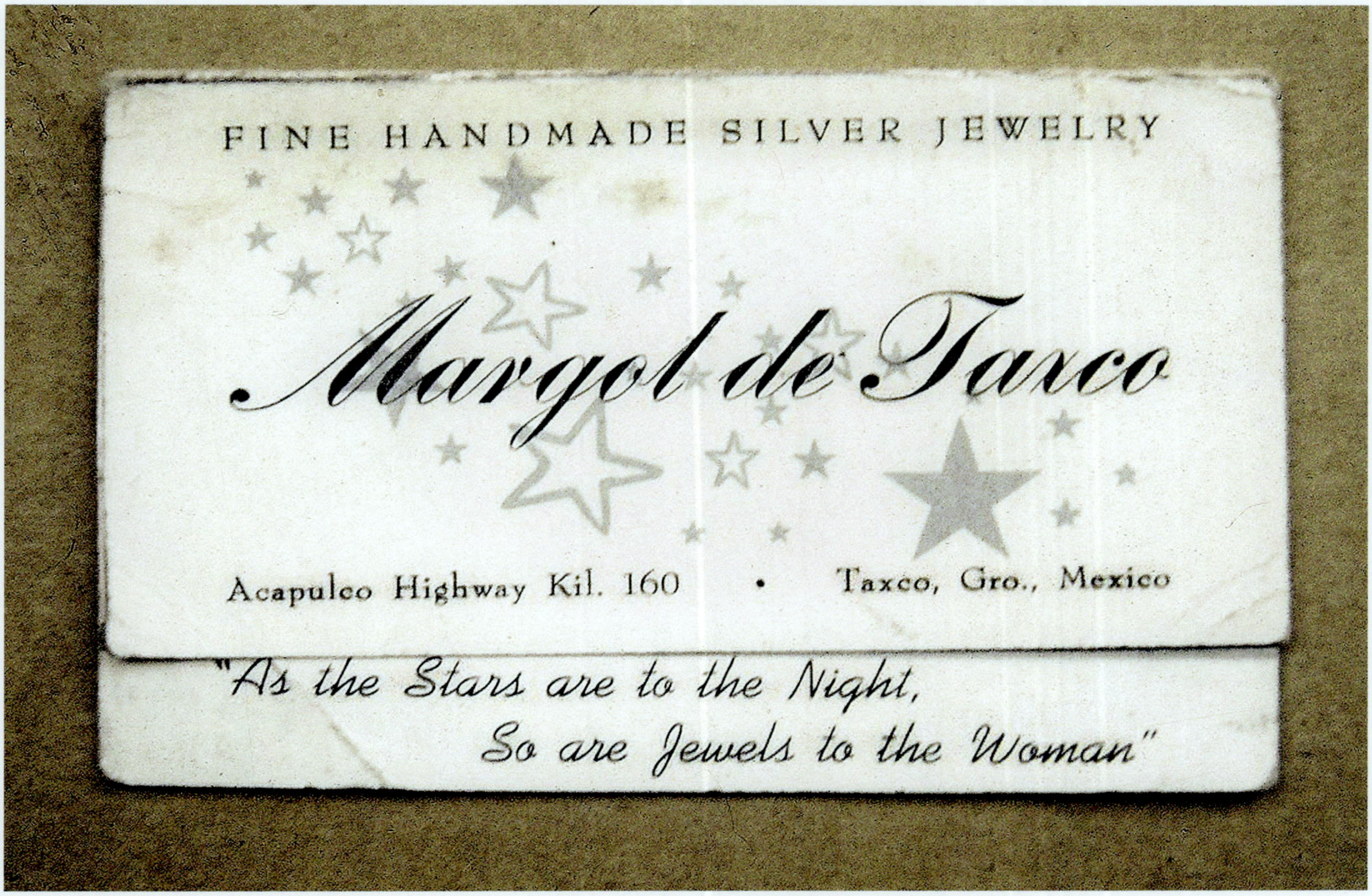

Plate VII-38 Margot de Taxco business card. Collection Jaime Quiróz Duque

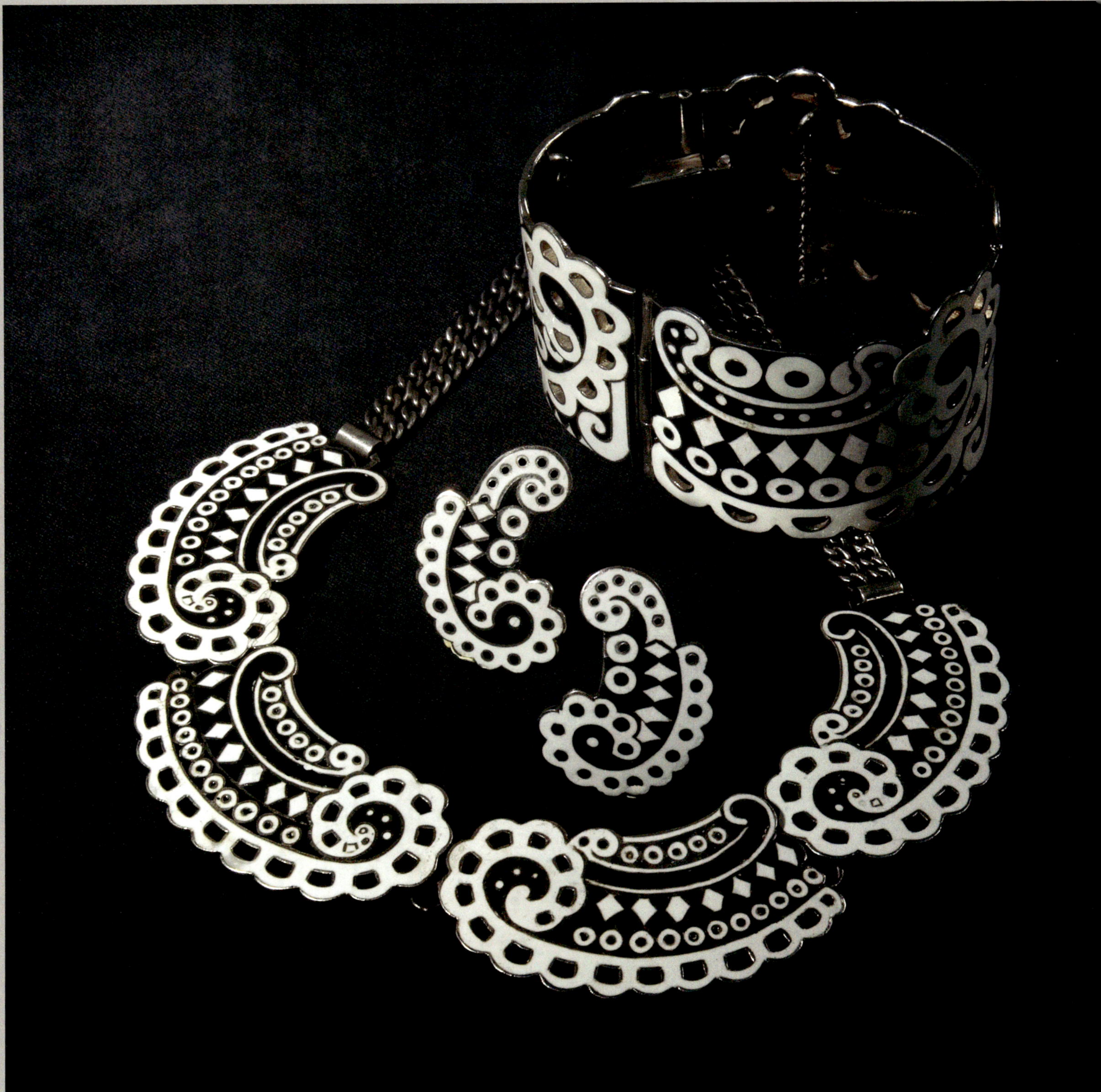

Plate VIII-1 Margot de Taxco 5552. *Diamonds and Dots Set.* necklace 14 x 1.25 in.; cuff bracelet 8 x 1.75 in.; earrings 1.38 x 1.125 in. Private Collection

Chapter VIII
Margot de Taxco Style

"Margot designed in circles," observed one of her associates.[150] The limitless energy of the spiral dominates Margot's designs, whether she was representing birds and flowers or pure form. Margot's approach towards her subjects could have been intuitive, an innate appreciation for the elegant simplicity of the spiral. On the other hand, the subtle complexity of her designs indicates another possibility, that, at some

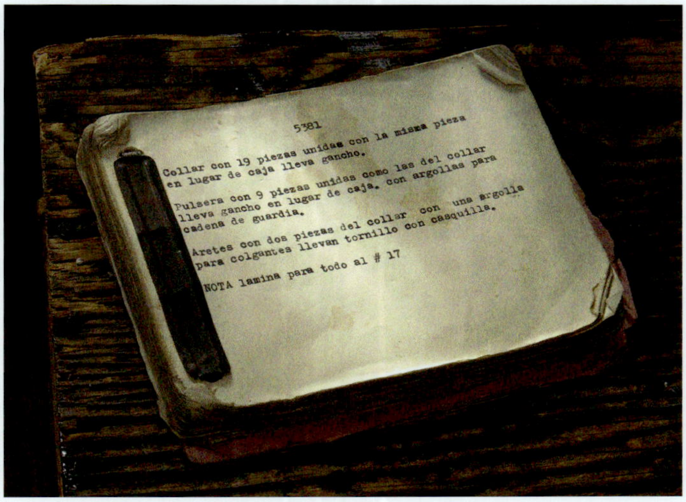

Plate VIII-2 Margot's Instruction Books. Tulane University Latin American Library, Spratling-Taxco Collection

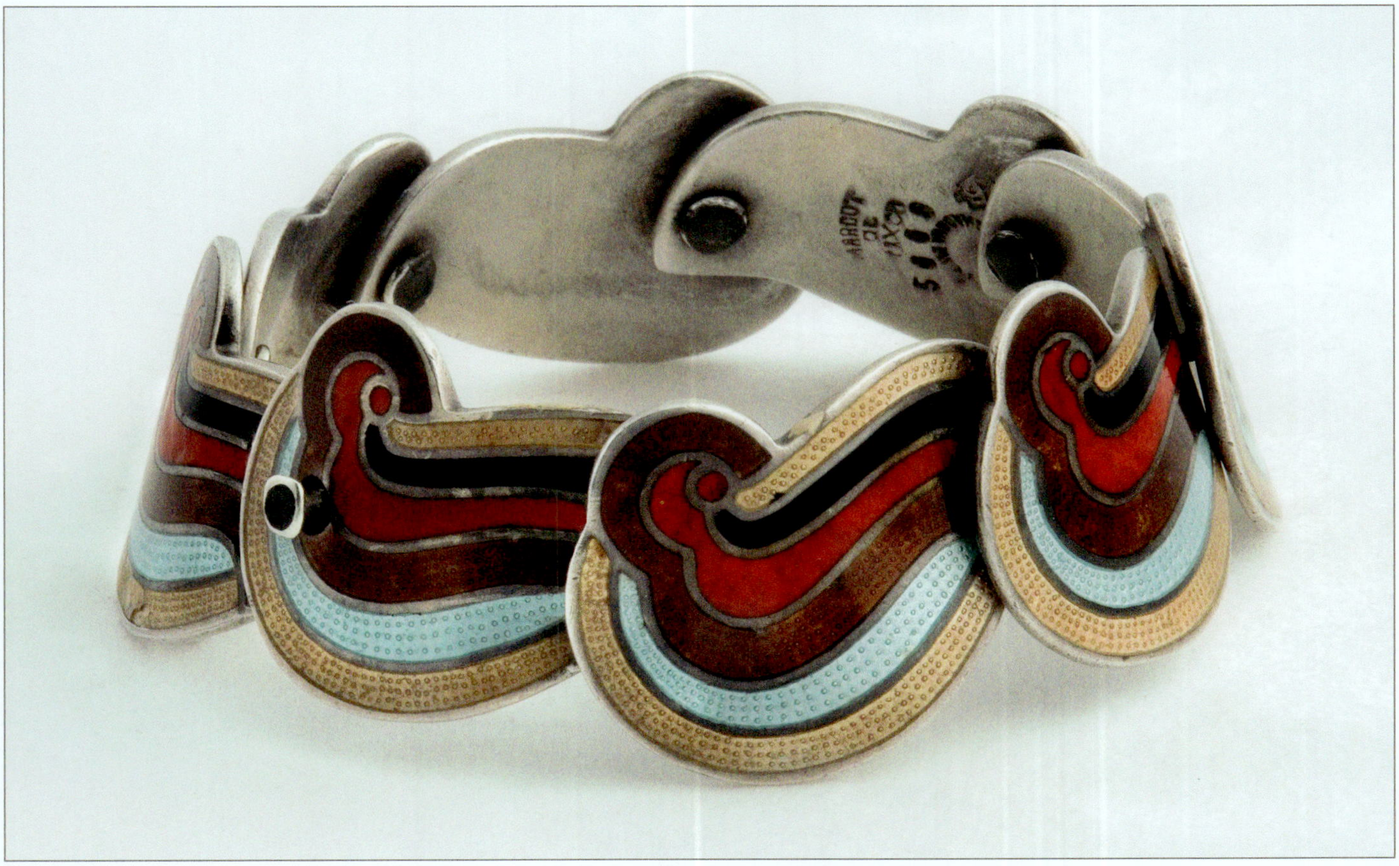

Plate VIII-3 Margot de Taxco 5668. *Multi-Color Curl* bracelet 7.5 x 1 in. Elizabeth Finch Collection

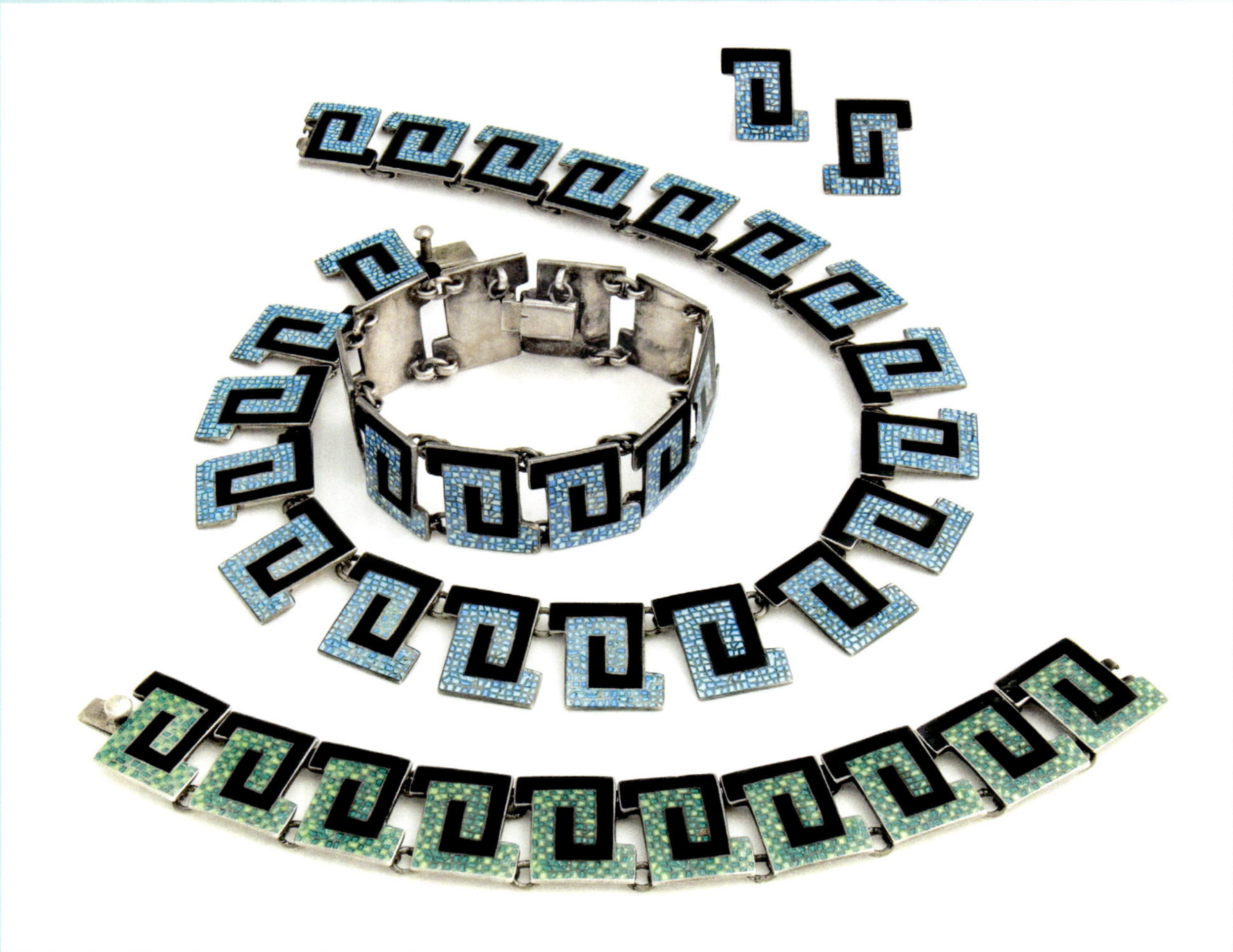

Plate VIII-4 Margot de Taxco 5442. *Greek Key Set.* necklace 14.5 in.; bracelet 7.25 x 1 in.; earrings 1 x 0.75 in. Private Collection

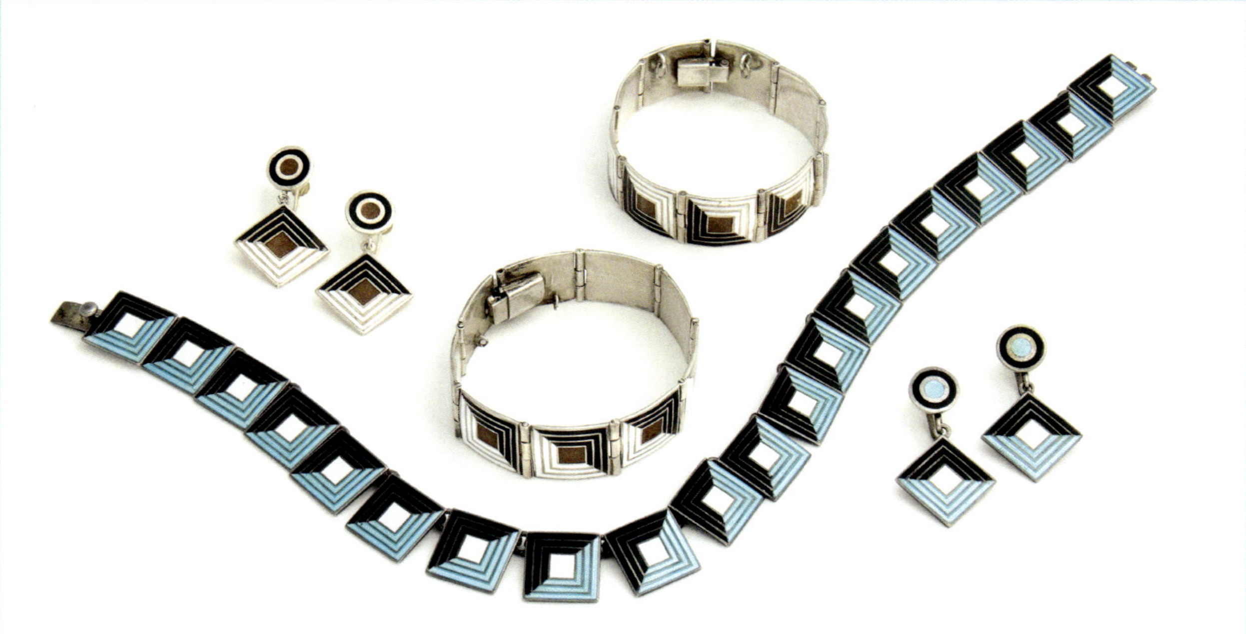

Plate VIII-5 Margot de Taxco 5461. *Shadow Box Set*. necklace 14 in.; bracelet 7.25 x 0.5 in.; earrings 1.5 x 0.5 in. Private Collection

point in her training as an artist, Margot encountered the golden mean. Also known as the divine proportion or the Fibonacci sequence, the golden mean is an irrational number, in that it produces an infinite set of numbers. The Greeks considered this sequence, especially as it was manifested in the golden rectangle, as an aesthetic form so close to perfection, that it was the desired architectural and artistic measure.

To begin the sequence, the golden rectangle is divided into a square and a smaller golden rectangle. In this construct, the diagonal of the original rectangle becomes perpendicular to the diagonal of the smaller rectangle. One mathematician has written, "Connecting vertices of this progression of golden rectangles with suitably curved lines will yield a spiral which converges at the intersection of the two diagonals. This same spiral can be constructed by connecting the acute vertices of the golden triangles in a progression of inflated triangles. This particular spiral is referred to as the equiangular or logarithmic spiral. The logarithmic spiral plays an important role in the growth and structure of certain biological systems." [151]

These Fibonacci spirals often occur in nature, in pinecones, sunflowers, and, most especially, in the shell of the chambered nautilus. The concept of perfection in nature is nowhere more evident than in this shell: "As the animal grows, it constructs larger and larger chambers in the form of a spiral, sealing off the smaller unused chambers. . . . The relative volumes of consecutive chambers are related to the golden ratio."[152] Herein lies the essential character of the golden mean that has captured the imaginations of architects, musicians, and artists for centuries.

Two cuff bracelets from Margot's years at Los Castillo and then at Margot de Taxco exhibit the spiral's infinite potential for movement (LosC154, 5192, Plate VI-21). The representation of the shell in silver must have been, for Margot, an ongoing exploration of the spiral's dynamics as a manifestation of the Fibonacci sequence. Margot achieved visual and tactile power in the two shell cuff bracelets when she combined heavily oxidized curling lines on the surface with the voluptuousness of *repoussage*.

In a very different manifestation of the spiral, Margot found inspiration for the more two-dimensional enamel shell set, number 5611 (Plate 25), in images from folios 75 and 80 of the pre-conquest Mixtec Codex Nuttall (Plate 24). Like many of the artists working in Mexico at the time, Margot was drawn to the pre-conquest aesthetic for its abstraction of natural form. This two-dimensional patterning has mathematical implications, and undoubtedly allowed ancient and modern artists to achieve balance with continuous movement.

In Margot's jewelry, plant forms swirl in overlapping color and line, as in numbers 5768, 5384, and

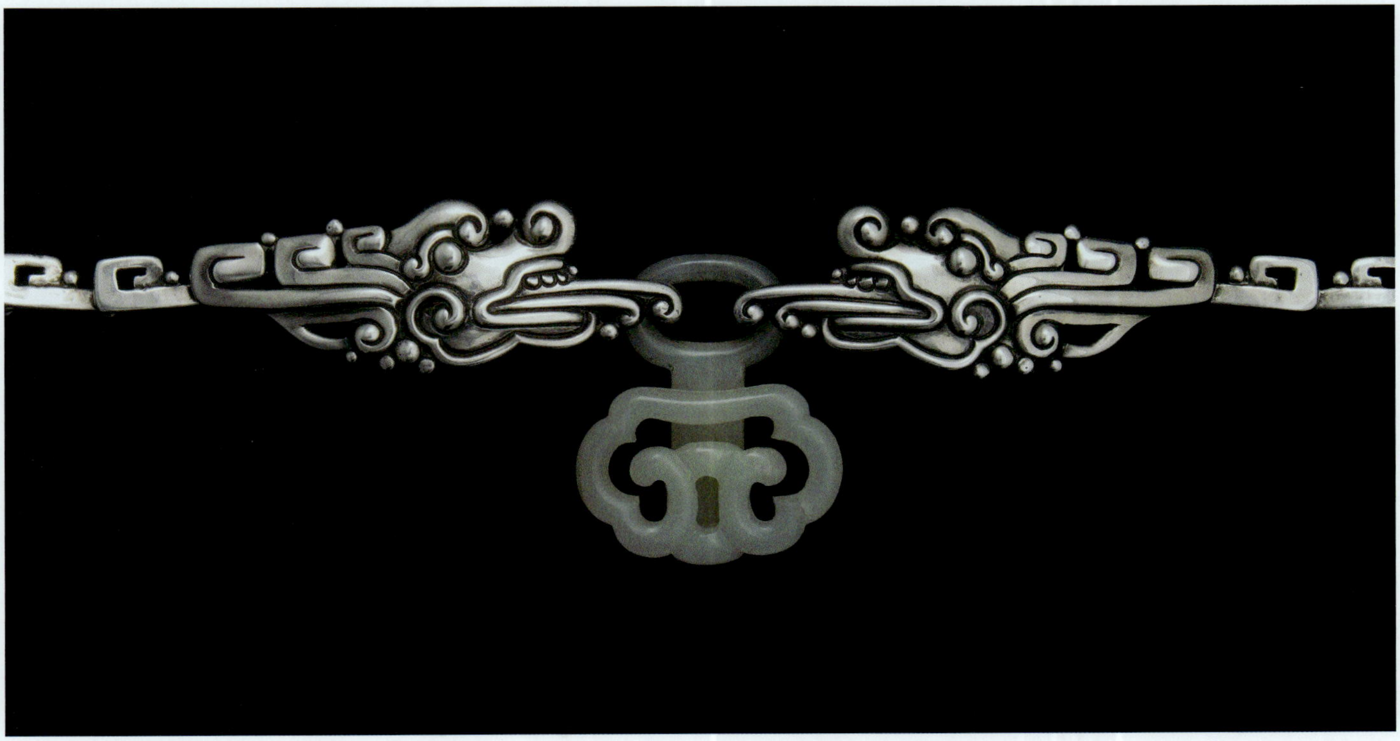

Plate VIII-6 Margot de Taxco 5243A. *Silver Dragons* necklace with Chinese Jade pendant 20 in., jade 2 x 1 5/6 in.15, central motif 1.75 x 2 in. Private Collection

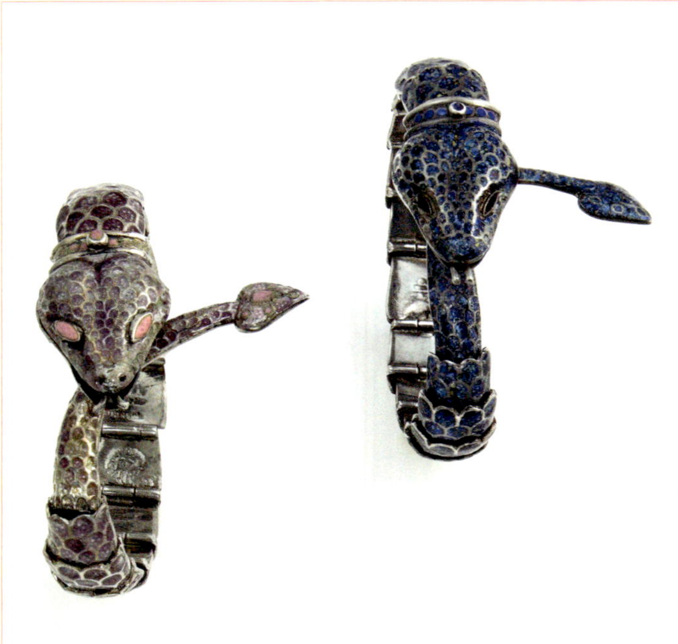

Plate VIII-7 Margot de Taxco 5554. *Serpents* in lavender and dark blue, bracelets 7.5 in. d. Private Collection

5559 (Plates XI-4, VII-33-36, 28). The use of color to emphasize and differentiate the spiraling forms in number 5907 (Plate VII-17) makes this pendant/brooch one of Margot's most successful pieces of jewelry. Even more important is the tension created, as the semi-circles in the central flower move clockwise and the flaming forms of the outer circle move flamboyantly counter-clockwise to strike the balance.

Both the quetzal bird (5755, Acknowledgments) and butterfly (5750, Plate VIII-27) are elegant adaptations of the spiral in nature. Combined with an emphasis on line, the expression of the qualities of the butterfly and bird becomes the central challenge. At the same time, Margot used color to enhance these lively creatures and she did the same with the pure form of the spiral. So many of her designs are focused on the expression of movement, which can be characterized as both continuous and in balance (LosC619, 5357, 5546, Plates VI-9, VII-37, 33). In one particularly successful example, Margot took the mirror image of one spiral form (5337, Plates III-3, III-5-6) and created three entirely different designs, one of which is a butterfly. Needless to say, the motions of unfolding and curling inward in Margot's necklaces and bracelets are amplified when a woman turns her head or gestures with her hand.

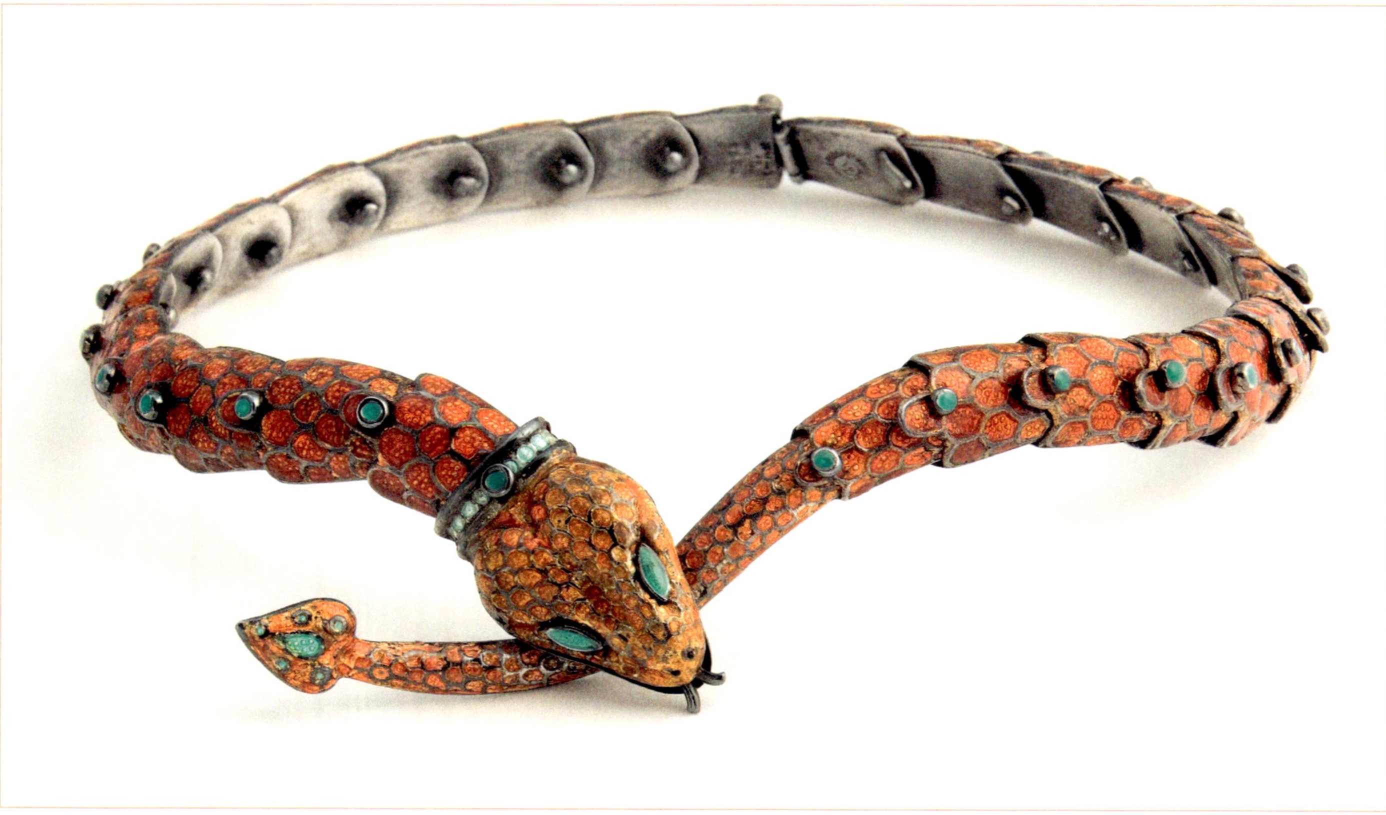

Plate VIII-8 Margot de Taxco 5554. *Serpent* in orange with turquoise, necklace 16 x 0.63 in. Private Collection

Pre-Columbian serpents and Chinese dragons prevail in Margot's work. The interconnectedness of these two themes is evident, even though they come from two different parts of the world. The Quetzalcoatl pendant (5753, Plate VII-9), with its spiral of swirling feathers, and the necklace with a Chinese jade pendant (5243A, Plate VIII-6) held by opposing dragons represent mythic creatures, both replicating the sinuous path taken by water and curling green vines. The jade pendant, admired on its own for its preciousness, is a replica in miniature of a Chinese metal belt hook, carved from a single piece of jade.[153] Among Mesoamericans before the conquest, carved jade was equally prized.

Margot's enamel serpent *parures* continue to be admired and avidly collected. According to her instruction book, now in the Spratling-Taxco Collection at Tulane University's Latin American Library (Plate VIII-2), the serpent necklace was made up of the head and tail, with twenty-five additional parts for the body. As in the early days at Los Castillo, silver pins or rivets were used to hold the pieces together, which allowed for the flexibility of the necklace. The mouth of the serpent was made to fit over the tail in both the brace-

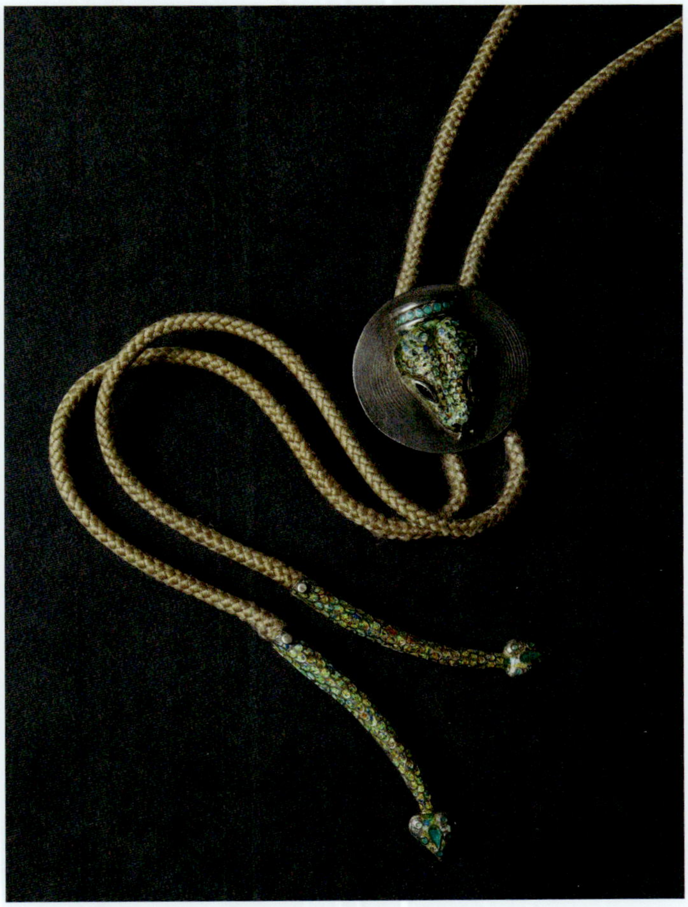

Plate VIII-9 Margot de Taxco 5554. *Serpent* bolo 40 x 1.5 in. d. Katherine Armitre Collection

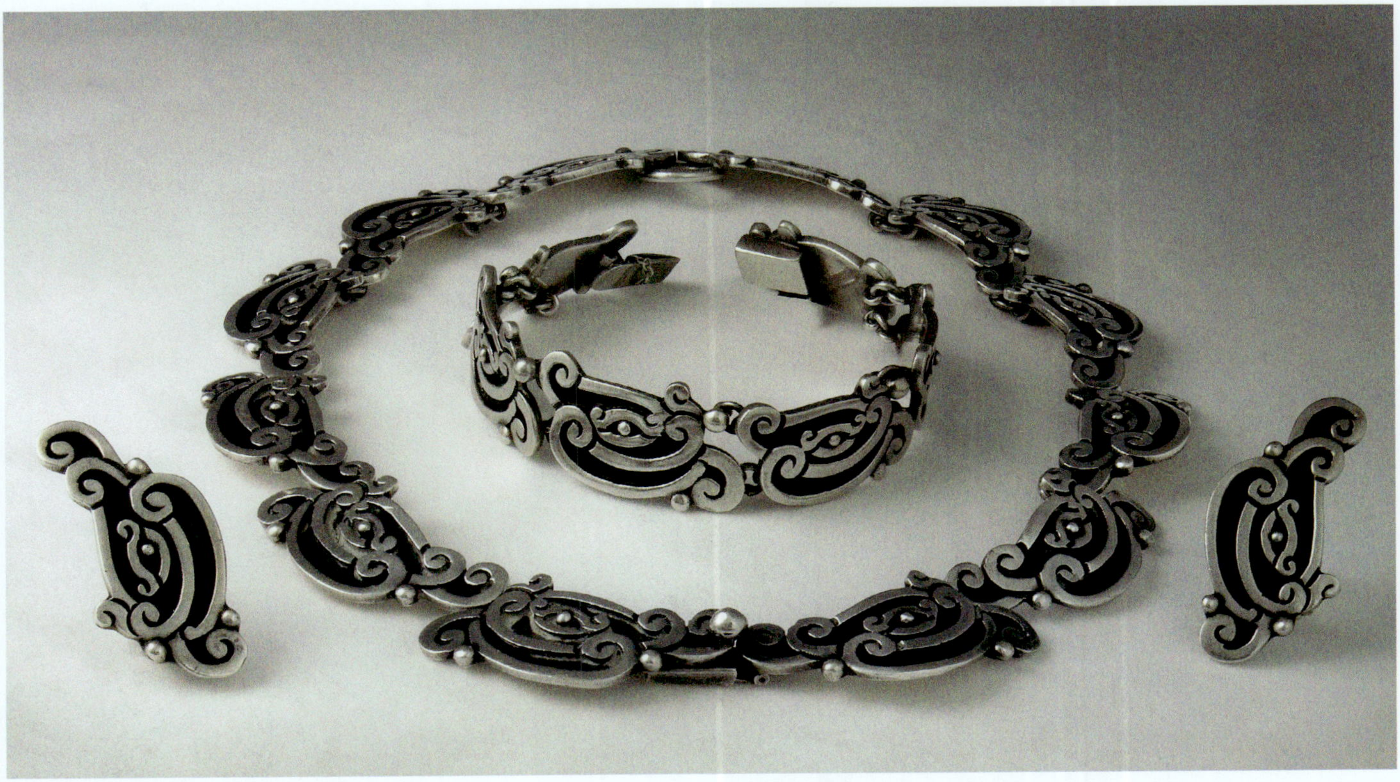

Plate VIII-10 Margot de Taxco 5109. *Linear Serpent Set.* necklace 14.5 x 0.75 in.; bracelet 7.5 x 0.75; earrings 1.63 x 0.75 in. Katherine Armitre Collection

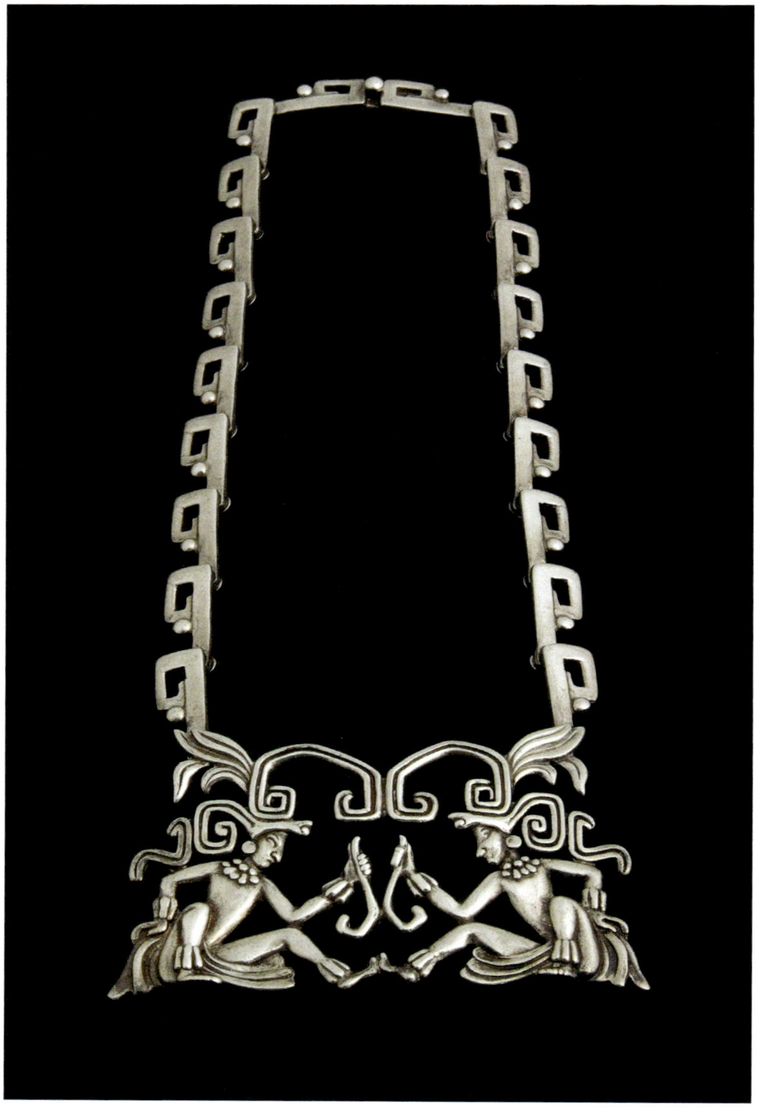

Plate VIII-11 Margot de Taxco 5605. *Maya Lords* necklace 16.25, central motif 2 x 3.13 in. Orlando Rigono Collection

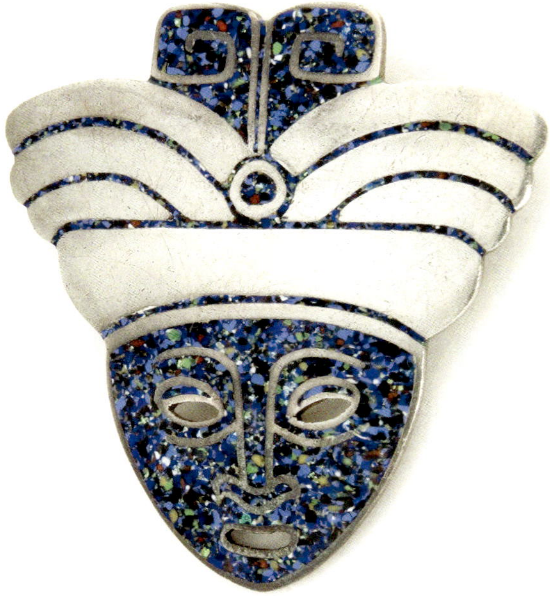

Plate VIII-12 Margot de Taxco 5403. *Mask* brooch in mélange 2 x 2 in. Private Collection

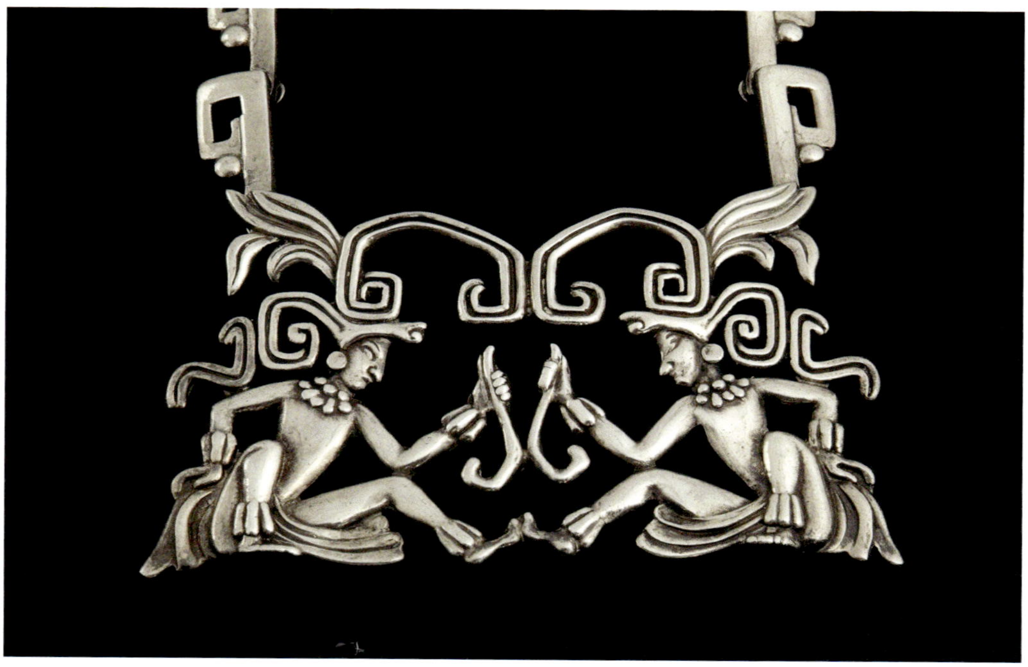

Detail of 5605, *Maya Lords* necklace.

let (of fourteen parts) and the necklace, with a single rivet holding the two together (5554, Plates VIII-7-8). All parts of the necklace were provided with a backing in silver except for the head, which was completely enameled. The closure at the back of the neck for the necklace is a disguised box clasp.

The most spectacular interpretation of the dragon motif is a set that includes a necklace, cuff bracelet, earrings, and ring (5255, Plate 11). The stone is potch blue opal.[154] The large carved pendant could be a representation of Buddha, with serene face and long earlobes. However, in traditional iconography, Buddha's eyes would have been half-closed and he would not have been wearing earrings. The opal, which could have originated in Mexico, was probably carved by a lapidary in Taxco who took a few artistic liberties. The stone's setting is equally exotic. The dragons compare to those in the necklace with the jade pendant (5243A, Plate VIII-6) and to the enamel design drawings for 5798 and 5905. These dragons are more like feathered serpents, again a wonderful interpretation of a Chinese dragon from a Mexican point of view. The bracelet, number 5905 (Plate 10), is a beautiful variation on the theme, taken from the drawing of the dragon. Margot must have used a mirror against the serpent scales to create an entirely new design for this stunning bracelet.

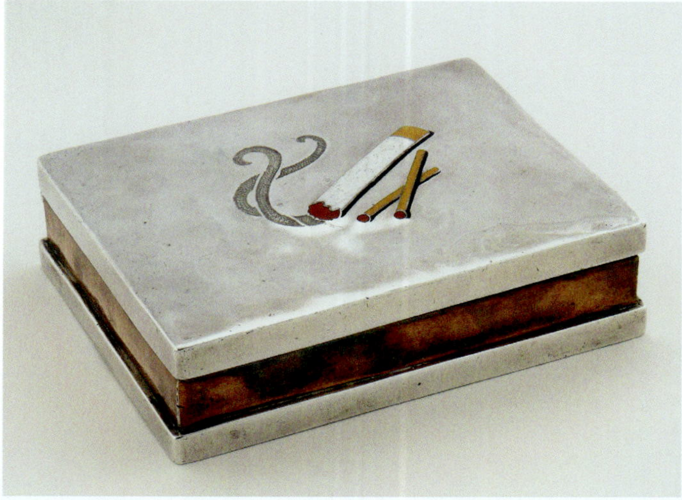

Plate VIII-13 Margot de Taxco 5592. *Lighted* cigarette box 1.38 x 5.25 x 4 in. Carole Berk Collection

When Margot's borrowings from pre-Columbian art were a matter of style or aesthetics, she was at her most successful. Several Margot de Taxco designs hark back to the Los Castillo days. The cuff bracelet

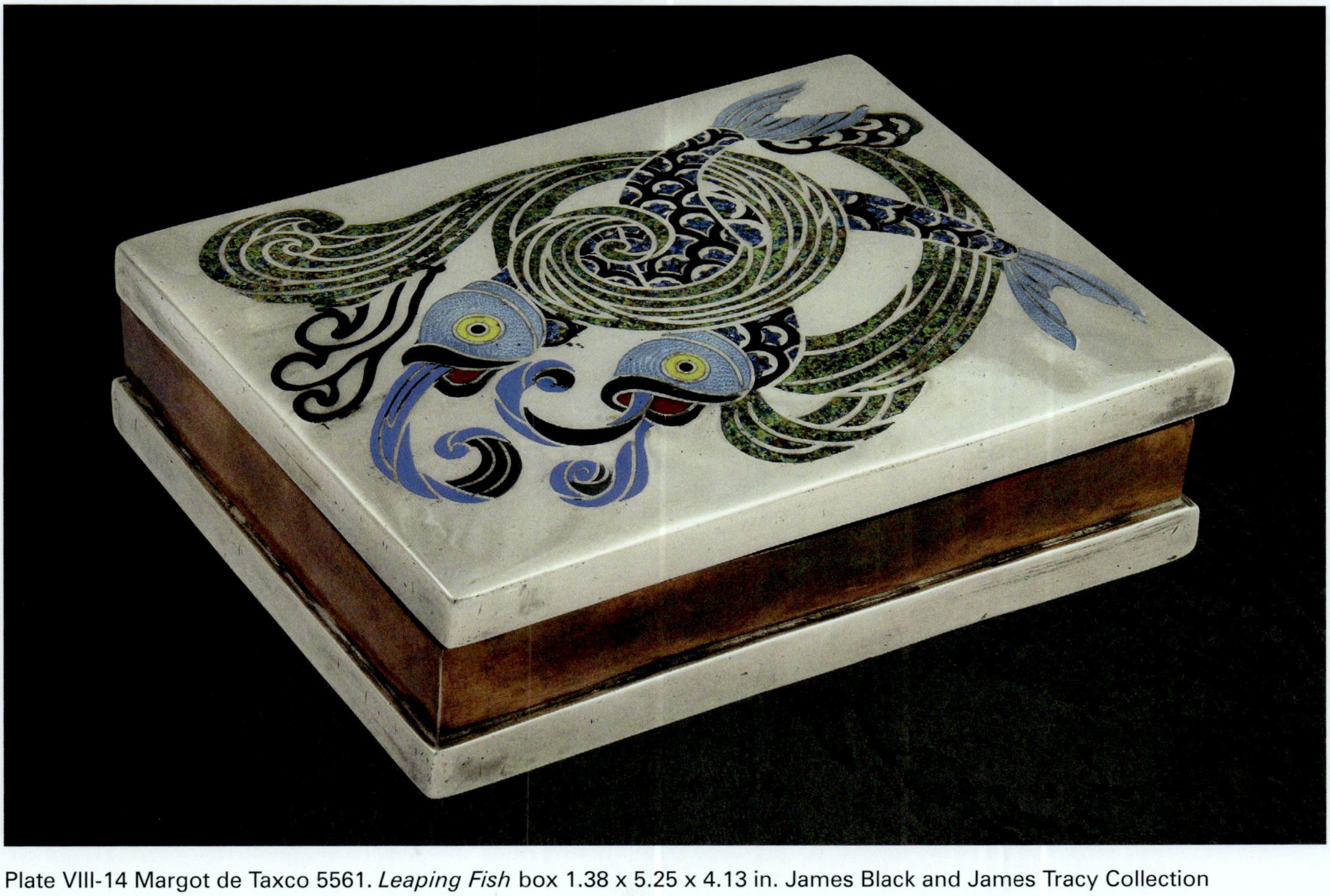

Plate VIII-14 Margot de Taxco 5561. *Leaping Fish* box 1.38 x 5.25 x 4.13 in. James Black and James Tracy Collection

(5415, Plate VII-31), hammered out in *repoussage* by one of the *maestros* in the workshop, is large-scale with opposing elements that imply three-dimensional abstract forms.

Most of the jewelry inspired by pre-Columbian art was figural and suggests a specific source. The facing Maya figures in a silver necklace by Margot (5605, Plate VIII-11) may have been based on a carved limestone throne from the Maya Usumacinta River region. This large-scale throne back was owned by Josué Sáenz (now in the Museo Amparo in Puebla), and Margot could have seen it published.[155] The enamel pendant with the profile of the Maya noble (5404, Plate VII-10) is in the style of Palenque and Yaxchilán, for which Margot had a particular affinity. However, Margot occasionally turned to central Mexico for design ideas. The enamel mask brooch (5403, Plate VIII-12) is symmetrical and formally abstract, resembling the stone masks of Teotihuacan.

The Eagle Knight box (back cover) is high drama. The warrior's rapid forward movement is enhanced by the exaggerated diagonal of the composition and the sweeping motion of the feather headdress and cape. The polychrome enamel used to depict the feathers on the cape glitters across the surface. One of the most interesting aspects of the design is in a minor detail. The small color squares that make up the interlocking "Greek key" in the shield are asymmetrical. What Margot has achieved with this slight change to a motif that is normally depicted in symmetry is to imply convexity to the shield's surface.

Margot did several figural brooches, bracelets, and necklaces derived from Asian models. These include the geisha (5670), fan dancer (5708), and Thai figure (5806, Plate 34). There was a precedent for these, the parure depicting Thai folk dancers she designed while at Los Castillo (LosC519, Plate II-14). A fascination with Japanese art led Margot to return over and over to several motifs, particularly the koi fish leaping in ocean waves. They appear in massive cuff bracelets in silver or blue enamel or on the cover of a cigarette box (5207, 5644, 5561, Plates I-14-15, VIII-14). This particular design was not new to Margot's oeuvre, for she had actually developed it at Los Castillo (LosC106, Plate VI-20).[156] In one example, she enhanced the bracelet with the addition of glass cabochons or spheres (LosC514, Plate VI-19).

In these special pieces by Margot, whether silver or enamel, the fish cavorting in the waves and worn round the neck or wrist impart a sense of joy and freedom. The curling waves, in the abstract, intersect with the curved forms of the fish to produce alternating lines of movement. The representation of the fish and waves is realistic and three-dimensional and, because it is compressed within a small surface, the effect is dynamic. One of the most beautiful examples of this theme is the silver necklace (and matching cuff bracelet) with tiny amethyst cabochons or "bubbles" coming from the fishes' mouths (5207, Plate I-14). These two faceted fish become brooches, while the necklace stands alone as a series of tiny abstract curling wave forms.

The chrysanthemum, an Asian theme depicted in several ensembles by Margot (5685, Plates 6, 7), has the same quality of expression. The extravagance of petals overlapping and flowing represents both a natural phenomenon and the abstract in the linear pattern that is seemingly in continuous movement. In another design, Margot confined the chrysanthemum to a circular format, producing even greater emphasis on line (5773, Plate I-8).[157]

Other floral designs by Margot are quite delicate in their appearance. During the rainy season in central Mexico, periwinkle blue morning glories appear in the valley south of Taxco. These are gracefully represented in a necklace and bracelet (5771, Ack-3) by Margot as opposing forms. Equally enchanting is a series of designs in which the jewelry is covered in tiny blue flowers, much like they might have appeared as a carpet of blue on a hillside (5600, 5607, Plates 31, 32).

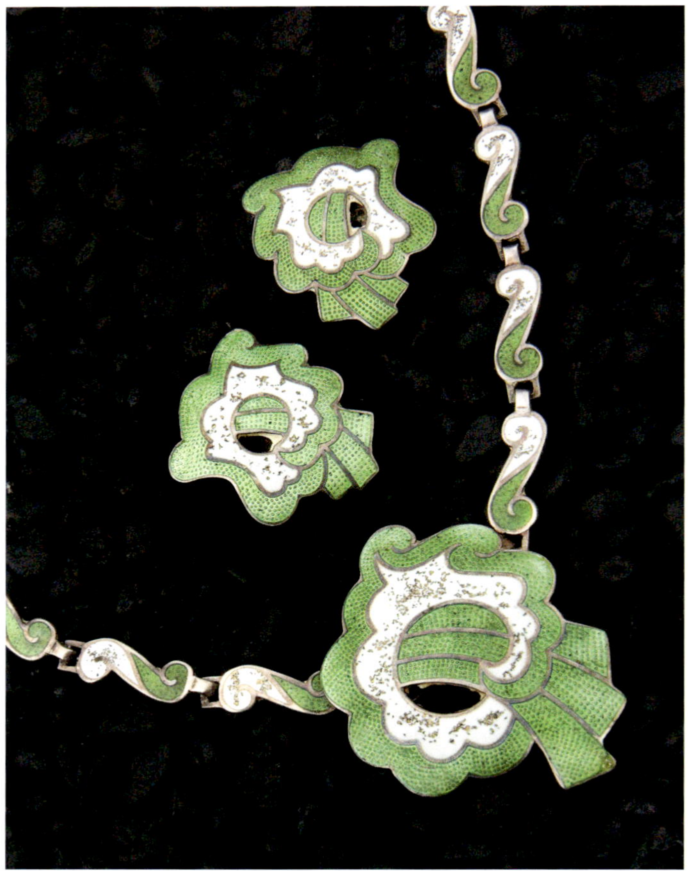

Plate VIII-15 Margot de Taxco 5896. *Floral Circle with Flanges Set*. necklace 16 in., central motif 2 x 1.5 in.; earrings 1.25 x 1 in. Private Collection

In contrast are the psychedelic colors of the flower brooch and earrings in design number 5408 (Plate 23). The gaiety of the composition, with its asymmetry, curling lines, and bright spots of color, makes it unique among Margot's enamel work. These enamel pieces resemble a pair of earrings in silver (Plate 22), making for an interesting comparison between the two different lines Margot produced.

Jaime Quiróz mentioned that Margot always carried a magnifying glass and would stop the car to look at flowers and plants while they were traveling.[158] Margot observed nature closely as an artist would, and found interesting ways to express form and color in her jewelry. Some of the most visually pleasing designs depict leaves in several shades of green (5405, 5906, F-2, F-5). In these works, she varied the level of color saturation and the opacity of the enamel to augment the variety of form and movement. Margot's depictions of leaves in silver are equally as elegant, as can be seen in design numbers 5474 and 5612 (Plates I-11-12).

One of Margot's more sophisticated adaptations of the leaf motif can be found in the openwork enamel necklace, number 5583 (Plates 4-5). The design reveals the wizardry of the silversmiths who worked with Margot. The floral pattern is slightly convex and held in place by a rim of silver that follows the contour of the necklace's central motif. The backing was oxidized to provide a shadow box effect for the floral pattern. The subtle choice of color, symmetry, and flow of line recall Renaissance interpretations of classical motifs. These elements enhance a design that is perfectly adapted to the quarter-moon-shaped format.

Margot took a trip to Italy in the sixties, but the art of ancient Rome (and Greece) was already an important influence in her work.[159] In an early design, Margot interpreted the Greek key

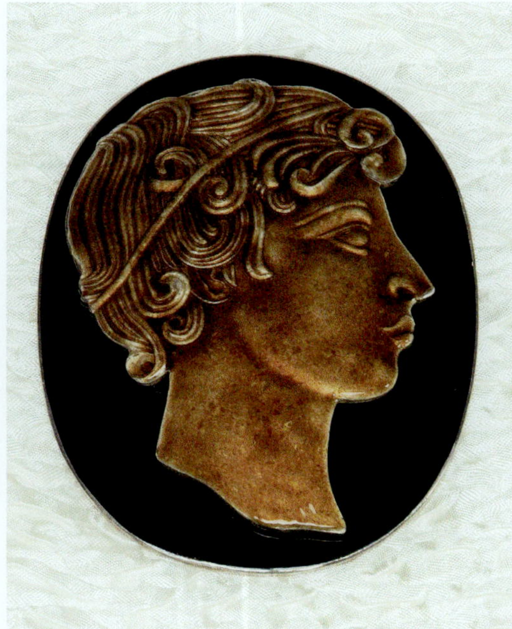

Plate VIII-16 Margot de Taxco 5826. *Roman Patrician* medallion/pendant 3 x 2.25 in. Maestros de Taxco Collection

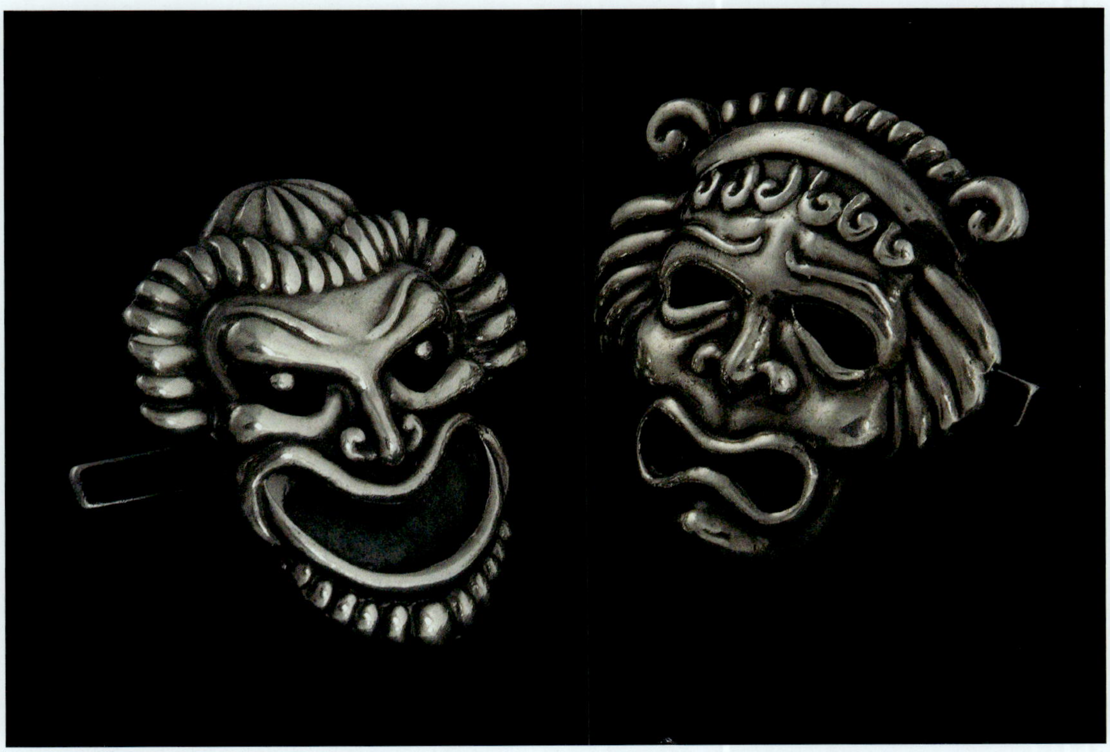

Plate VIII-17 Margot de Taxco 5458. *Tragedy/Comedy* cufflinks in silver. 1.25 x 1 in. Private Collection

or fret motif in a number of color combinations (5442, Plate VIII-4). The later profile portrait of a Roman patrician (5826, Plate VIII-16) is in a format that resembles a coin. This piece reveals an implied three-dimensionality, with the use of transparent color over gold leaf, in contrast with a dark opaque background. In an innovative use of enamel, the color and luster of the portrait head combine to suggest the patina that would have been found on an ancient coin or medallion.

The hippocampi were the horses of the sea known to have drawn the chariot of Neptune.[160] Margot depicted a hippocampus (5360, Plate 35) rising in a swirl of water on the lid of a box. The effect of this phantasmagorical creature is elegant and

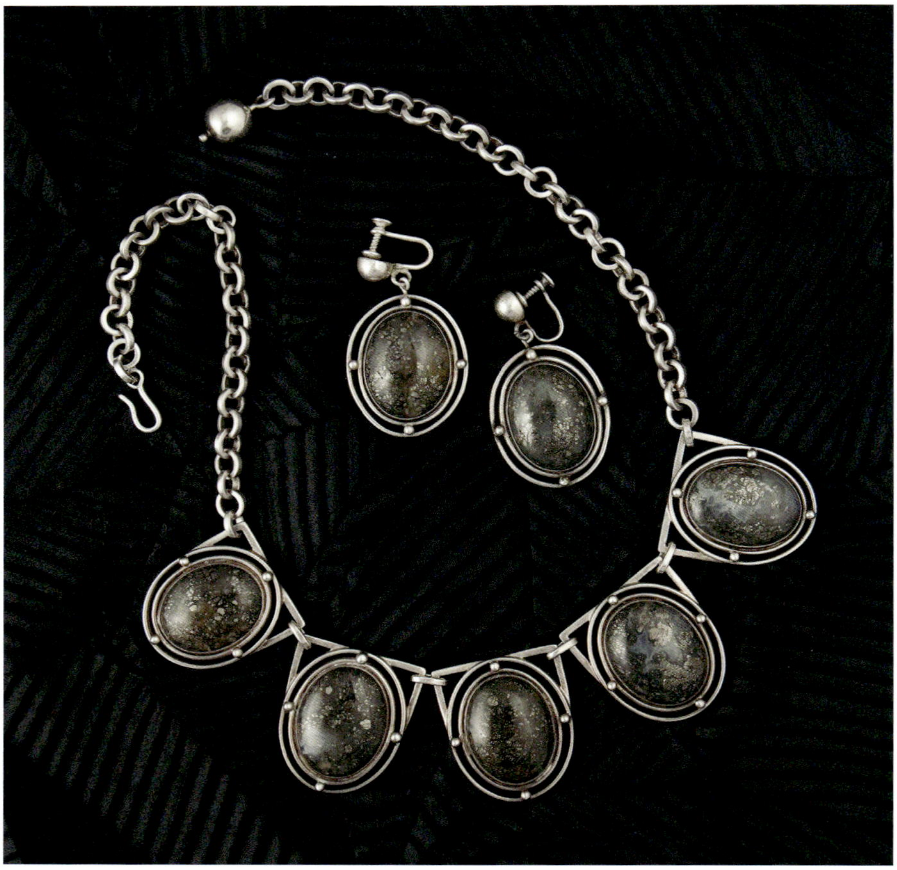

Plate VIII-18a & b Margot de Taxco 5630. *Moss Agate in Silver Set*. necklace 16 in., stone setting 1.25 x 1 in.; bracelet 7.25 x 1 in.; earrings 1.75 x 1 in. Private Collections

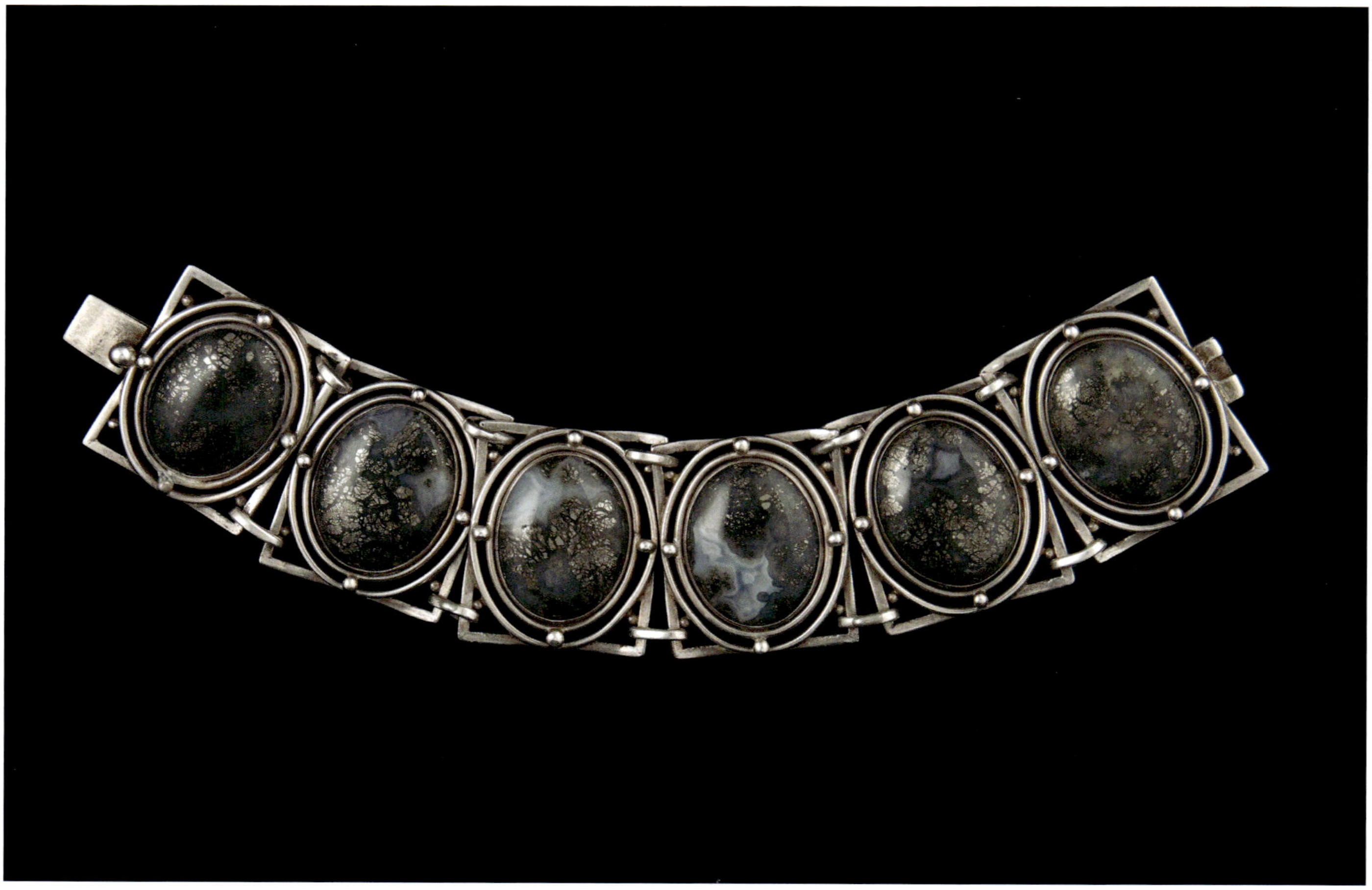

appealing. The box recalls a large cuff bracelet Margot designed at Los Castillo (LosC354, Plate VI-4). Two lion griffins face each other as formal heraldic figures with mythic references, much like the wonderful hippocampus of later date.[161]

Margot allowed function to add meaning to a design. For example, the masks of Tragedy and Comedy as cufflinks (5458, Plate VIII-17) reflect, with some humor, the duality of life's choices at either hand. These miniature masks in silver appear to be have been lifted from a classical relief sculpture. Another intriguing choice of motifs was the recumbent nude as a tie bar (5295, Plate VIII-25). Margot had adapted the female nude in a similar fashion once before. At Los Castillo, she designed goblets with stems in the form of the female body. Some of her most attractive nude figures appear in her zodiac charms (5225, Plate VIII-23). Their bodies are stylized and gracefully adapted to the contours of the circular, slightly convex pendants.

Egyptian Revival had arrived in the wake of the discovery of Tutankhamen's tomb in 1922, and remained a stylistic reference for Margot. The profile head of Cleopatra was a very early enamel design (5242, Plate 21) that was produced continuously by the workshop until its closing. Margot added an intriguing detail to this pendant (indicated in the drawing), the tiny chains that embellish the headdress.

Another overtly Egyptian borrowing is the "papyrus" parure (5587, Plate VIII-19), in which Margot suggests overlapping leaves by adjoining three colors. She made yet another innovative use of enamel in design number 5603 (Plate 18). Heavy applications of purple and green enamel in small areas on the surface of the earrings resemble precious stones. This same design as a necklace in brilliant green against black enamel is stunning. The reference to Egyptian jewelry also appears in design number 5758 (Plate 20), which seems to be a marriage of numbers 5603 and 5587 (Plates 18, 19, VIII-19). This dynamic rendition of the petals spiraling away from the central "jewel," is once again evidence of Margot's fascination with the visual power of the spiral.

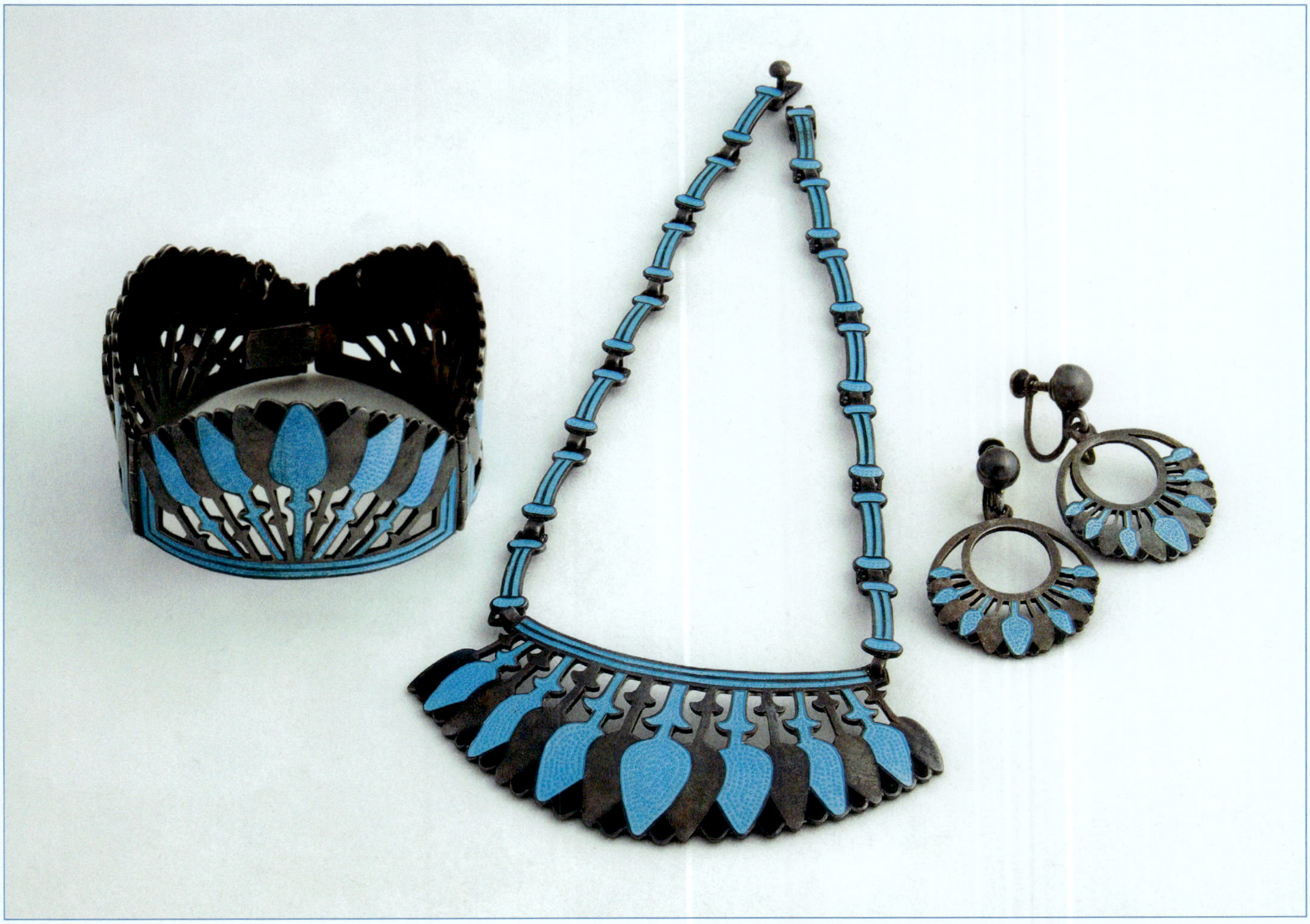

Plate VIII-19 Margot de Taxco 5587. *Papyrus Set* in blue, gray, and black. necklace 15 in., central motif 1.5 x 3.63 in.; bracelet 7 x 1.5 in.; earrings 1.63 x 1.25 in. Elizabeth Finch Collection

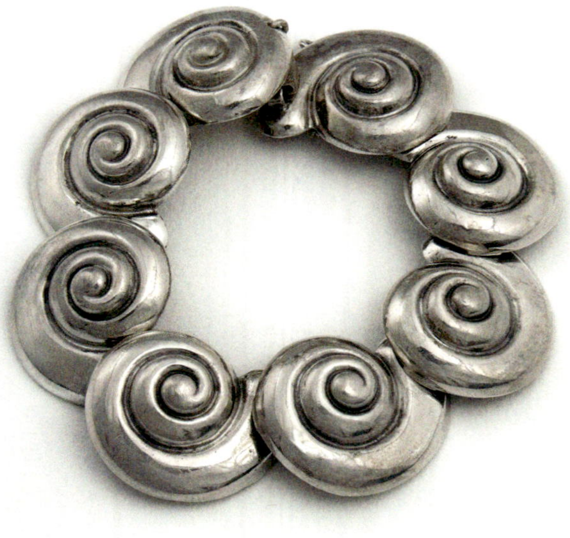

Plate VIII-20a & b Margot de Taxco 5104. *Rounded Shell* necklace with 2 extensions 20 x 1 in.; bracelet 7 x 1.25 in. Private Collections

Margot found inspiration in the Art Deco style, an influence that combined well with other design elements of interest to her. Easily adapted to jewelry, the emphatic line and strong color contrasts resulted in exceptional work (5225, 5717, 5743, Plates VIII-23, V-10, IX-1). Margot, like many of her contemporaries in Taxco, gave a Deco interpretation to pre-Columbian motifs, as in number 5285 (Plate VII-27). This necklace, with its geometric two-dimensionality that is reiterated by the line of beading, recalls jewelry from the Los Castillo era. The enamel mask brooch, inspired by the masks of Teotihuacan (5403, Plate VIII-12), is reminiscent of Picasso's cubist interpretation of the African mask. Another Art Deco motif, the zigzag, is integral to the design for the Aquarius zodiac as well as in design number 5914 (cover). The black and white geometric stepped forms in the necklace and bracelet number 5743 (Plate IX-1) conjure up the sounds and setting of the Jazz Age. A particularly creative and elegant set is number 5696 (Plate 13), in which Margot looked closely at the layering in cut stone and discovered a Deco rhythm.

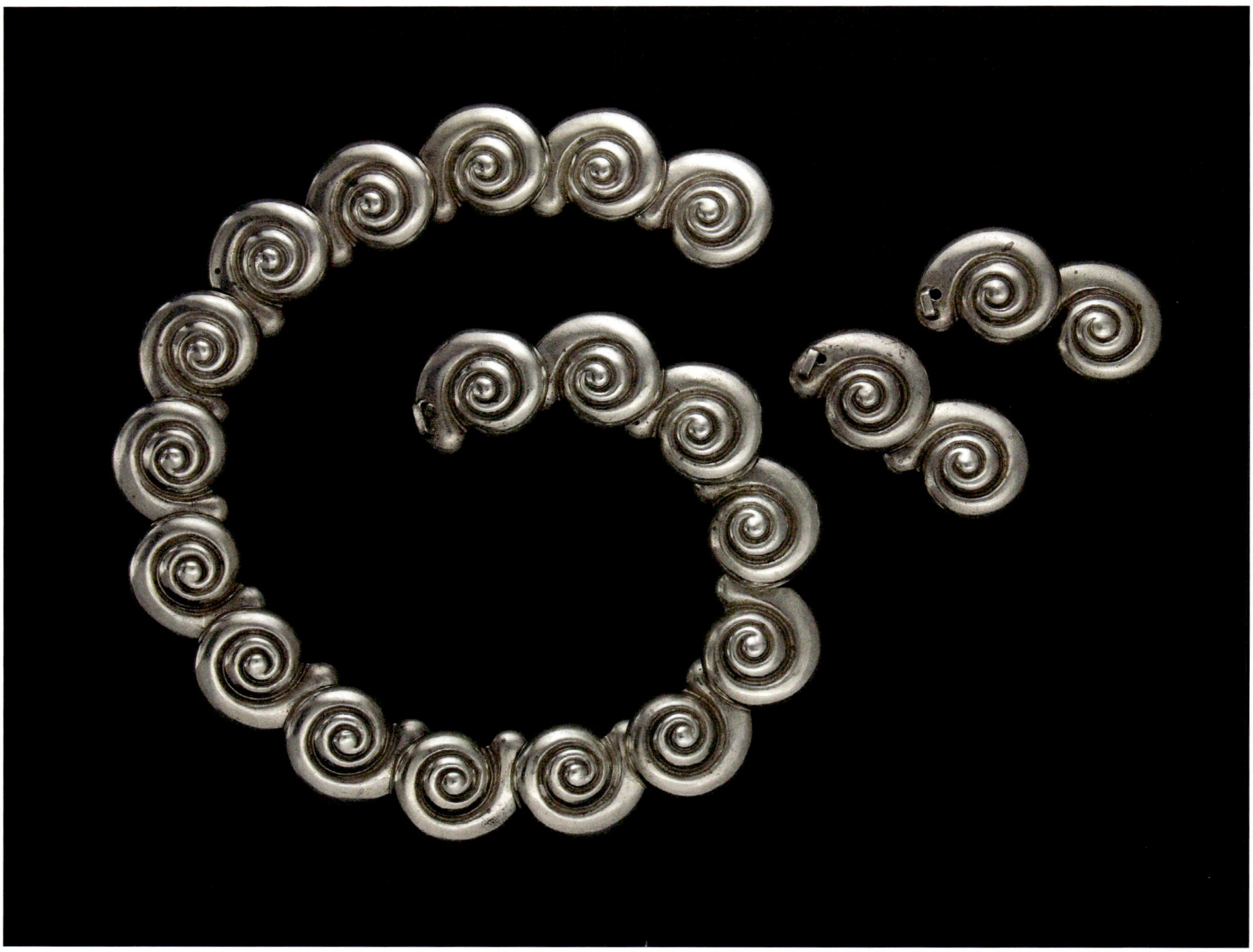

Margot had worked with Juan González to produce several one-of-a-kind pieces. One of the most spectacular examples is the Portrait set (5780A, Plate 3). The miniature painting is, according to scholars at the National Portrait Gallery, "European, possibly French, from the Directoire period or early Napoleonic period. The woman's Empire costume and hairstyle can be dated to circa late 1790s-1800's. The woman's hair and headband follow the antique Grecian style, which was revived during this neoclassical era." [162]

Margot could have found this miniature portrait in Mexico. After Napoleon III's incursion into Mexico in 1861-1867, known as the "Maximilian Affair," and again, during Porfirio Díaz's presidency, the Porfiriato (1876-1911), the Mexican elite avidly collected French art and antiquities. This portrait could have been placed with a Mexico City antiques dealer where Margot would have discovered and purchased it. There are other possibilities, that Margot found the portrait while in Paris or that this remarkable set was a special commission. In this unique work, Margot used color to enhance the portrait. The enamel and colored stones emphasize rather than detract from the coloration in the miniature painting, creating a totally integrated composition.

A comparison can be made between the Portrait set and several variations Margot produced from design 5389 (Plates 1, 2). Margot's instructions indicate that a brooch could be produced as a pendant. An element from the bracelet could become an earring.[163] What is apparent is that Margot could decide to apply gold leaf, add semi-precious faceted stones, or vary color combinations in order to create quite distinctive works.

Margot was a painter and colorist, so it has been quite satisfying to study the impact that her color choices had on any one particular motif. Margot, in the drawing for number 5545, required the enamelists to paint purple grapes, grey-green leaves, and to use gold leaf beneath the yellow-green branches for Model 1, as it appears in the drawing. For Model 2, the directions in Margot's handwriting call for the leaves to be in two colors and the grapes to be yellow with gold leaf, as can be seen in the actual brooch (Plate VIII-28). In this work, the difference color makes is dramatic, as in number 5774 (Plates 8, 9), where gold leaf beneath the yellow enamel replaces periwinkle blue. Color transforms the floral spirals of design number 5384 (Plates VII-33-36), and results in a masterpiece in the polychrome fish necklace (5693, Plate 17). Margot's enamel serpents (5554, Plates VIII-7-9, cover) defy natural coloration; actually, the choice of color charges the creatures with meaning. The green mélange enamel recalls leafy plants bordering a meandering river. The purple serpent bracelet with pink eyes is exotic and just a bit menacing, while the orange serpent with turquoise eyes glitters with the preciousness of gemstones.

The quality of the feminine is pervasive in the work produced at Margot de Taxco. A cherub holds a mirror (5441, Plate I-20), and ballerinas twirl in their circular skirts (5200, Plate II-13). The necklace and earrings

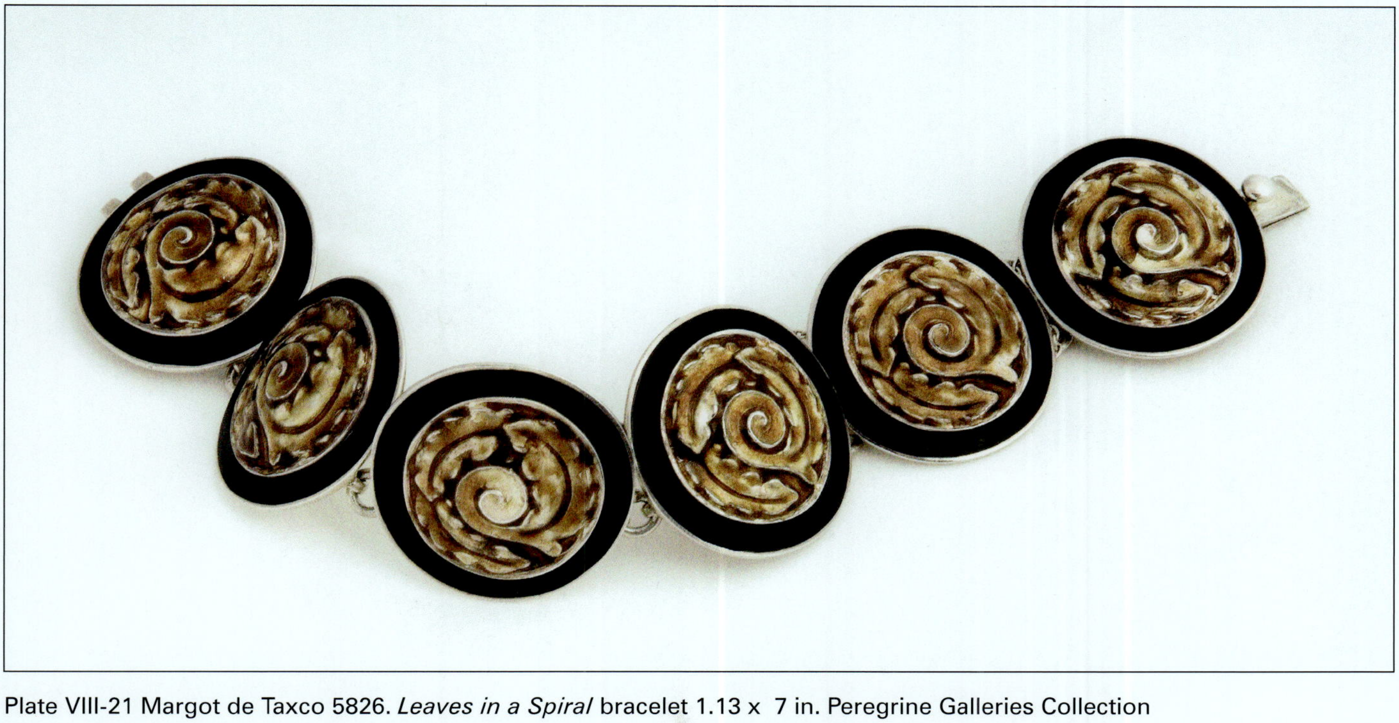

Plate VIII-21 Margot de Taxco 5826. *Leaves in a Spiral* bracelet 1.13 x 7 in. Peregrine Galleries Collection

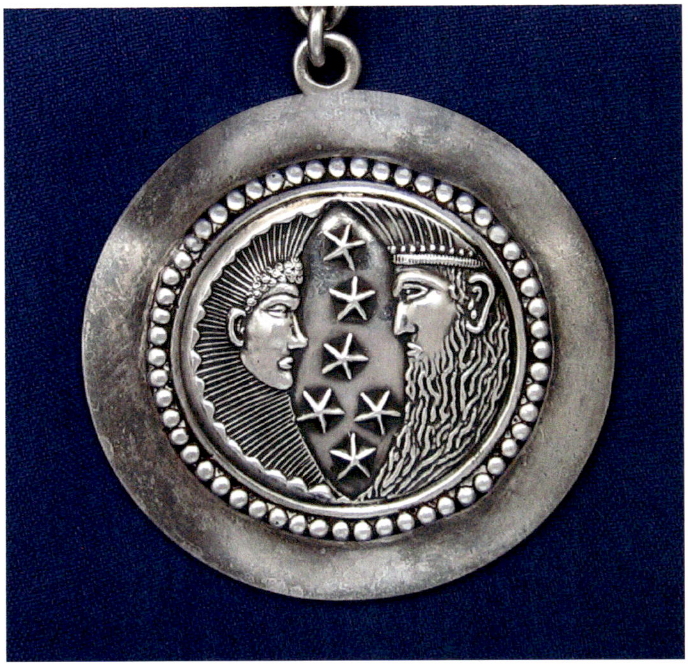

Plate VIII-22 Margot de Taxco 5485. *Neptune and Amphitrite* pendant 2.5 in. d. Photograph provided by a collector. Silver Huntress Collection

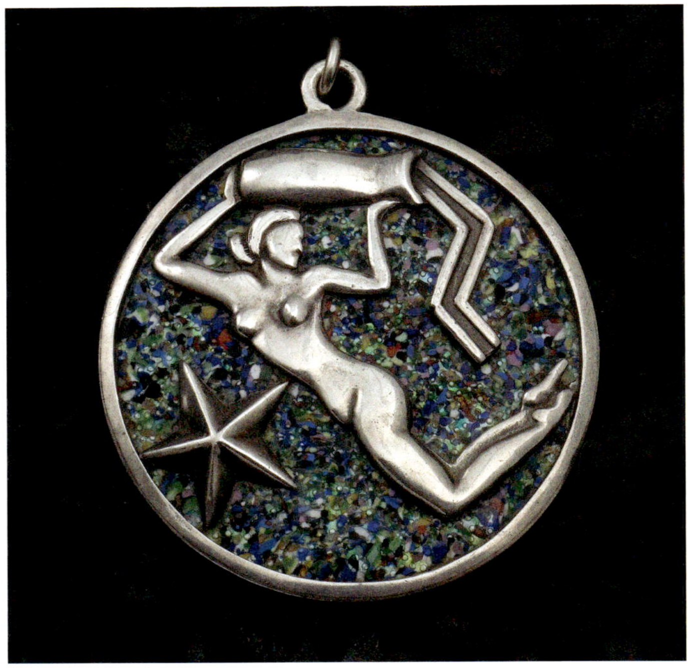

Plate VIII-23 Margot de Taxco 5225. *Aquarius* pendant in mélange enamel 1.63 in. d. Private Collection

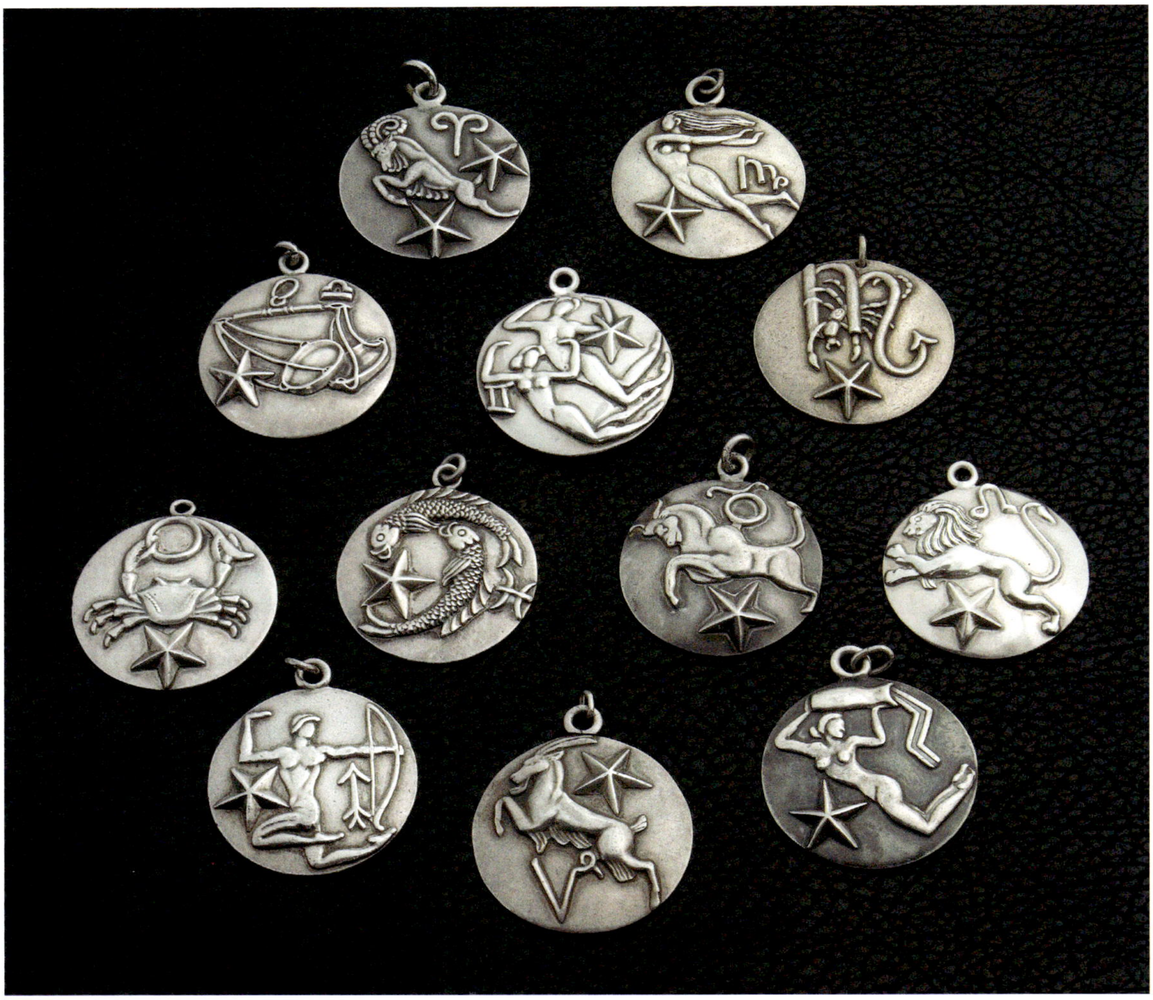

Plate VIII-24 Margot de Taxco 5220-5232. *Twelve Zodiac Signs* 1.5 in. d. Carole Berk Collection

of miniature roses (5871, Plates IX-5-6), the doily pattern of design 5552 (Plate VIII-1), and the tiny folk art doves (5868, Ack-4) have a sweetness about them that contrasts with the more robust designs of the Los Castillo period. Comparing the early pieces to what was produced over two decades at Margot de Taxco reveals an expanding number of themes with a greater variety in the choice of sources and influences. Margot, now independent and successful, was reaching out to an ever-expanding audience. Committed to technical experimentation, Margot added color, texture, and the use of gold leaf in her designs. She set herself apart from everyone else in Taxco. Very few designers were doing the kind of innovative work that was coming out of the Margot de Taxco workshop.

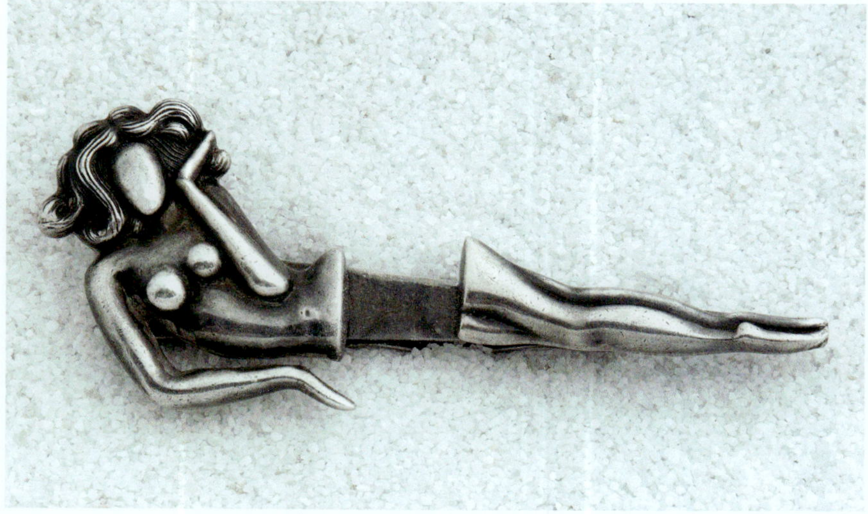

Plate VIII-25 Margot de Taxco 5295. *Female Nude* tie bar in silver 1.13 x 2.75 in. Orlando Rigono Collection

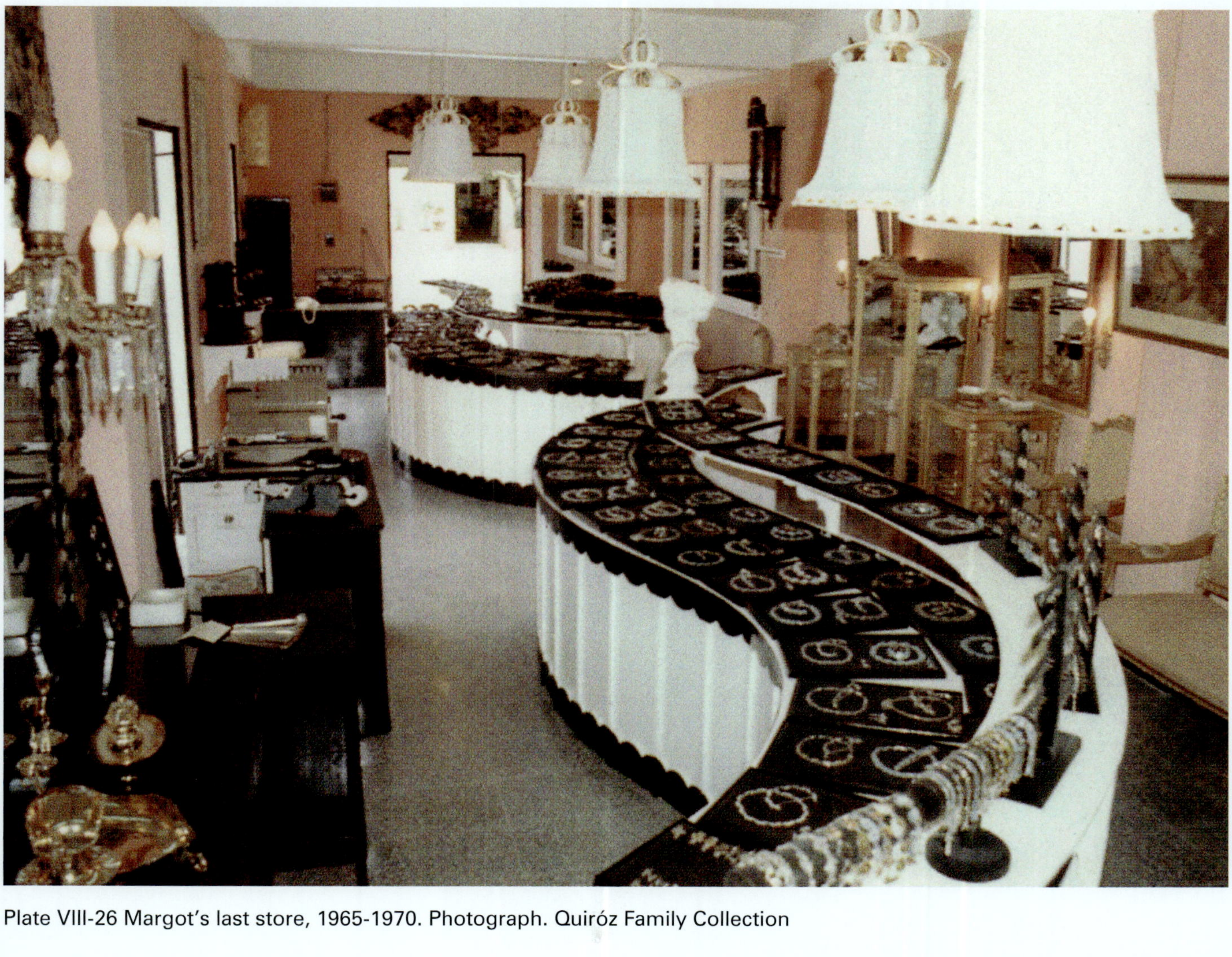

Plate VIII-26 Margot's last store, 1965-1970. Photograph. Quiróz Family Collection

Plate VIII-27 Margot de Taxco 5750. *Mirrored Butterfly Swirls* brooch 1.63 x 2 in. Private Collection

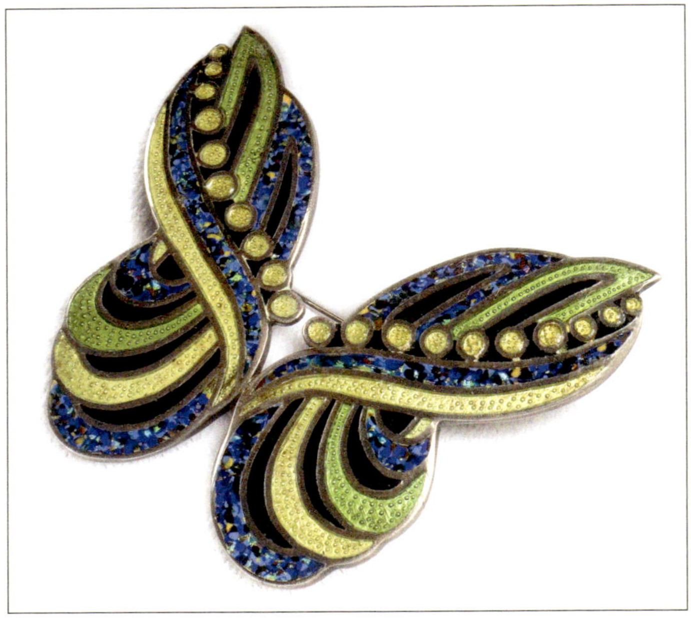

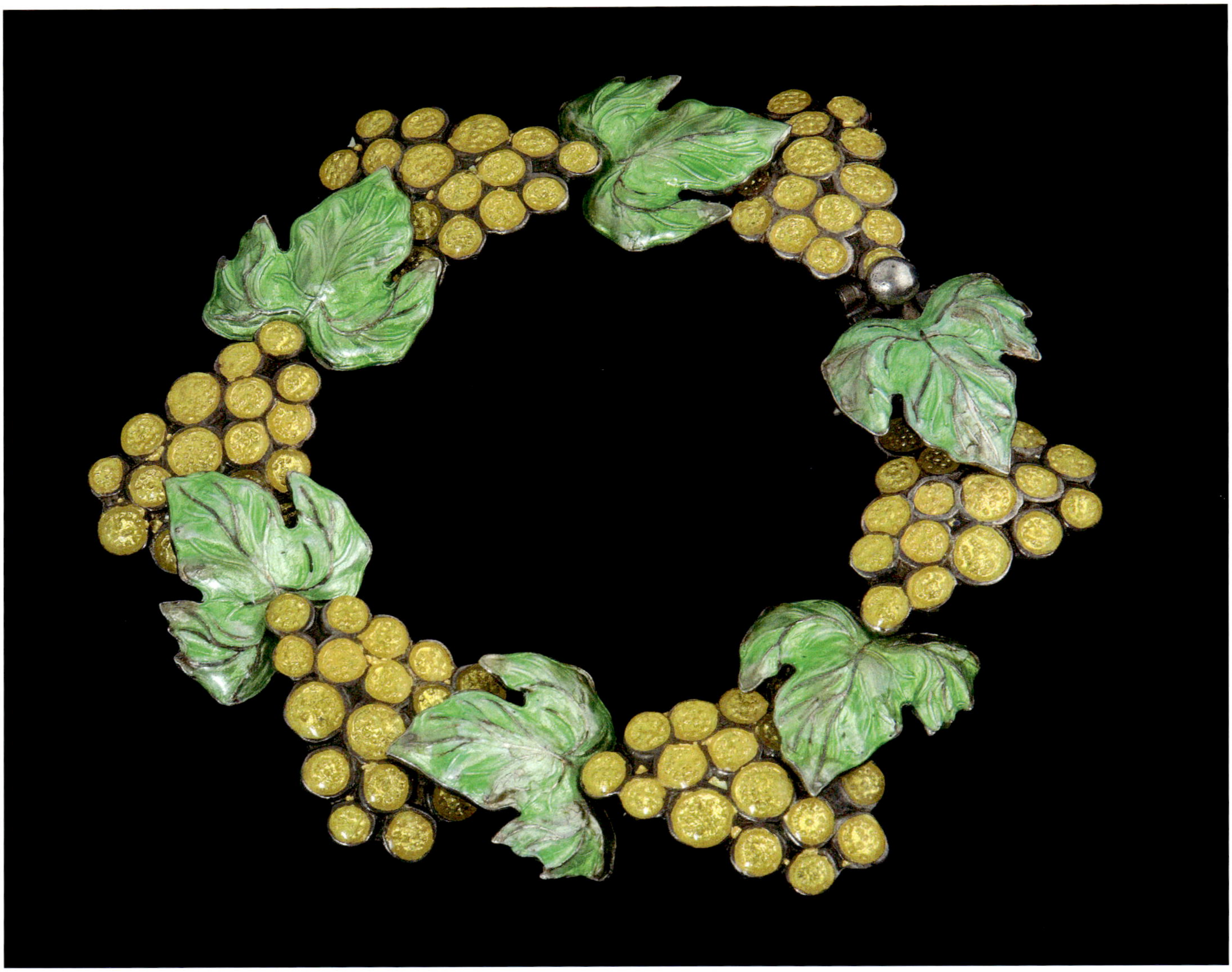

Plate VIII-28 Margot de Taxco 5545. *Yellow Grapes* bracelet 7 x 0.88 in. Elizabeth Finch Collection

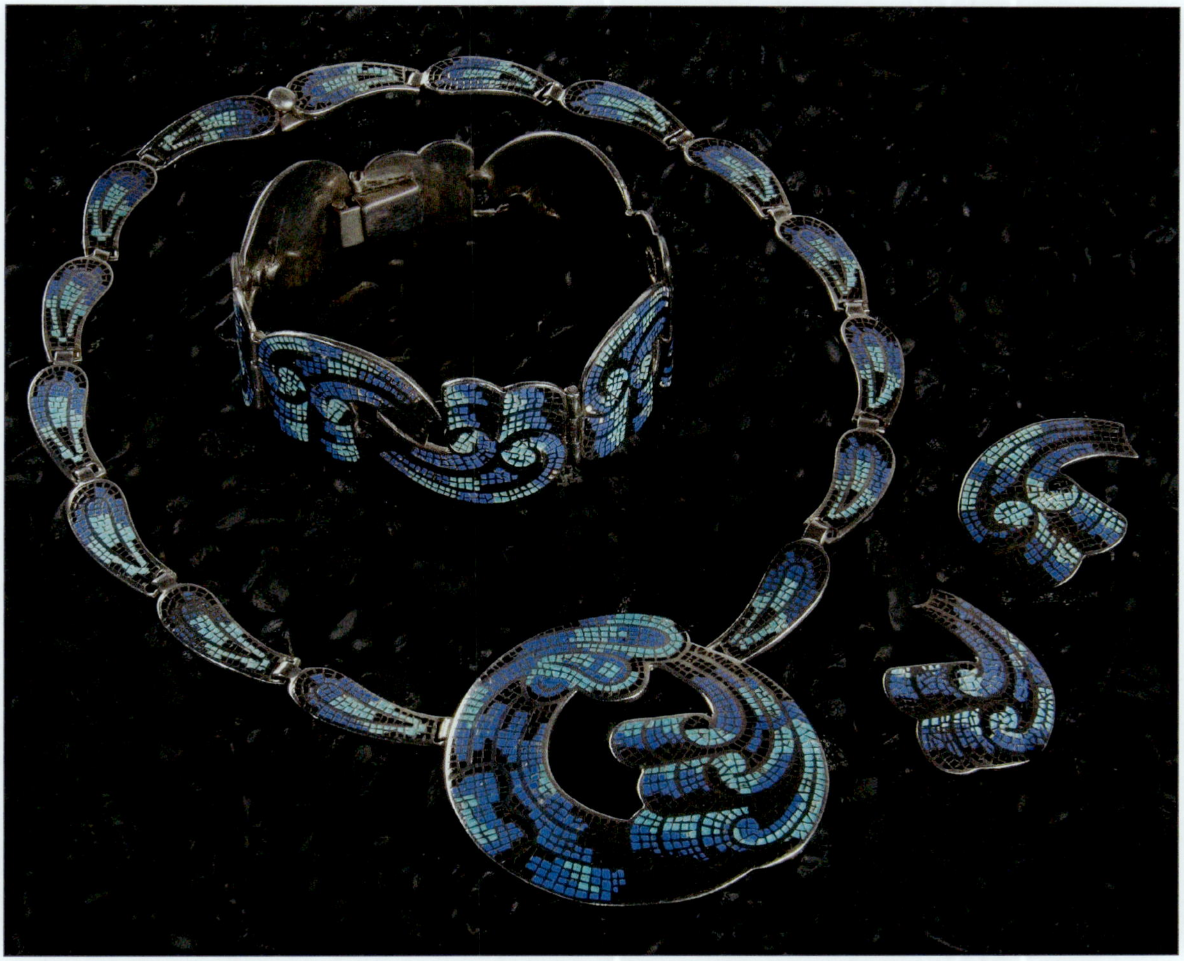

Plate VIII-29 Margot de Taxco 5547. *Wave Circle Set.* necklace 15 x 0.5 in., central motif 2 in. d.; bracelet 7 x 1 in.; earrings 1.25 x 0.75 in. Stella and Fred Krieger Collection

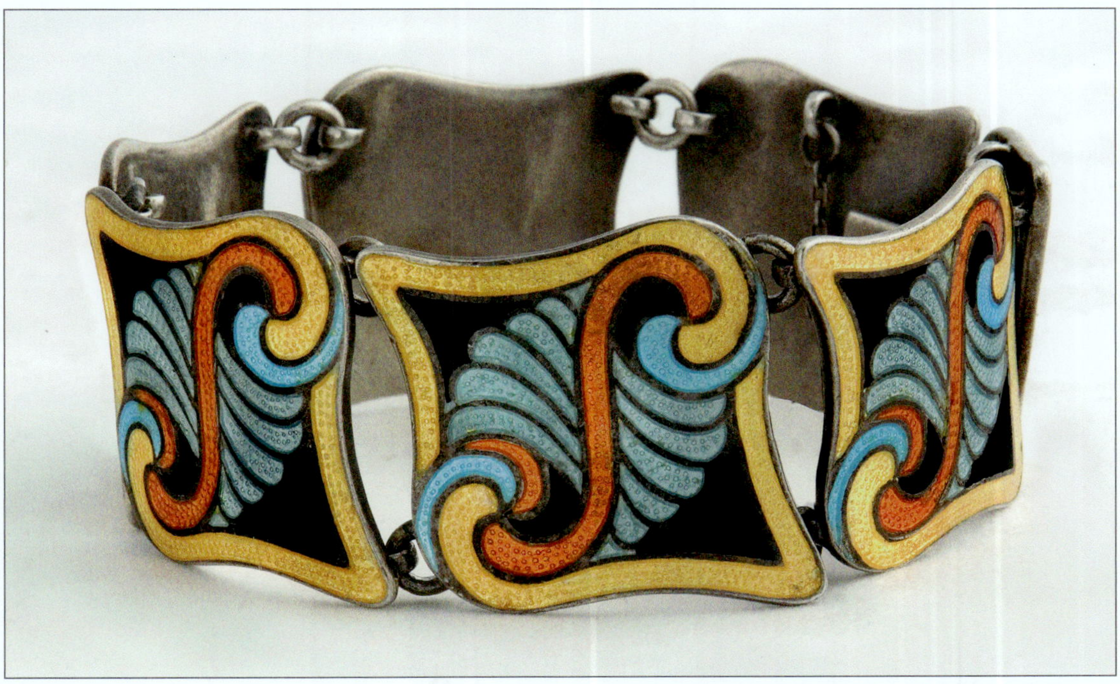

Plate VIII-30 Margot de Taxco 5468. *Egyptian S-Curve* bracelet 7 x 1 in. Elizabeth Finch Collection

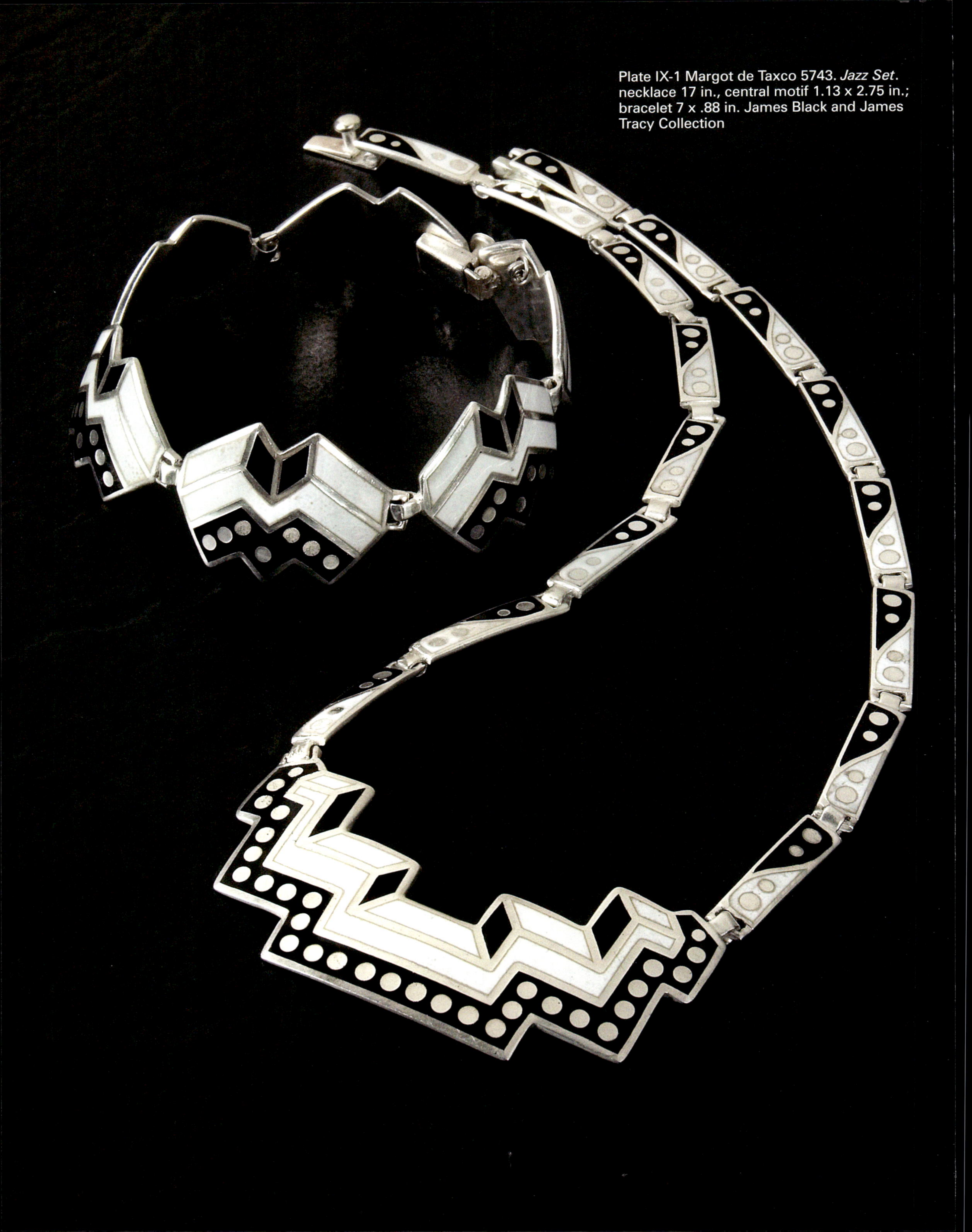

Plate IX-1 Margot de Taxco 5743. *Jazz Set*. necklace 17 in., central motif 1.13 x 2.75 in.; bracelet 7 x .88 in. James Black and James Tracy Collection

Chapter IX
The Last Years

Fortunately for many of the silver designers, Hollywood had come to Taxco. The town was occasionally a movie set but, more often, a destination for the stars. Lana Turner came every year to see Margot, and Jack Palance, John Wayne, Tony Curtis, and John Derek all shopped at Margot de Taxco.

Margot continued to leave Taxco to travel and this probably kept her from becoming stale as an artist for, as María Pineda said, Margot always returned from her trips with new ideas. Margot and Jaime Quiróz traveled together to the state of Yucatán and visited the Mayan archaeological sites, Chichén Itzá and Uxmal, and on the way, toured La Venta, an Olmec site in Veracruz. Margot made drawings of what she encountered while on this trip.[164]

Margot very rarely mentioned her own family or family's history to the people in Taxco. In 1972, she had a visit from her cousin Elizabeth Freeman, Robert Chabot's granddaughter. On another occasion, Elizabeth's aunt came to Taxco and reported to her niece

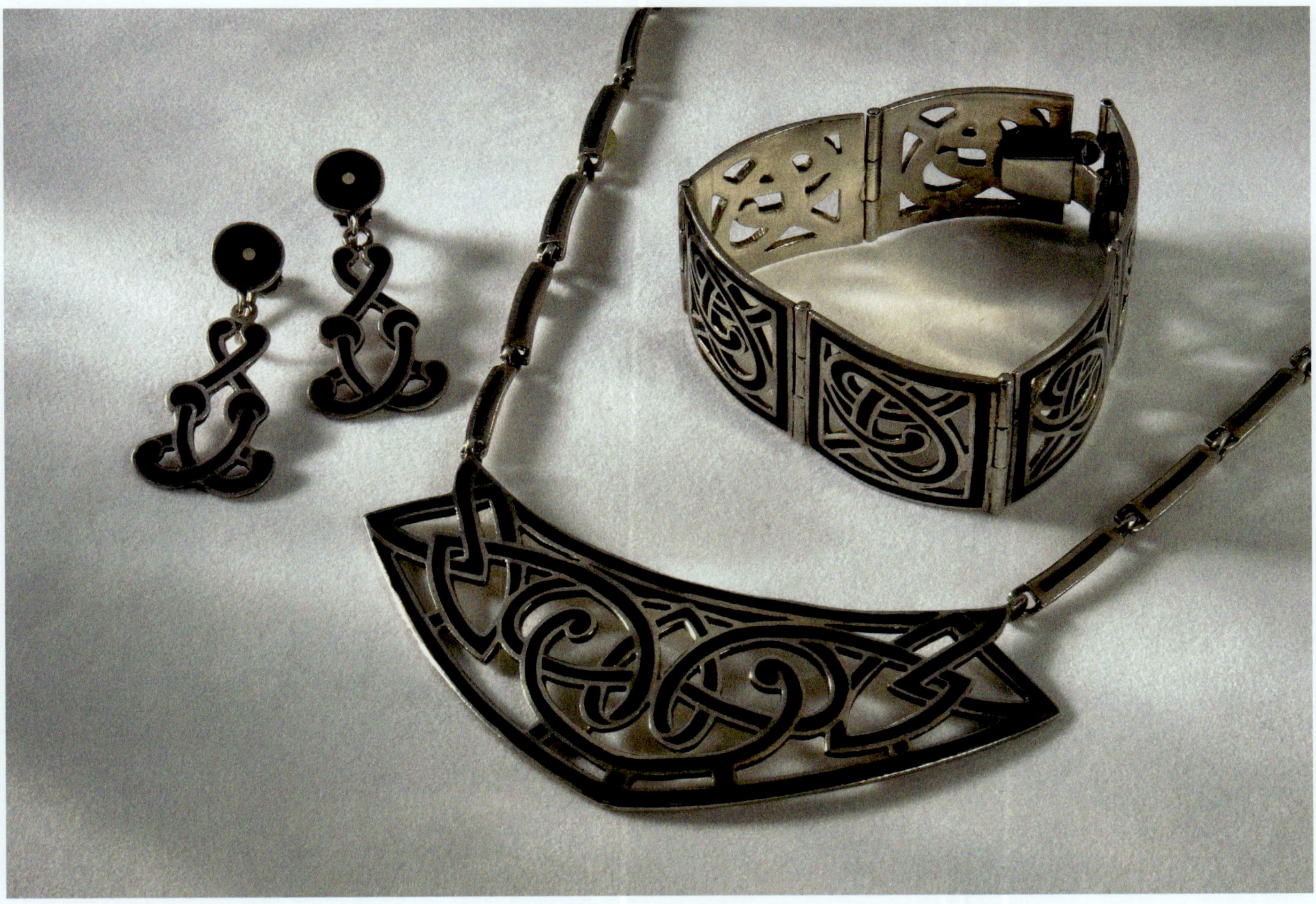

Plate IX-2 Margot de Taxco 5391. *Scrolling Linear Patterns Set*. necklace 14.5 in., central motif 2 x 4 in.; bracelet 6.5 x 1.13 in.; earrings 1.75 x 0.88 in. Elizabeth Finch Collection

how beautiful she thought the store was and that she had bought a necklace and bracelet from Margot.[165] As if in an effort to create the semblance of family, Margot became close to Jaime Quiróz's children. She did a drawing of his first child Alba in 1969 and then sculpted a portrait head of the little girl (Plate IX-9). Carl Pappe, an artist in Taxco, helped Margot find someone to do the casting and bronzing.

A saying that Margot incorporated into a brooch (5754), "Más que ayer, menos que mañana," [More than yesterday, less than tomorrow], refers to a love that grows and increases over the years, and it seems to be reflective of Margot's romanticism. According to Jaime Quiróz, there were those in Taxco who became close to Margot, who won over her trust, and then took advantage of her.

In 1972, Melecio Rodríguez discovered that the administrator, stock manager, and the accountant at Margot de Taxco were falsifying the records and stealing hundreds of thousands of pesos from the company. Rodríguez confronted the administrator and accused him of taking commissions for sales that he had never made. Margot became enraged and Rodríguez left to work out of his home. After three months, Margot came to him and asked Rodríguez to return, and his response was that he would not come back until she

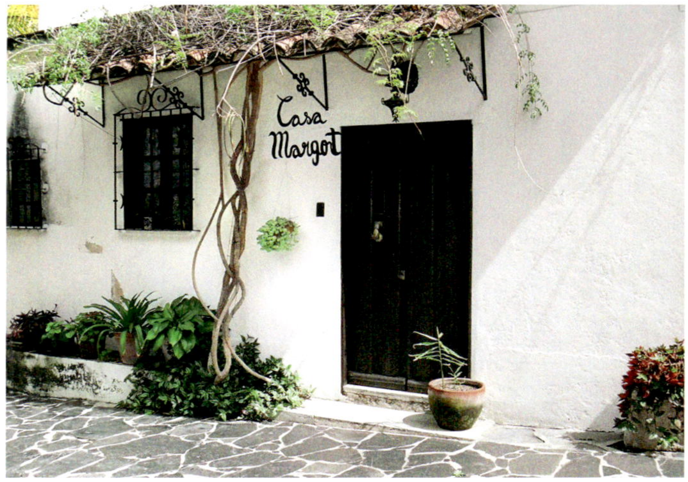

Plate IX-3 Casa Margot. Photograph by author, 1993

had run off the thieves. In a decision that revealed poor judgment, Margot did not fire the people she should never have trusted.

Like so many of the workshops in Taxco, Margot de Taxco was beset by problems with the silversmiths' union. The Taller Borda had gone bankrupt and closed its doors in the mid-sixties. During this same period, Antonio Castillo had attempted to organize ownership of Los Castillo as a workers' cooperative,

Plate IX-4 Margot de Taxco 5860. *Flowers in a Circle Set.* necklace 21 in.; pendant brooch 2 x 1.63 in.; earrings 1 x 0.88 in. Elizabeth Finch Collection

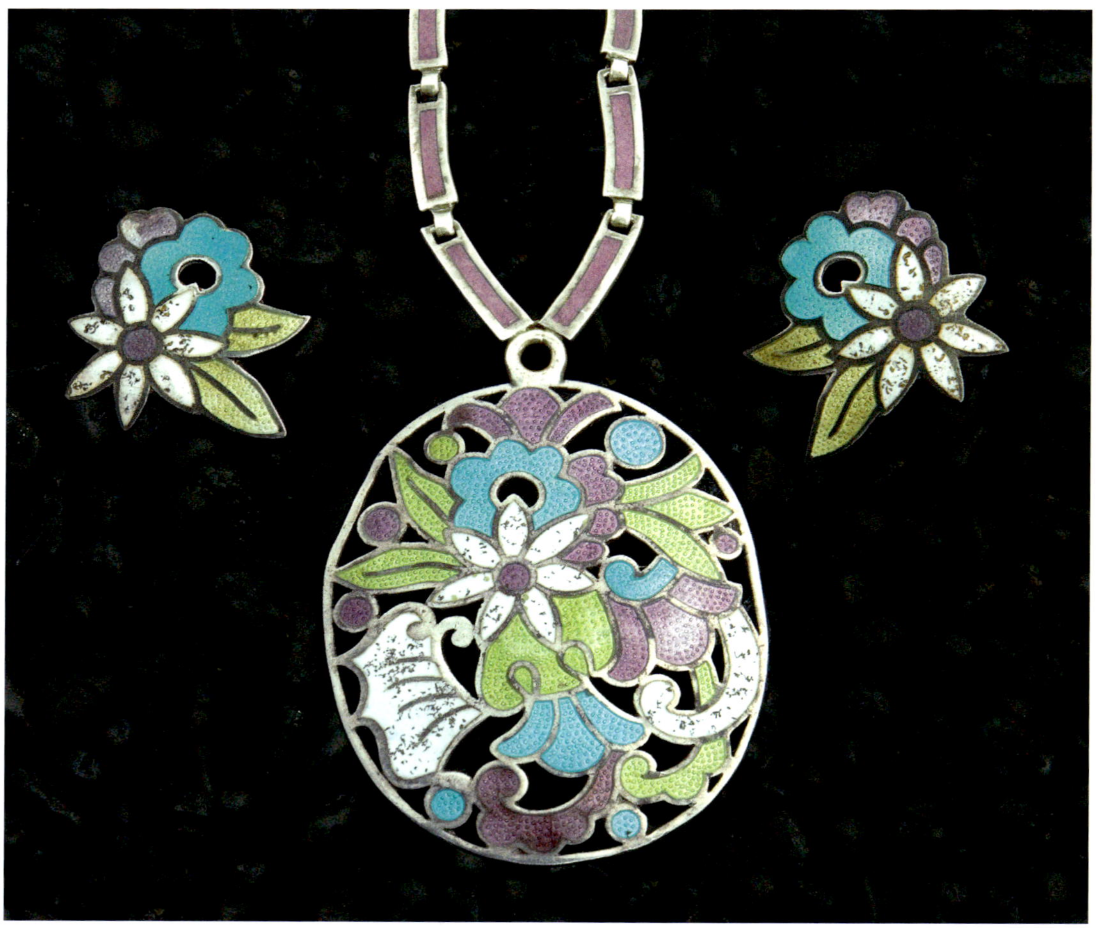

which ended in the embezzlement of funds on the part of the cooperative's administrator.[166] By 1974, the silver industry had seen a major transformation. The cost of keeping a workshop operational had become so prohibitive, that more than two-thirds of the silver production in Taxco was taking place in small family workshops.[167]

After 1958, the government began requiring the silver workshops to pay social security (health and social services) for their workers, along with an additional deposit in each worker's account towards a place to live. Most workshops went into bankruptcy in order to avoid paying out more than they had. The costs accelerated with greater demands from the union, and Margot could not keep everyone working. The drop in production as a result of fewer silversmiths in the workshop led to a decrease in available funds. A strike in 1974 closed Margot de Taxco. In order to get the silversmiths back into the workshop, Margot agreed to pay the workers for the month the workshop was closed down, along with higher wages. At this point, she began borrowing from the bank in order to cover labor expenses. Her debts became insupportable and, in 1977, Margot was forced into bankruptcy.

The government's labor authority intervened and auctioned off all of Margot's assets. She lost everything — her house and the entire contents of the workshop. With the money from the auction, the workers were paid off for the hours they had worked and given additional payments related to the number of years they had been with the company. During the fiasco, the administrator, stock manager, and accountant walked away with large sums of money and silver. Margot's car went to one employee and the machinery and

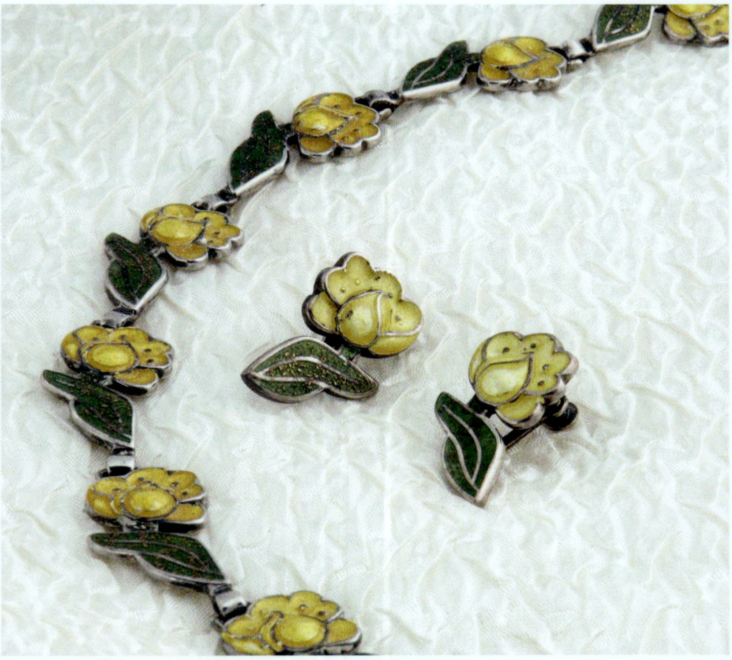

Plate IX-5 Margot de Taxco 5871. *Tiny Flowers Set* in yellow. necklace 16.5 x 0.88 in.; earrings 1 x 0.63 in. Elizabeth Finch Collection

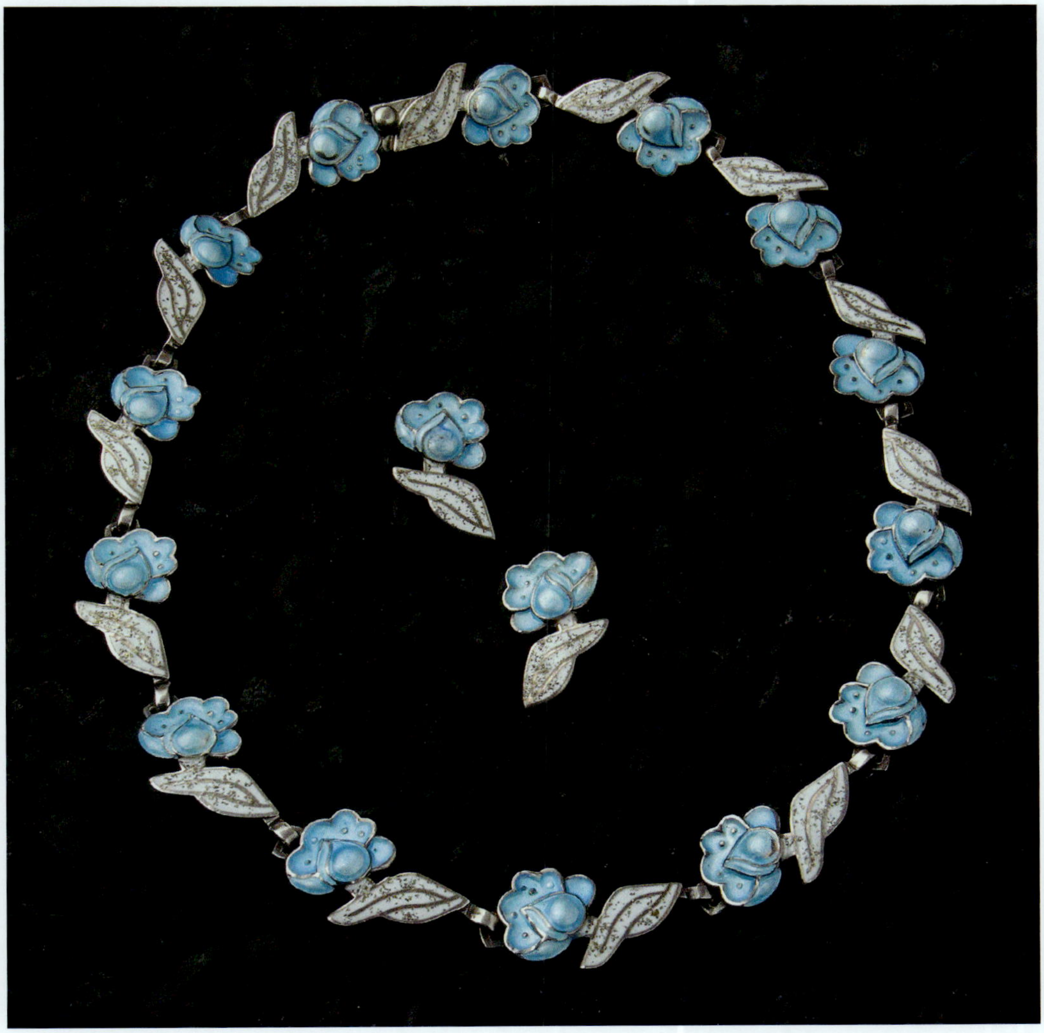

Plate IX-6 Margot de Taxco 5871. *Tiny Flowers Set* in blue. necklace 15 x 0.88 in.; earrings 1 x 0.63 in. Private Collection

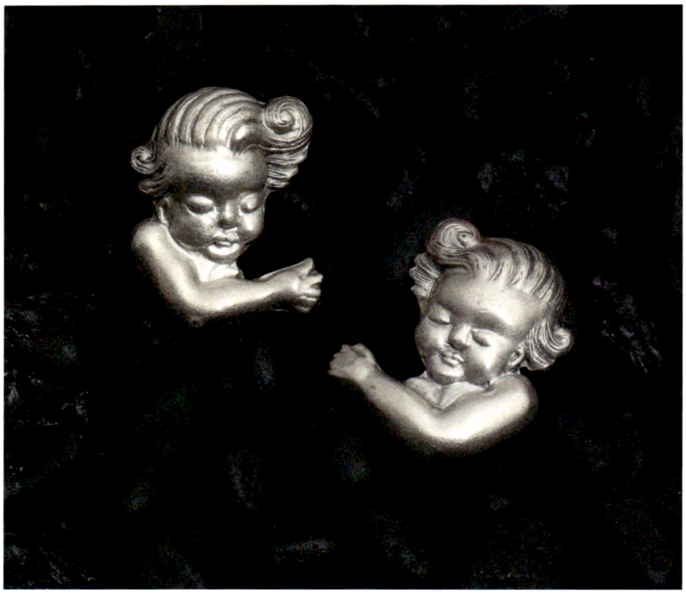

original dies were scattered all over Taxco. Pedro Pérez bought the mirrors, antique vitrine, and display cases.[168]

Most of the dies for Margot's designs in silver or enamel were bought by Manuel Quinto, Melecio Rodríguez and Benjamín Santarriaga. Santarriaga had worked for Valentín Vidaurreta in Mexico City and, after 1941, went to Los Castillo in Taxco, where he worked for a couple of years. He later joined the Margot de Taxco workshop.[169]

Over the last few years, Melecio Rodríguez has been buying the dies where he finds them and has accumulated a fairly large collection. He and his sons and grandson have been producing reproductions of Margot's designs with the family's mark. Jaime Quiróz also worked in enamel for several years, marking the pieces "Alba," his eldest daughter's name.[170]

Plate IX-7 Margot de Taxco 5447. *Silver Cherub* brooches 0.75 x 0.75 in. Private Collection

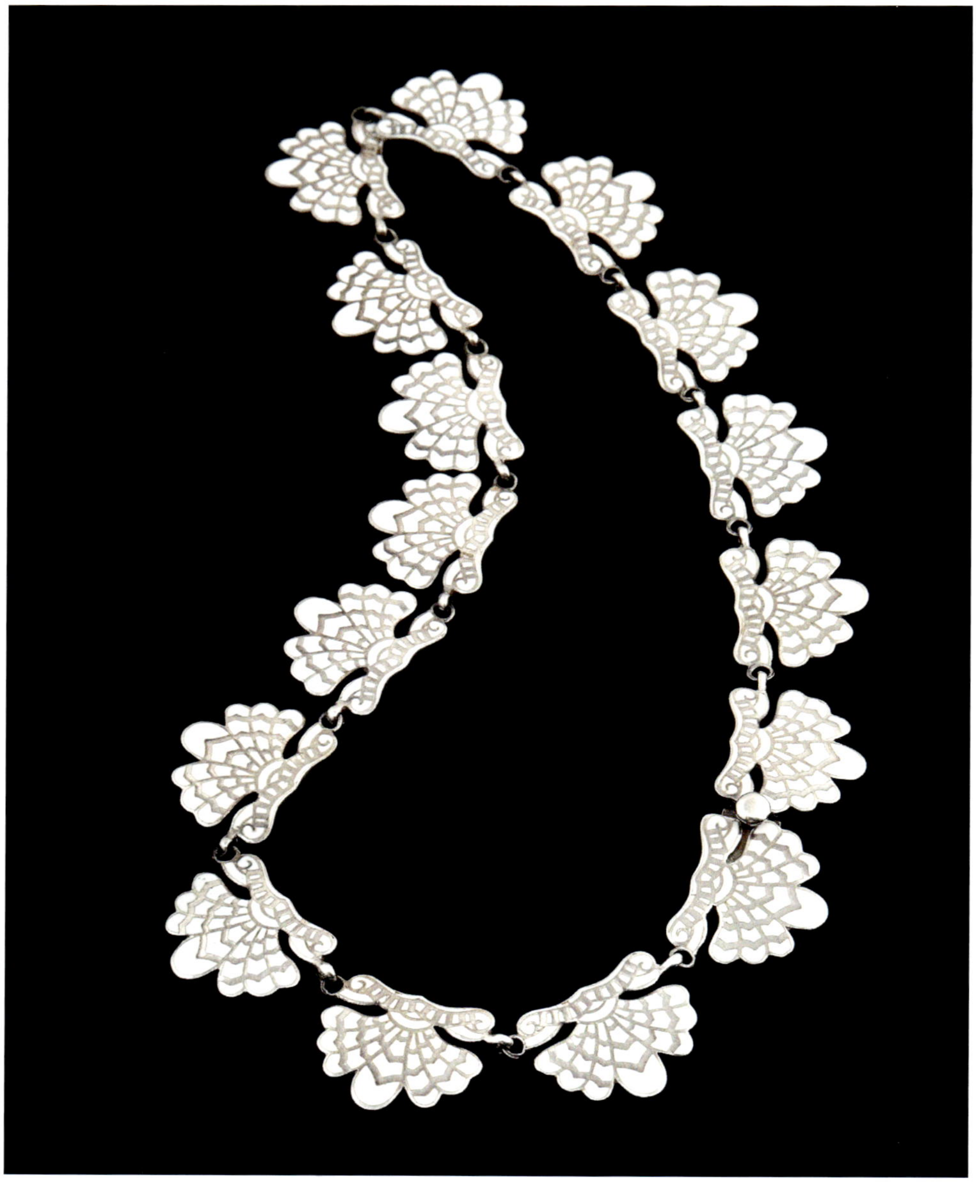

Plate IX-8 Margot de Taxco 5380. *Frills in White* necklace 16.25 in. Mary Jane Fletcher Collection

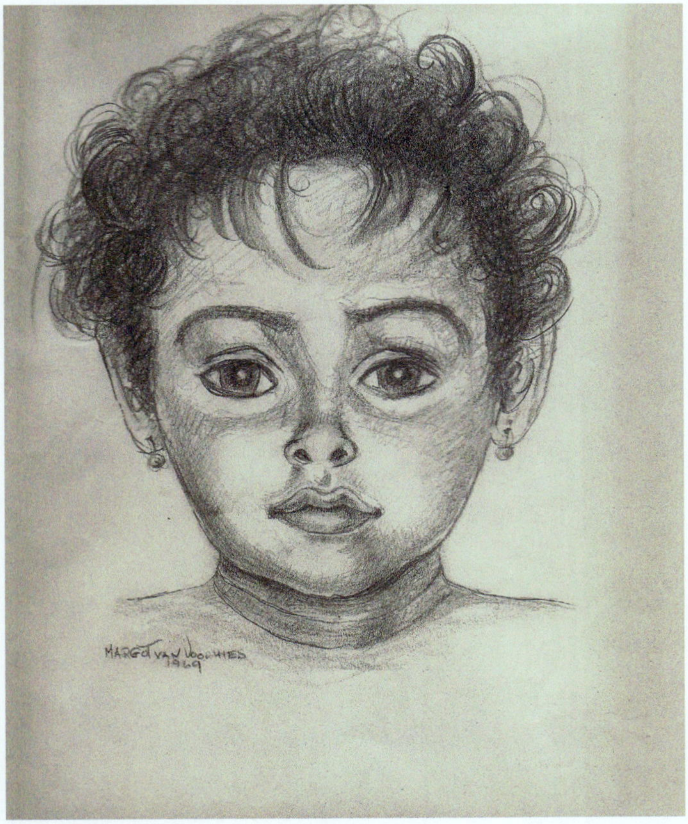

Plate IX-9 Margot van Voorhies. *Portrait of Alba Quiróz*. 1969, pencil sketch. Quiróz Family Collection

Margot was left with a few pieces of furniture, her books, and clothing. Once again, her beloved friend María Castillo stepped forward and, in 1978, at the age of 82, Margot went to live on the grounds of the small hotel owned by Reveriano and María Castillo. Margot spent the rest of her days in the Casa Margot (Plate IX-3).

On April 26, 1985, Margot had a fall and went the hospital. Her friends, John Anderson and Linda Bennett, paid for two weeks of her hospital stay, and at that point, Margot's adopted son Rafael stepped forward to assist with her care. He had government health benefits and he signed Margot onto his account as his mother.

Margot died on July 26, 1985, just weeks before her ninetieth birthday. At the time of Margot's death, Rafael inherited a few objects with personal meaning, as did María Castillo. The manuscript of her trip to Japan has never surfaced. The Casa Margot still stands, a testament to the goodness of a dear friend. Margot's grave remained unmarked until recently, when an anonymous donor commissioned a gravestone engraved with her name.[171]

Plate IX-10 Margot de Taxco 5686. *Poodle* brooches 2.38 x 1.38 in. Margot de Taxco 5708. Fans in Black from *Fan Dancer* bracelet 7.5 x 0.75 in. Margot de Taxco 5542/5543. *Mask and Fan* sweater brooches, mask 1 x 0.5 in.; fan 1.75 x 2 in. Margot de Taxco 5419. *Mexican Lady* brooch 1.88 x 0.88 in.; *Mexican Man* brooch 1.88 x 1.25 in. Margot de Taxco 5422. *Peter Rabbit* brooch 1.75 x 1.13 in. Things Finer in Santa Fe Collection

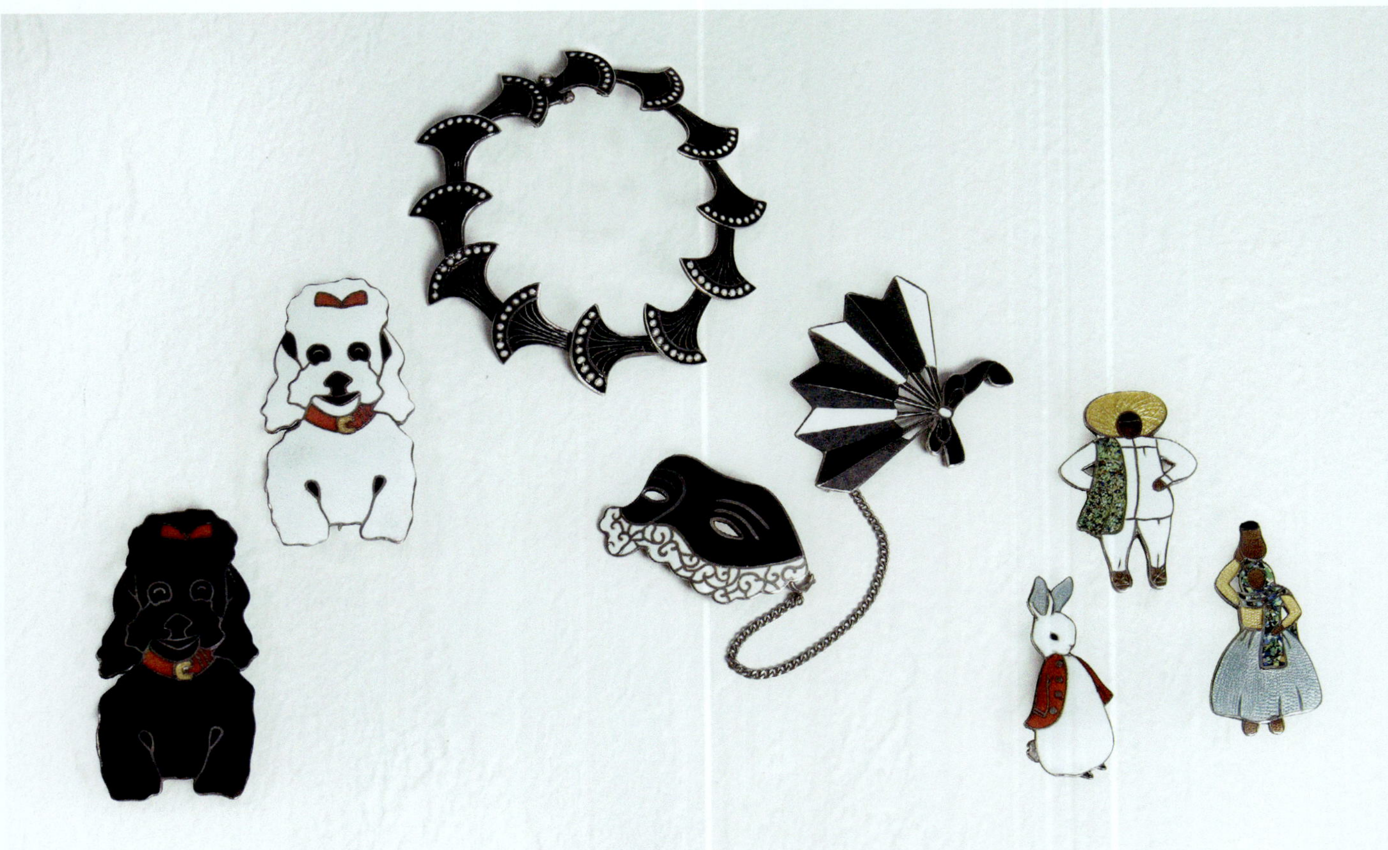

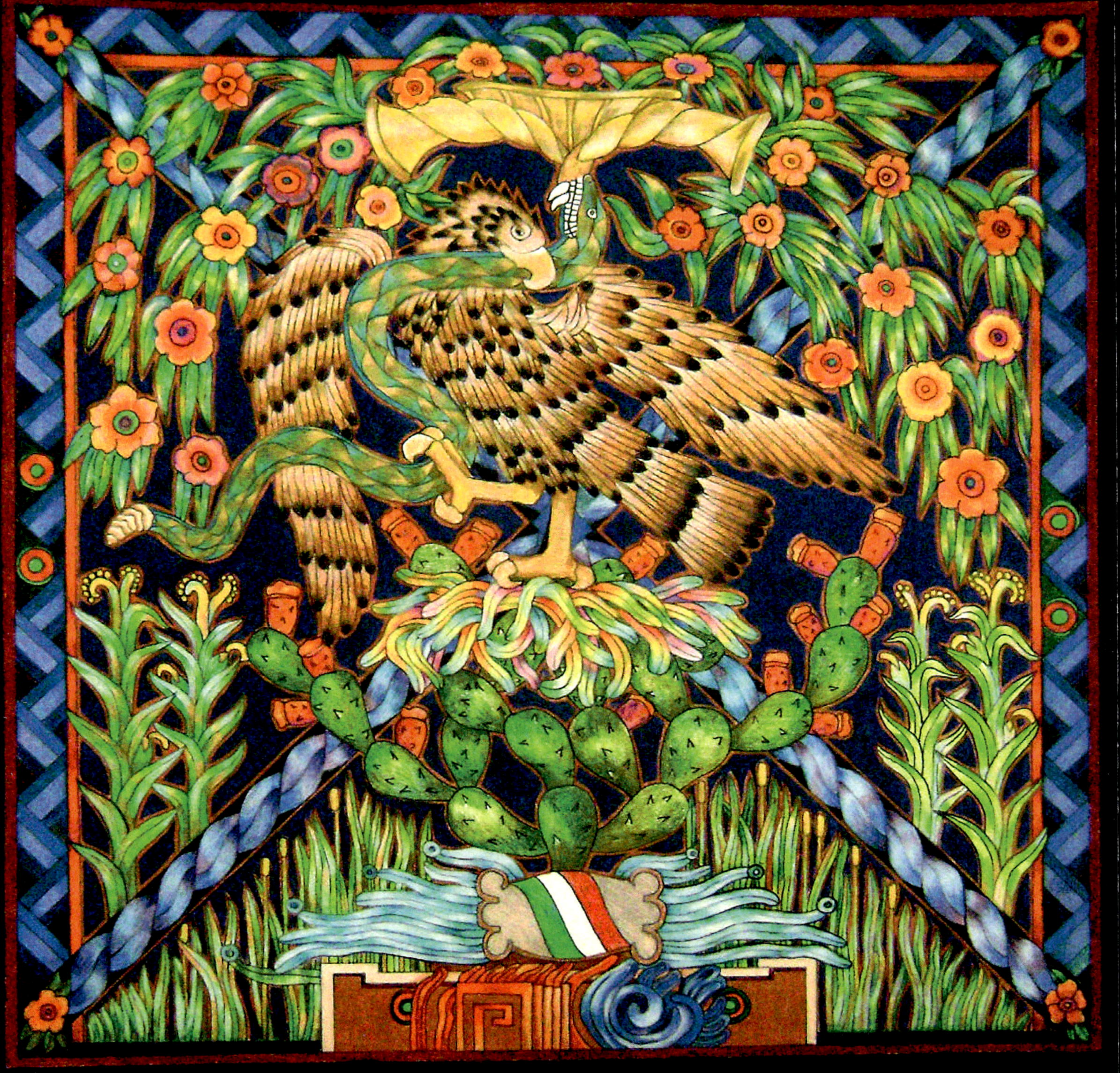

Plate X-1 Adrian Brun. *The Eagle and The Serpent* mural. 7 x 7 m. Centro Médico Nacional Siglo XXI. Photograph provided by Adrian Brun

Chapter X
Contemporary Enamelists In Mexico

Adrian Brun

Adrian Brun was born in Mexico City and spent his childhood in Actopan, Hidalgo, where he showed a propensity for art at an early age. Fortunately, he was able to follow his inclinations, leading to studies at the national university and the Instituto Nacional de Bellas Artes (INBA). For a year and a half, he apprenticed with Diego Rivera until the artist's death in 1957.

In 1972, Brun was invited to teach design to students in the enameling workshop at INBA. In order to inspire the students to attempt more ambitious work, Brun collaborated with them to create a large mural for the newly constructed College of Architects building. Brun's success both as an artist and innovative teacher led to his election as director of INBA's School of Design and Applied Arts.

To date, his most ambitious work is the enamel mural at the Centro Médico Nacional Siglo XXI (Plate X-1-2). Measuring seven meters square, the work was the result of a partnership with the architect, Joaquín Alvaréz Ordoñez. According to Brun, "The architect created a distribution of spaces and I was given the opportunity as the artist to embellish them [vestir los espacios]. A work of art in a building is in a constant process of change in terms of who will view the work of art and how they will respond. Will the work

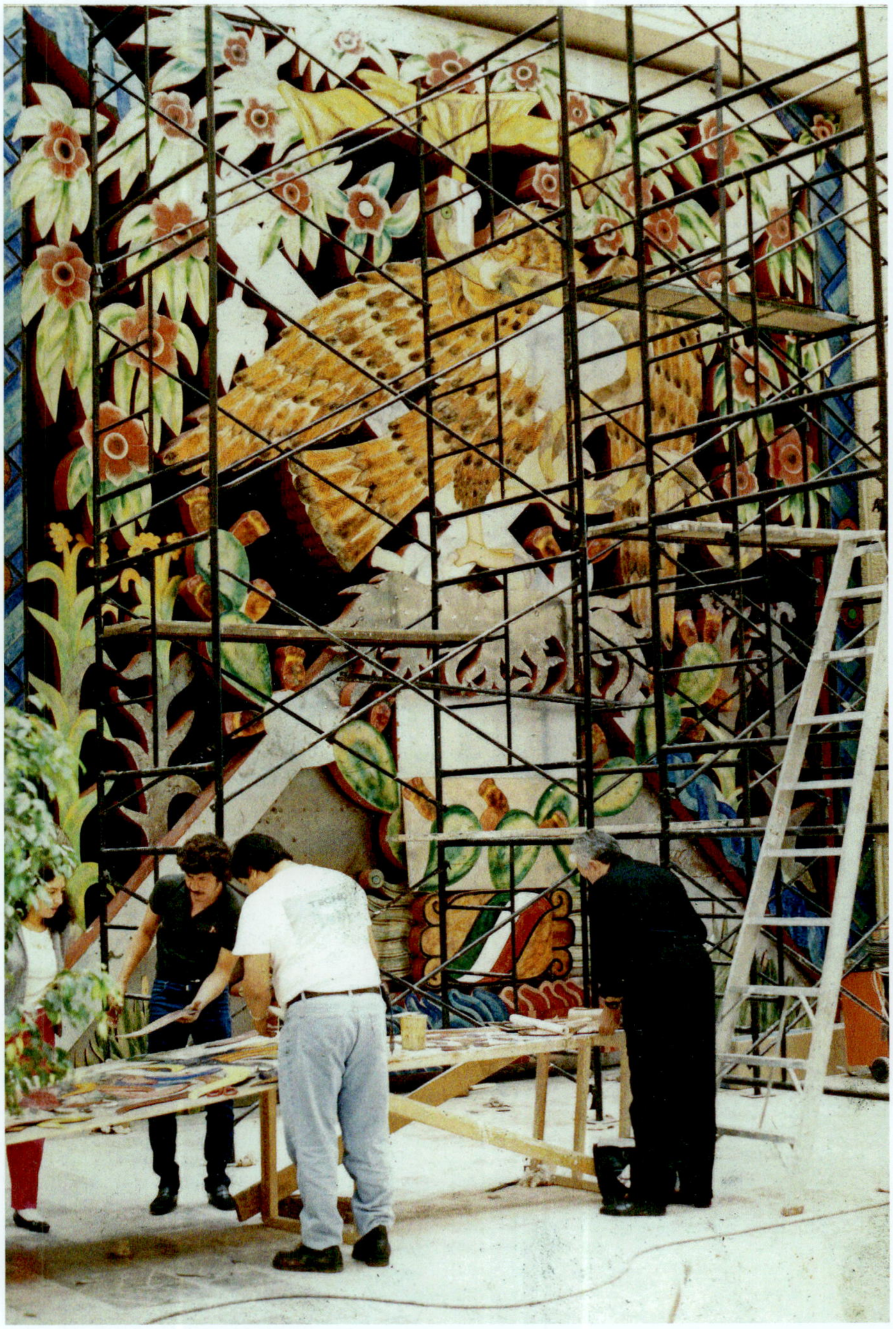

Plate X-2 Adrian Brun, with assistants, installing *The Eagle and The Serpent* mural. Photograph provided by Adrian Brun

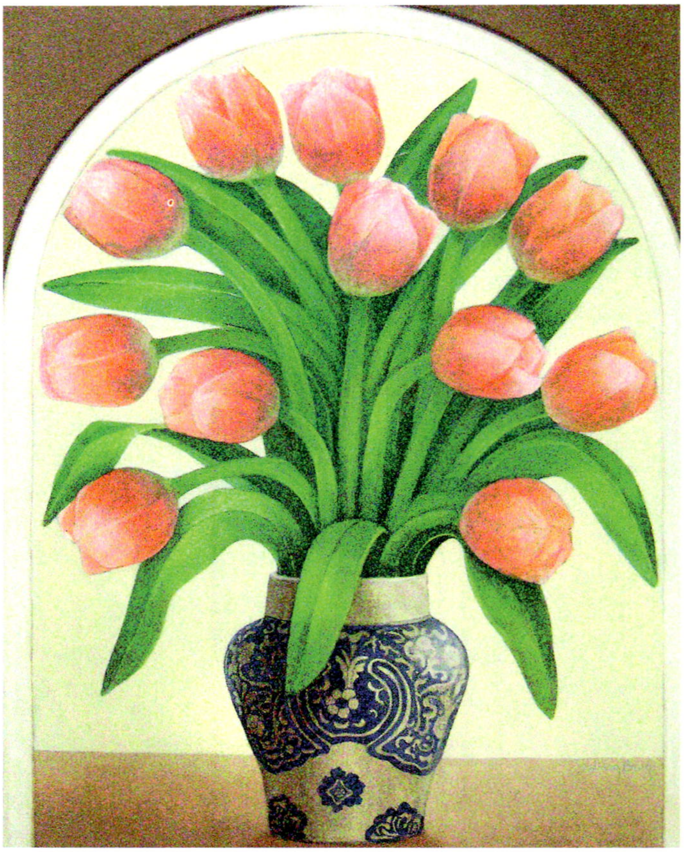

Plate X-3 Adrian Brun. *Flowers in a Niche*. enamel on copper. 80 x 60 cms. Adrian Brun Collection

be considered tranquil or exciting, formal or heroic? Will it illustrate the character of the place?"

Adrian Brun has felt strongly the influence of Diego Rivera, in that he believes art should be public and accessible for people. When Brun was director of the school of design, he focused on the concept that even utilitarian objects must be elegant and balanced, and that the quality of art is in its form. Brun believes that in all cultures, beauty is considered a necessity, and that people in all parts of the world search to express themselves in beauty.

According to Brun, Diego Rivera and William Spratling reinterpreted the symbols of Mexico, and these carry great importance for him as well. At the same time, Brun is passionate about world art, particularly the intention of the artist to enlarge upon life's experience. Because all artists extend the quality of life, for Brun, it seems logical that there would be an affinity among artists of all periods and places. Artists represent the best of what it is to be human. Brun senses that affinity; he puts forth the example of the grand Aztec sculpture of Coatlicue, in which fifteenth- and twenty-first-century observers can discover beauty within a monstrous sculpture. While we cannot relate to human sacrifice, we can appreciate the work of the artist, particularly his attempt to express something deeply meaningful.

For Adrian Brun, "the idea of an enamel mural is a surprise because it is unique, and so it inspires conversation. The mural is elegant, one might say, jewel-like, because enamel resembles the colors of jewels." The mural at the Centro Médico is a fine example of the technique that Brun employs. He applies enamel to copper in pieces that resemble a jigsaw puzzle. The surface of the copper must be clean or the enamel will not adhere. The level of the enamel is crucial: too

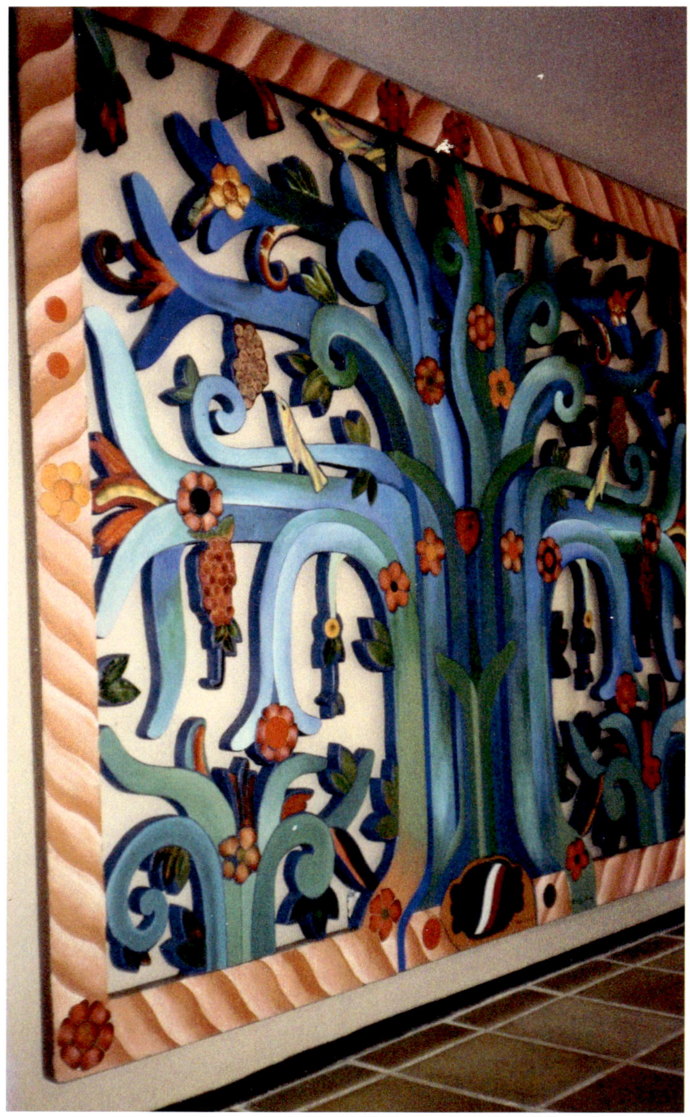

Plate X-4 Adrian Brun. *Tree of Life* mural, enamel on copper. 2.45 x 1.84 m. Photograph provided by Adrian Brun. Collection Dr. Jorge Aguilar, Cuautla, Morelos

much and the enamel will separate from the copper, too little and the enamel will have dark indentations. The enameled segments are placed into the kiln at about 800 C. degrees before being reassembled on-site. Colors have different values in heat. In Brun's mind, the beauty of the material is in its unpredictability and instability, so that the artist is never completely sure what will emerge from the kiln.

Brun's experiments with the material have included the use of gold beneath translucent enamel, just as Margot did in her jewelry. He has also put enamel over sand to create texture. He is emphatic that he does not want the enamel mural to resemble a painting but, instead, to be true to its jewel-like characteristics.

Brun works constantly. "Enamel is my passion — it is an incredibly beautiful material, always surprising. Enamel has allowed me to express my feelings as an artist." He currently is working on a series he has entitled, "Visitors to the Garden." He will represent the plants, trees, and flowers as he imagines they would be experienced by butterflies, birds (especially hummingbirds), and animals. Another project, which is more iconographic than representational, is a collection of thirty variations on the symbol of Mexico — the eagle with the serpent.[172]

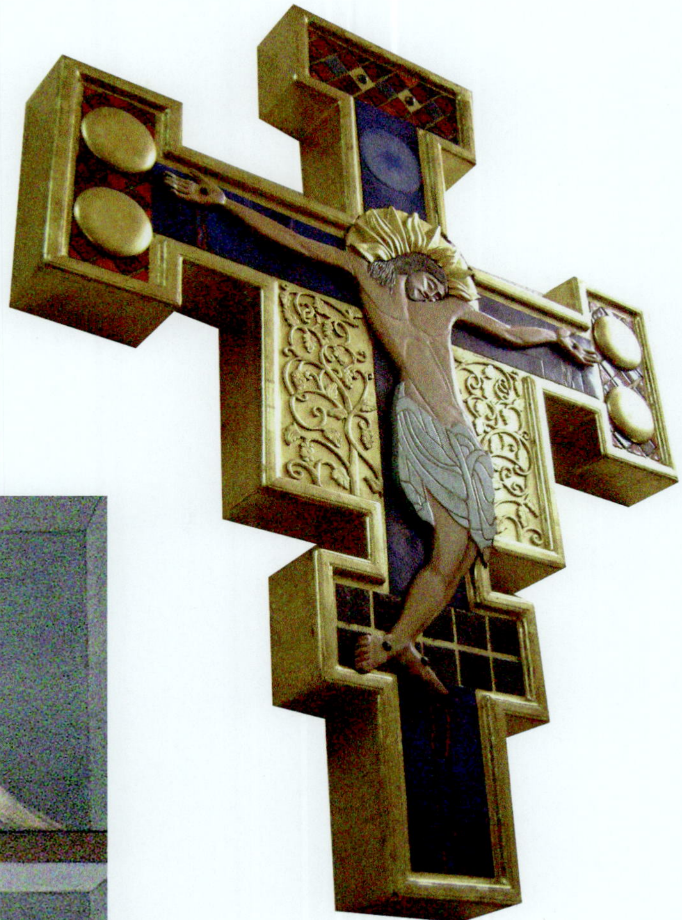

Plate X-5 Adrian Brun. *Crucifix*, enamel on copper. 1 x 1.5 m. Photograph provided by Adrian Brun. Adrian Brun Collection

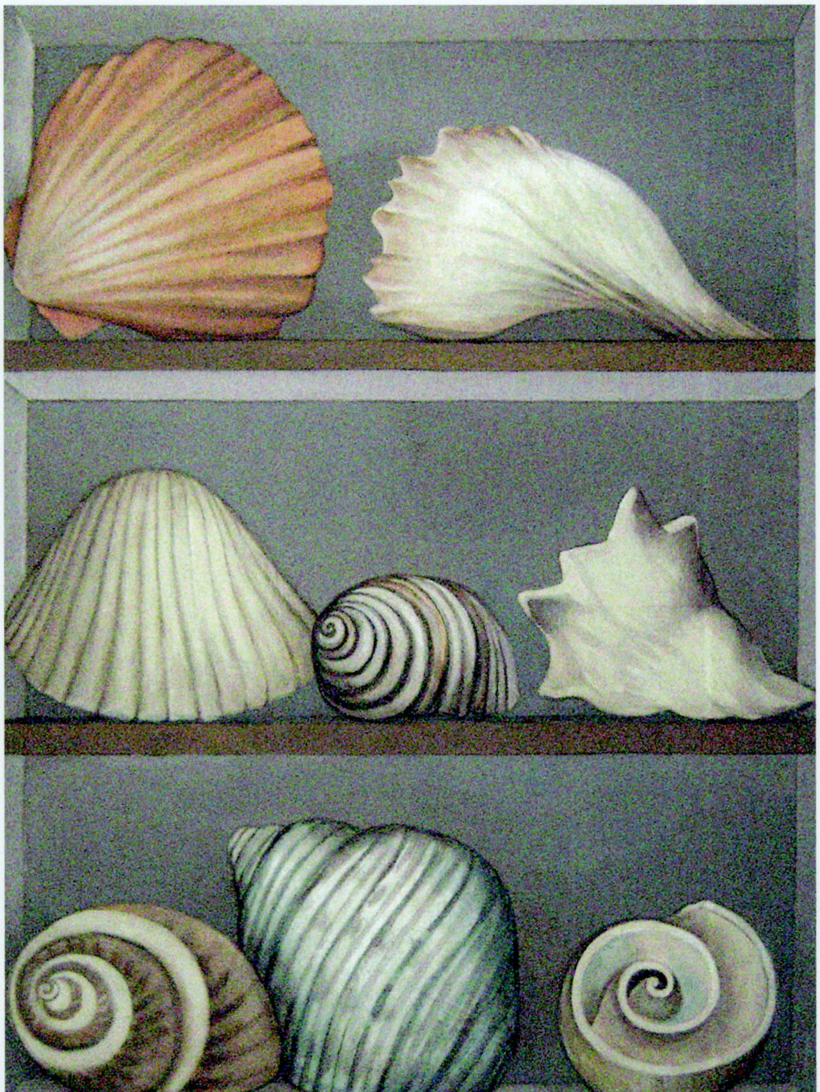

Plate X-6 Adrian Brun. *A Display of Shells*, enamel on copper. 80 x 60 cms. Photograph provided by Adrian Brun. Adrian Brun Collection

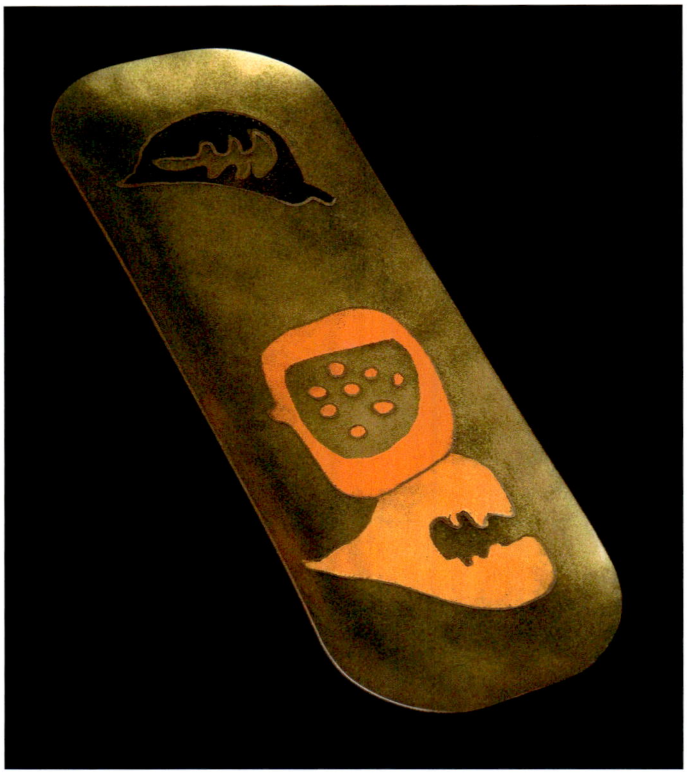

Plate X-7 Maggie Howe. *Fruit and Leaf* tray 3.75 x 10.38 in. Rocio Sierra-Honigmann Collection

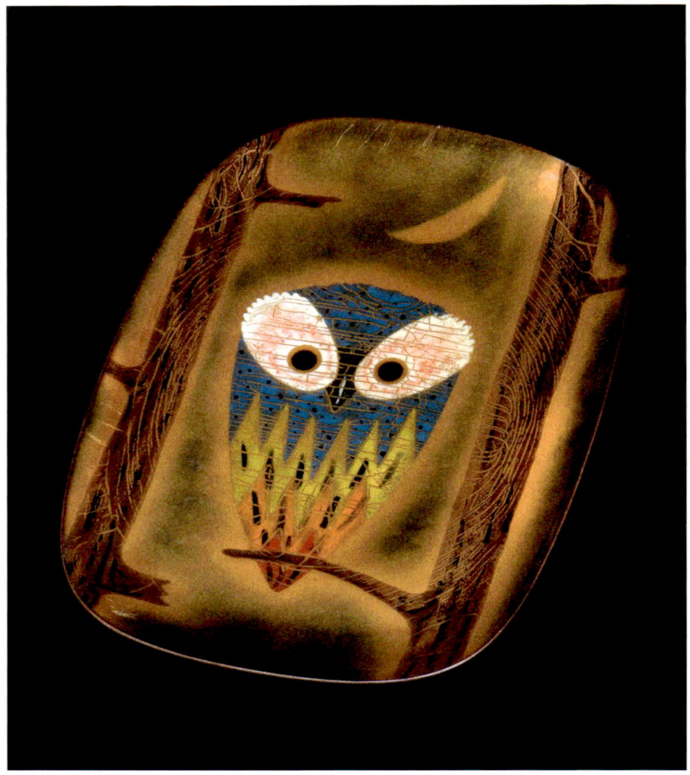

Plate X-8 Miguel Angel Pineda with Los Castillo mark. *Owl* tray 0 8.5 x 6.38 in. Rocio Sierra-Honigmann Collection

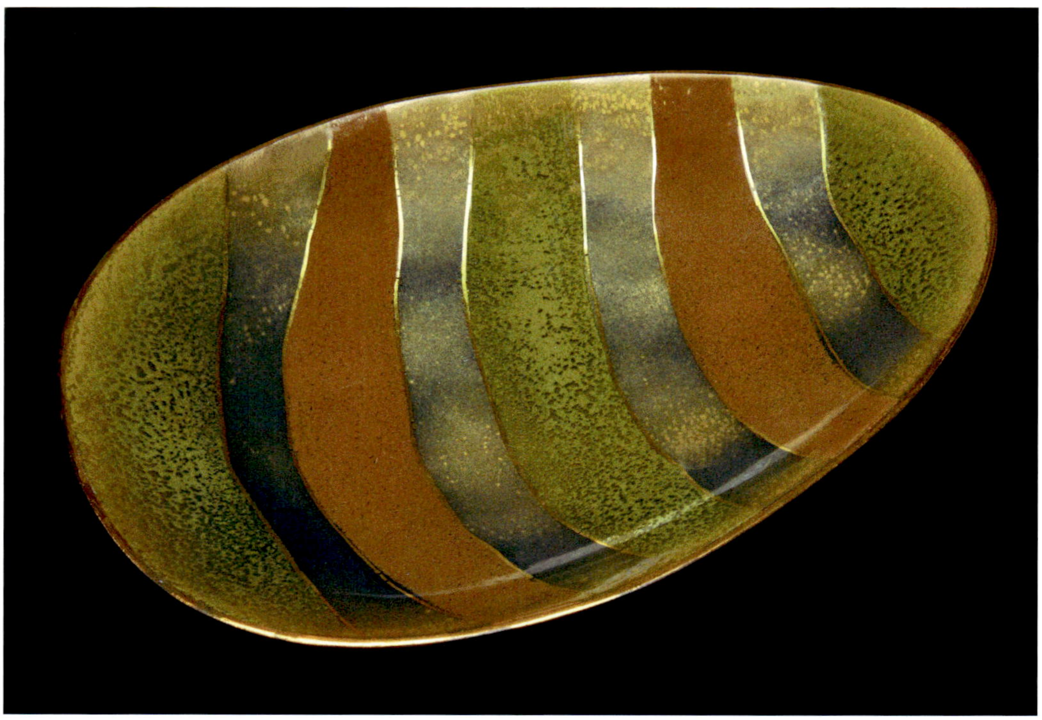

Plate X-9 Maggie Howe with Los Castillo mark. *Striped Egg* plate 4 38 x 5.75 in. Rocio Sierra-Honigmann Collection

Miguel Angel Pineda Godoy

Miguel Pineda devotes his life to art, believing that work should bring pleasure. ["¡El trabajo debe ser un placer!"] Like other enamelers, he is a painter first, with a love for color. Aesthetically, he is particularly drawn to the Byzantine and early Medieval European approach to form. He is equally passionate about the art of Mexico's past, with a special appreciation for the Maya. [173]

From 1958 to 1960 Pineda studied with Maggie Howe, an American enamelist who worked and taught in Cuernavaca. With a group of other visionaries, Howe assisted in the establishment of the Bazar del Sábado in San Angel Coyoacán, a revived colonial section of Mexico City. Howe produced trays, bowls, and other decorative pieces in *champlevé, cloisonné, basse-taille*, and other techniques, which she sold in the Bazar (Plate X-7).[174] Pineda acquired his technical skills from Howe and was influenced by her style. However, Pineda has remained committed to experimentation, which has led him to work in an ever-increasing variety of techniques and styles.

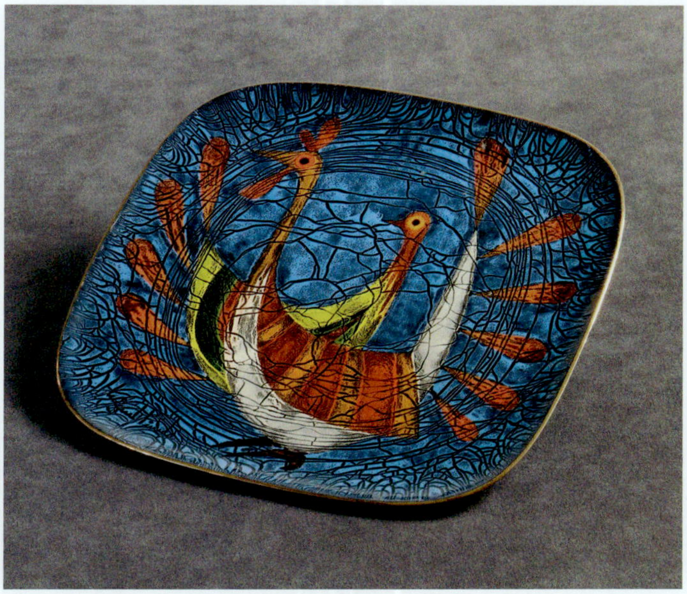

Plate X-10 Miguel Angel Pineda. *Birds* tray 6.5 in. sq. Rocio Sierra-Honigmann Collection

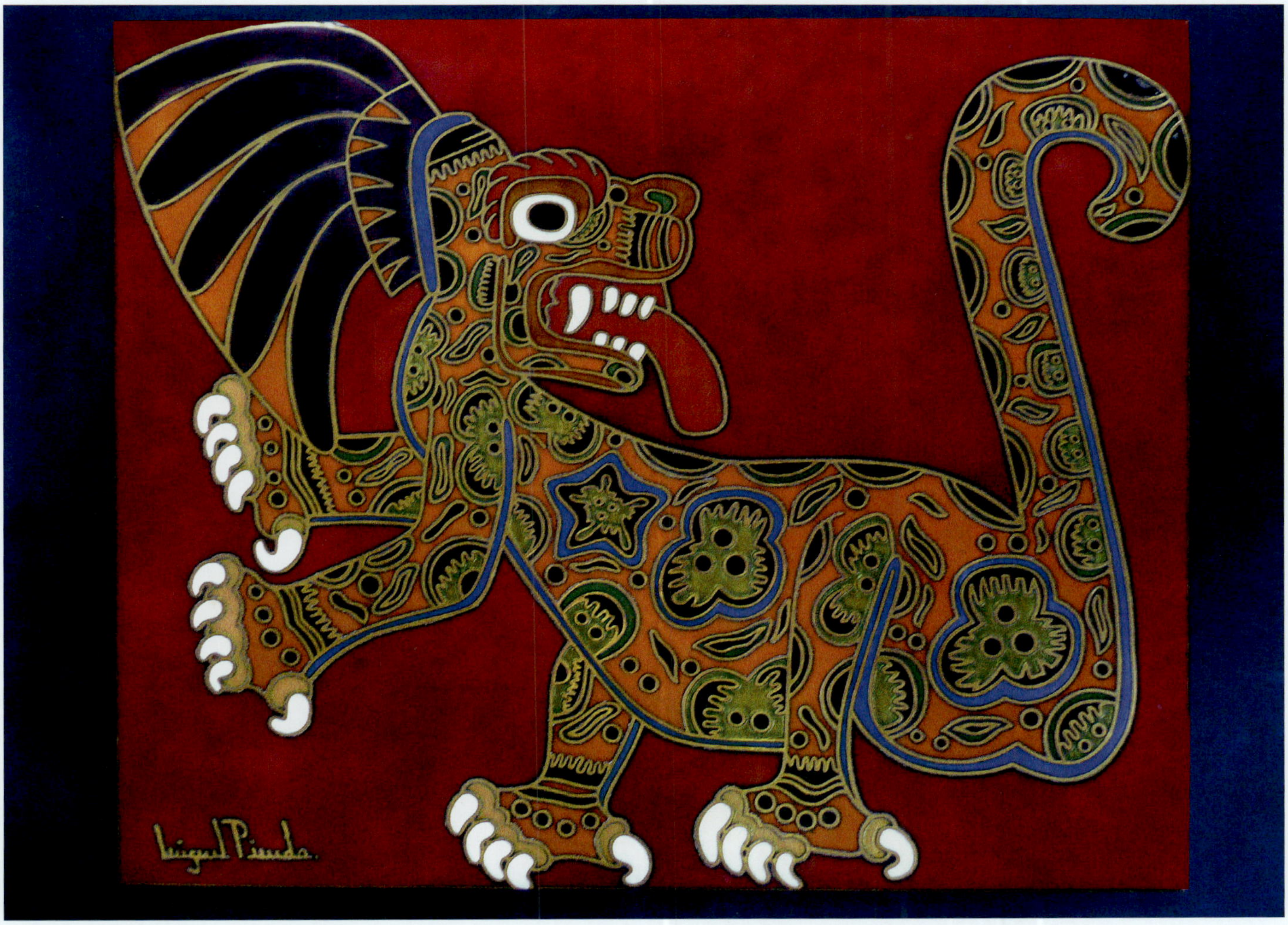

Plate X-11 Miguel Pineda. *Aztec Tiger* plaque. Photograph provided by a collector. Patrick Kapty Collection

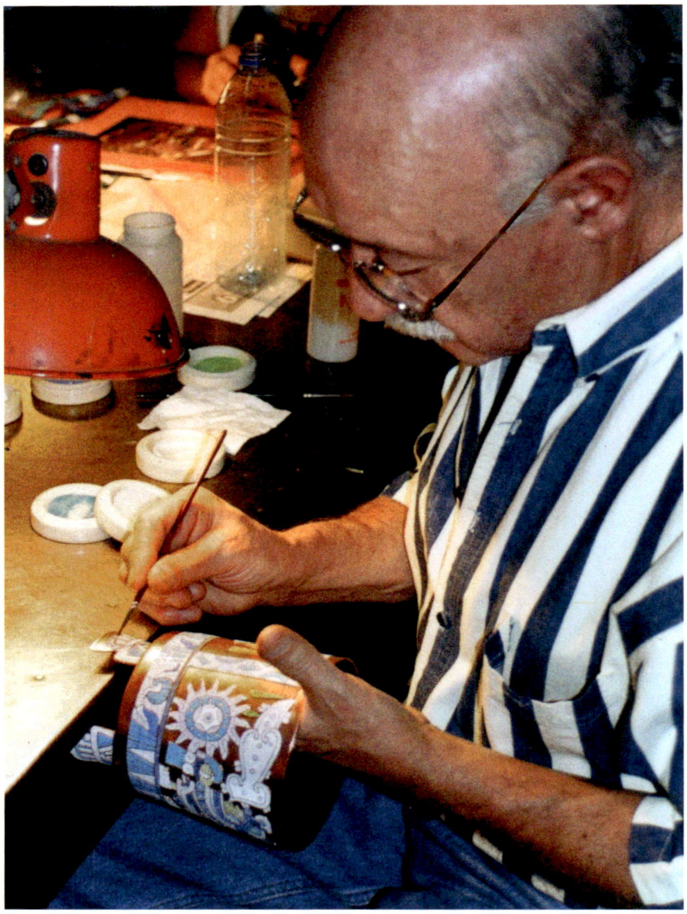

Plate X-12 Miguel Pineda applies enamel to copper in his workshop. Photograph provided by Miguel Pineda

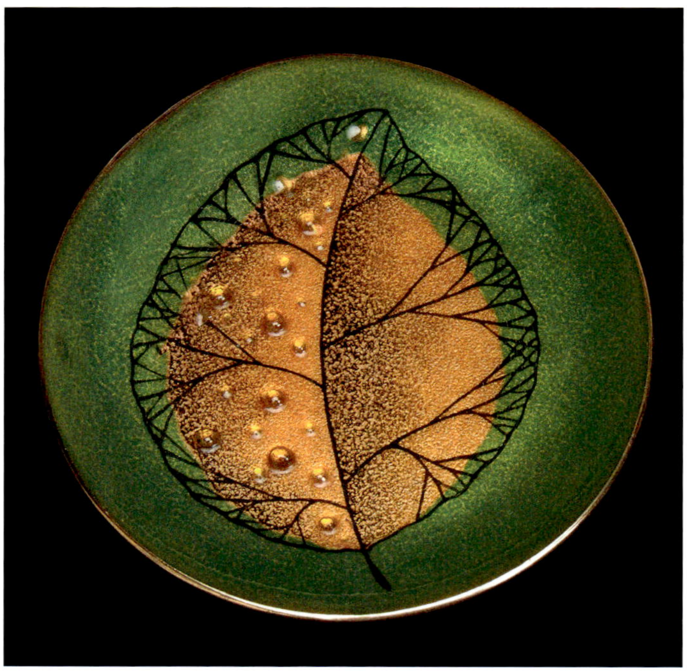

Plate X-13 Miguel Angel Pineda. *Leaf* bowl in orange and green 4.25 in. Rocio Sierra-Honigmann Collection

Success came to Pineda and Maggie Howe, who both sold their enamelwork at the Bazar del Sábado.[175] These two artists captured Antonio Castillo's attention and he brought them into the Los Castillo fold. Pieces can be found with the double signatures (Pls. X-8-9).

In the past, Pineda has purchased copper vases from Santa Clara del Cobre in Michoacán and applied enamel to the outer surfaces of the vases (Plate X-15). His attention to form inspires him to develop unique designs, bringing importance to the work of the coppersmith as well. Techniques he has used on copper in a number of contexts include stencil, wet-packing or wet inlay, painting with acid [grabado con ácido, pintura que corosiva], and *sgraffito*. His favorite technique continues to be *champlevé*. In his workshop, Pineda and his assistant Blanca de la Cruz paint color according to his designs (Plate X-12).

Pineda has named his company, Arte y Esmalte Pineda, S.A. He sells from his workshop and at the Bazar del Sábado, where his nephew Mariano Pineda has responsibility for the sales. He appeals to a wide range of sensibilities, as he continues to produce plaques, trays, bowls, and jewelry in a number of different styles.

The Bonampak vase is a technical tour-de-force (Plate X-14). The background has been given texture with acid and the individual figures are in *champlevé*. Pineda carefully studied the murals to achieve an accurate but unique interpretation of the late eighth-century tomb paintings. We sense the movement and excitement of the moment as the costumed participants march in procession around the vase.

The plaque, *Man's Face and Nude Woman*, recalls the work of modernist Georges Rouault, in the use of heavy black outlines surrounding two-dimensional color fields (Plate X-16). The isolation of the figures suggests a sense of the tragic. The man seems to be wearing the mask of a clown, and his separation implies loss. The woman displays her sexuality while remaining unaffected by the presence of the man.

Like Adrian Brun, Pineda has produced large murals with cut and assembled enamel on copper pieces. At the same time, because Pineda has a large kiln, he is able to take on large-scale commissions. He produced a work on copper for the Sagrario in Monterrey and is working now on reliquaries of the saints for the Clínica San Rafael. He created the *champlevé* panels for the doors of the Shrine of the Virgin of Guadalupe, a work that took three years to design and produce.[176] In considering the monumental in enamel, Pineda's words define his mastery: "The enameling technique is complicated — one error and the entire piece is lost. It is such a rich medium, and can be applied to any setting."[177]

Plate X-14 Miguel Pineda. *Bonampak Mural* vase 6.5 x 19.38 in. Private Collection

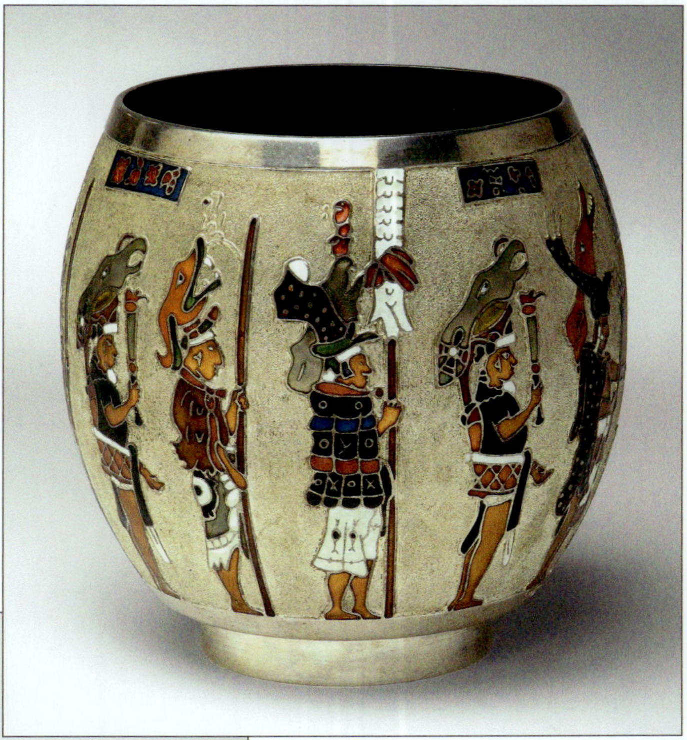

Plate X-15a & b Miguel Pineda. *Modernist Design* in enamel on Santa Clara copper vase. 13 x 28.5 in. Rocio Sierra-Honigmann Collection

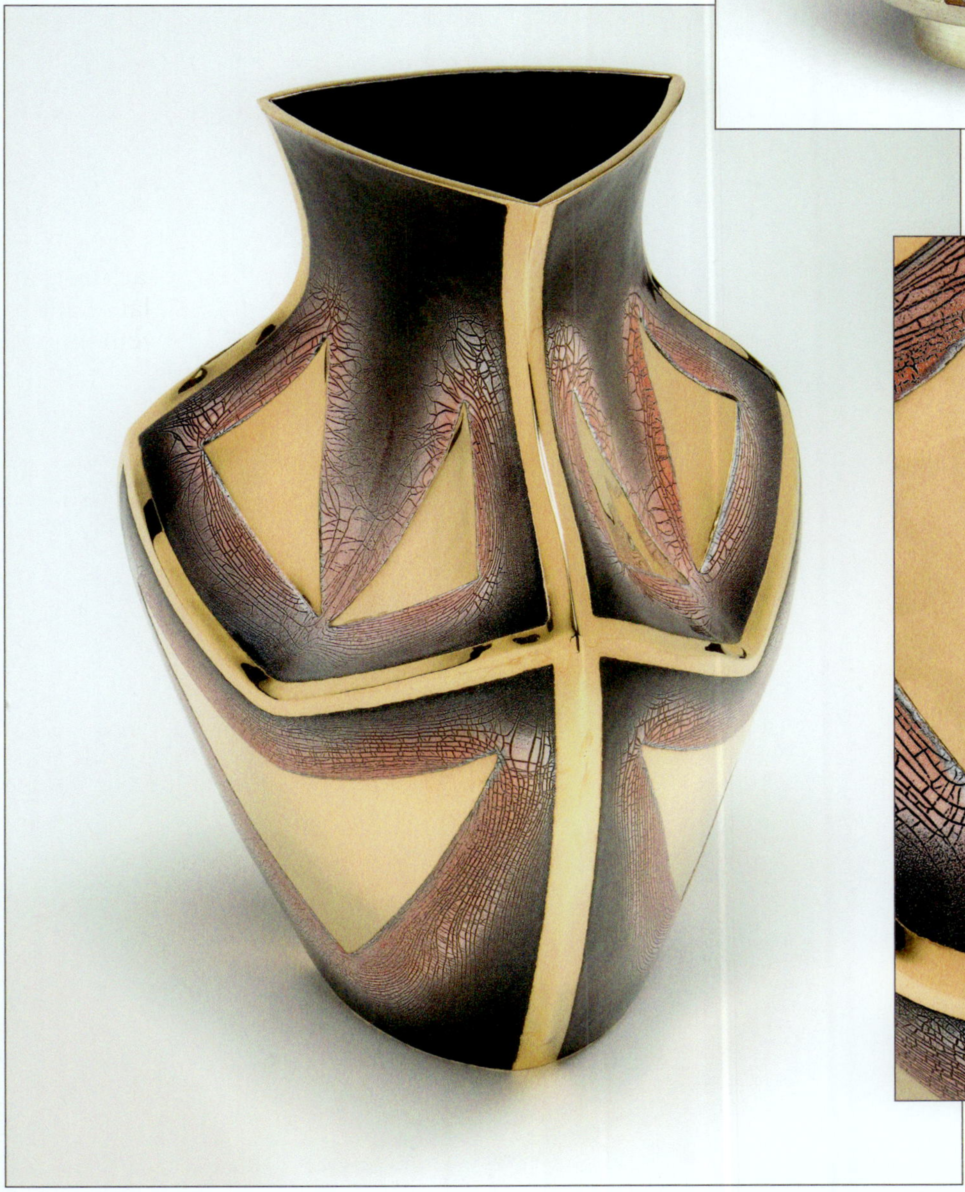

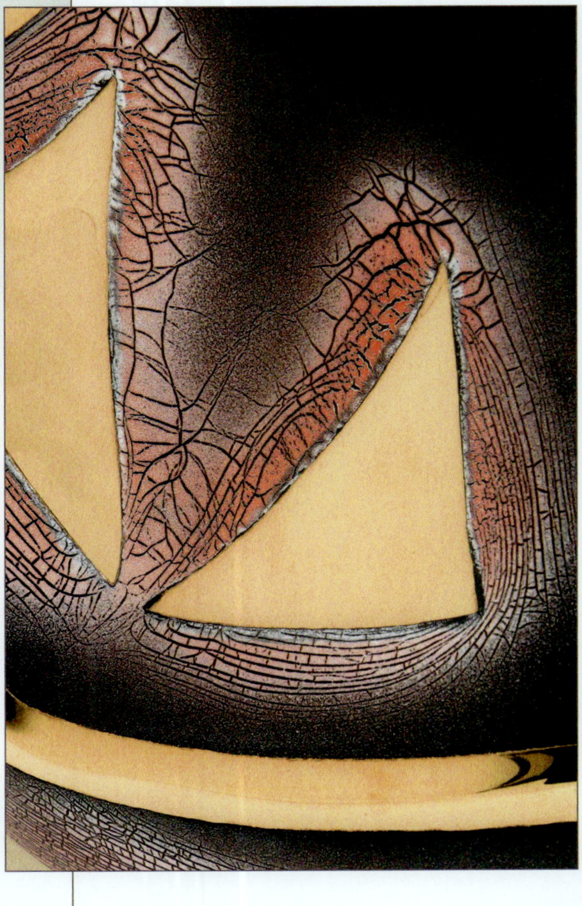

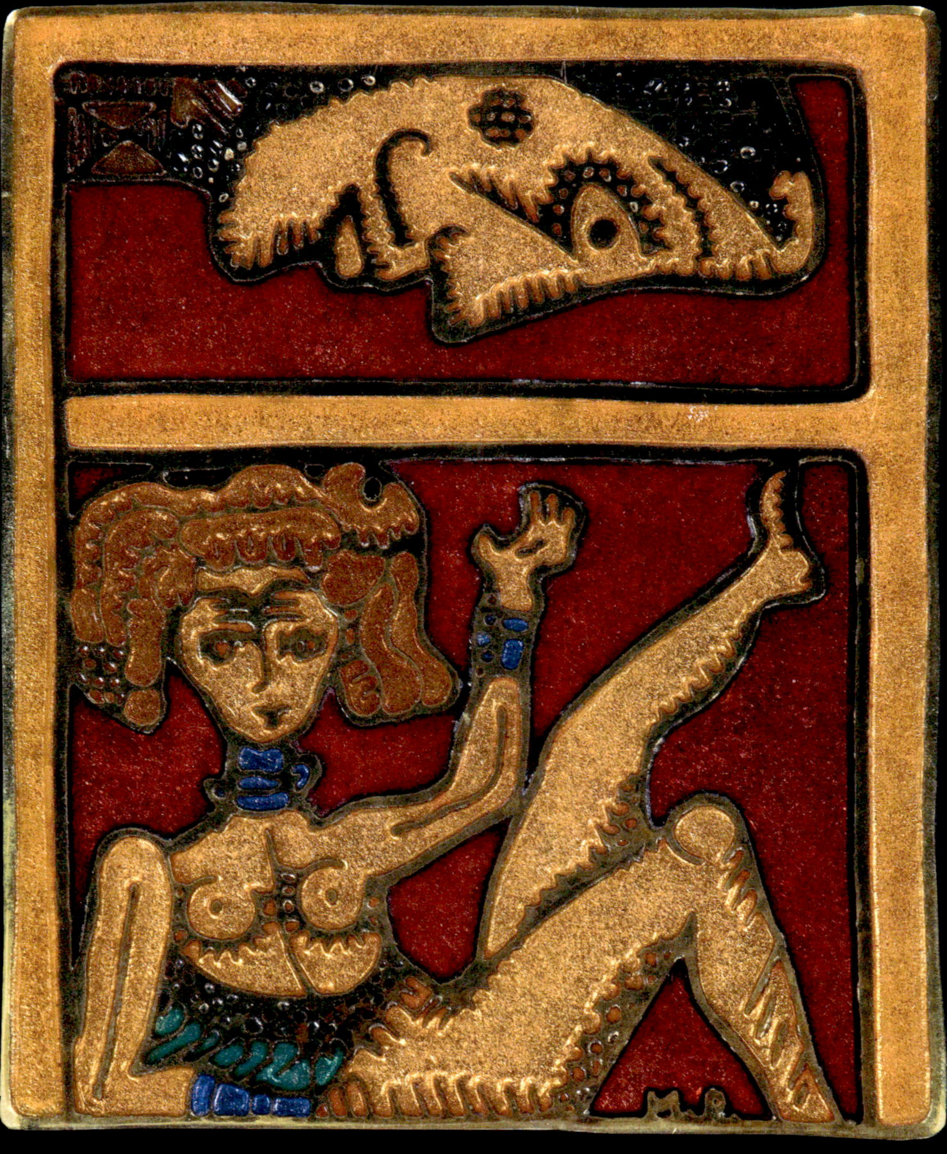

Plate X-16 Miguel Angel Pineda. *Man's Face and Nude Woman* plaque 4 x 3.25 in. Judy P. Steele Collection

Priscila Maria Canales Sanchez Leon

Priscila Canales studied painting at the Instituto Regional de Bellas Artes de Cuernavaca in Mexico before getting her degree in painting in 1992 at the Accademia di Brera in Milan, Italy. Upon her return to Mexico, Priscila made a significant change in the direction of her training when she chose to study silversmithing and enameling from 1994 to 1996 at the Centro Artesanal Independencia in Mexico City.

Priscila collaborates with her husband José Dionisio León in the workshop. She is one of the few designers to actually produce her own work (Plate X-17). Her jewelry combines *repoussage* with semi-precious stones or enamel. She writes, "Our jewelry is totally fashioned by hand. All parts are assembled one by one. Although designs are repeated, each is individually produced and unique. Silver is our principal medium of expression and we enjoy giving color by adding semi-precious stones and enameling. I believe that the techniques one applies must relate directly to life changes one encounters, an expansion of learning related to experience. Thus, silver evolves into a form of communication, as *repoussage*, chasing, and lost wax become elements of a language one learns to 'speak.'"

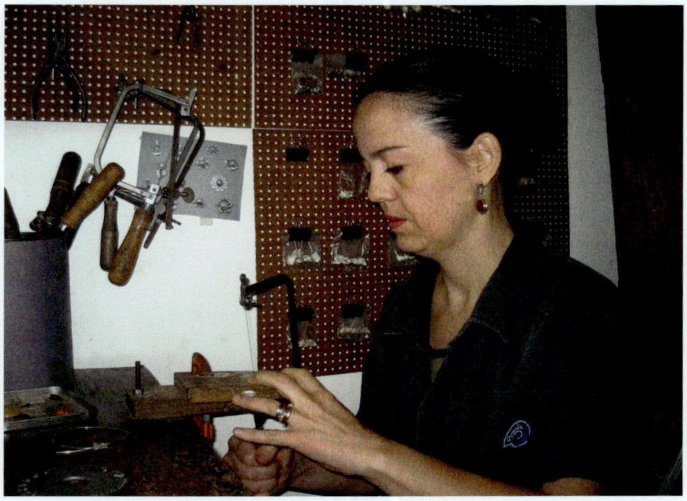

Plate X-17 Priscila Canales in the workshop. Photograph provided by Priscila Canales

In the case of Priscila's design, *Bugambilia*, the painterly application of enamel on an oval dome implies the natural coloration of a stone (Plate X-18). In contrast, the butterfly's wings fully extend to emphasize their symmetry and the large areas of solid color (Plate X-19). This brooch/pendant has a graphic quality as opposed to the feminine elegance of the necklace. While Priscila's work embodies a youthfulness of spirit and a focus on nature, her stylistic approach unquestionably emanates from the strength of Mexico's artistic cultural heritage.[178]

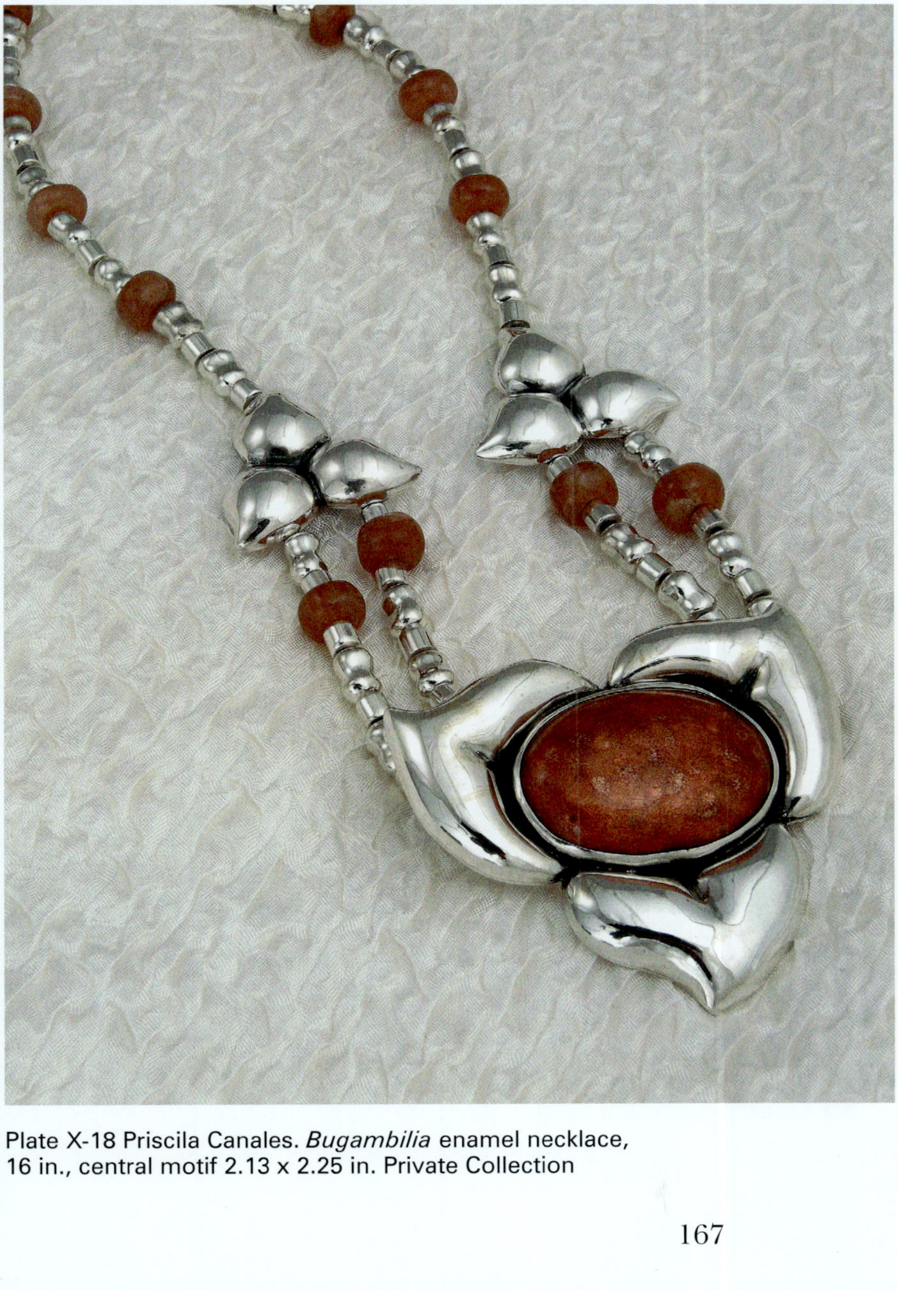

Plate X-18 Priscila Canales. *Bugambilia* enamel necklace, 16 in., central motif 2.13 x 2.25 in. Private Collection

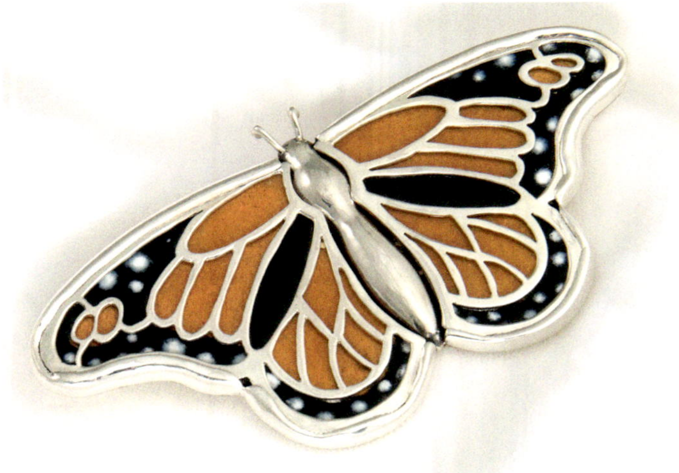

Plate X-19 Priscila Canales. *Butterfly* enamel pendant/brooch, 2 x 1.88 in. Private Collection

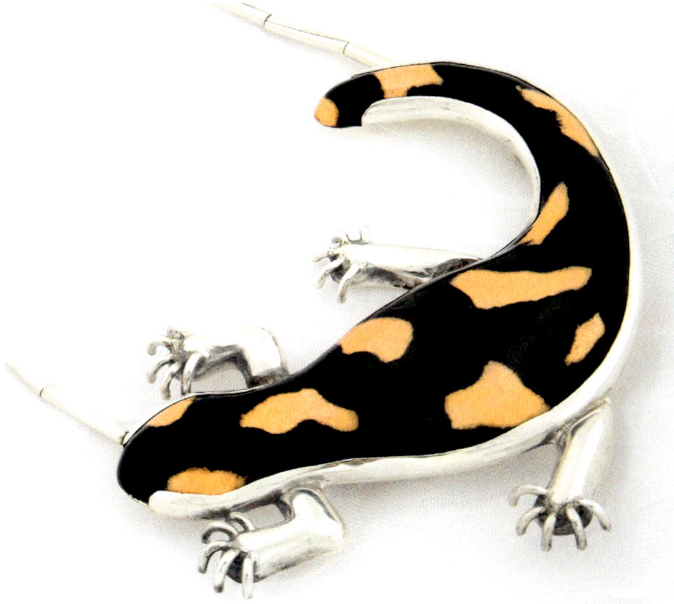

Plate X-20 Priscila Canales. *Lizard* pendant, enamel on silver, 3 x 3 in. Private Collection

Siguifrido Pineda

Sigi Pineda has worked in many different media and has encountered just about everyone in the Mexican silver world (Plate VII-8). It is awe-inspiring to consider that, with only minimal formal education, Sigi, like Antonio Castillo and many of his peers, has achieved so much in his lifetime. He is self-taught, whether as a silversmith, enamelist, guitar-maker, draftsman, or designer.

After leaving Margot de Taxco, Sigi started his own workshop in an old building next to Piedra y Plata, the workshop of Felipe Martínez. An American from Los Angeles who was in the film industry arrived in Taxco and proposed a collaborative venture. This man took Sigi's silver designs to Mexico City with the promise of sales but then did not return to Taxco with the money or the silver. This outrageous theft forced Sigi to close his store and workshop.

Sigi's next offer was one he tried to refuse. Sigmund Mezger wanted to do business with Sigi but, with the memory of the recent fiasco fresh in his mind, Sigi had no enthusiasm for the project. Sigmund kept coming until Sigi relented, and "pavonado" was born. The pearl-like oxidizing finish on silver sold particularly well for Sigi in Europe.

Over the years, Sigi has opened several stores in Taxco and Mexico City. One of his enterprises was particularly ingenious. He had a box made up with compartments and a glass top. Every day he and a helper placed fifty different pieces of jewelry in the compartments and the box was circulated to a different beauty salon in Mexico City each day. The salon owner received one third of the profit for the sales, another third went to Sigi, and a third covered cost of materials and labor. The project, which lasted a couple of years, was financially quite successful.

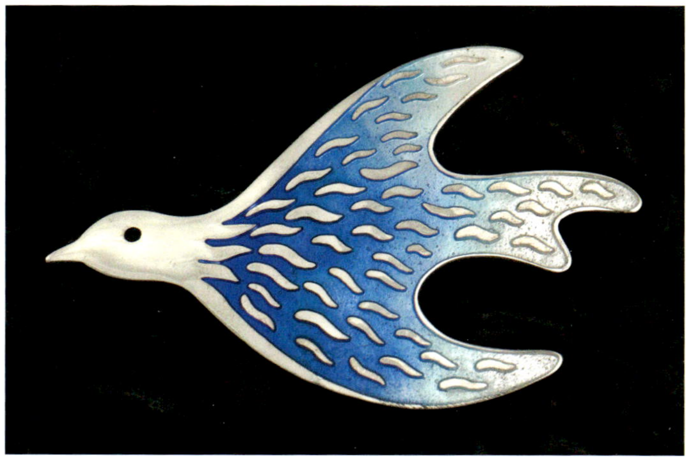

Plate X-21 Siguifrido Pineda. *Blue Bird* enamel brooch, 1.63 x 2.5 in. Luisa di Pietro Collection

Recently, Sigi has returned to enamel on silver and his experiments are at once whimsical and elegantly simple. Using the lost wax method, Sigi has created the silver form of the bird as a two-dimensional dynamic shape, with the feathers widening towards the tail (Plate X-21). The enamel, painted on the bird's body, changes hue. Together these aspects of the design imply flight.

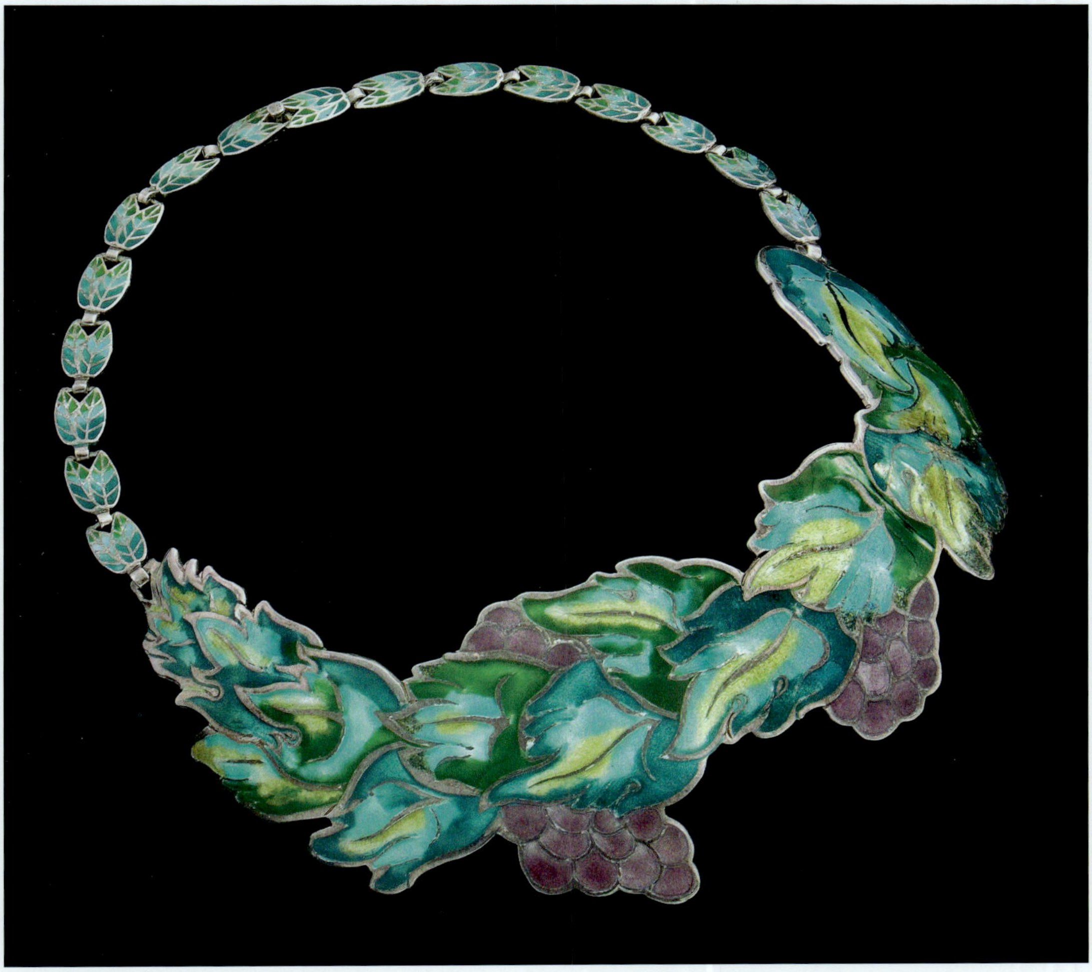

Plate X-22 *Floral Spray* necklace. Antonio Pineda crown mark. Stella and Fred Krieger Collection

Enamel Jewelry In Yucatán

The village women of the eastern and southeastern part of Yucatán favor the incorporation of enamel in their jewelry, which they buy in the market. The majority of enamelists in Yucatán work in Valladolid, with a few in Mérida.[179] These artisans paint opaque enamel onto silver or gold. The enamel jewelry takes various forms as pendants, bracelets, earrings, and rings. Among the shapes are the rhombus, flower, star, fan, peacock, heart, bead, and bell. Tiny areas of color are often interrupted by a word, "cariño" or "amor."[180] This work has its own charm and is produced according to tradition.

The Grapevine Necklace

Antonio Pineda's maker's mark appears on the back of a large-scale enamel necklace with painterly qualities (Plate X-22). Since his name is not synonymous with enamel on silver, another jeweler must have been responsible for the enameling. The first person asked to solve this puzzle was Jaime Quiróz. Imagine the surprise to discover that he had actually done the work himself!

The piece was originally a crown meant for the sculpted figure of a saint. The chased outline of the grapevine brought the crown to the attention of Pedro Domecq, owner of Casa Pedro Domecq Wines. Domecq bought the crown about twenty-five years ago, and asked Antonio Pineda to make something of the piece.

Pineda brought the crown to Jaime Quiróz, with the intention of enhancing the silver surface with enamel. To keep the work in the family, Quiróz had Bruno Pineda (Antonio's brother) flatten out the curve of the crown. Both Quiróz and Vicenta Rodríguez, who had been an enamelist in the Margot de Taxco workshop, applied the enamel according to the chased lines that were still visible on the crown. Using the die for Margot's design number 5839A, Quiróz made the chain for the pendant and Vicenta applied the enamel.[181] Thus, produced in c. 1990, the necklace is the stunning work of two people who learned their artistry thirty years earlier in the Margot de Taxco workshop.

Appendices

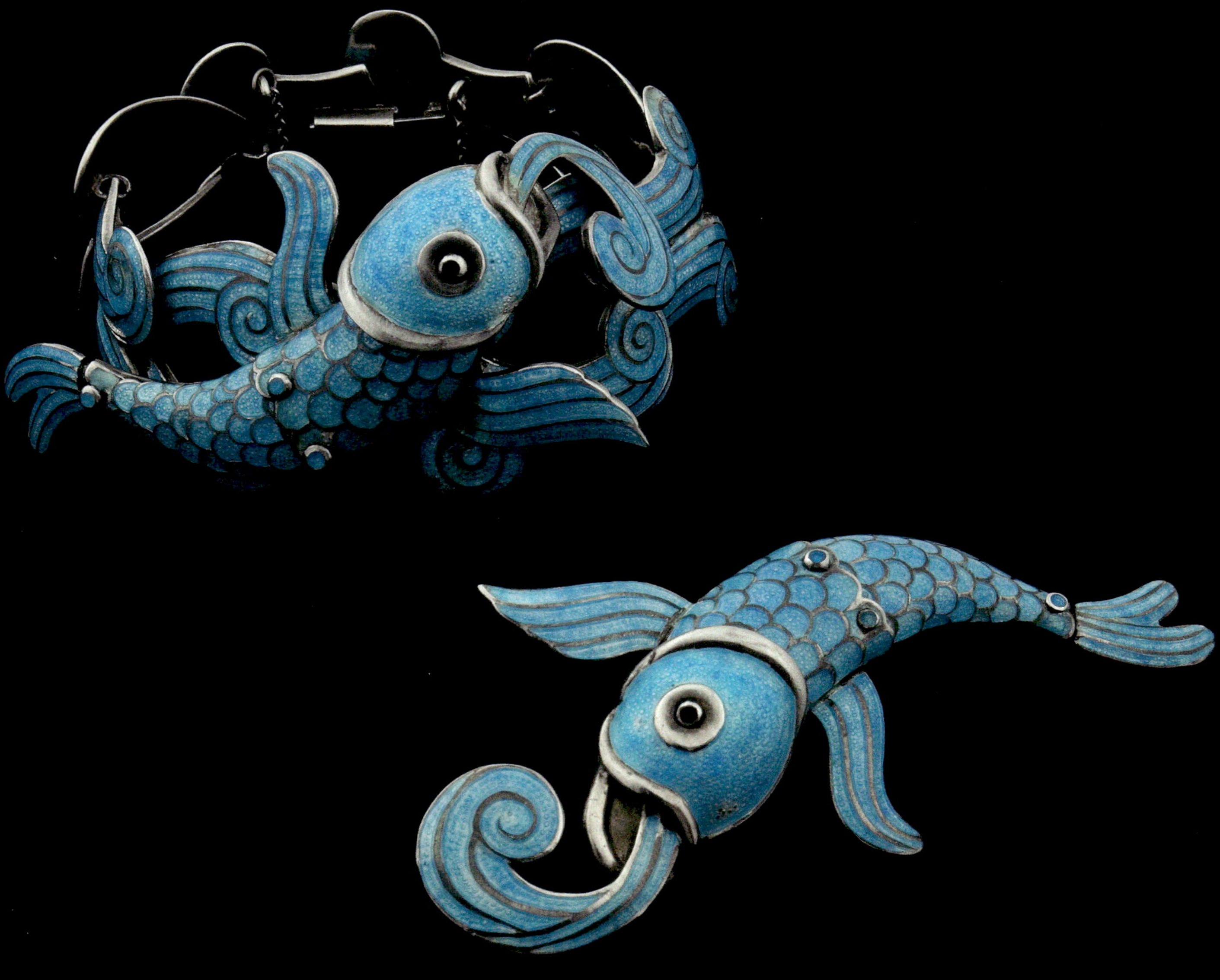

Margot de Taxco 5644. bracelet 7 x 1 in.; brooch 1.75 x 3.5 in. Stella and Fred Krieger Collection

Appendix 1.
Margot's Design Drawings

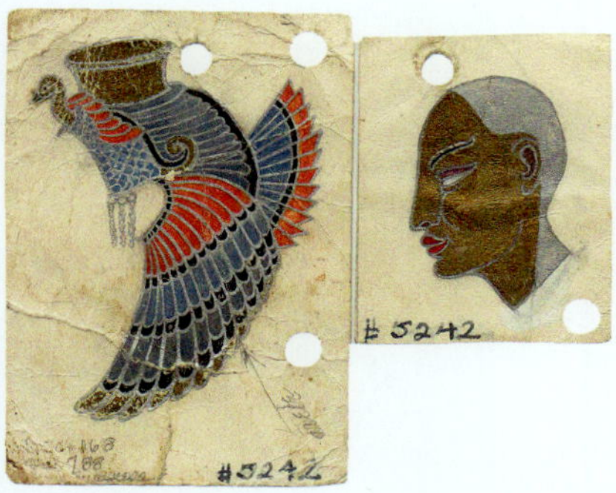

5242 Cleopatra

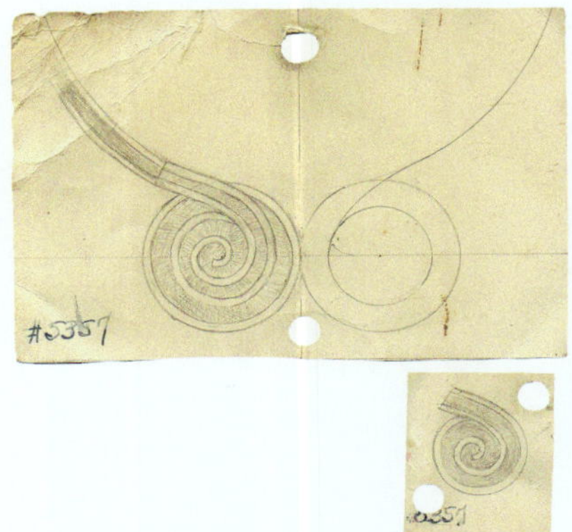

5357 Double Scroll

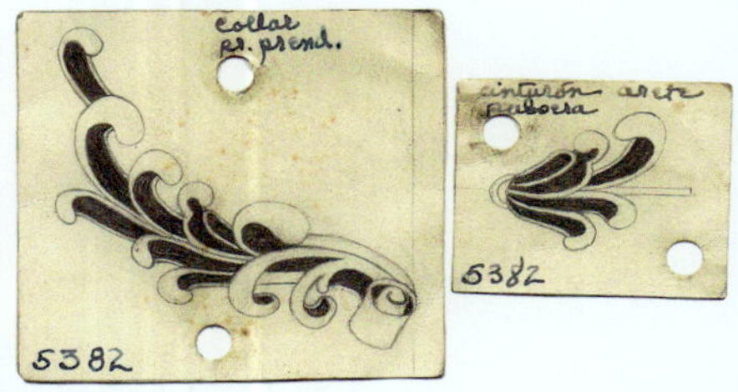

5382 Curled Leaf

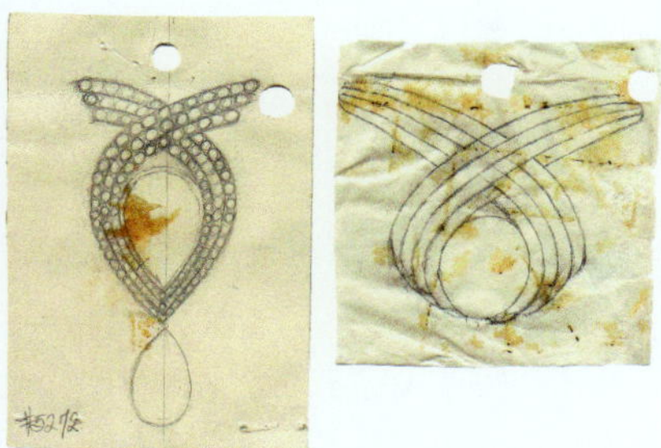

5272 Crossed Curve

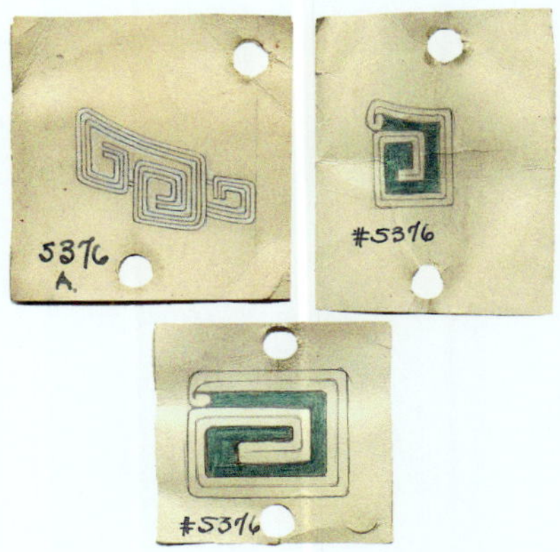

5376 Square Scroll

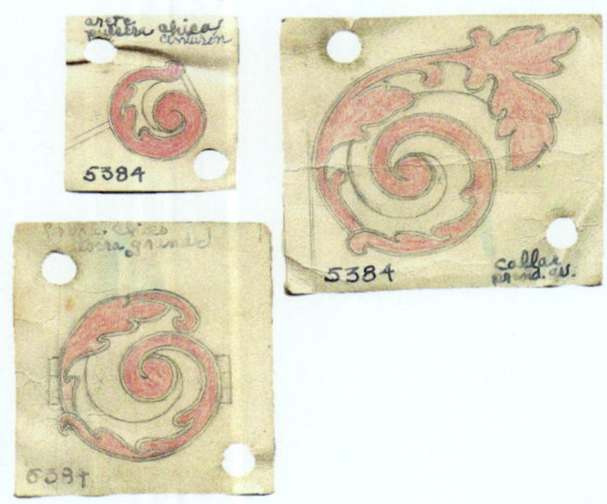

5384 Leaf Scroll

5280 Pre-Columbian Design

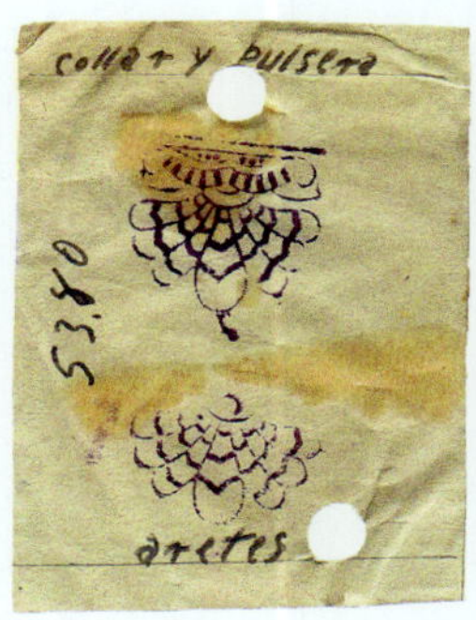

5380 Frills

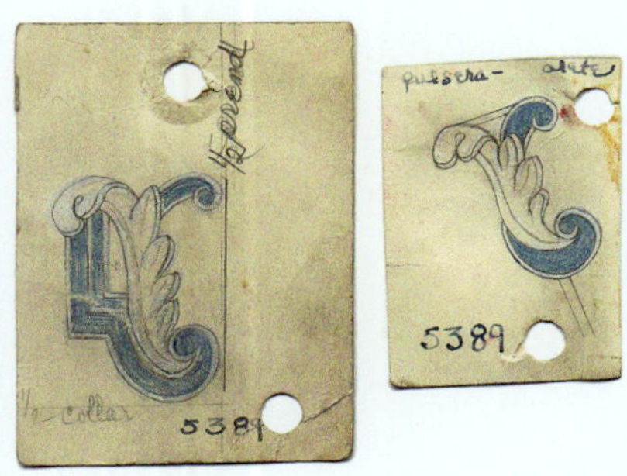

5389 Leaf & Scroll

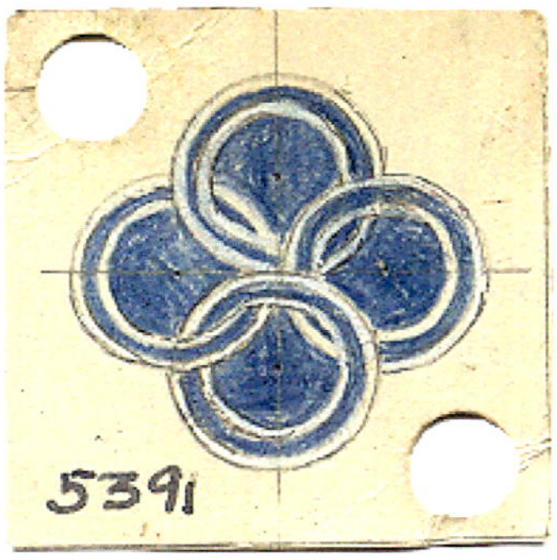

5391 Four Circles

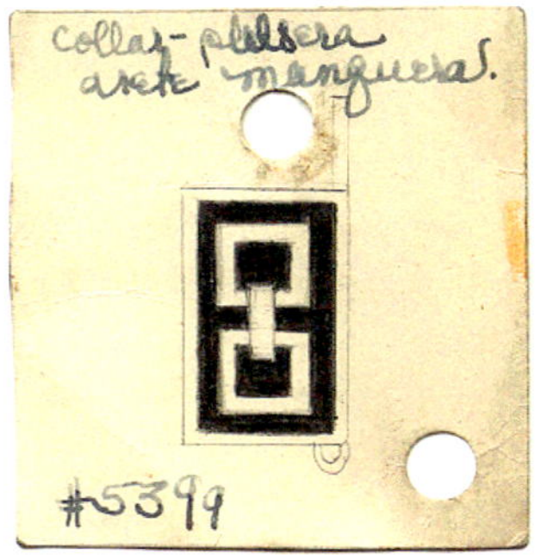

5399 Greek Key

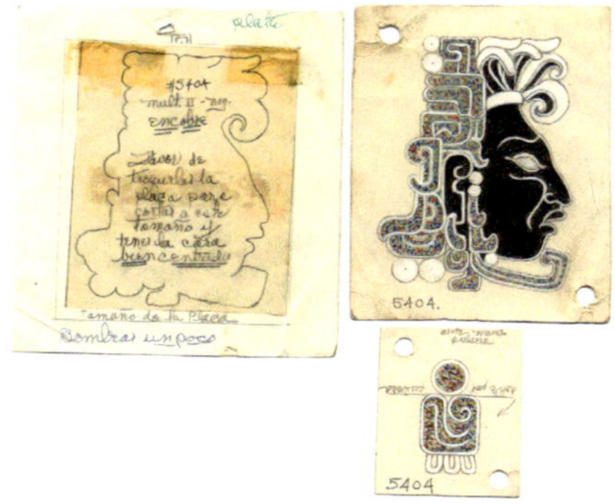

5404 Noble

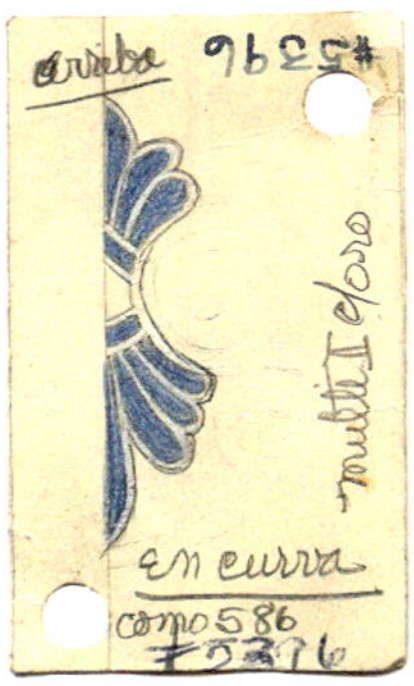

5396 Flared Bow

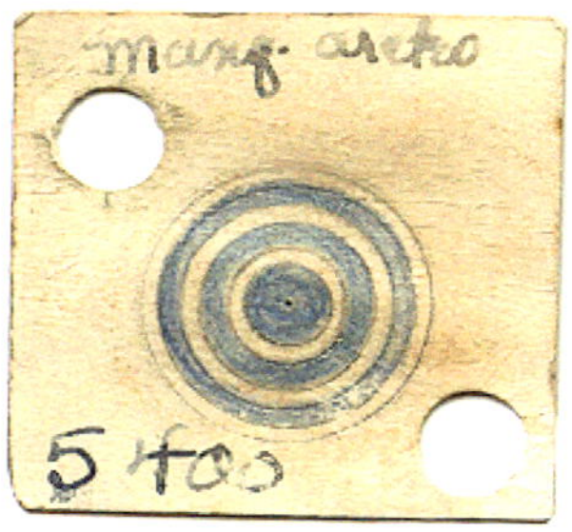

5400 Bull's eye

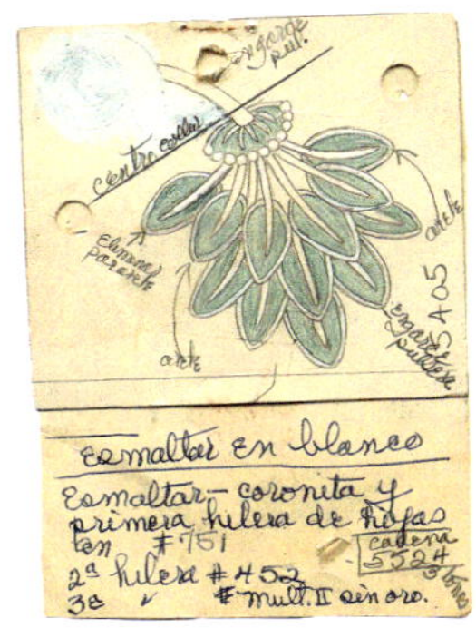

5405 Leaf Spray

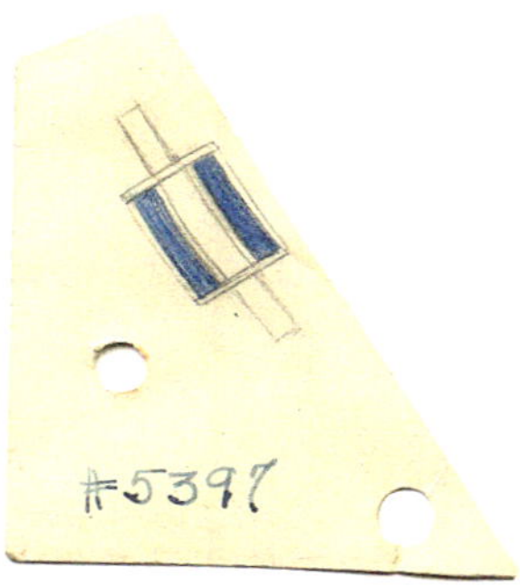

5397 Parallel Lines

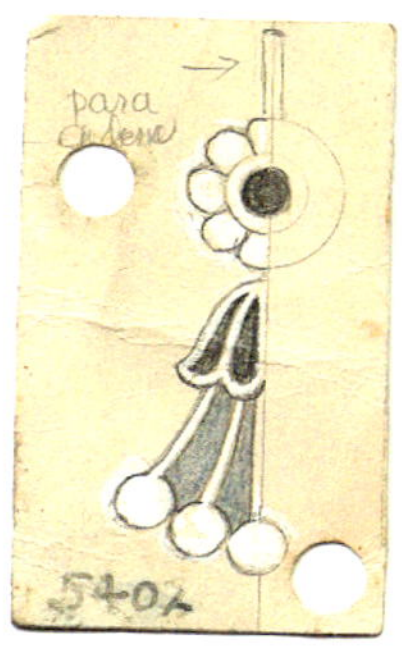

5402 Flower & Spray

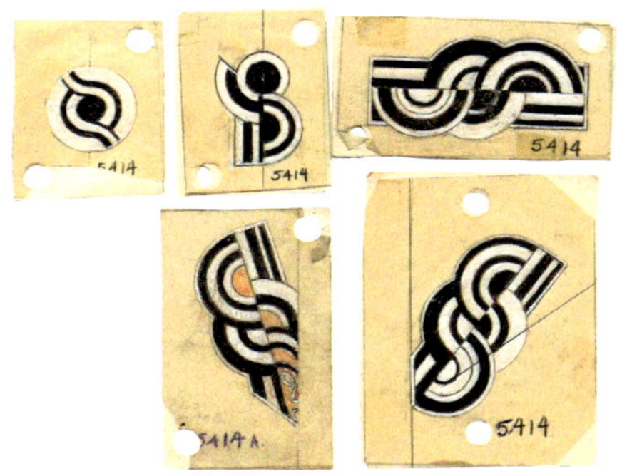

5414 Pointed Face

5398 Multi-color

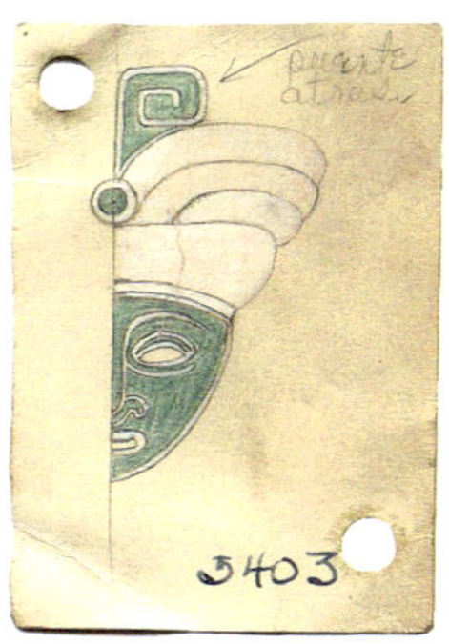

5403 Mask

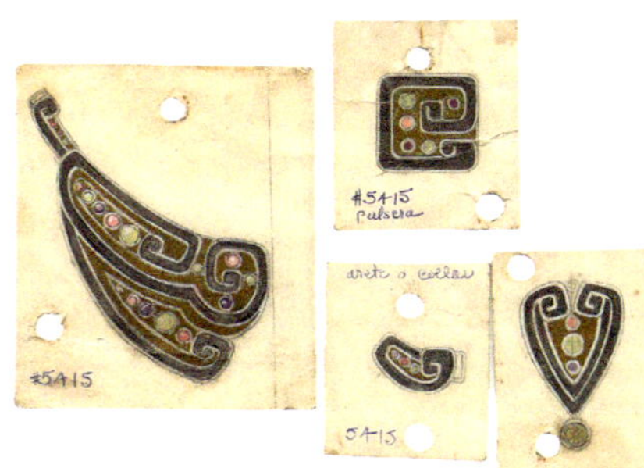

5415 Jeweled

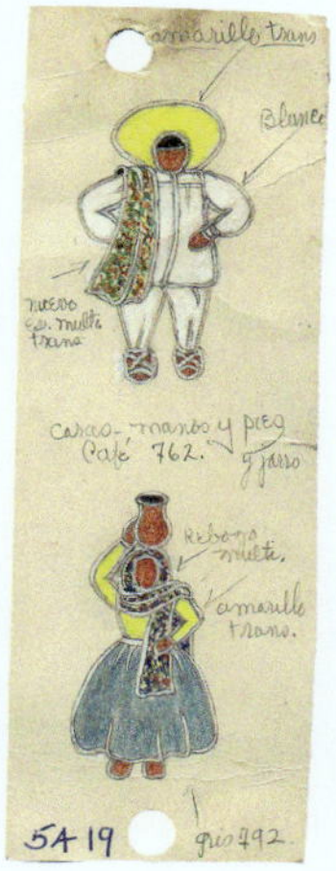

5419 16A5

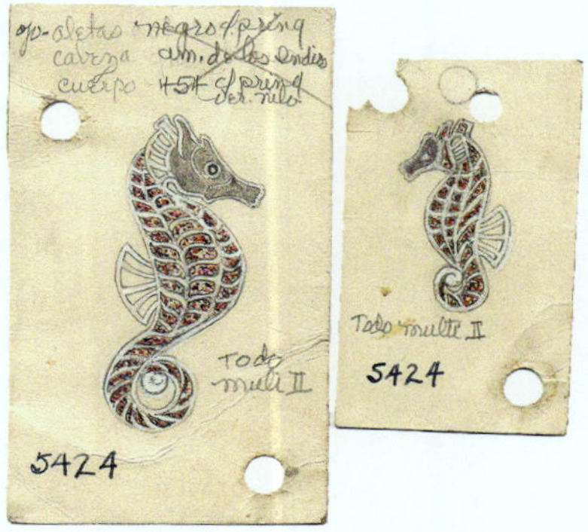

5424 Seahorse

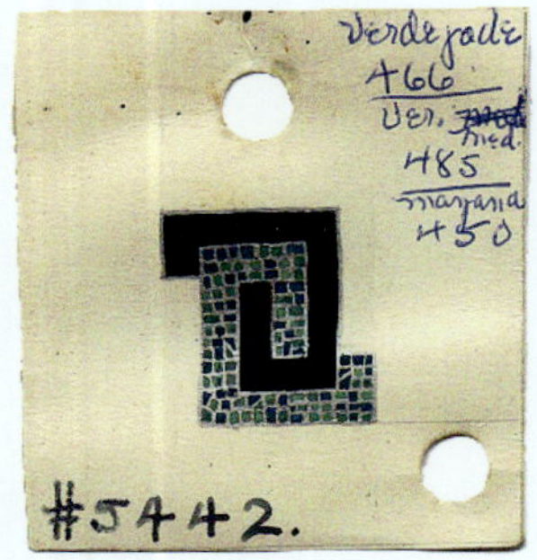

5442 Greek Key

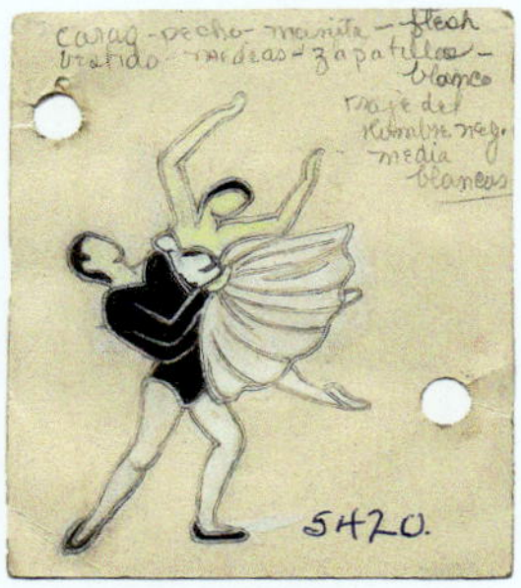

5420 Dancers

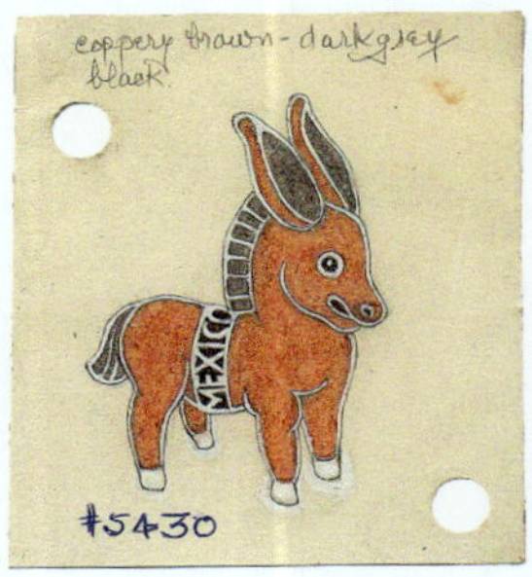

5430 Donkey

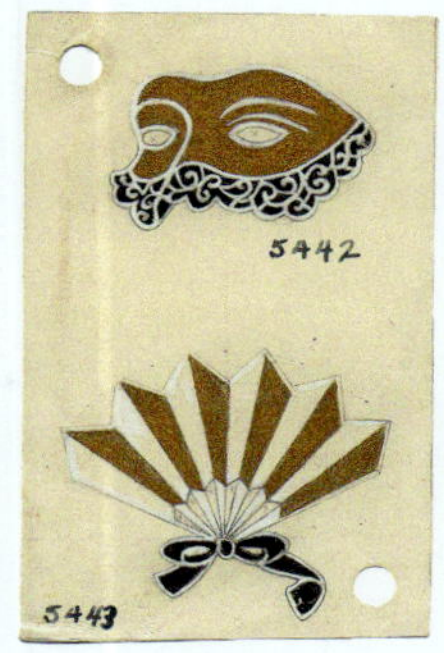

5442 Mask & 5443 Fan

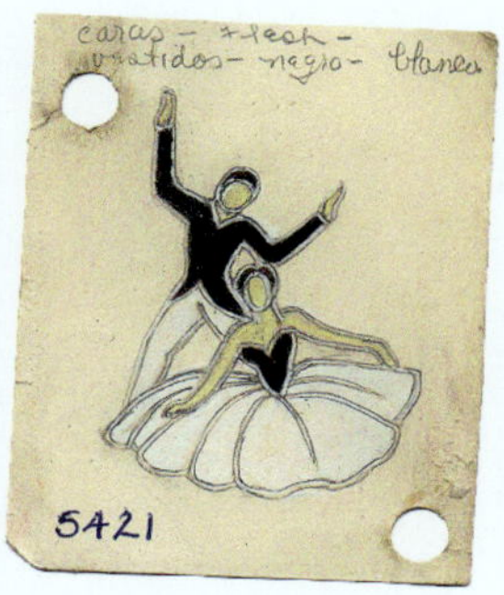

5421 Ballet Dancers

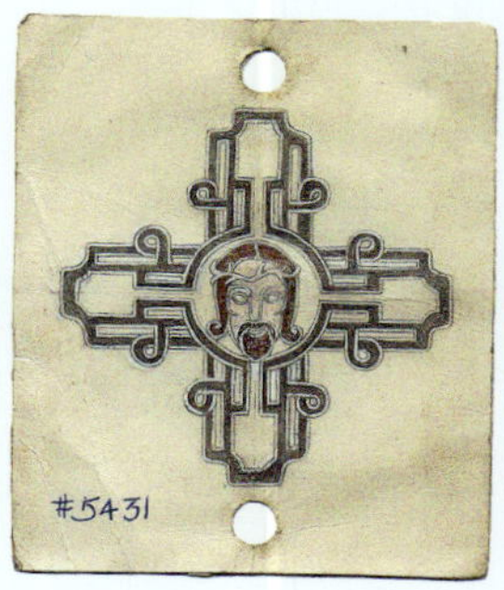

5431 Cross

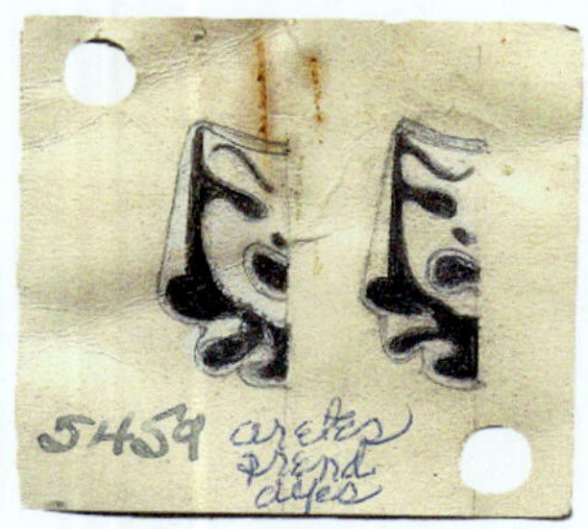

5459 Mask

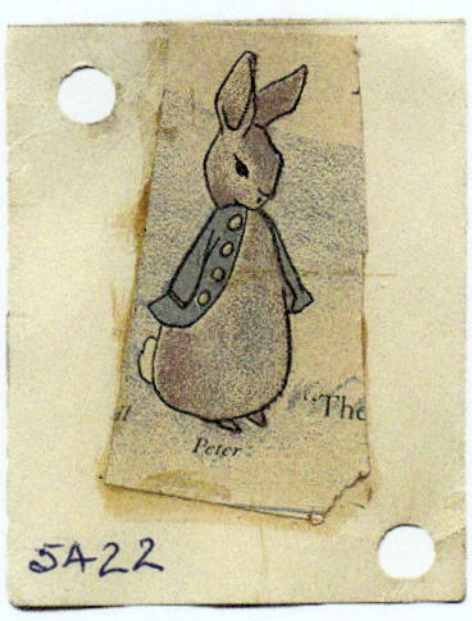

5422 Peter Rabbit

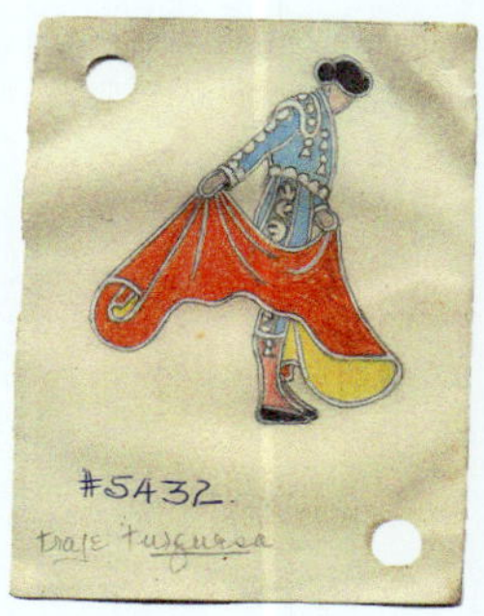

5432 Bullfighter

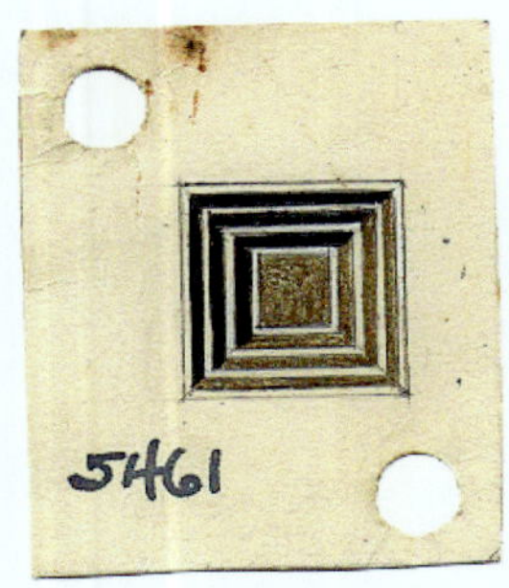

5461 Shadow Box

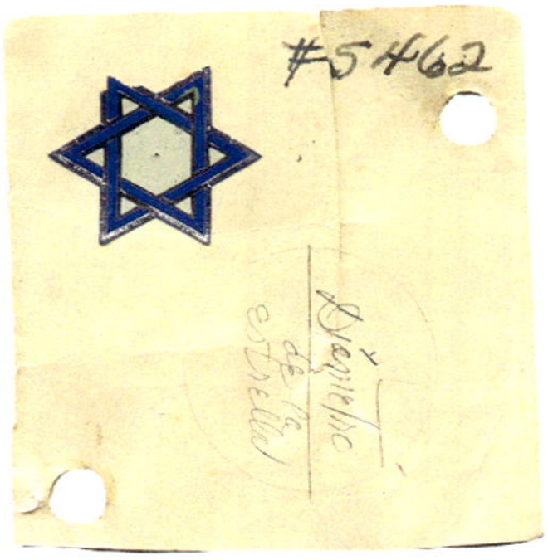

5462 Star of David

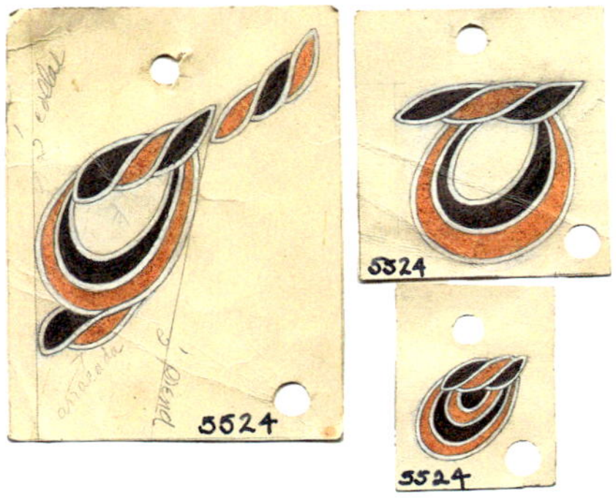

5524 Tricolor Torque

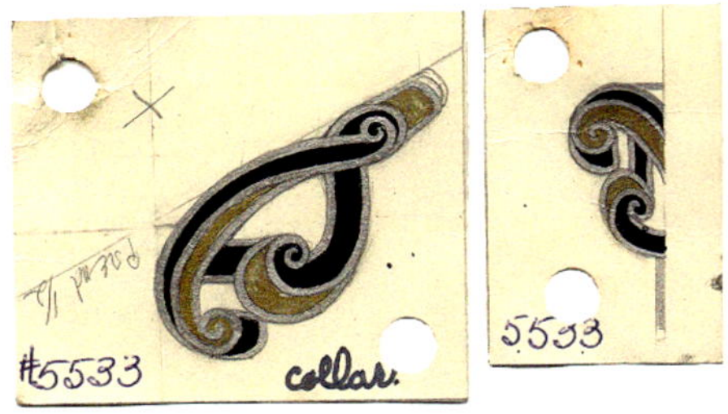

5533 Open Scroll

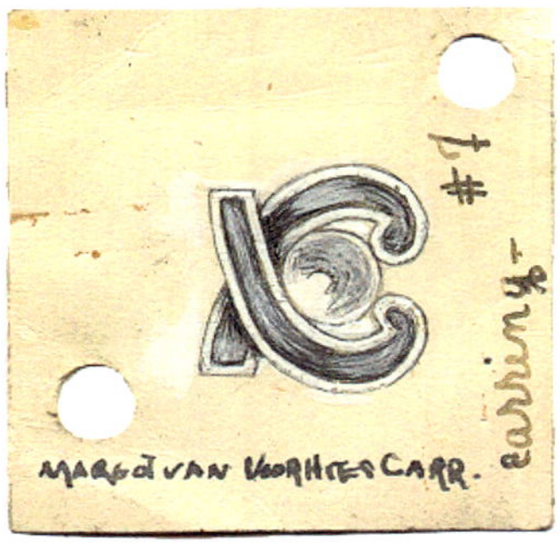

5463 Enfolded Bead

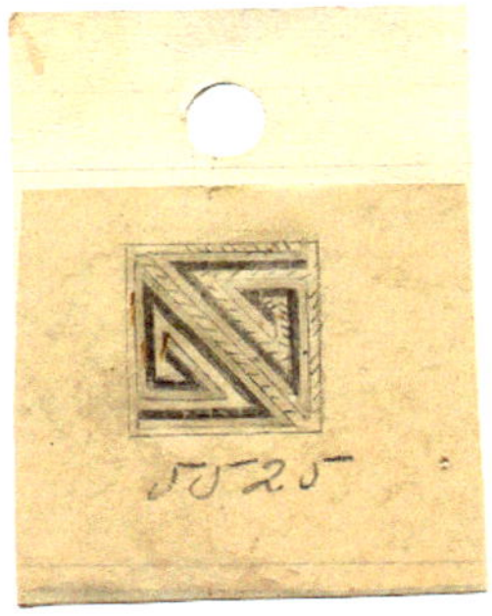

5525 Square Triangle

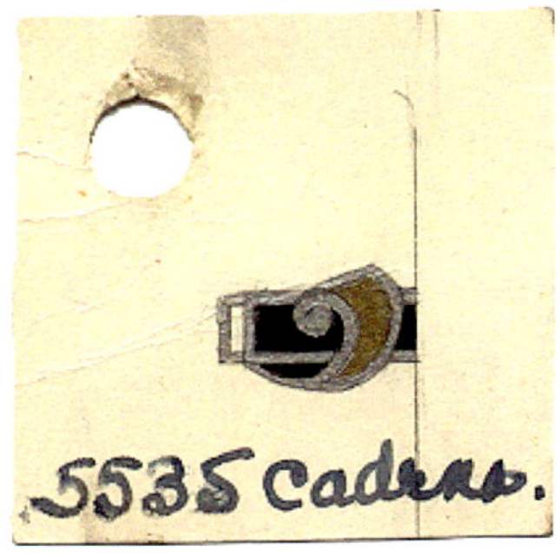

5535 Chain

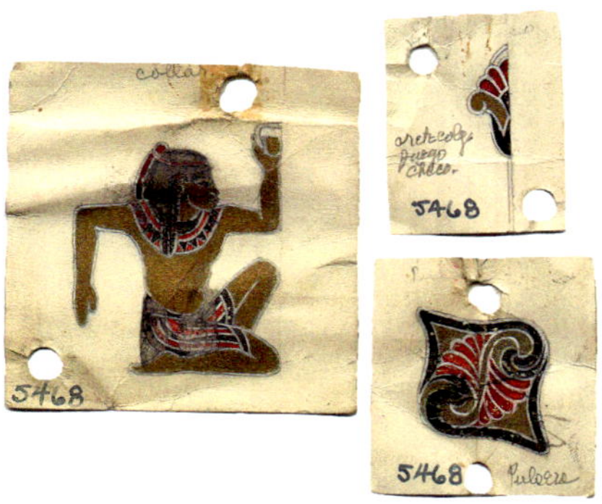

5468 Egyptian

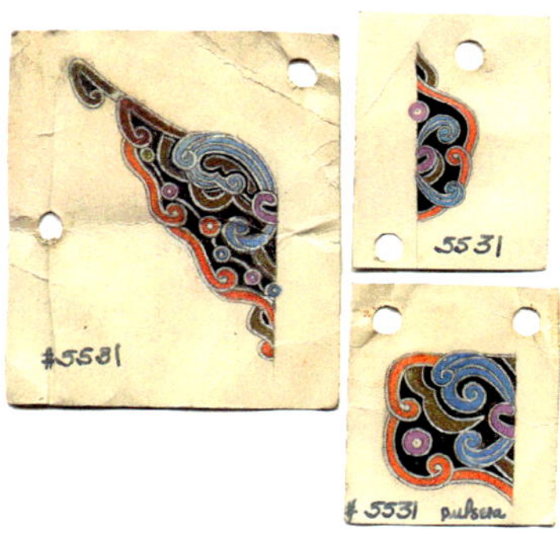

5531 Arabesque

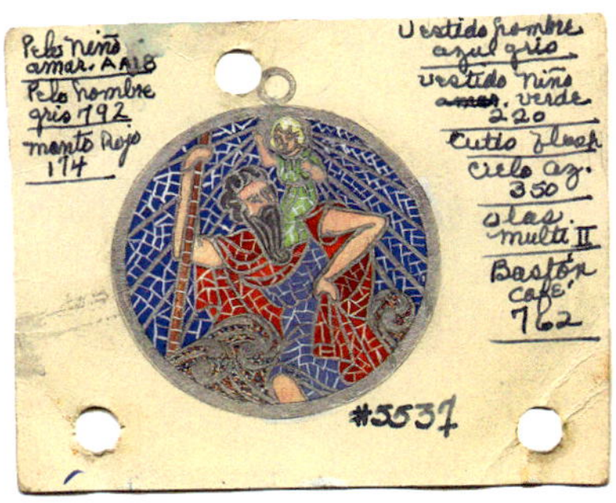

5537 Saint Christopher

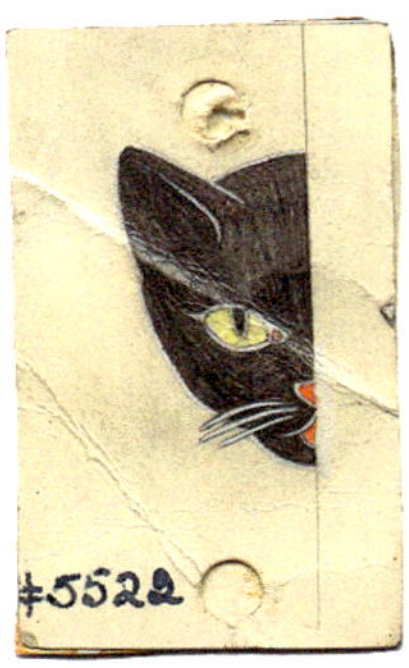

5522 Cat Face

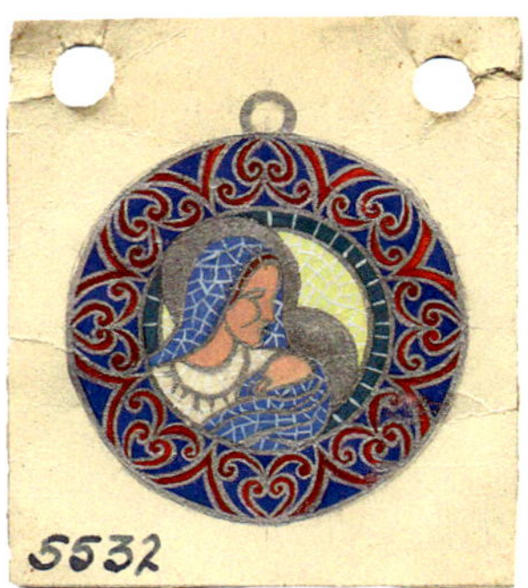

5532 Virgin and Child

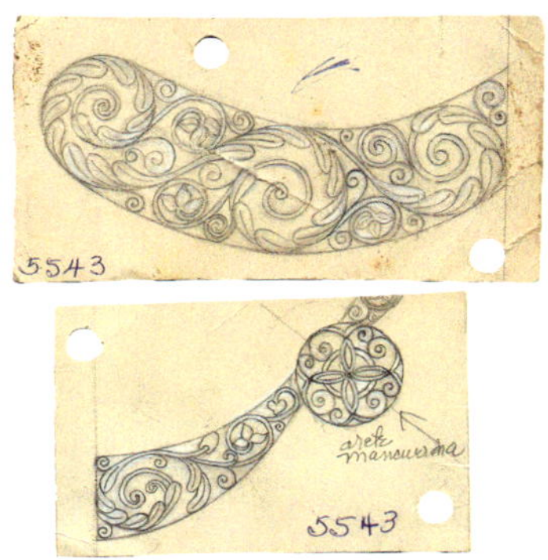

5543 Swirls

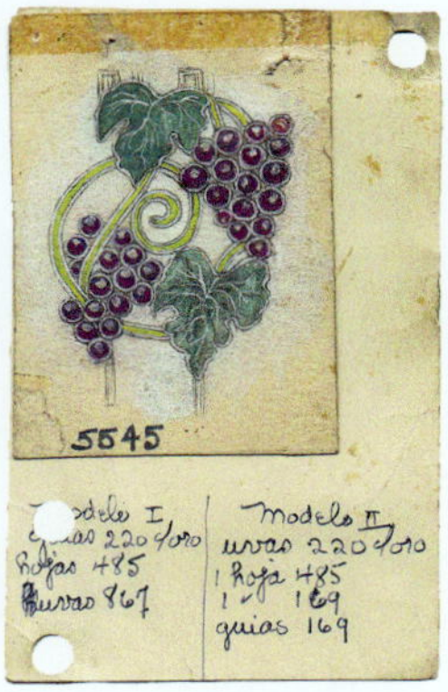

5545 Grapes

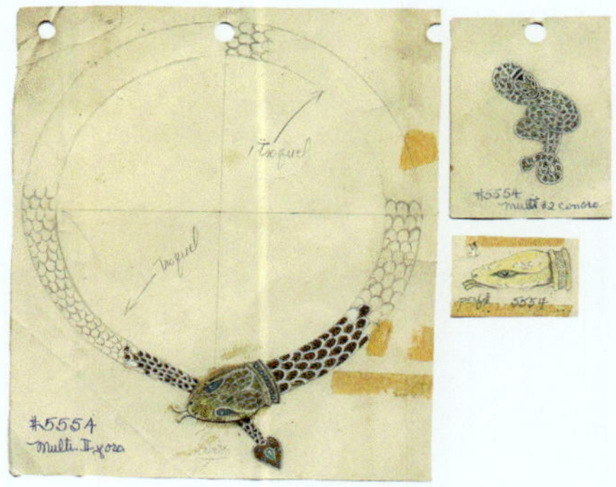

5554 Serpent

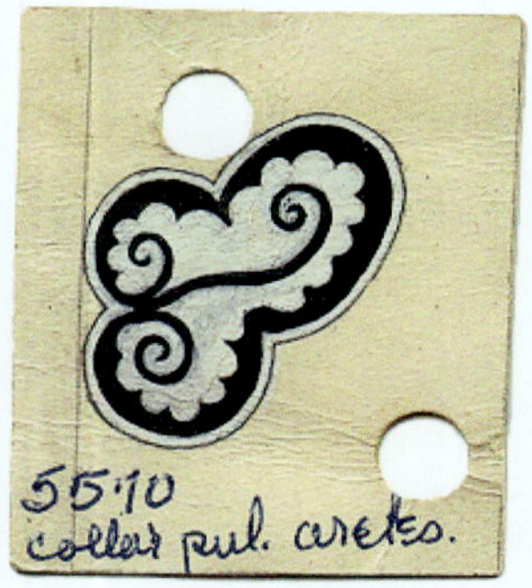

5570 Doily Pattern

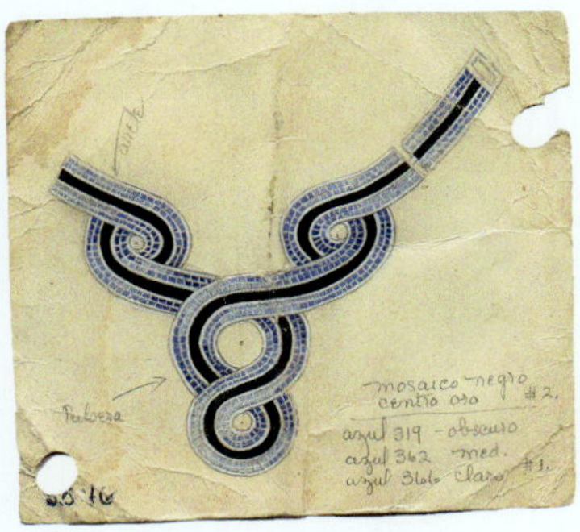

5546 Twirled Lines

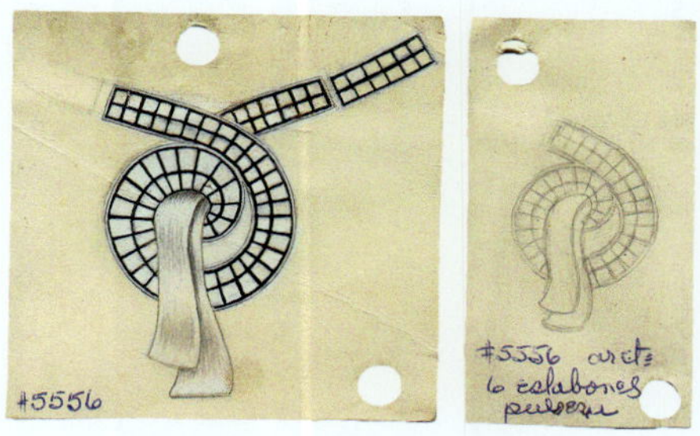

5556 Ribbons from Swirl

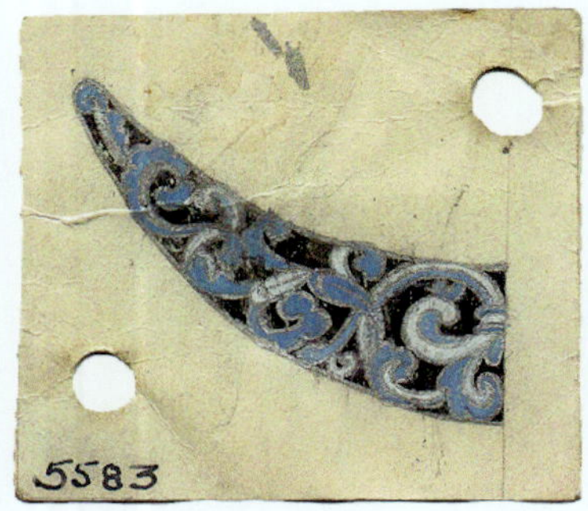

5583 Leaf Curls

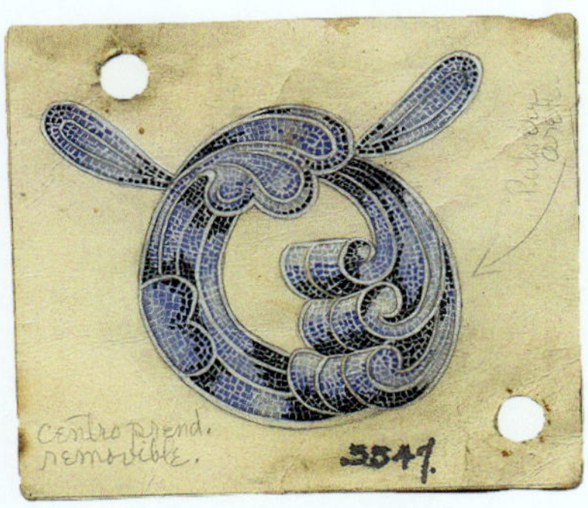

5547 Wave Circle

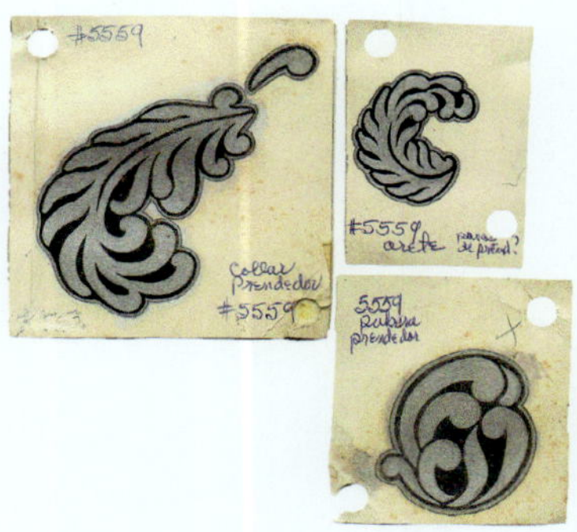

5559 Feathery Leaf

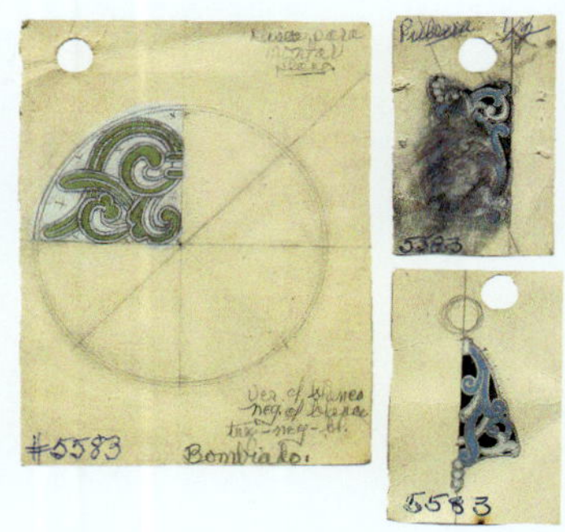

5583 Scrolls In Circle

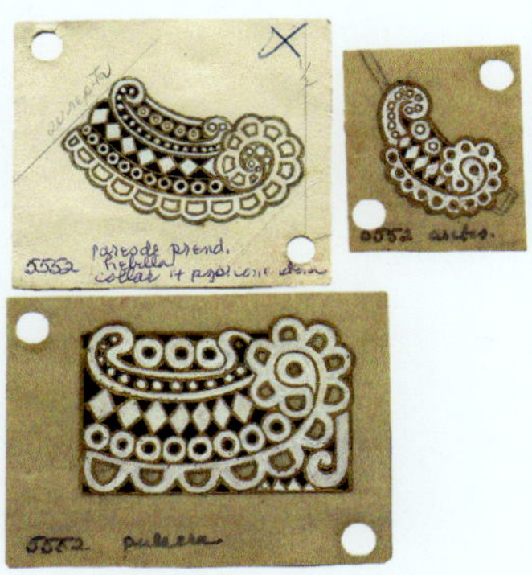

5552 Diamonds & Dots

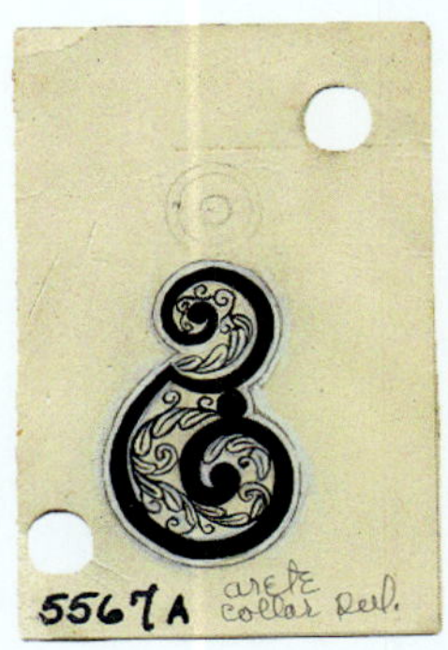

5567A Music Scroll

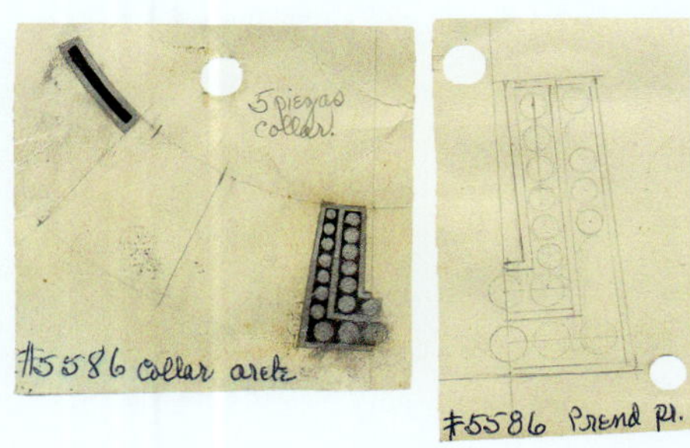

5586 Modern Dots

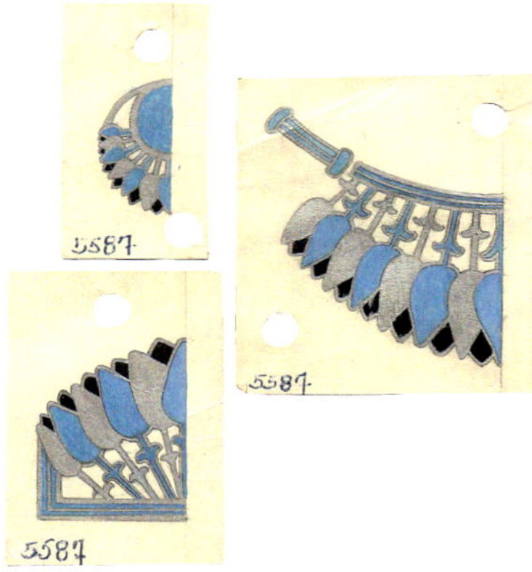

5587 Papyrus

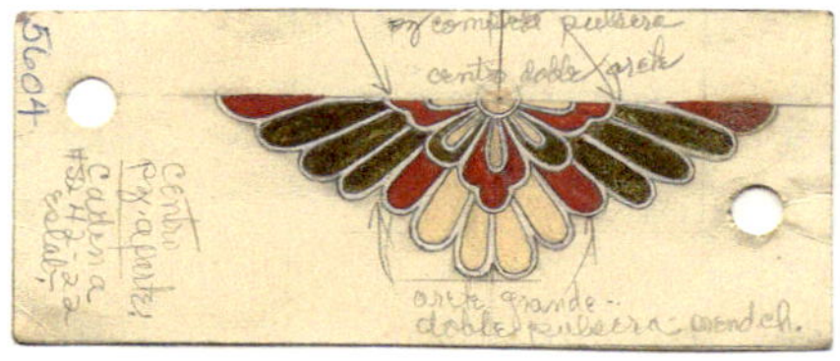

5604 Fanned Out

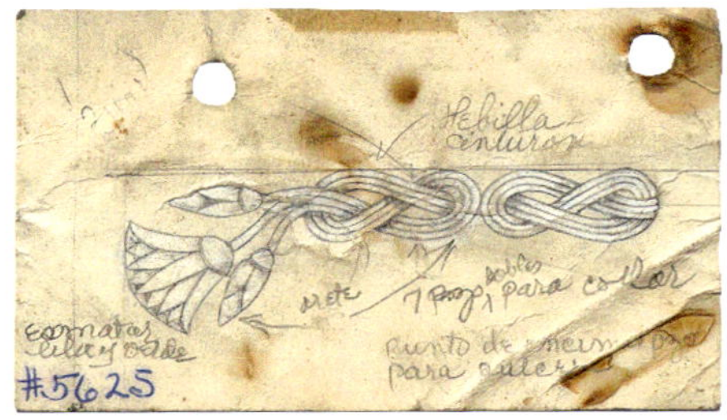

5625 Twist & Lotus

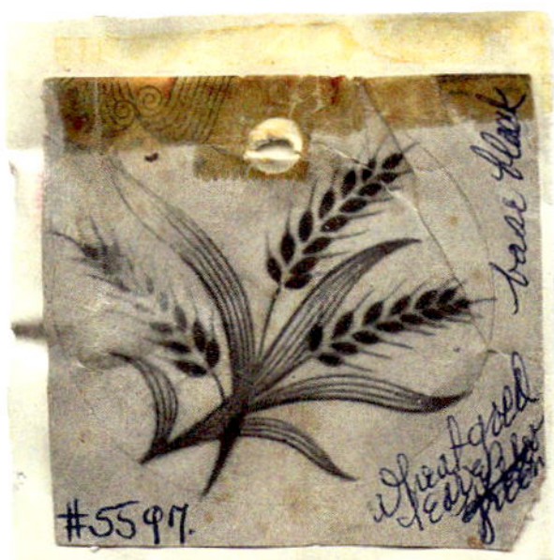

5597 Wheat Sheaf

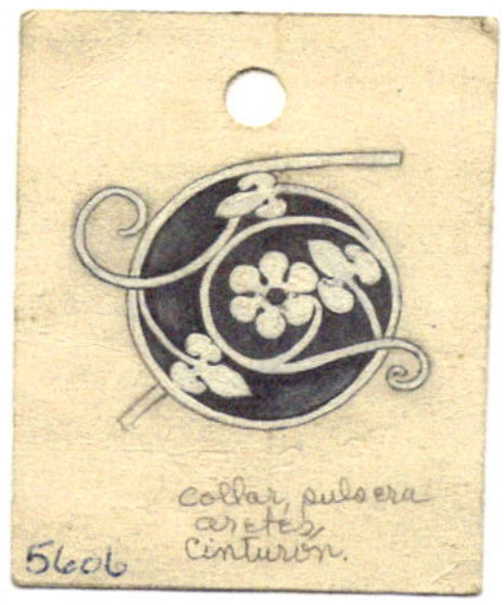

5606 Flower In Circle

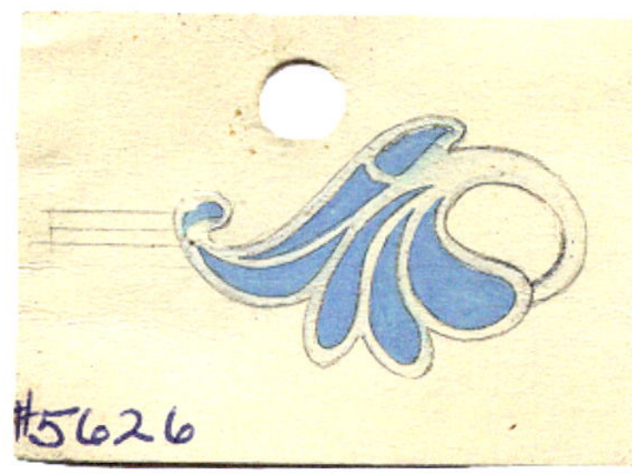

5626 Small Curled Leaf

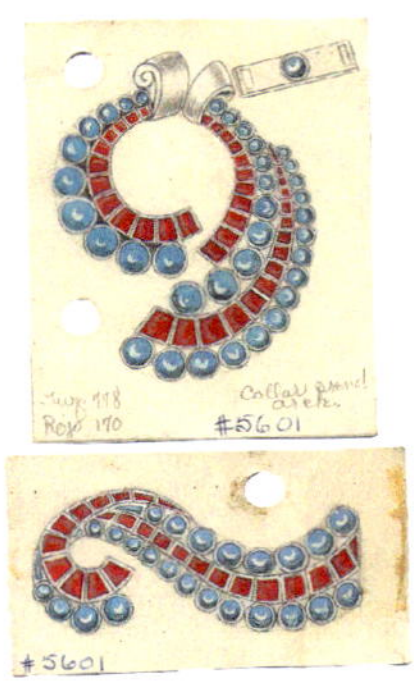

5601 Dot & Square Scroll

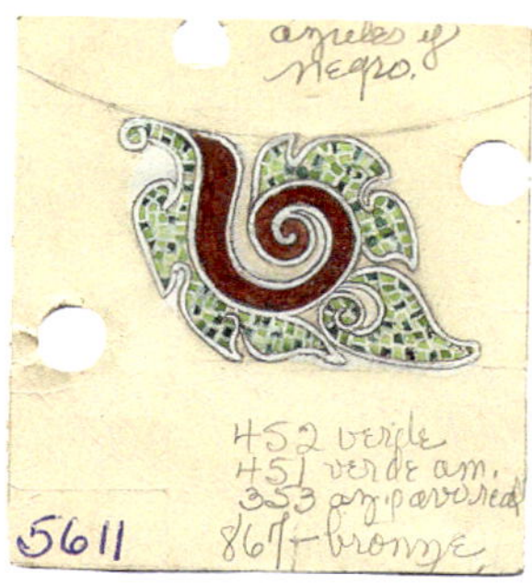

5611 Leaf Shell

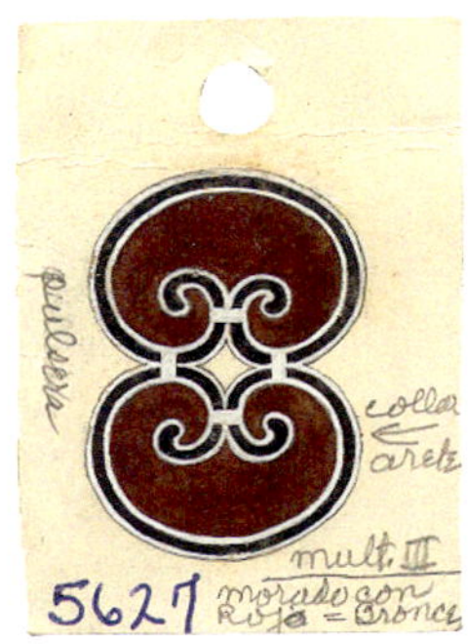

5627 Curled In Oval

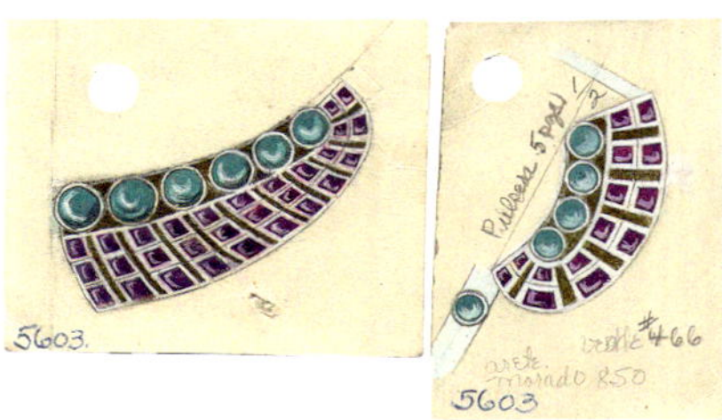

5603 Jeweled Circles

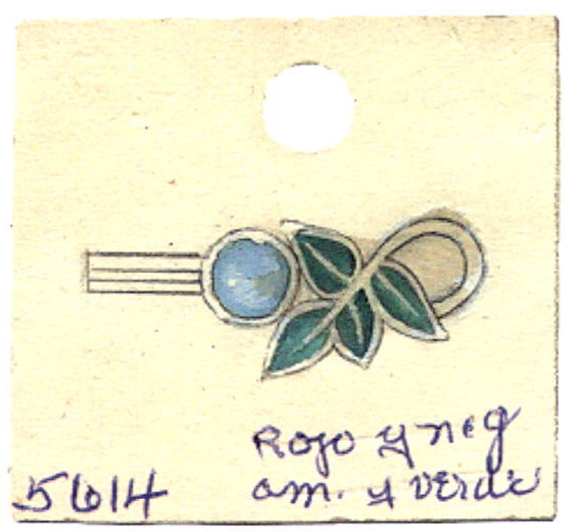

5614 Fruit & Branch

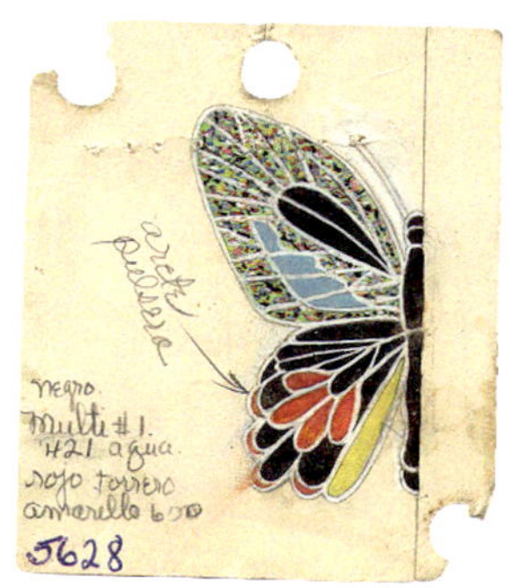

5628 Butterfly

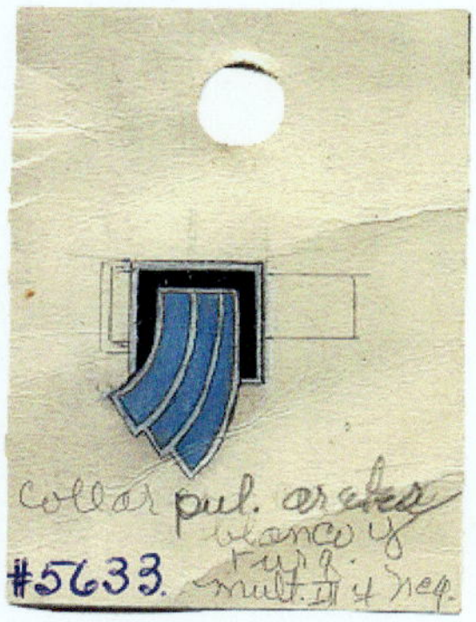

5633 Flanges In Square

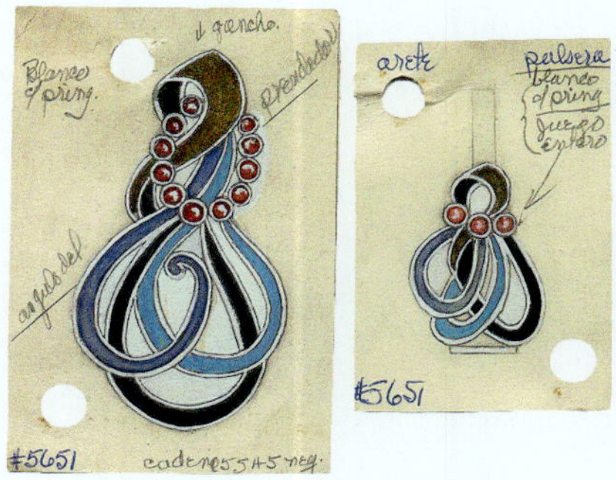

5651 Gathered Swirls

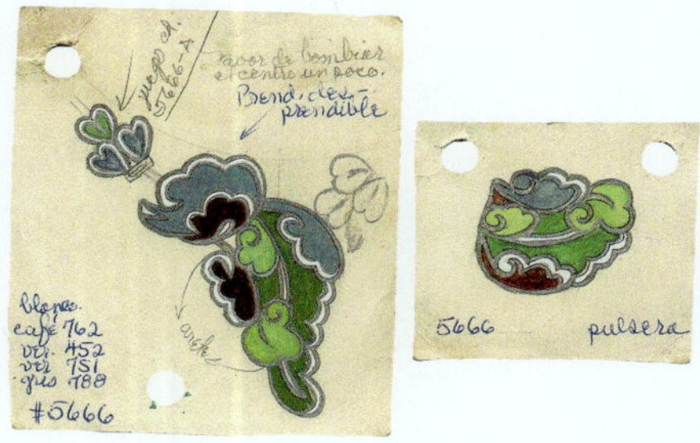

5666 Scallop-edge Leaf

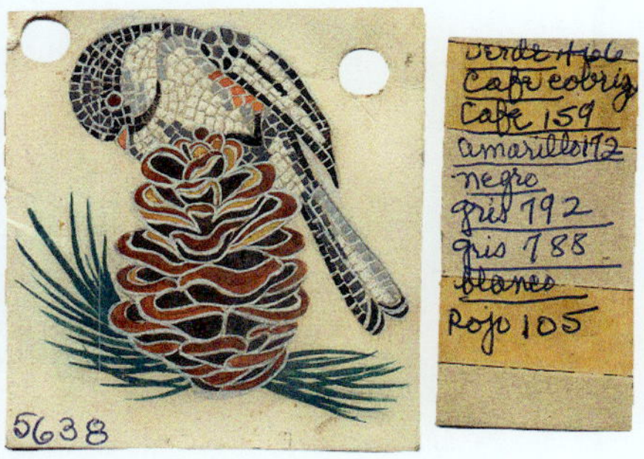

5638 Bird & Pinecone

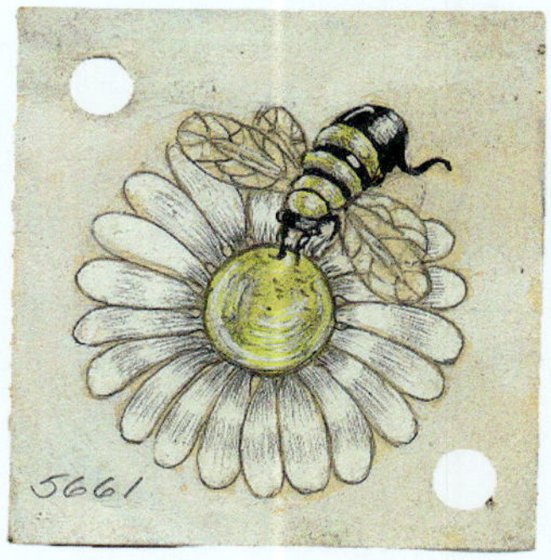

5661 Bee & Flower

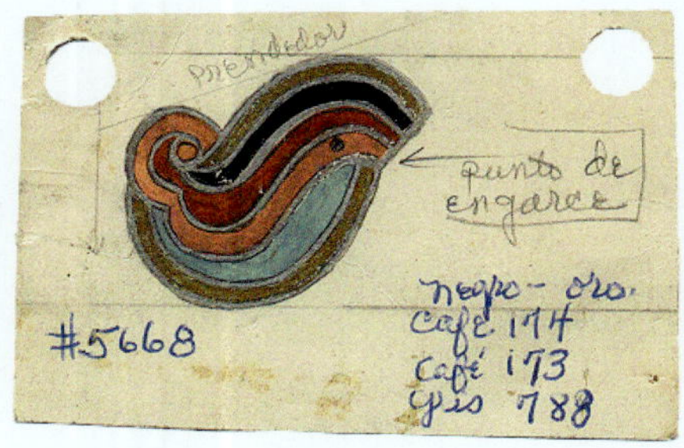

5668 Multicolor Curl

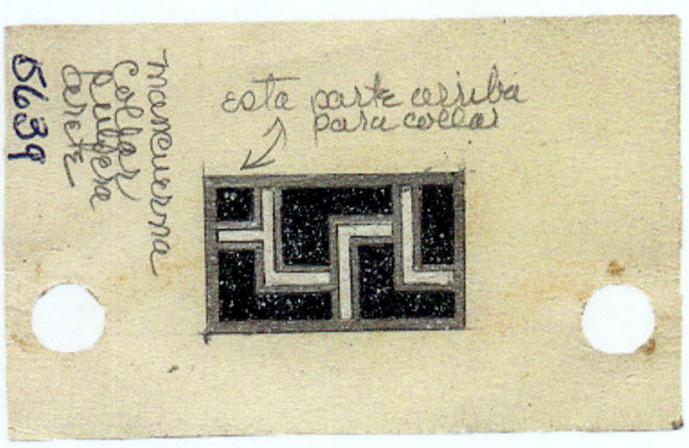

5639 Interlocking Lines

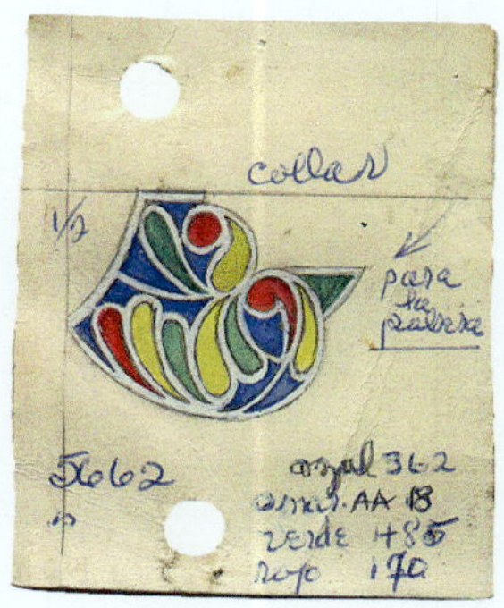

5662 Bright Multicolor

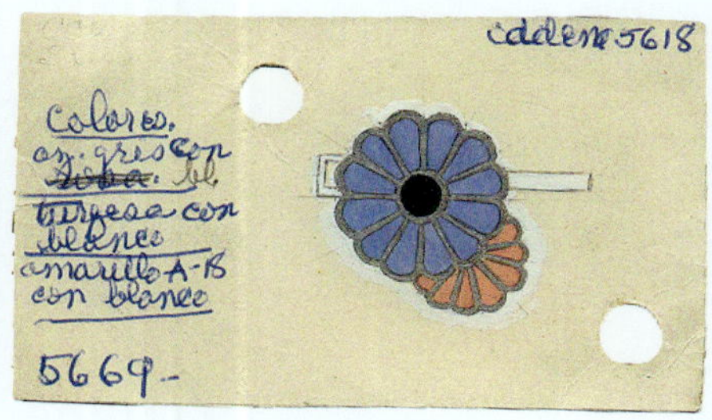

5669 Daisy Chain

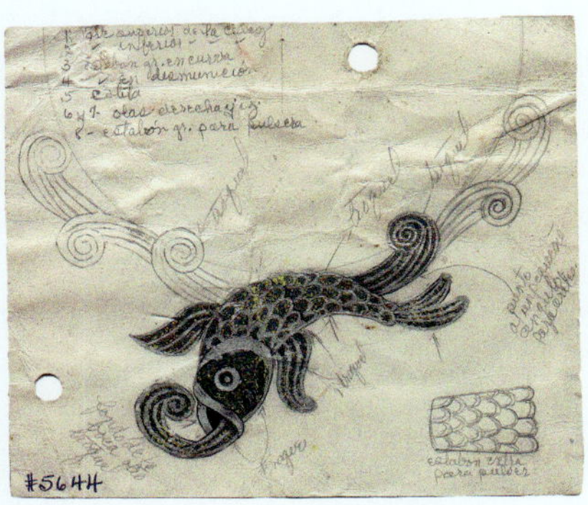

5644 Fish

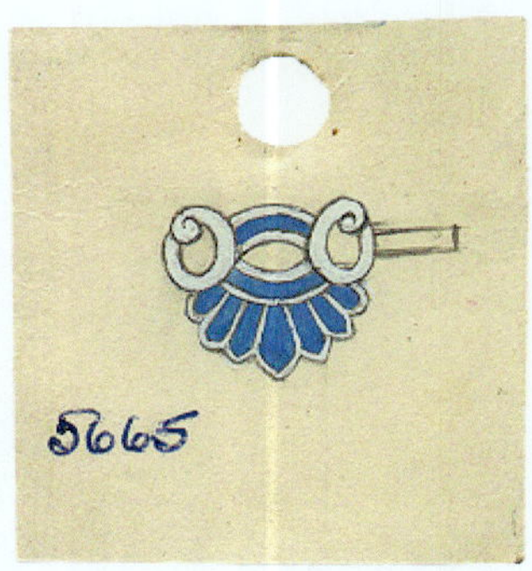

5665 Small Formal Fan

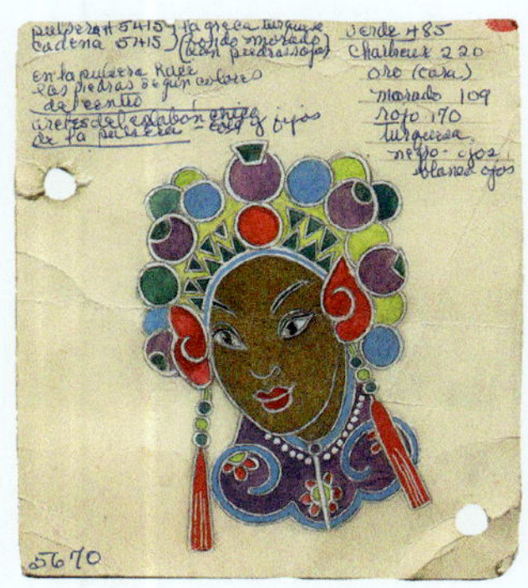

5670 Geisha Face

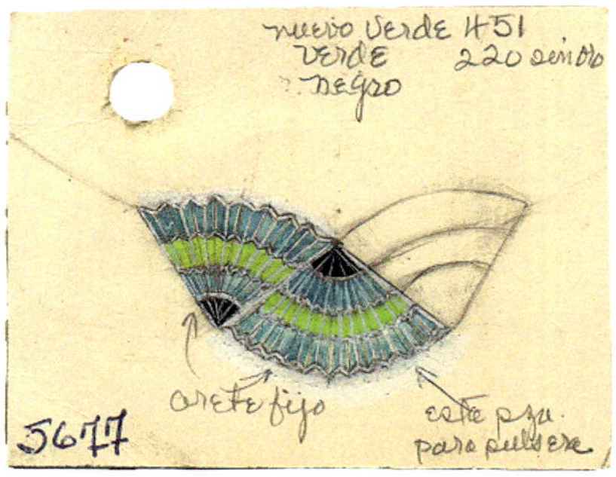

5677 Alternating Fans

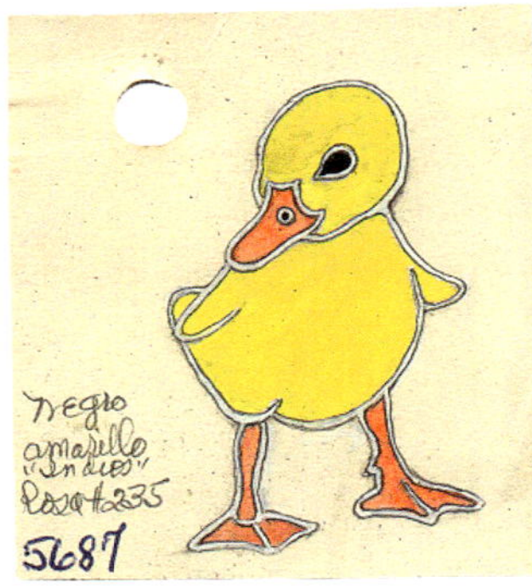

5687 Duck

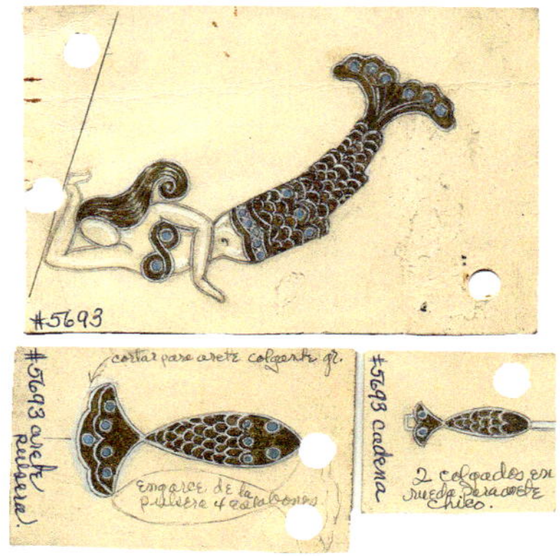

5693 Mermaid & Fish

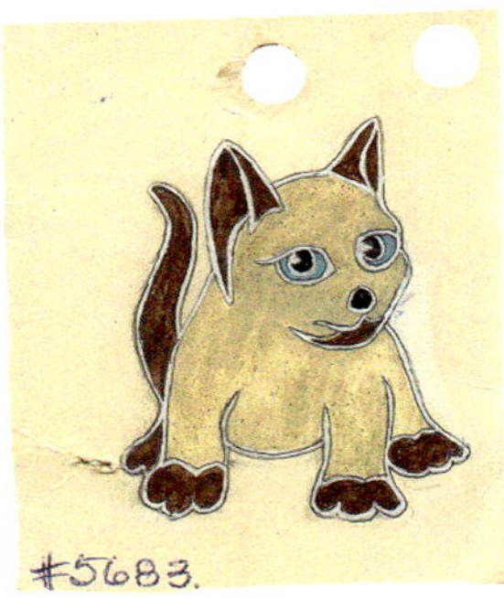

5683 Cat

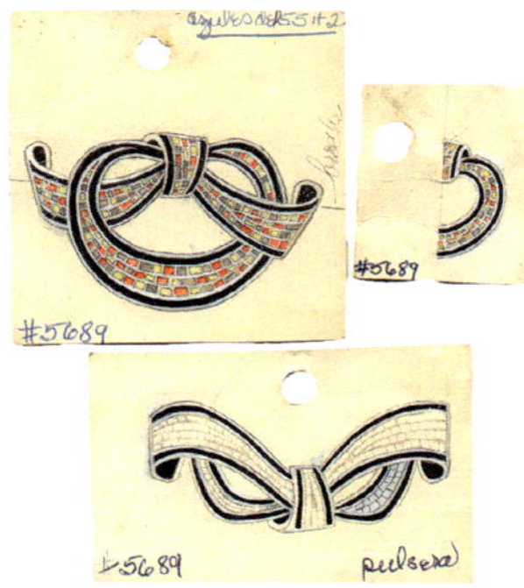

5689 Circle Tied Bow

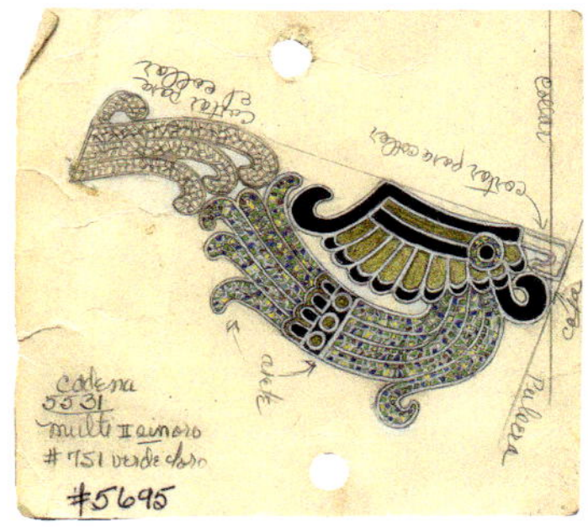

5695 Quetzal Feathers

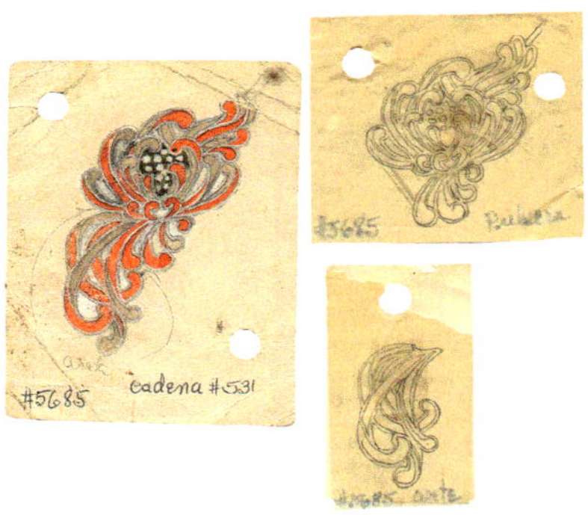

5685 Chrysanthemum

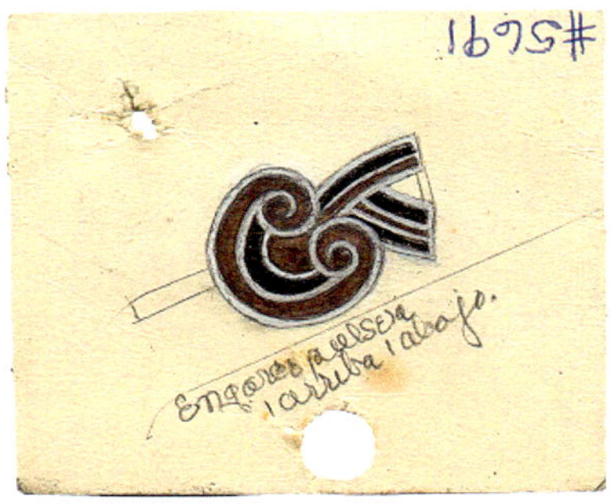

5691 Tricolor Curled In Circle

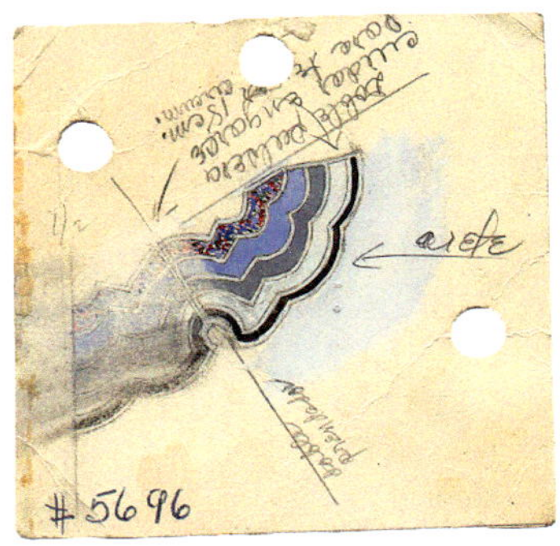

5696 Scalloping

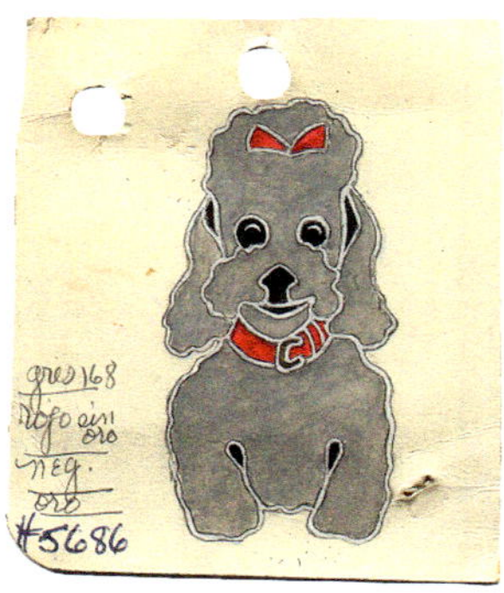

5686 Poodle

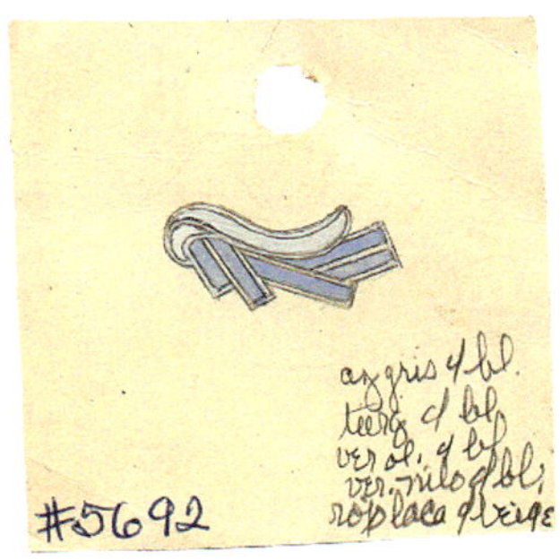

5692 Lines From Scroll

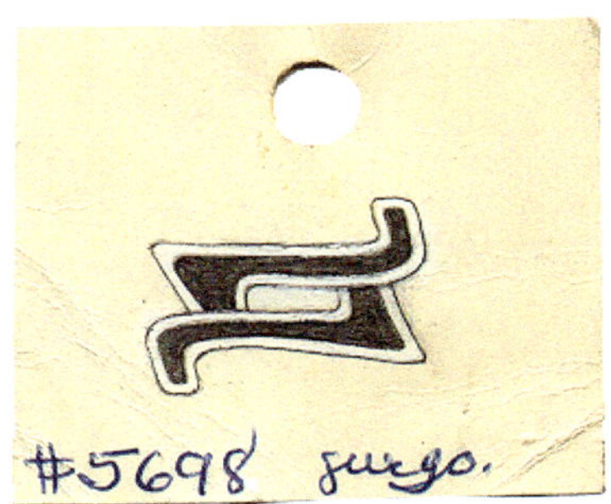

5698 Parallel

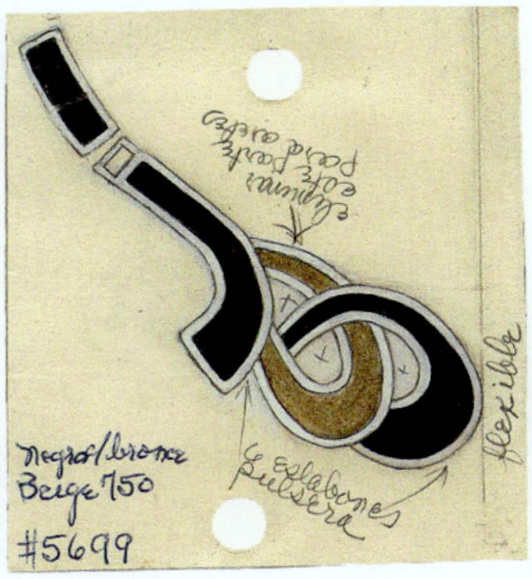

5699 Double Loop

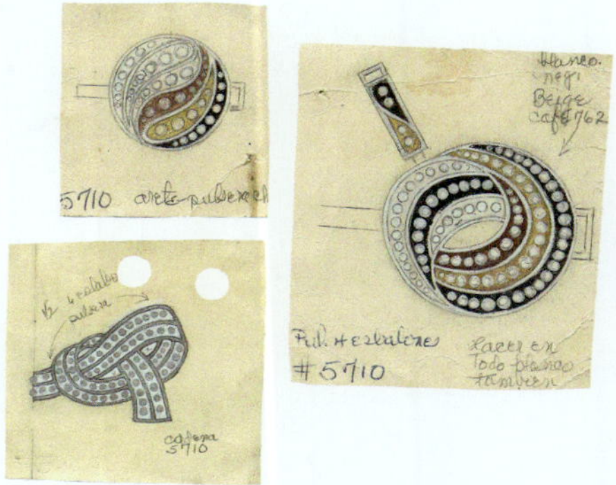

5710 Circle Curves

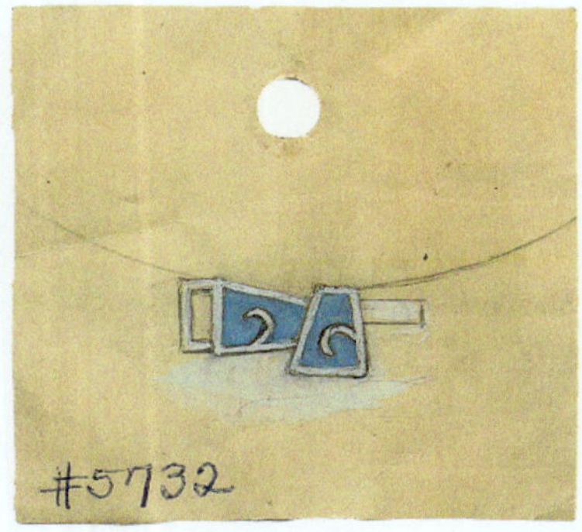

5732 Right Angle Mini Pre

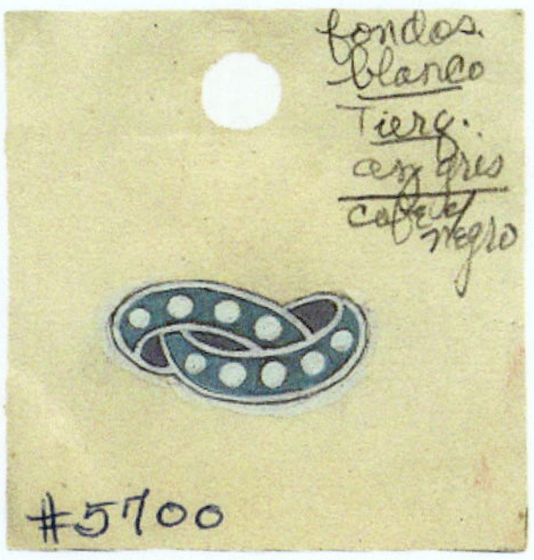

5700 Dots On Knot

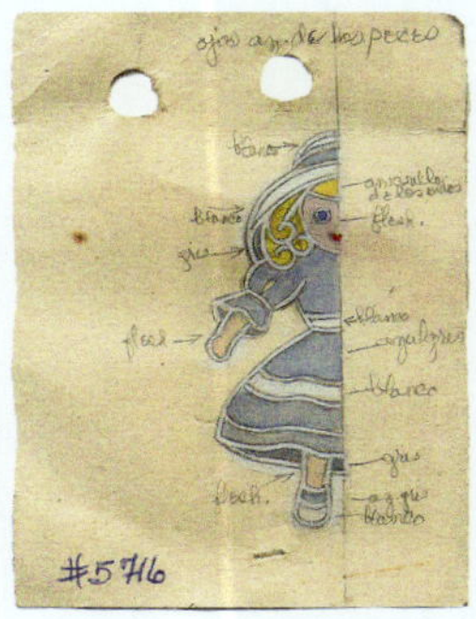

5716 Doll

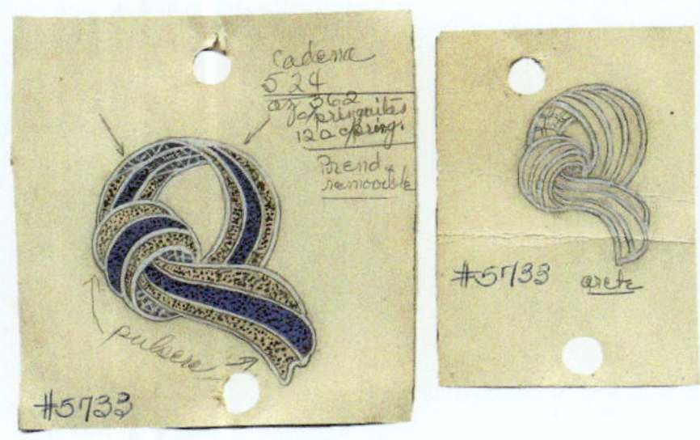

5733 Elegant Knot

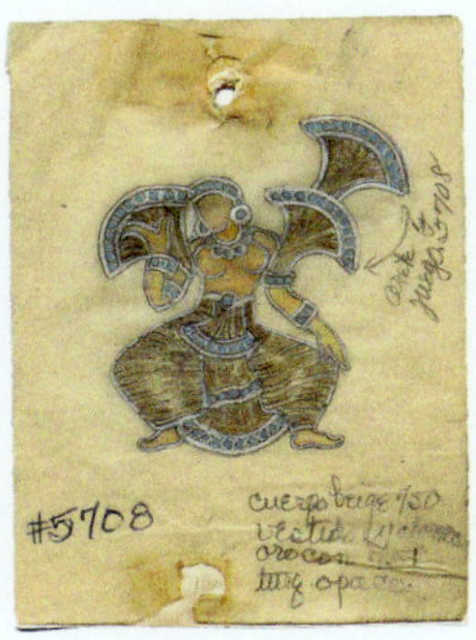

5708 Fan Dancer

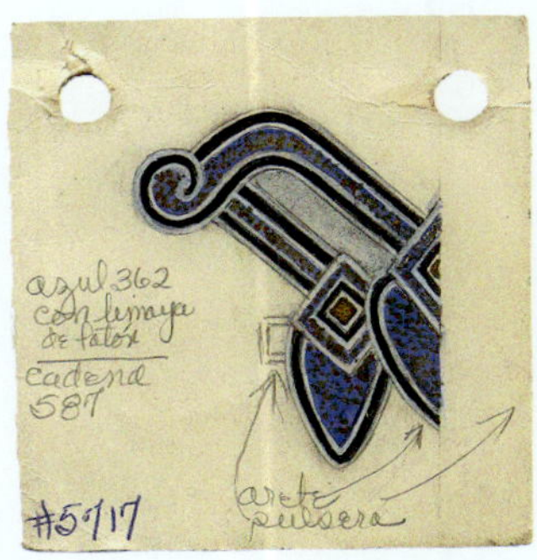

5717 Deco Diamonds

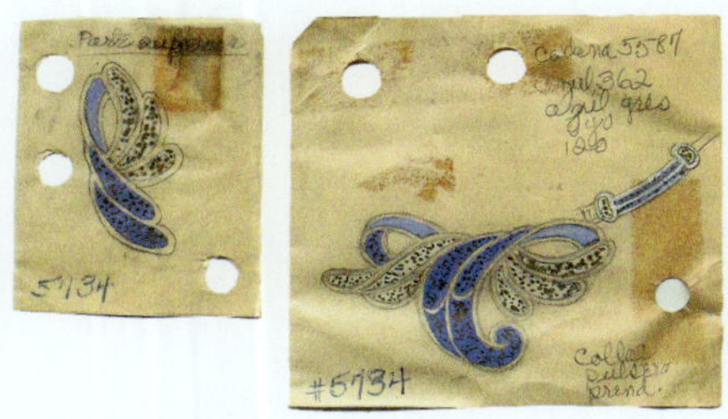

5734 Overlapping Curls

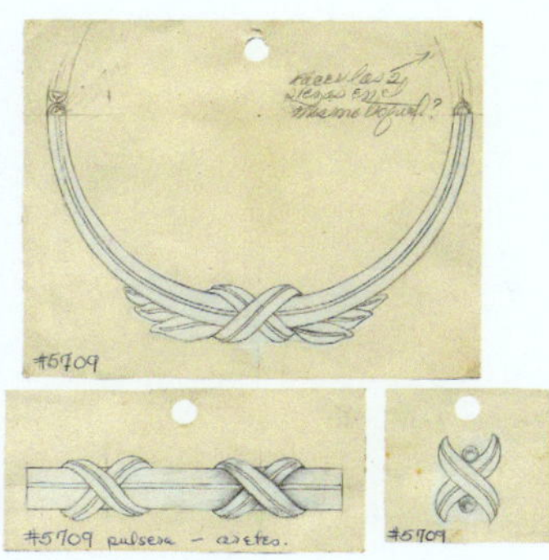

5709 Crossed Bands

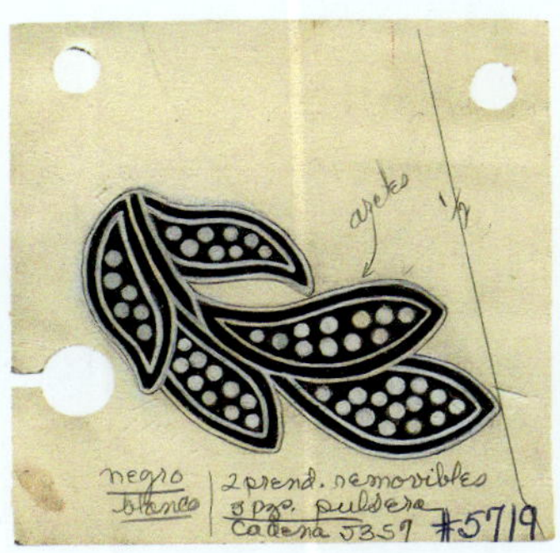

5719 Leaves With Dots

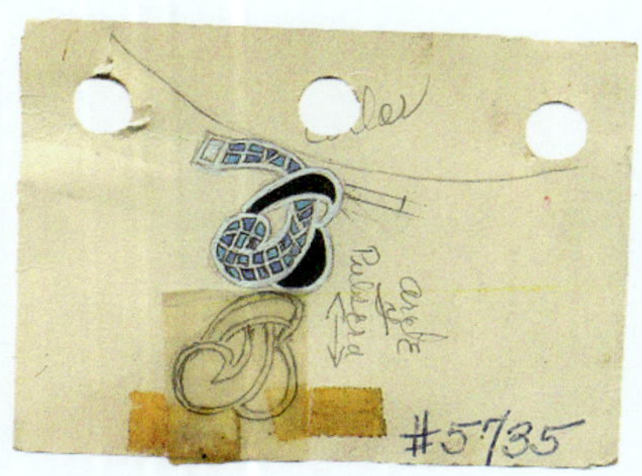

5735 Small Swirl

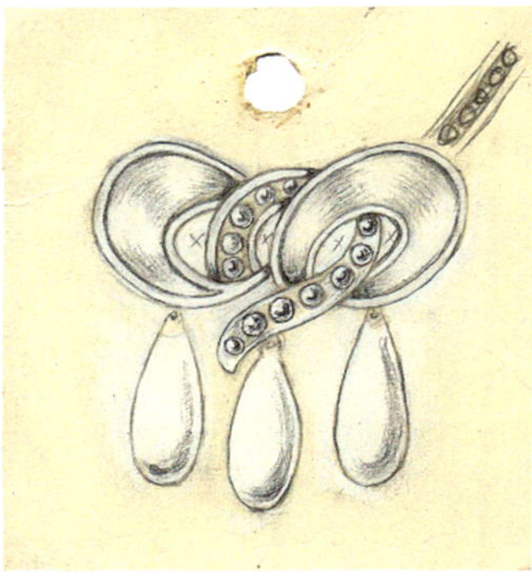

5740 Triple Drop

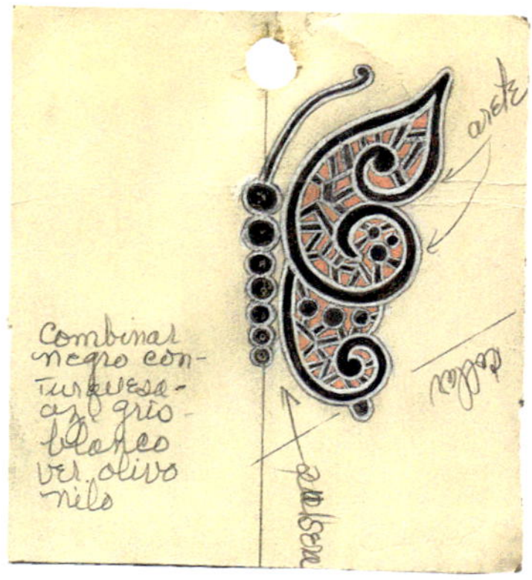

5748 Elegant Butterfly

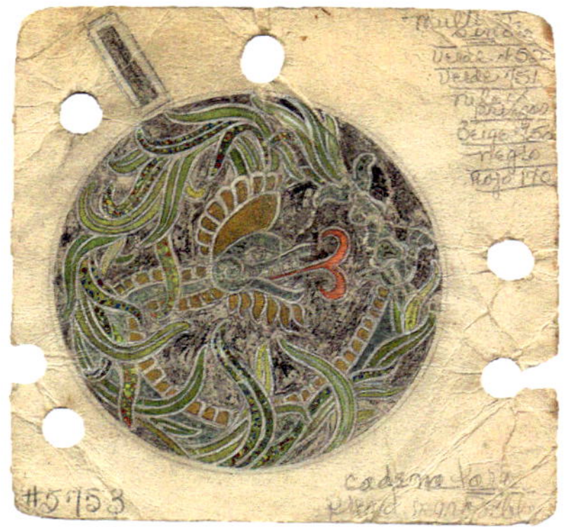

5753 Quetzalcoat

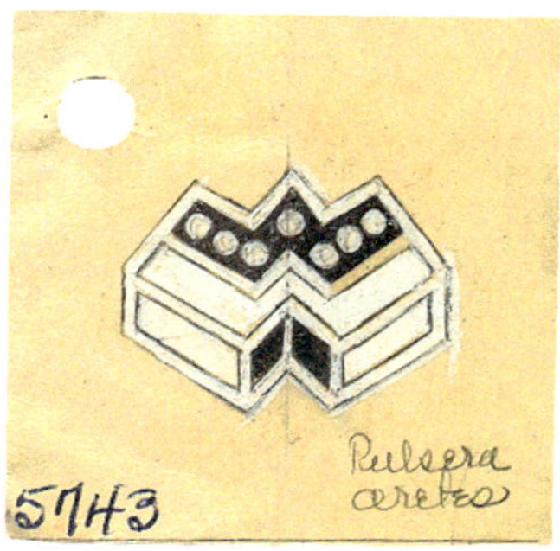

5743 Three-dimension W

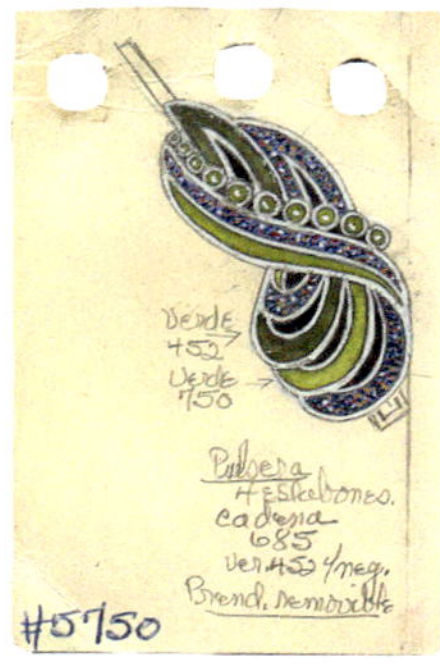

5750 Butterfly Swirls

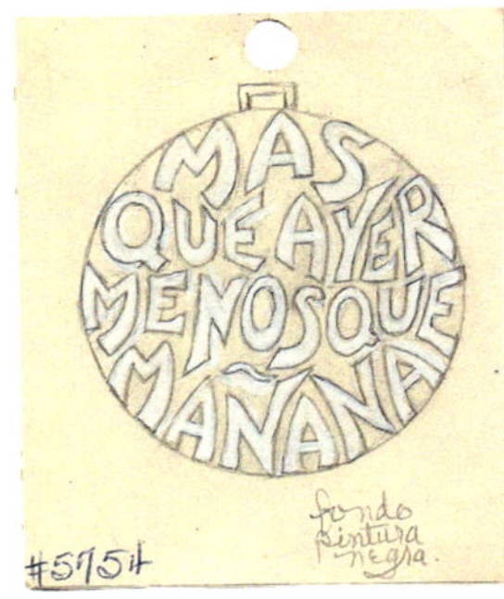

5754 Mas Que Ayer

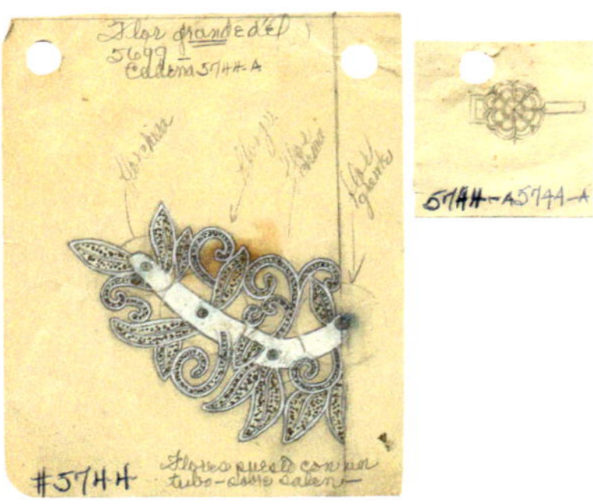

5744 Wild Leaves With Flowers

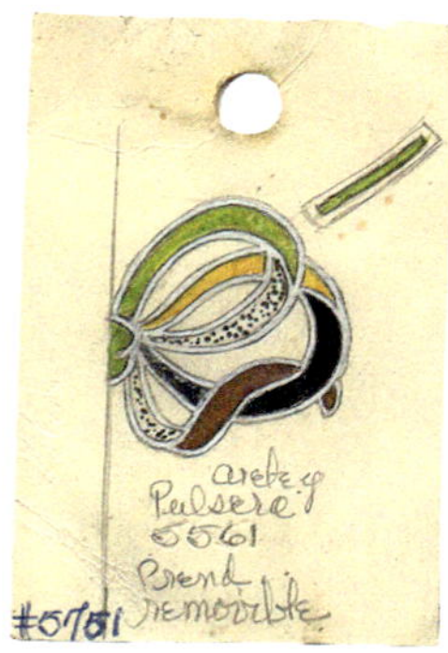

5751 Open Bow

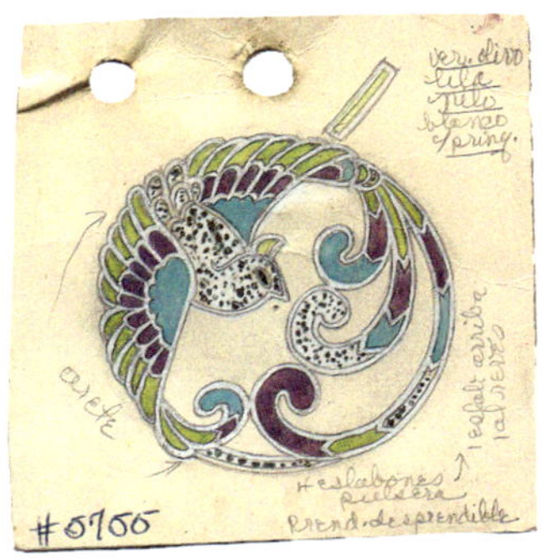

5755 Quetzal Bird

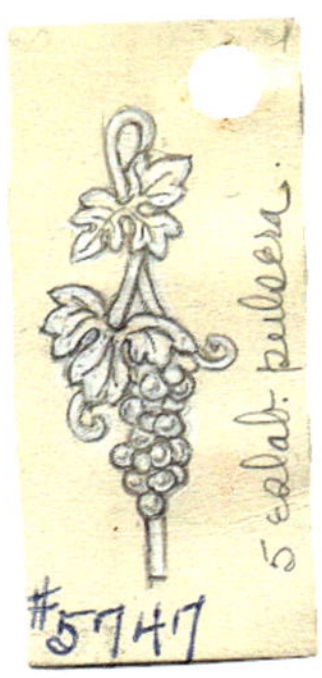

5747 Grapes

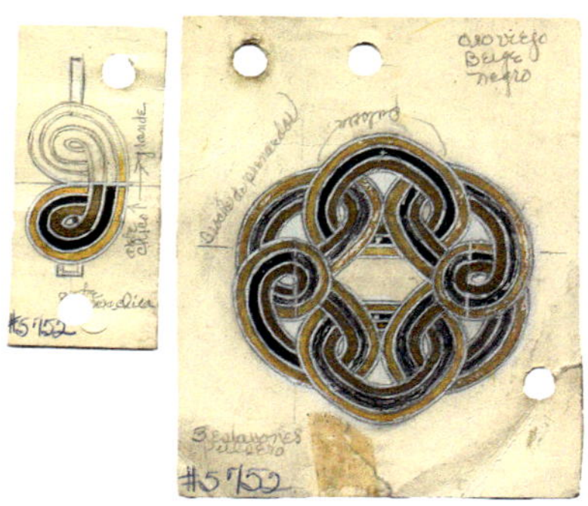

5752 Spiraling Circle

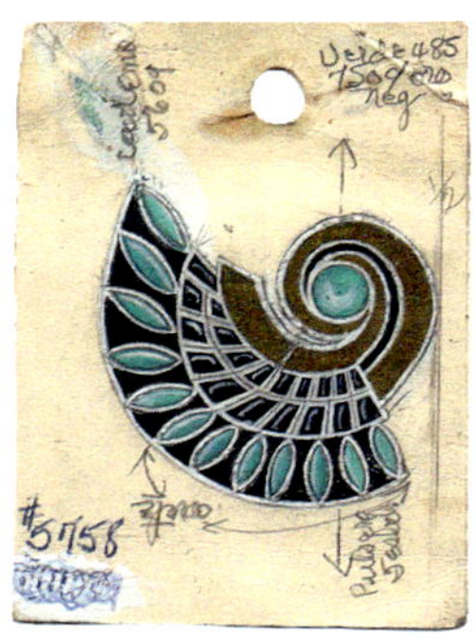

5758 Egyptian Scroll

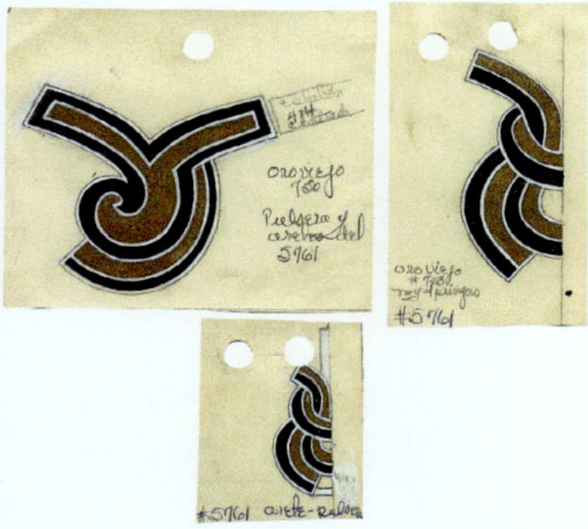

5761 Gathered Curves

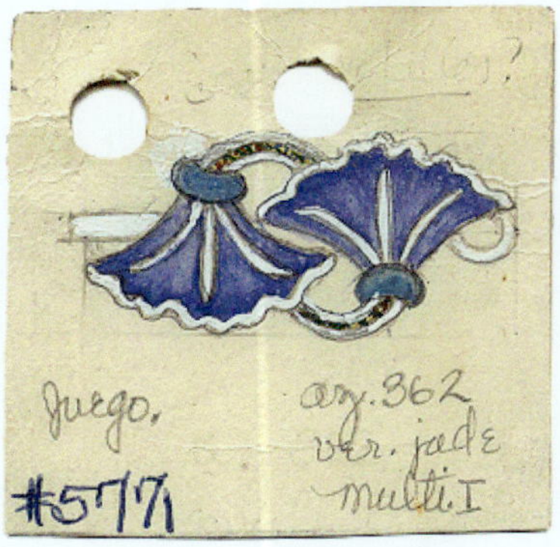

5771 Morning Glory

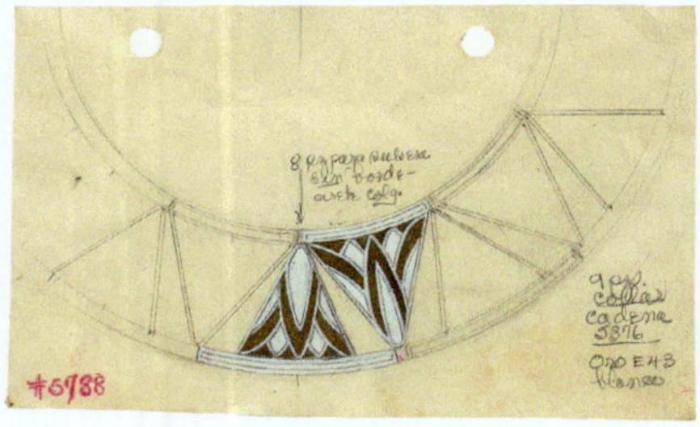

5788 Egyptian Collar

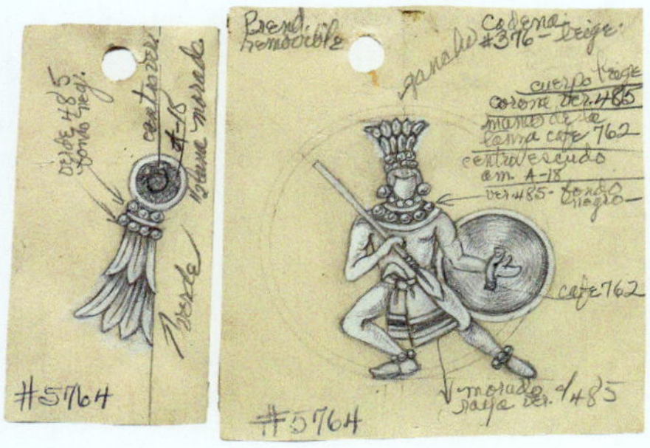

5764 Warrior

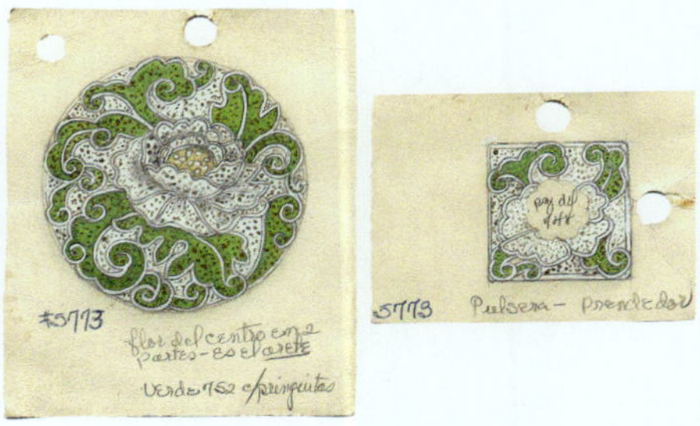

5773 Florid Mum

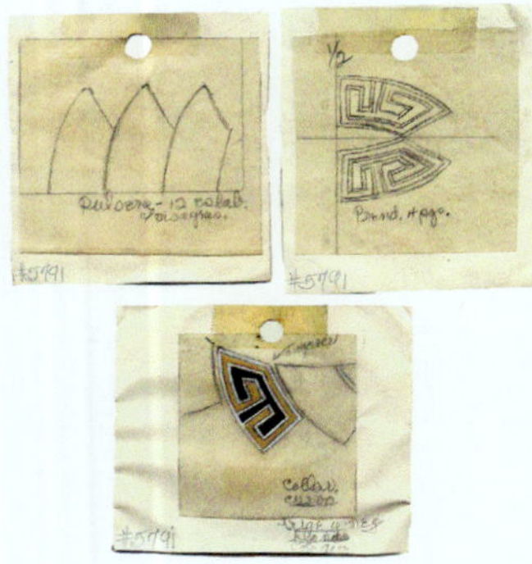

5791 Pre-Columbian Greek Key

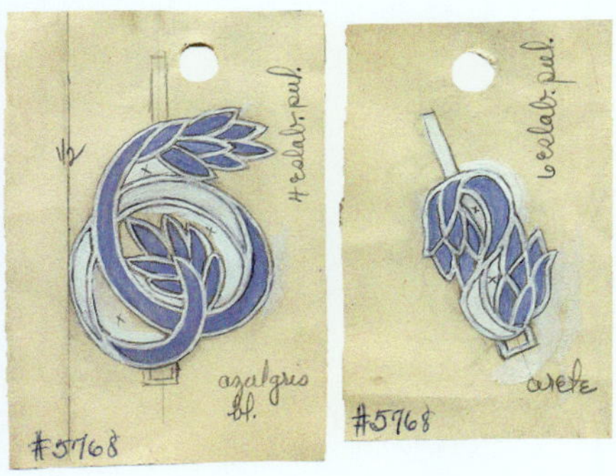

5768 Flowers In Curls

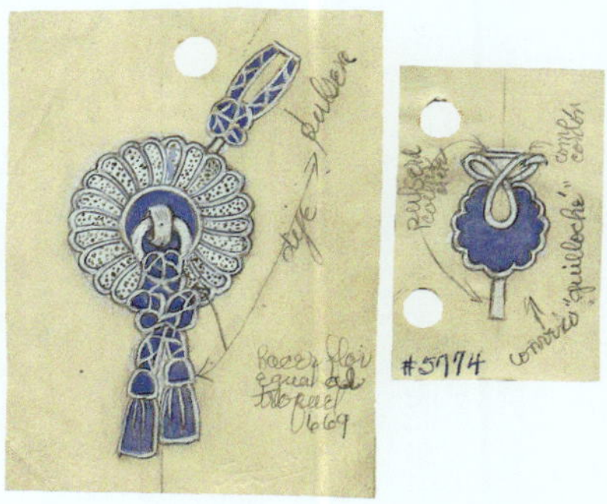

5774 Tassels

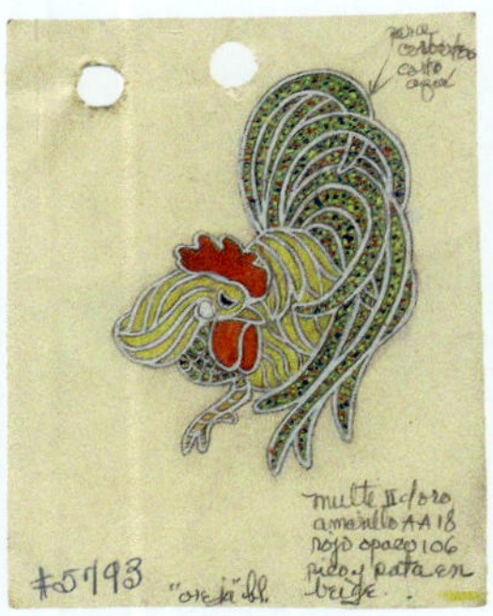

5793 Gallo

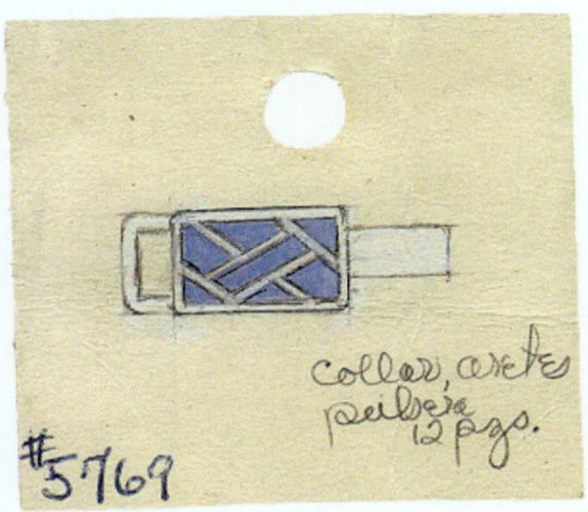

5769 Small Petate

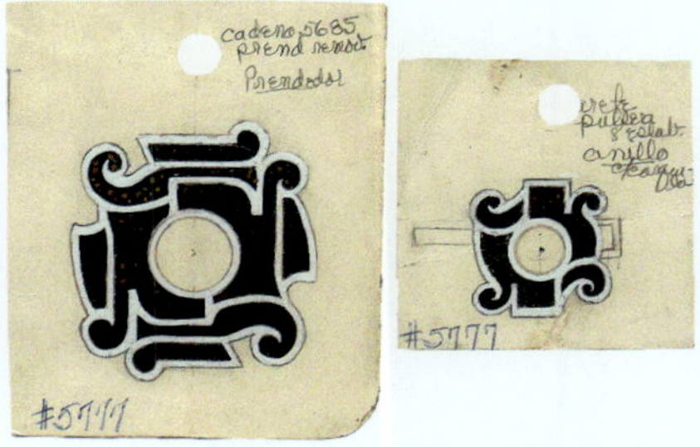

5777 Curled-edge Square

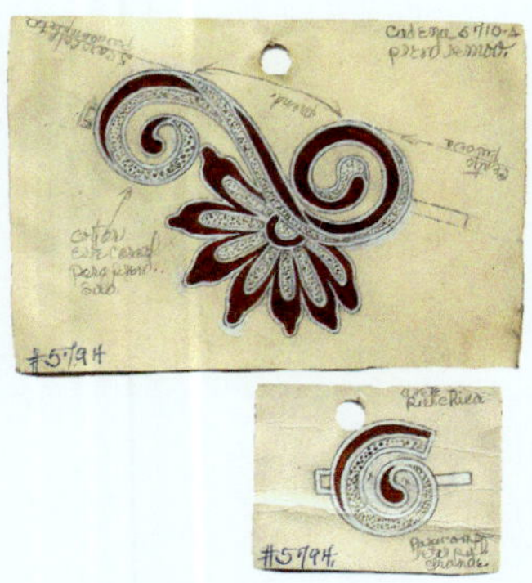

5794 Floral Flanges & Curls

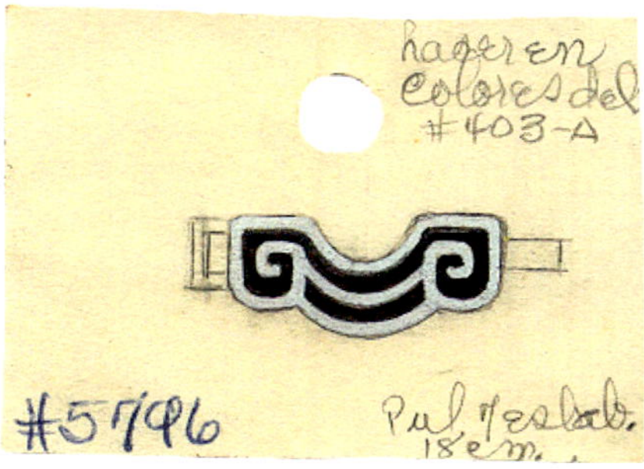

5796 Classical U

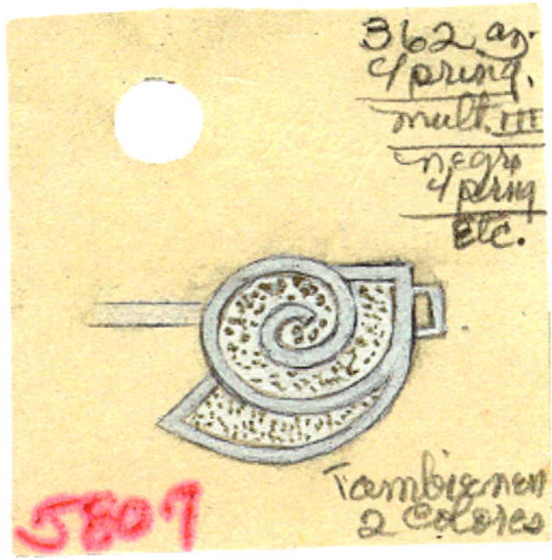

5807 Caracol Coil

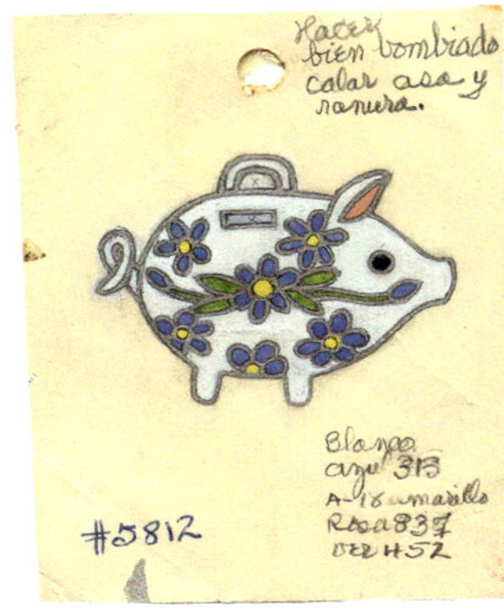

5812 Marranito

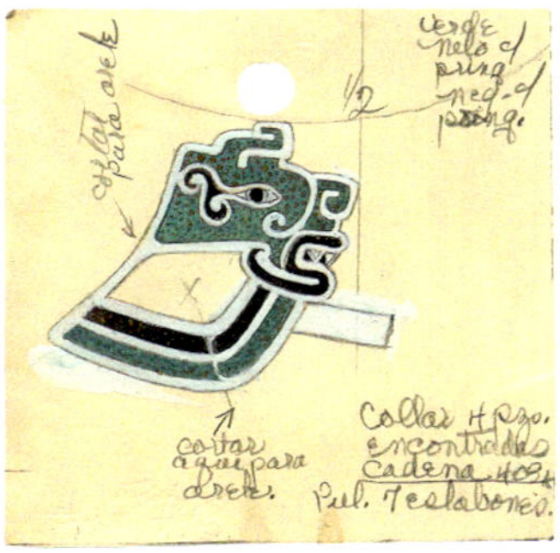

5798 Dragon Head

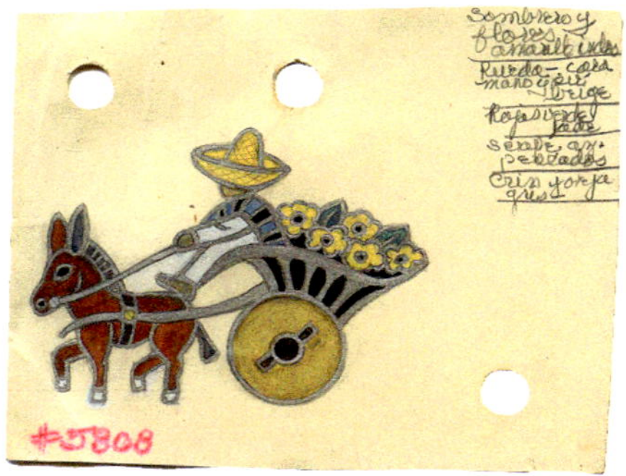

5808 Flower Cart

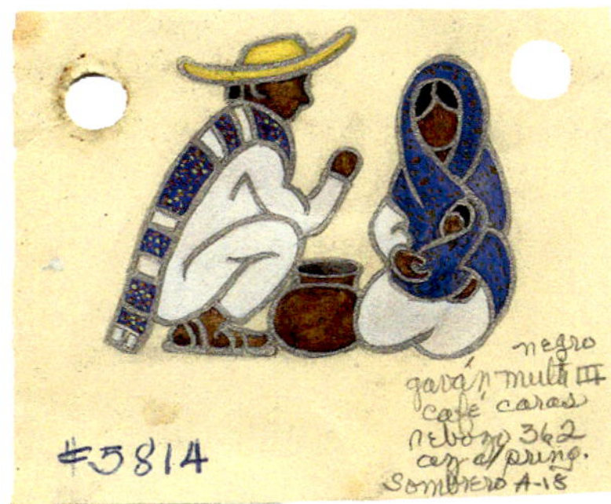

5814 Seated Man & Woman

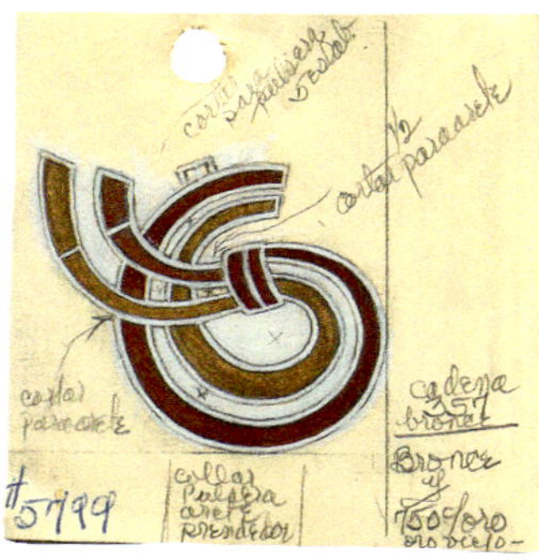

5799 Twist & Turn

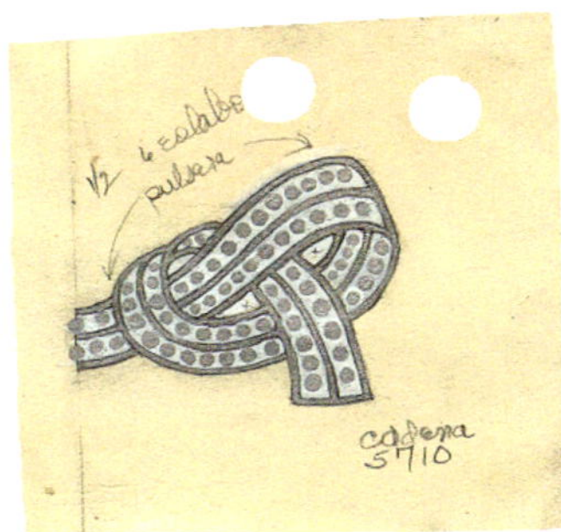

5809 Dotted Knot

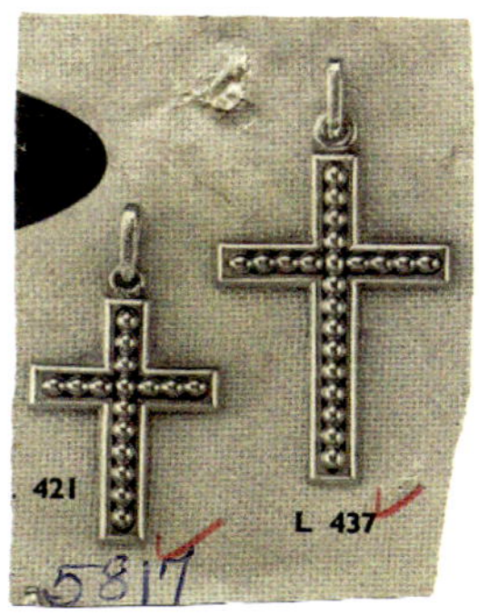

5817 Beaded Crosses

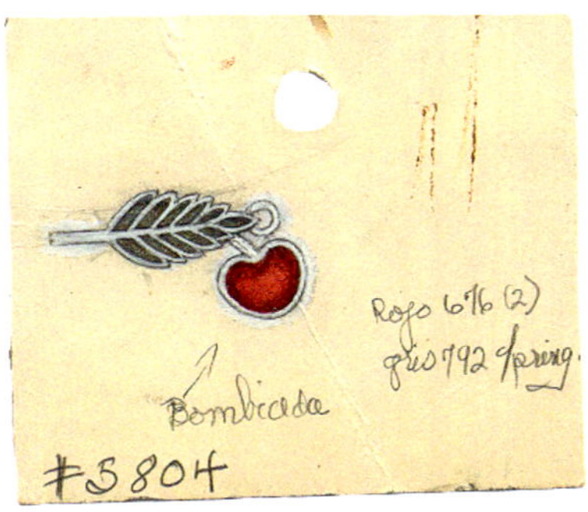

5804 Apple & Leaf

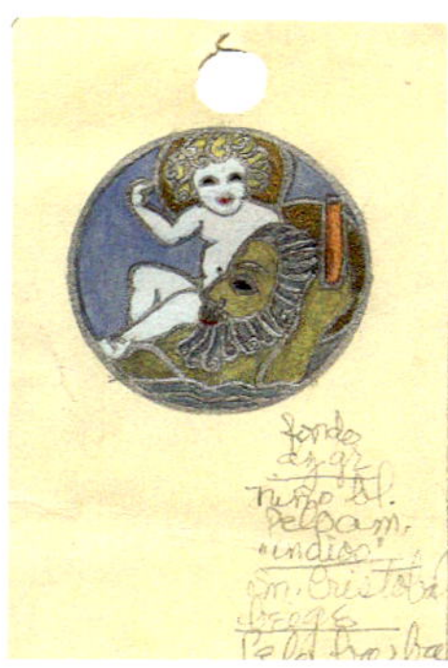

5810 Christ Child & Saint

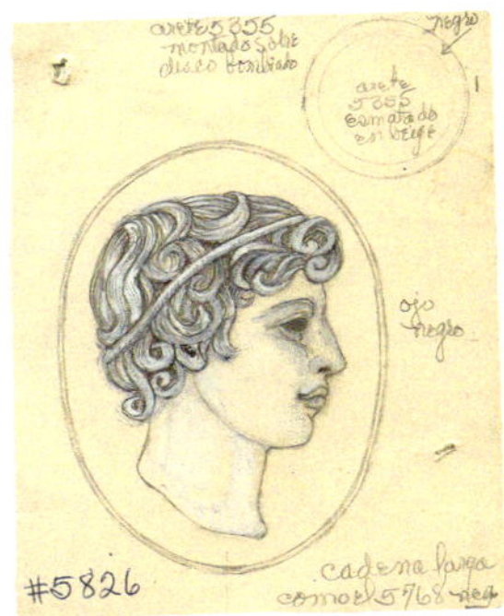

5826 Roman Head

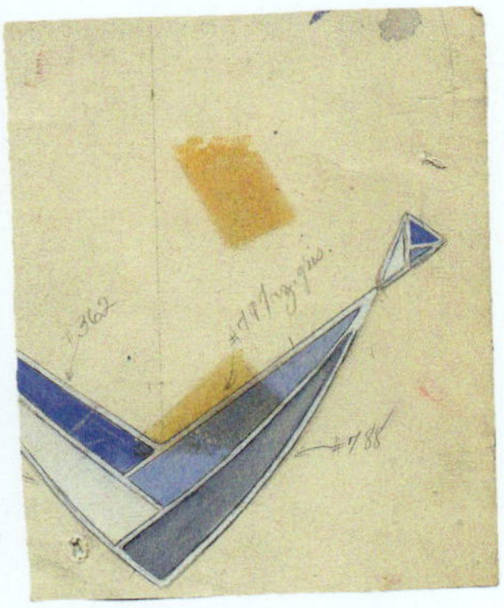

5826 Back

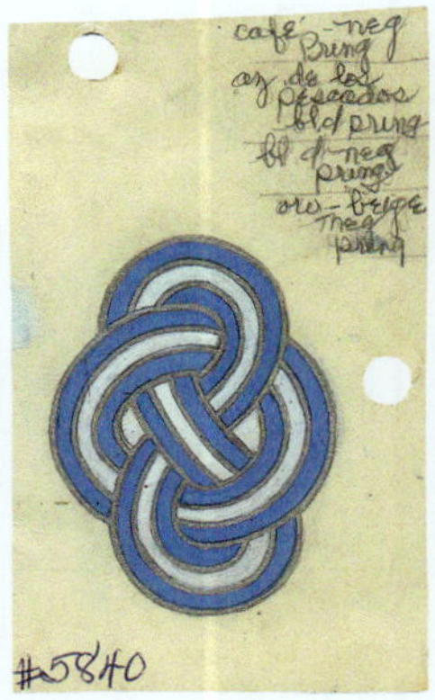

5840 Intricate Knot

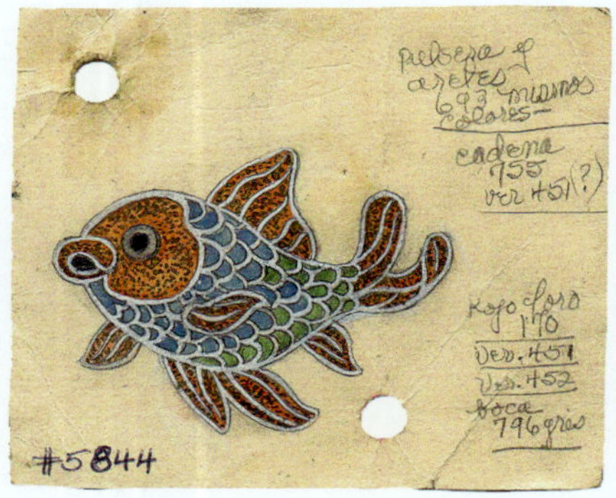

5844 Plump Fish

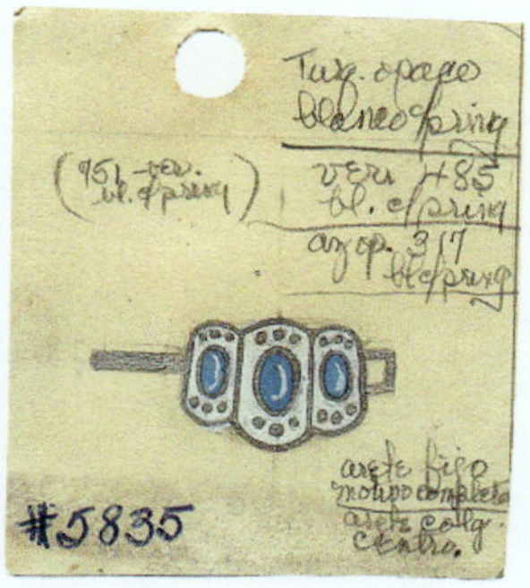

5835 Semi-precious

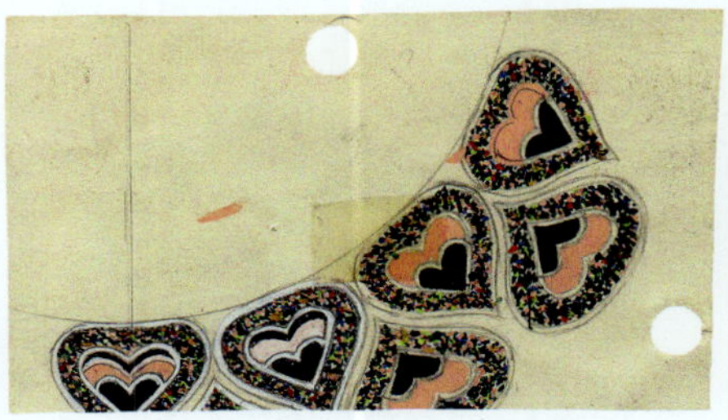

5840 Back

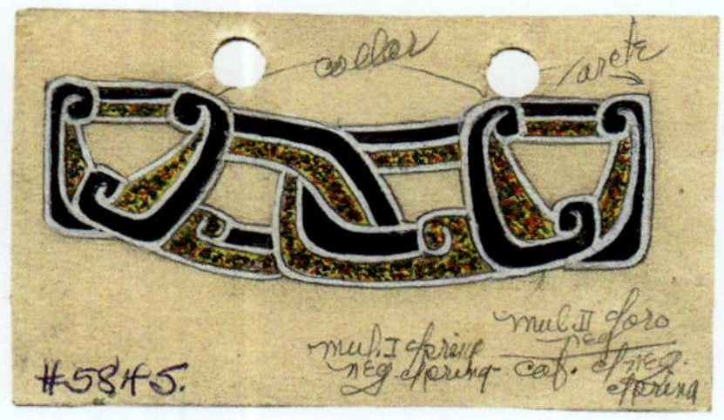

5845 Links

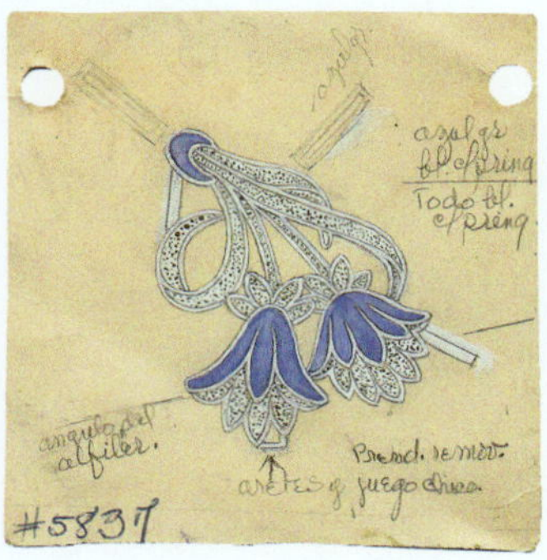

5837 Hanging Flowers

5842 Mum In Circle

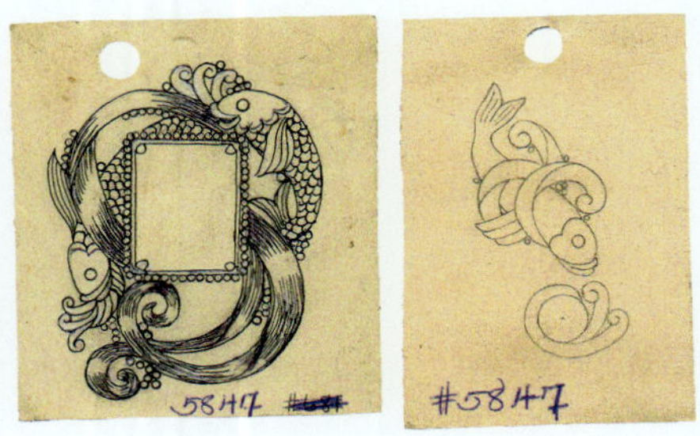

5847 Fish Pin & Earring

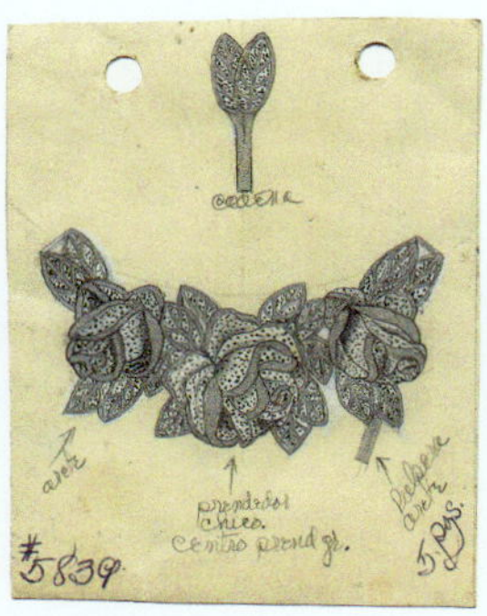

5839 Roses

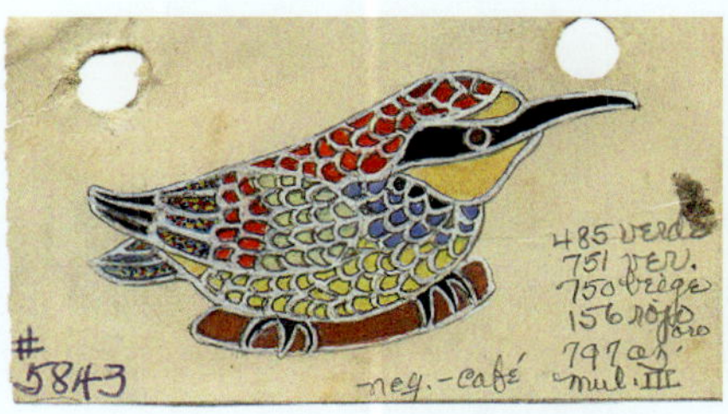

5843 Bird

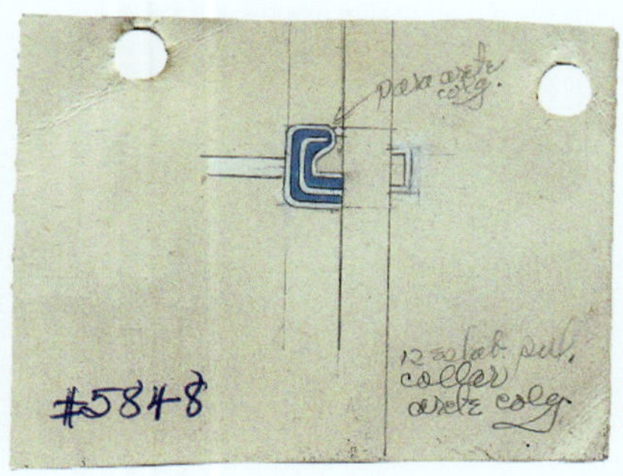

5848 L-link

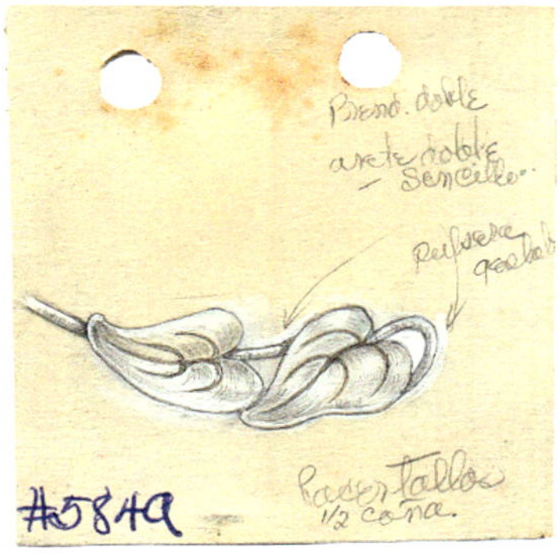

5849 Leaves

5861 Fragile Colors Butterfly

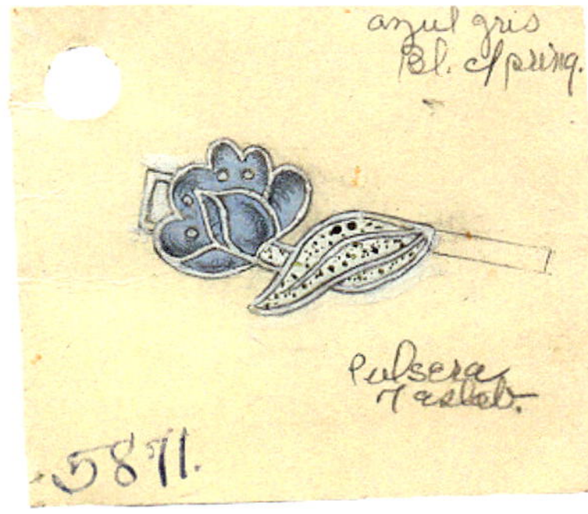

5871 Small Flower & Leaf

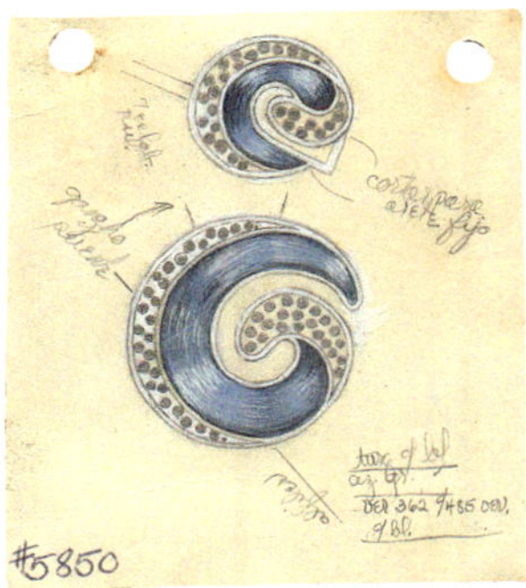

5850 Dots Spiral

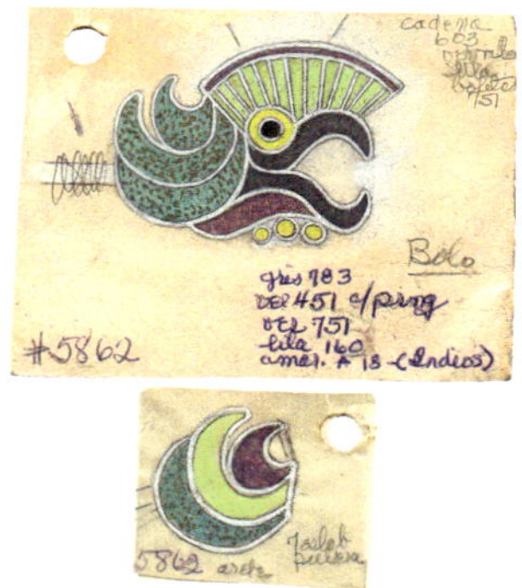

5862 Parrot

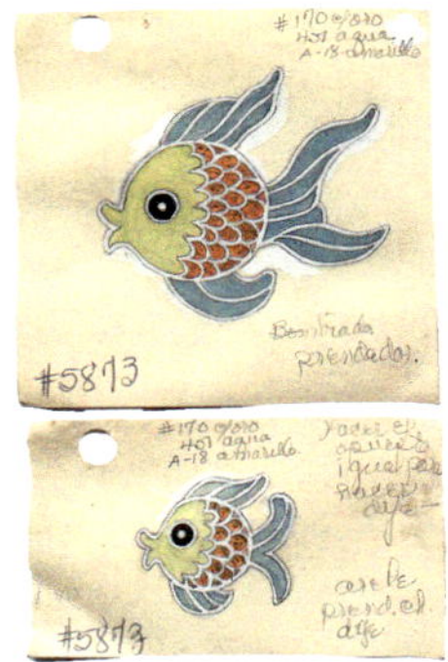

5873 Circle Fish

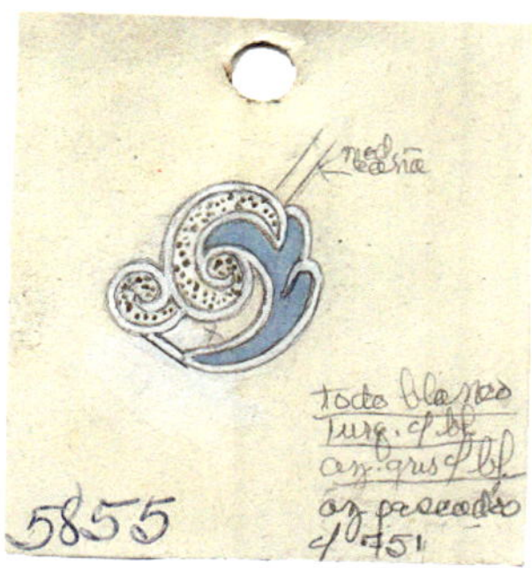

5855 Interconnected Leaves

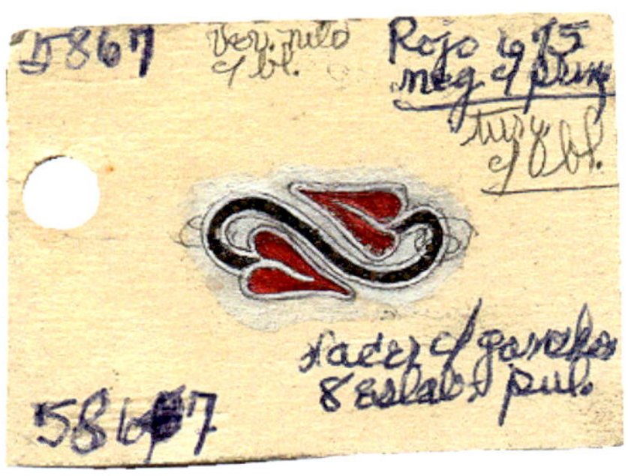

5867 Opposing Leaves

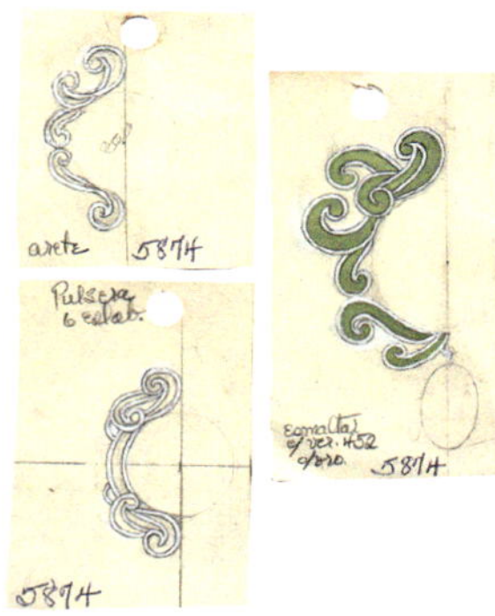

5874 Curling Leaf Lines

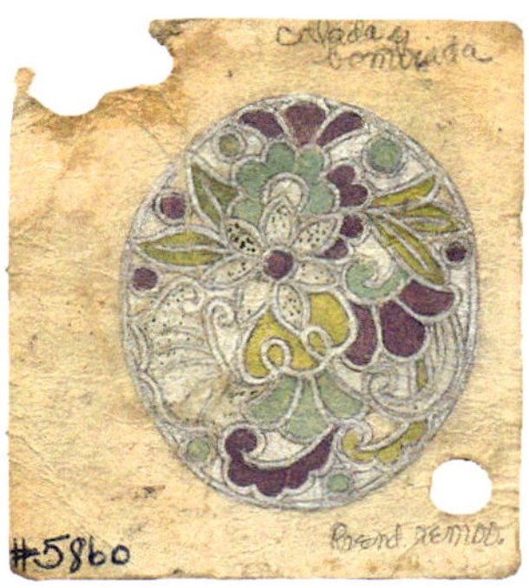

5860 Multi-floral circle

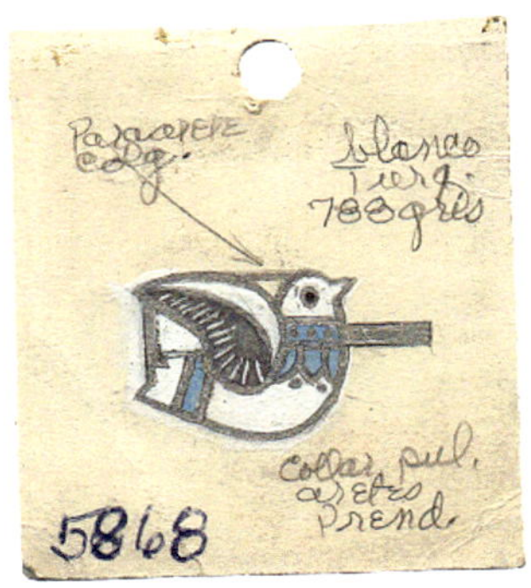

5868 Folk Art Bird

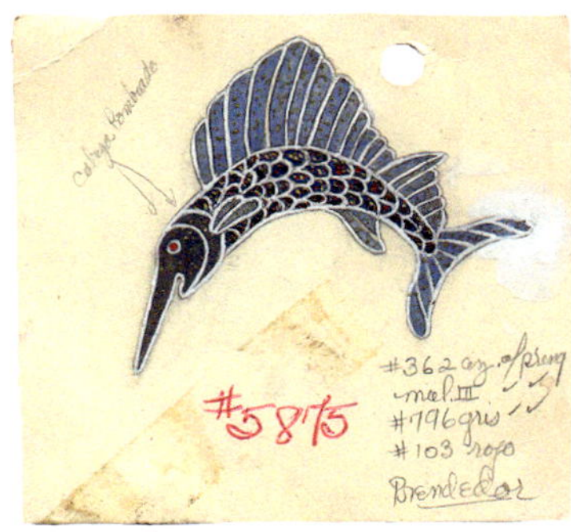

5875 Sailfish

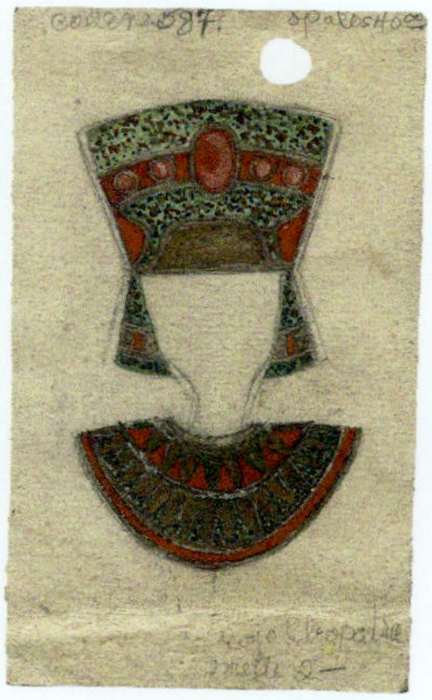

5877 Cleopatra

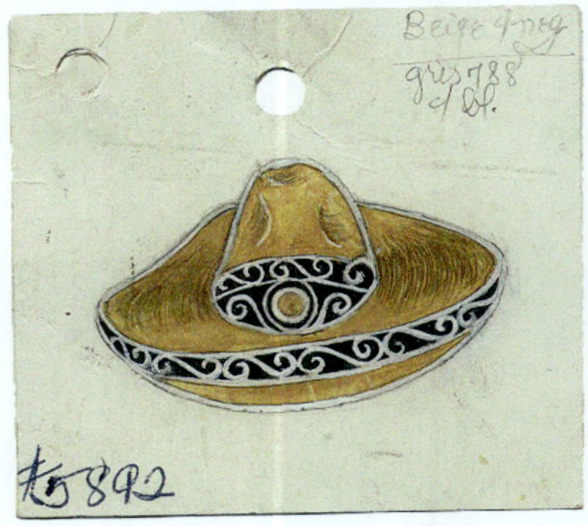

5892 Sombrero

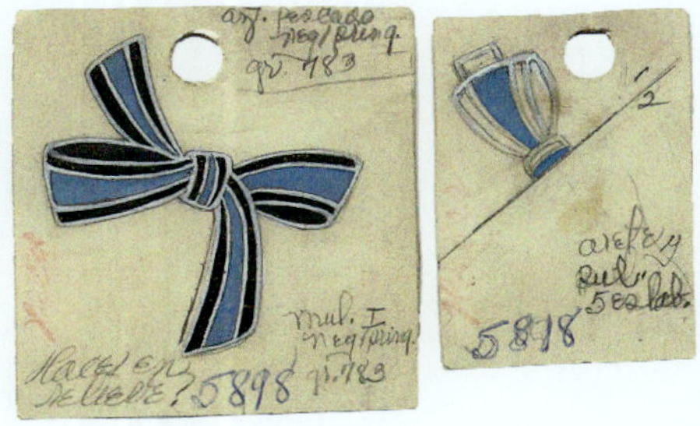

5898 Bow Tied

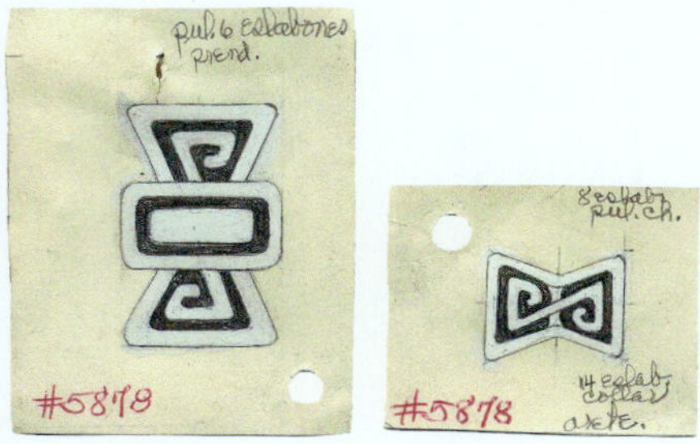

5878 Pre-Columbian Z Shape

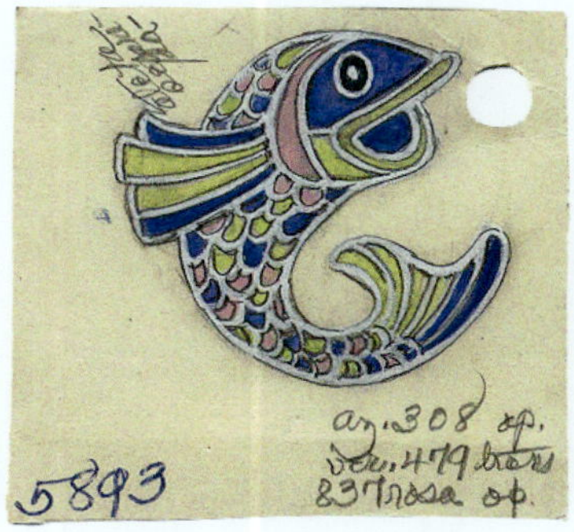

5893 Smiling Fish

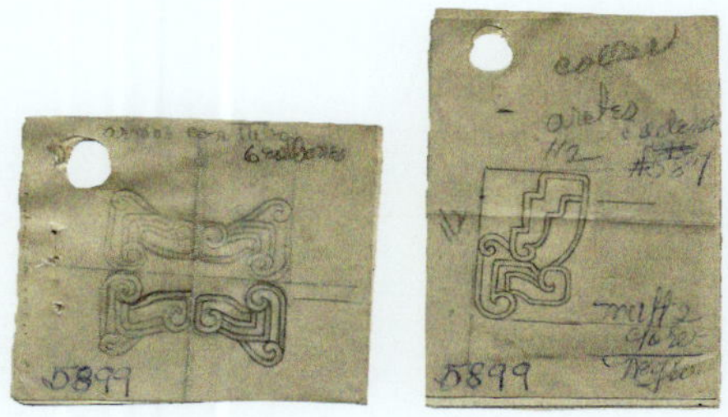

5899 Stepped Curl

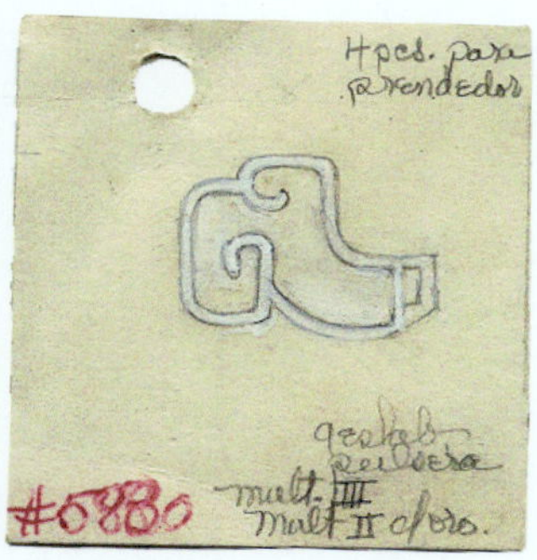

5880 Three-dimensional Curl

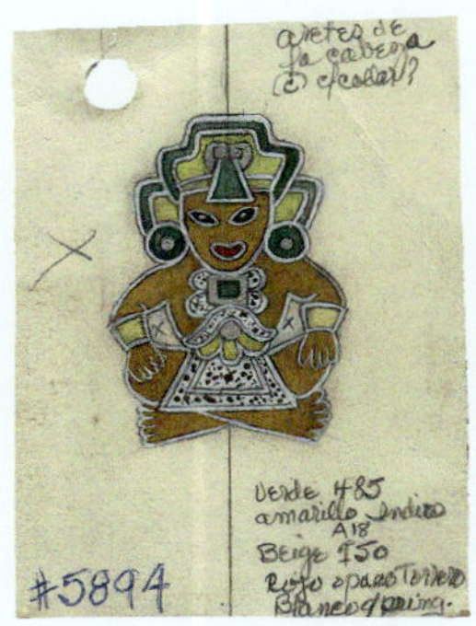

5894 Seated Pre-Columbian Figure

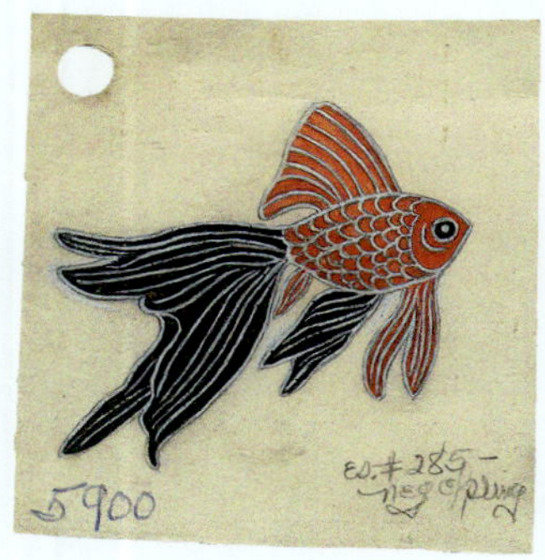

5900 Flowing Fish

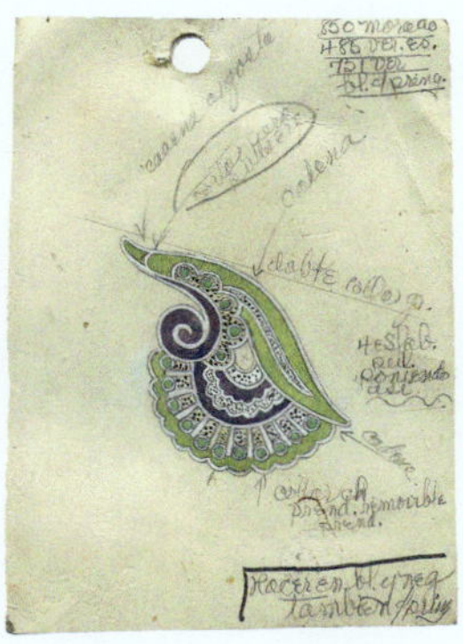

5891 Emerging Floral

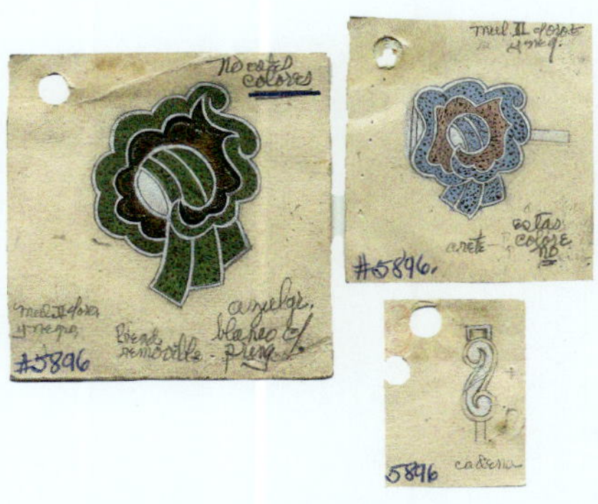

5896 Three Flanges & Floral Circle

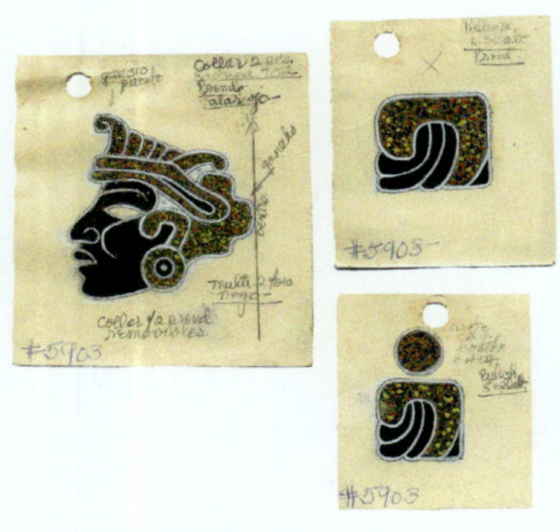

5903 Profile Maya

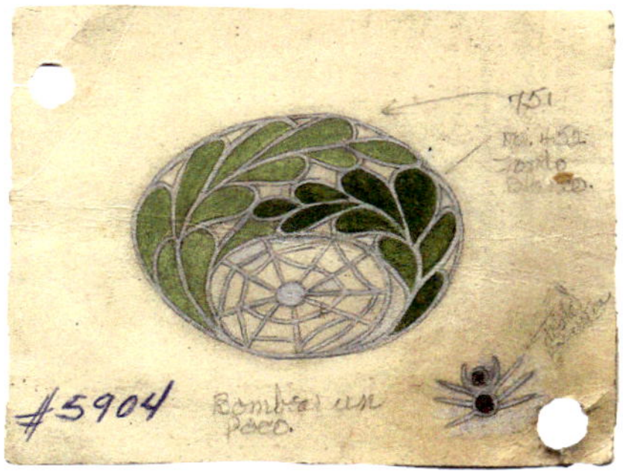

5904 Spider & Web

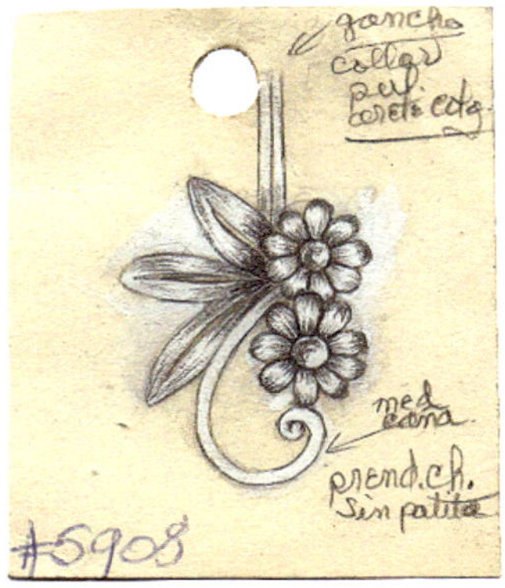

5908 Tiny Flowers & Leaves

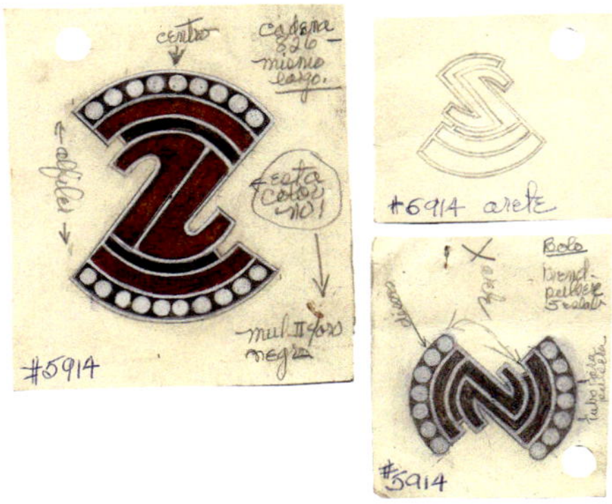

5914 Dots With Rocking Z

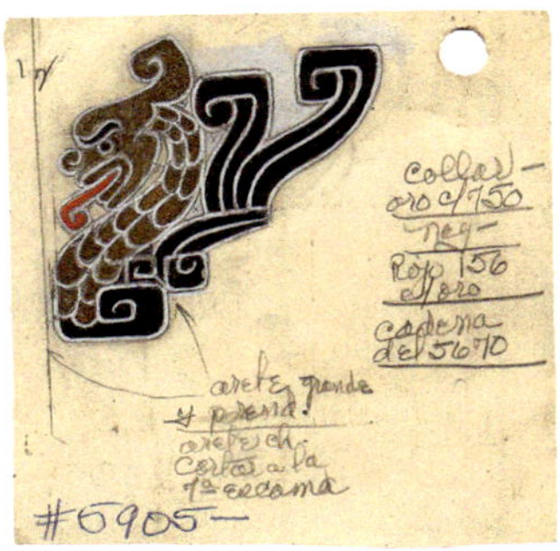

5905 Dragon

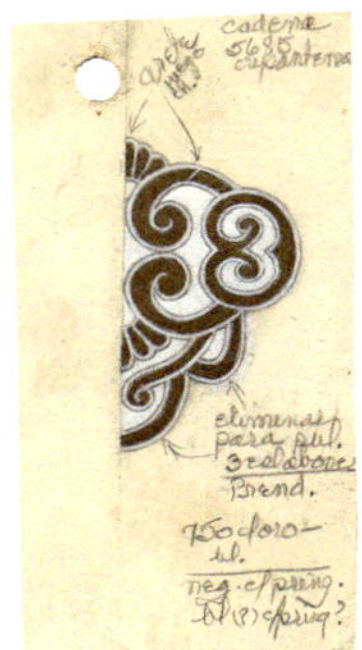

5910 Double Inward Curl

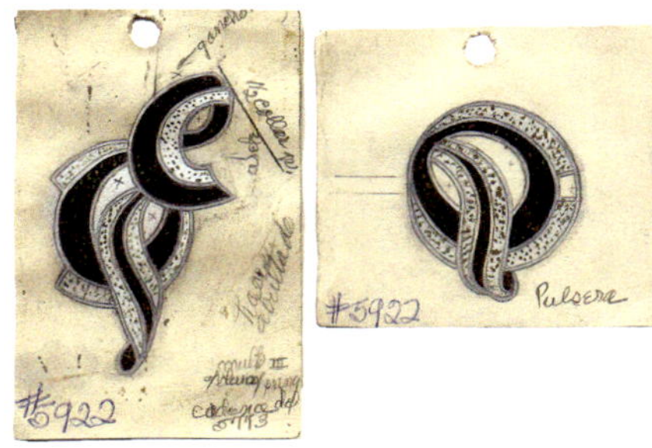

5922 Refined Curled Circle

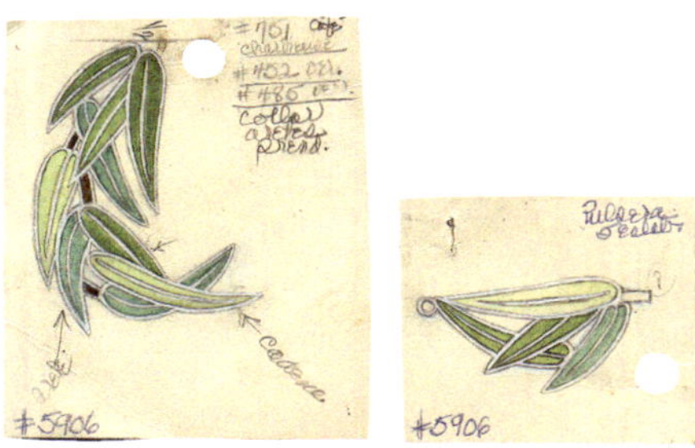

5906 Narrow Leaves

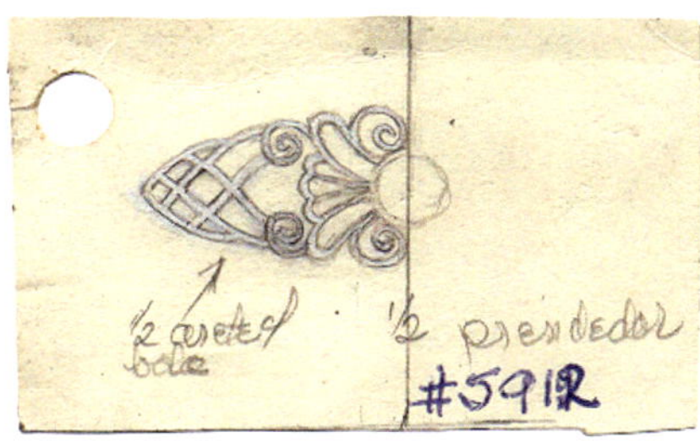

5912 Interlaced Point

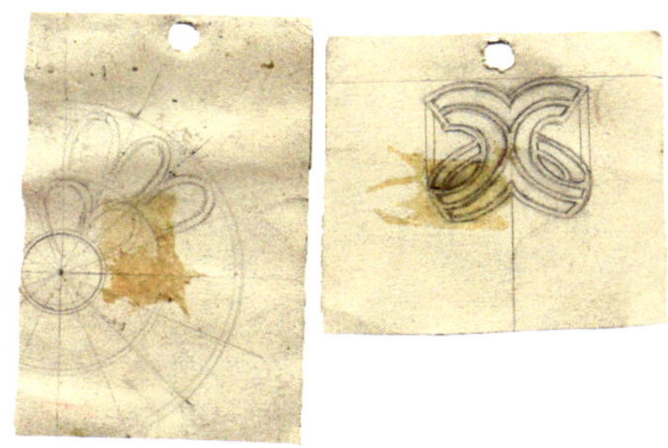

5922 Back

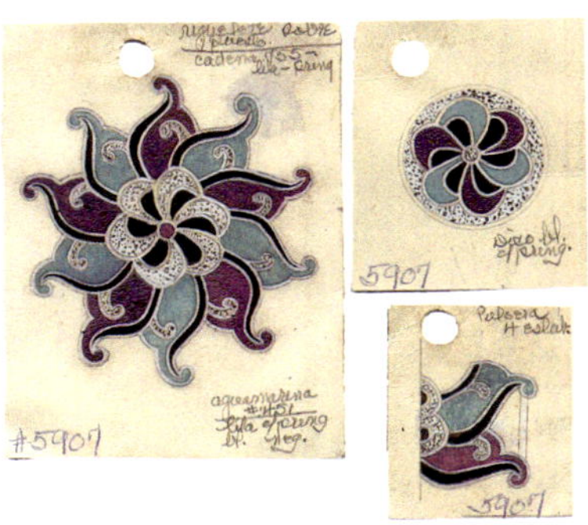

5907 Spiraling Spiral

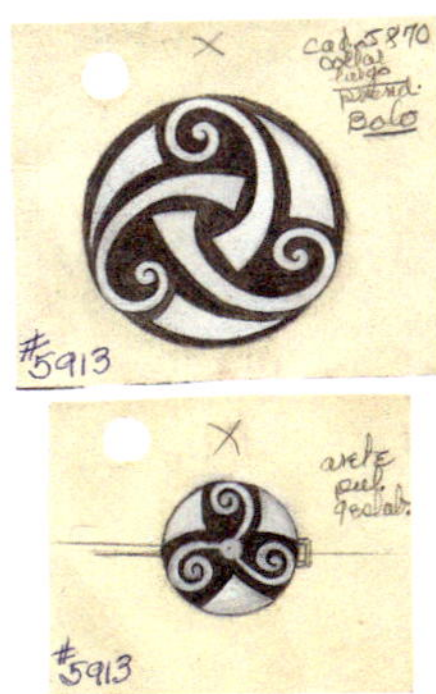

5913 Triple Spiral Circle

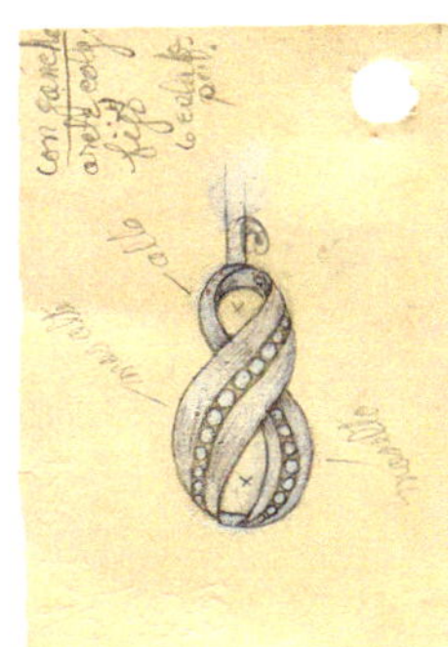

5923 Figure Eight

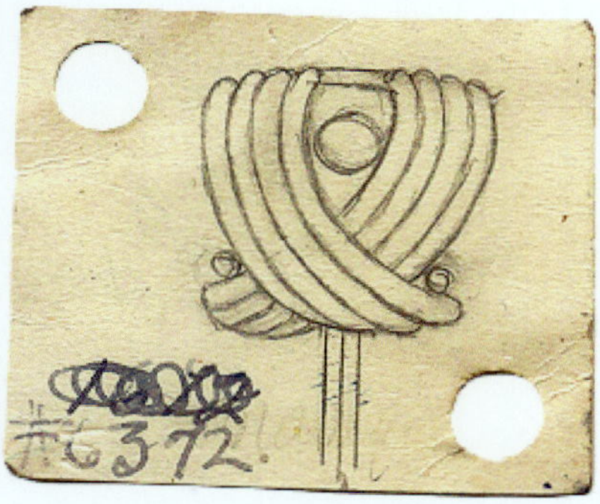

6372 Wrapped Bead

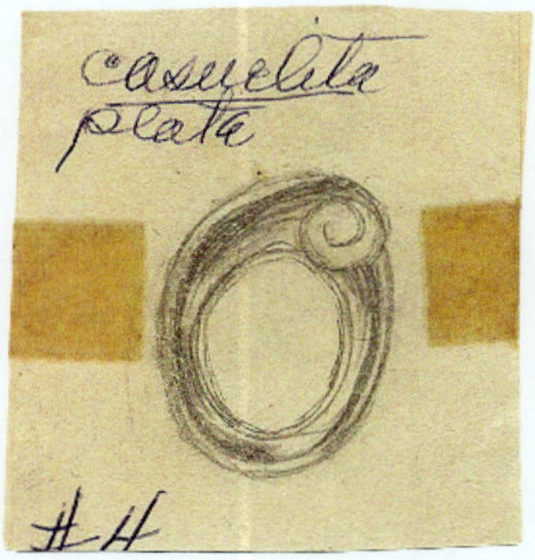

Casuelita

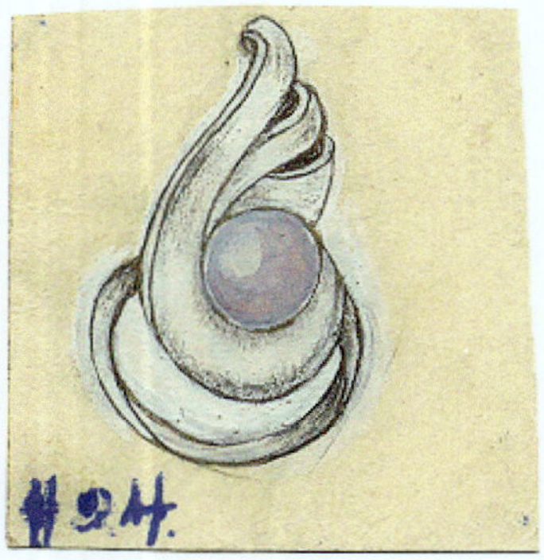

Pearl Ring

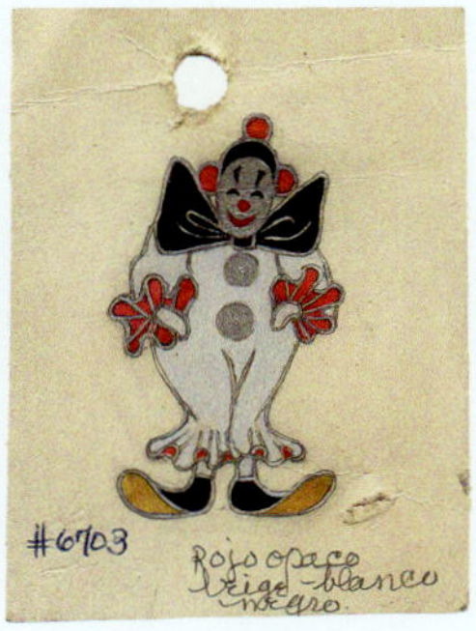

6703 Clown

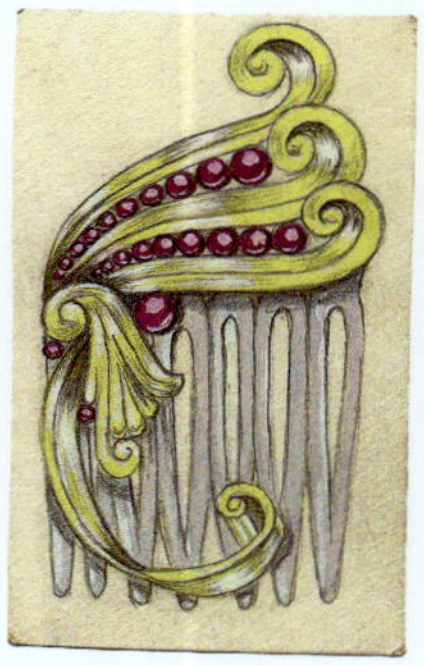

Haircomb

Appendix 2.
Makers' Marks

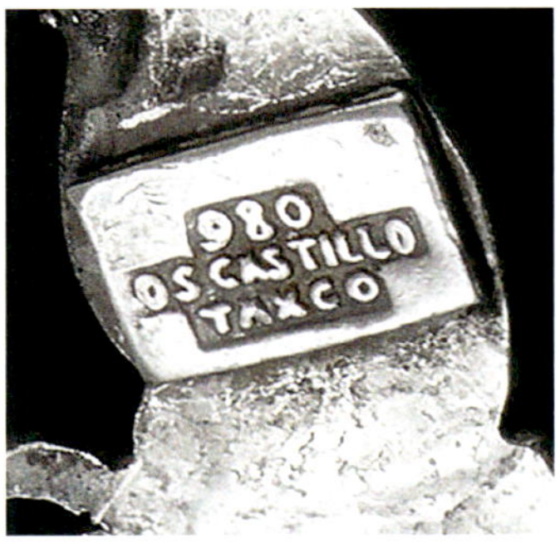

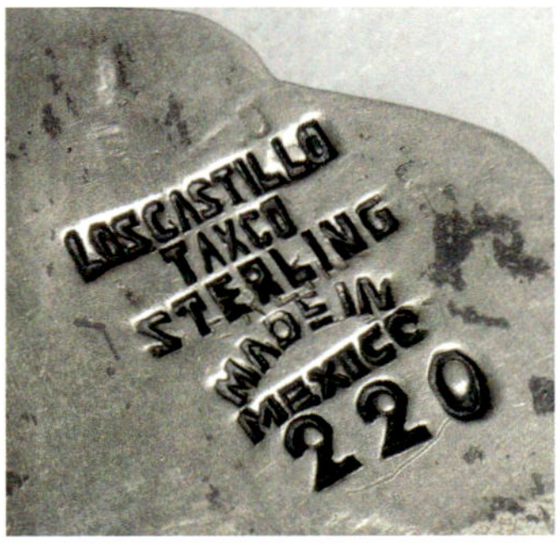

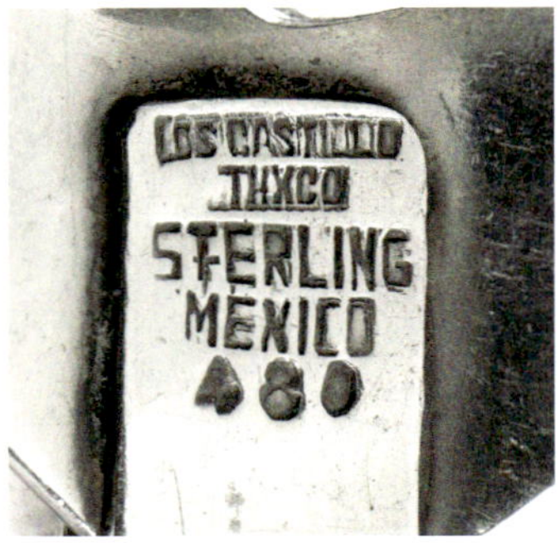

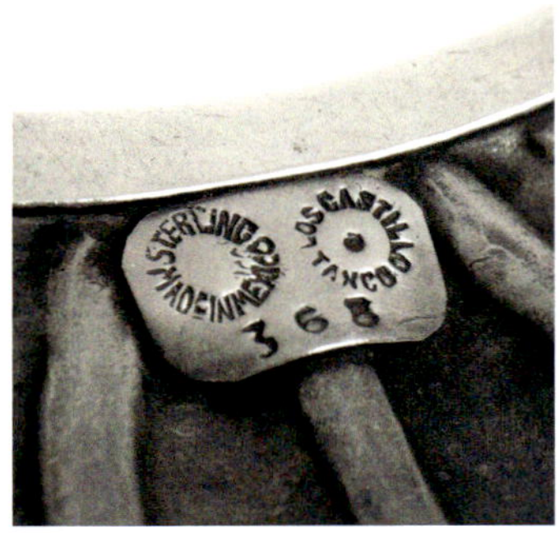

Los Castillo

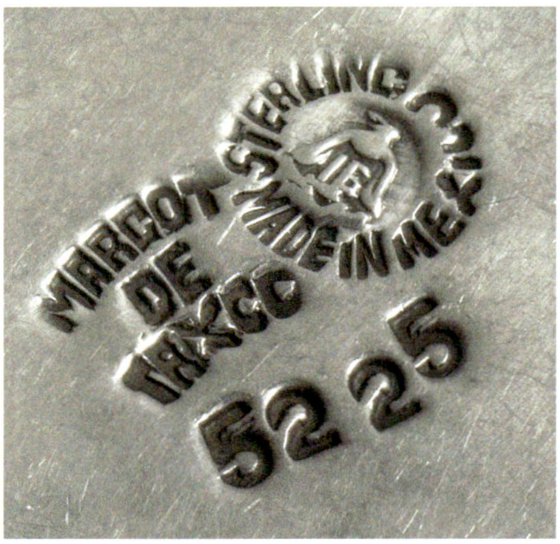

Margot de Taxco

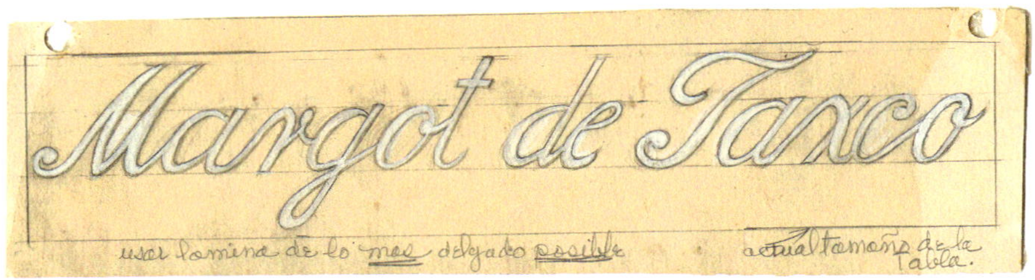

Margot's Sign

Contemporary Enamelists

Adrian Brun

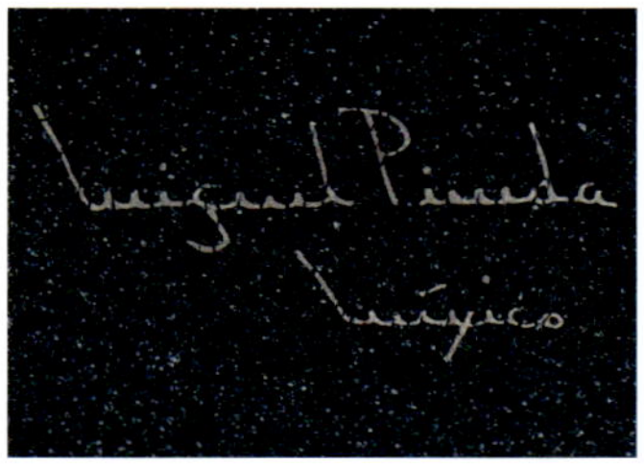

Miguel Pineda

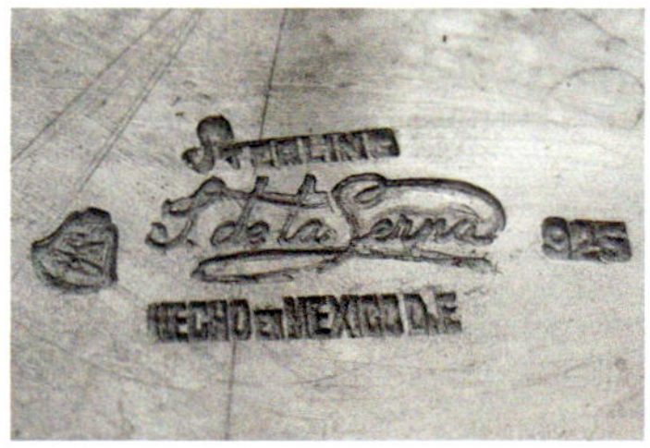

S. de la Serna

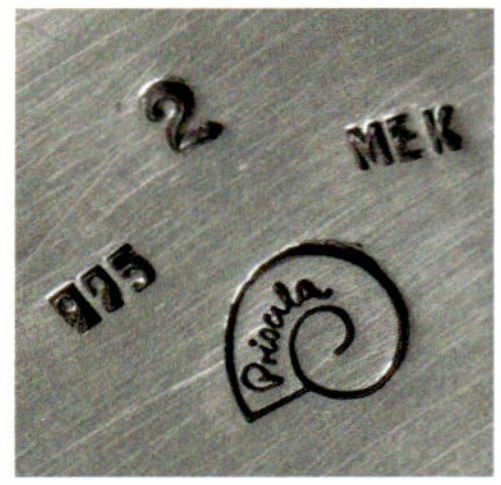

Priscila

Appendix 3
Timeline for Margot Van Voorhies — Margot de Taxco

May 27, 1847
Marriage of Albert Voorhies to Marie Leontine Durand at Pine Alley, St. Martinsville, Louisiana
November 23, 1860
Birth of Albert A. Voorhies, possibly Margot's father, to Albert and Marie Voorhies
October 9, 1861
Birthdate of Marie Albine Chabot, Margot's mother, in St. Simon, Quebec, Canada
1871
Albine Chabot, age 9, living in Bagot, Quebec with her father Joseph Chabot and mother Cecile Chabot
1877
Albina's brother Robert immigrates to the United States. Robert works for his uncles Remi and Anthony Chabot (Joseph's brothers) in Oakland, California.
1880
Albine Chabot immigrates to the United States. Albert A. Voorhies is living with his parents Albert and Marie Durand in New Orleans.
1883
Robert develops cranberry bogs for his uncle Anthony Chabot in Ilwaco, Washington.
1889
Death of Anthony Chabot
1890
Death of Remi Chabot. Remi's wife Emelie receives controlling interest in cranberry company and sells shares.
1891
Albina Chabot is working at the Lawrence Hosiery Company and living in rooms at 113 Salem in Lowell, Massachusetts.
1892
Robert Chabot leaves Ilwaco to establish his own cranberry bogs in Washington State.
1893
Albina Chabot is a dressmaker in Lowell, MA and living in rooms at 59 Read. Albert A. Voorhies is living with parents at 74 Orleans and working as a clerk at the New Orleans Board of Health.
1894
Albert A. Voorhies is living at 776 Claude, New Orleans, LA and working as a printer.
1895
Albert A. Voorhies is working for printer Sendker & Wessel at 228 Chartres in New Orleans.
1896
Albert A. Voorhies is working for Sendker & Wessel and living at 2028 Dauphine, New Orleans.
1894-1896
Albina Chabot probably arrives in San Francisco sometime between 1894 and 1896. Albina's marriage to Albert Durand probably takes place in 1896.
August 1896
Birth of Margot Durand to Albina and Albert Durand
1897
Although Albert A. Voorhies is listed in the New Orleans directory as living at 812 N. Liberty, he may have already left Louisiana for San Francisco, where he apparently changed his identity to Albert Durand (his mother's maiden name). The Durand family (Albert, Albina, and Margot) is living at 2207 ½ Pine in San Francisco.
1898
Birth of Maurice George Durand, Margot's brother
1899
Death of Maurice George Durand. The Durands move to 1109 ½ Leavenworth Street in San Francisco.
1900
Margot, age 3, is living in San Francisco with father Albert (Allie A.) Durand and Albina Durand.
1900-1903
Durand family at 1109½ Leavenworth Street
January 5, 1903
Death of Allie Durand. Allbert (Allie) Durand is buried at Holy Cross Cemetery in Colma next to his son Maurice George Durand. Albert A. Voorhies's death is recorded as January 4, 1903.
1903
Mrs. Allie Durand (widow) and daughter living at 1109 ½ Leavenworth. Albina is working as a hairdresser.
August 7, 1903
Death of Albert A. Voorhies's mother, Leontine Durand Voorhies
1905
Mrs. Allie Durand is now Mrs. A. Durand Voorhies and living at 2132 Fillmore with her daughter.
1906
San Francisco earthquake and fire
January 21, 1913
Death of Judge Albert Voorhies, Margot's grandfather
1915
Albina Voorhies and Margot have a beauty salon at 2131 Center and are living at 2135 Center in Berkeley, California. Margot now listed as Margot Van Voorhies.

1919

Albina Voorhies (widow of Albert) and Margot return to San Francisco and are living at 1677 Sacramento. Margot is a clerk with Travelers' Insurance Company.

1920

Margot is working as a bookkeeper and living at 2417 Washington.

1921

Margot is a stenographer and living at 2417 Washington.

1922

Margot is not listed in San Francisco directory. Albina Voorhies is living at 1008 Larkin and is unemployed.

1923

Albina is employed as a hairdresser and living at 1008 Larkin.

1924

Albina D. Voorhies is living at 1008 Larkin and is possibly employed by Margaret Clayes whose company manufactured art goods inspired by Asian art.

1925

Mrs. Albina D. Voorhies is at 1008 Larkin, without a listing of employment. Margot van Voorhies, hairdresser, living at 1460 Leavenworth.

1926

Mrs. A. D. Voorhies is listed as an artist and living at 1008 Larkin. Margot van Voorhies is listed as a clerk and living at 1460 Leavenworth.

1927

Mrs. A. D. Voorhies is listed as artist and living at 1008 Larkin. Margot van Voorhies is listed as hairdresser and residing at 1460 Leavenworth. Margot marries Francis Evans Carr.

1928

F. Evans Carr is listed as a student, with Margot at 145 Judah.

1929

Ada [sic] Voorhies is listed as grocer at 1513 48th Avenue; residence at 1515 48th. F. Evans Carr and Margot are living at 1355 Willard. Margot has separate listing as hairdresser.

1930

Margot and Evans Carr are living at 1355 Willard in San Francisco. Margot is a cosmetologist at a dry goods store and Evans is still a student. Albina D. Voorhies is a grocer at 1515 48th Avenue.

1931

F. Evans and Margot Carr are living at 1355 Willard. Albina Voorhies (widow Albert) is a grocer at 1515 48th Avenue. Albina is murdered on February 16, 1931.

1932

Evans and Margot Carr are living at 1150 Union. Evans is listed as a dentist; Margot is unemployed.

1933

Evans, a dentist, and Margot Carr are living at 1150 Union. Margot has a separate listing as a beauty operator.

1934-1935

Evans, a dentist, and Margot Carr are at 1150 Union. Margot is working at the United Artists Beauty Salon at 133 Geary Blvd.

1936

Evans and Margot Carr are living at 770 California. Evans is listed as a dentist at 450 Sutter. Margot is working at United Artists Beauty Salon. Margot and Evans Carr divorce in July 1936.

1937

Evans is listed as dentist at 450 Sutter. Margot is working at 133 Geary and living at 770 California. Margot moves permanently to Mexico and, in August 1937, is living in Mexico City at Chopo #139. Margot is employed at Hugo Brehme's photography gallery at Madero 1 in Mexico City. Margot becomes an accomplished artistic photographer. Margot befriends María de los Angeles. Margot and friends travel to Taxco.

1938

Margot marries Antonio Castillo de Terán.

1939

Margot and Antonio Castillo establish Castillos en el Aire, which becomes Los Castillo, with Antonio's brothers Jorge and Justo and cousin Salvador Terán.

1940

Margot and Antonio Castillo travel to the United States, crossing in Laredo, Texas.

1943

Margot travels to the United States.

Early 1940s

Margot adopts Rafael, possibly Antonio's illegitimate son.

1946

Margot and Antonio Castillo divorce. Margot goes to live with María de los Angeles.

1947

Margot van Voorhies Carr establishes Margot de Taxco.

1948

In 1948, Margot has two hundred artisans in the Margot de Taxco workshop.

1949

Margot, with Sigi Pineda as manager, establishes the enamel workshop.

1950

María de los Angeles marries Reveriano Castillo and establishes Reveri.

1952

Margot marries Alberto Cuevas.

1953

Fire in the Margot de Taxco workshop and the tragic death of two enamellists. Margot moves to a workshop and store away from the plaza and on the highway.

1954

Margot divorces Alberto Cuevas.

1960

Another fire in the workshop

before 1964

Margot moves Margot de Taxco to another smaller building on the highway across from Rancho Alegre owned by Pedro Pérez.

1974

Strike at Margot de Taxco

1977

Margot is forced into bankruptcy. The government takes her car, house, furnishings, workshop tools and dies to pay off workers and unpaid taxes. Margot moves into a small house where she lived for eight years, supported entirely by her friend María Castillo.

July 26, 1985

Margot's death

Notes

Chapter I. Early Childhood

1 Jaime Quiroz was in charge of sales for Margot from 1963 until the company collapsed in 1974 (Morrill 2007, 173).

2 *Border Crossings: From Mexico to U.S., 1903-1957*, Margot Van Voorhiss [sic] De Castillo October 12, 1940, with husband Antonio Castillo, at Laredo, Texas.

3 *United States Federal Census*, San Francisco, California, 1900. Listing for Allie Durand (wife Albina, and daughter Margot), 1109 ½ Leavenworth Street.

4 U. S. Federal Census, San Francisco California, 1900. For Albina, her age "39" was marked out and replaced with "37."

5 *United States Federal Census*, San Francisco, California, 1900.

6 Baptismal record for Marie Albine Chabot at the parish of St. Simon de Bagot. *Quebec Vital and Church Records*, Oct. 9, 1861. According to Sigi Pineda, Margot was thought to be of Jewish origin. The documentary evidence, however, points to a Roman Catholic heritage along all of her ancestral lines.

7 Joseph Chabot and Cecile Jared were married in the parish of La Présentation, Quebec in 1842 (*Quebec Vital and Church Records*). In the baptismal record for Albine, her father Joseph is referred to as "cultivateur."

8 United States Federal Census, San Francisco, California, 1900.

9 Burgess 1992, 148-52. This uncle assisted Anthony Chabot with the introduction of cranberry cultivation in the Northwest. Unfortunately, Burgess does not name Anthony Chabot's brother-in-law.

10 Blewett 1990, 73.

11 Marshall M. Hayden, Photographer, at 120 Merrimack, appears in the Lowell Business Directory for 1882. This photograph is owned by Elizabeth L. Forrest Freeman, Robert's granddaughter and Albina's grand-niece.

12 Hofstadter, Miller, and Aaron 1967, 256-7; Morison and Commager 1962, I: 505; II: 16-17. The difficult life of the female factory worker is almost unimaginable today in the United States (although these conditions continue to exist in other parts of the world). Hymowitz and Weissman (1978, 236-7) write, "For ten to twelve hours a day a woman factory worker might 'concentrate mind, body, and soul, itself, literally on the point of a needle,' or sit on a stool feeding shoes, one by one, into a machine that punched lace holes or stand at a table and press covers on thousands of thousands of jars or glue linings into an endless number of boxes." They worked by the piece, so the women had to work swiftly and carefully because, as one woman recalled, "'The machines go like mad all day, because the faster you work the more money you get. . . . All the time we are working the boss walks about examining the finished garments and making us do them over again if they are not just right.'"

13 Lowell, Massachusetts city directories, 1891-1893.

14 Information about *L'Etoile* appears on the Library of Congress's website, *Chronicling America*: http://chroniclingamerica.loc.gov/lccn/. Lowell, Massachusetts city directories, 1881, 1883, and 1890.

15 Burgess 1992, 145, 154-5.

16 Burgess 1992, v-vi, 153-65. Anthony Chabot's choice of philanthropic projects may have been influenced by his sister-in-law Emelie, whom he respected and trusted. Emelie took a leading role in many of Oakland's charitable institutions, and served on the boards of those that were founded and funded by her brother-in-law.

17 Burgess 1992, 151-2,167, 173.

18 Hofstadter, Miller, and Aaron 1967, 518-23.

19 Burgess 1992, 174-5.

20 San Francisco city directories, 1897-98.

21 Burgess 1992, 174. Ellen also served on the board of a charitable institution founded by her father.

22 U. S. Federal Census, San Francisco, California, 1900.

23 San Francisco, California, city directories, 1899-1902, list Allie Durand at 1109 ½ Leavenworth Street. In the directories for 1900 to 1902, he appears as Allie A. Durand.

24 *San Francisco Chronicle*, Jan. 7, 1903. The Chronicle for Jan. 8, 1903 announced: "DURAND

— In this city, January 5, Allie, beloved husband of Albina Durand, and father of Margot Durand, a native of Missouri, aged 36 years, 1 month, 23 days. Friends and acquaintances are respectfully invited to attend the funeral Friday, January 9, at 10:30 o'clock, from his late residence, 1109 ½ Leavenworth street. Interment, Holy Cross Cemetery, by 11:30 o'clock train from Third and Townsend streets."

25 I am grateful to Nancy Peterson and Jane Knowles Lindsey of the California Genealogical Society for providing this information.

26 San Francisco city directory, 1903.

27 Elizabeth Finch interview of Elizabeth Freeman (daughter of Elizabeth Forrest), 2008.

28 San Francisco city directory, 1905. Albina and Margot were now around the corner from Sacramento and two blocks from the original Pine Street address.

29 Border crossing, 1940. According to Alice Goyette's 1918 passport application, she was a French Canadian, born in 1882, thus, fourteen years older than Margot. During the 1930s, Alice lived at 1590 Sacramento and was at this address when she wrote the citizenship letter for Margot in 1940. Miss Goyette was a stenographer and bookkeeper, like Margot. Thus, Margot could have befriended Alice as a neighbor or co-worker, or as a friend of her mother's, since both were French Canadians (U. S. Passport Applications, 1906-1925; California Voter Registrations, San Francisco, 1930-1940).

30 Judge Albert Voorhies's obituary in the *Daily Picayune*, Jan. 21, 1913. *Biennial Report of the Louisiana State Board of Health, 1884-1885*.

31 *Louisiana: A Guide to the State 1945*, 624. The house is no longer standing. The wedding of Marie Leontine Durand and Albert Voorhies, Margot's grandparents, may have actually taken place on May 27, 1847 (Ancestry.com, family tree for Marie Leontine Durand). Also legend has it that $150,000.00 in gold coins was buried on the grounds of Pine Alley Plantation during the Civil War (Mills 2002, 19).

An interesting Voorhies cousin by marriage was from another prominent New Orleans family. Judge Albert Voorhies's sister Isabelle Voorhies married Bernard Avegno, Jr. and his brother Joseph William Voorhies married Bernard's sister Lucie Marie Avegno. Lucie and her brother Bernard, Jr. were first cousins of Virginie Amélie Avegno Gautreau, the famous Madame X, whose portrait by John Singer Sargent caused a scandal in Paris. When Sargent offered to sell the painting to the Metropolitan Museum of Art the year after it was displayed at the 1915 Panama-Pacific exposition in San Francisco, he stated that he considered it his finest work (U. S. Federal Census, New Orleans, LA, 1880; New Orleans, LA Marriage Records Index, 1831-1925, For William Voorhies and Lucie Avegno, Vol. 5, p. 380, for Bernard Avegno and Isabelle Voorhies, Vol. 8, p. 642; Ormond and Kilmurray 1998, 113-18; Official Catalogue of Exhibitors 1915, 40).

32 U. S. Federal Census, Fausse Point, St. Martinville, LA 1850; New Orleans LA 1860, 1870.

33 Christoph 2003, II, 803-4. The entry in the genealogical record reads, "b. 24 Nov. 1860 LA; d. 4 Jan. 1903 at age 42." A great effort has been made on the part of Elizabeth Finch and the researchers at the California Genealogical Society, Nancy Peterson and Jane Knowles Lindsey, to discover the source of the Voorhies family's information concerning Albert A.Voorhies's death date. Unfortunately, this particular mystery may never be solved.

34 New Orleans city directory, 1893; U. S. Federal Census, New Orleans, LA 1880. In 1880 Albert A. Voorhies was nineteen years of age and living with his parents. He was employed as a clerk in a store.

35 In 1894, Albert A. Voorhies was a printer, his residence at 776 St. Claude; in 1895, a printer for Sendker & Wessel at 228 Chartres. He continues to be a printer but, in 1896, his residence is at 2028 Dauphine, and in 1897, at 812 N. Liberty. Albert is a travel agent in 1898, living at 1426 N. Derbigny, but the following year, this exact listing is given for Albert's cousin, Arthur Voorhies. Thus, we can assume that the name should have been Arthur in the 1898 directory. New Orleans city directories, 1893-98.

36 U. S. Federal Census, San Francisco, 1930.

37 *San Francisco Chronicle*, Jan. 8, 1903.

38 Elizabeth Chabot Forrest's daughter, Elizabeth L. Forrest Freeman, now owns the photograph.

39 Christoph 2003, II, 803-4.

40 New Orleans, LA Death Records Index, 1804-1949. Leontine Voorhies died on August 7, 1903 at the age of 71.

Chapter II- The San Francisco Earthquake and Fire

41 Smith 2005, 3. For more on the San Francisco earthquake, see *San Francisco Chronicle* website, *The Great Quake: 1906-2006* (http://sfchronicle.us/greatquake/), Carle Nolte, Chronicle Staff Writer, 2006.

42 Smith 2005, 52-6, 78.

43 Smith 2005, 222.

44 Watkins, 1906.

45 Smith 2005, map on viii-ix.

46 Smith 2005, 127.

47 Knox family papers, letter addressed to Uncle Wales Knox from Anna Sprague Cox, 1906, The Barcroft Library, University of California, Berkeley. Lola Clayes Monroe and her sister Margaret Clayes may have befriended Albina and Margot at a later date.

48 Some evidence exists that Margot and her mother Albina left a burning building. Margot told Jaime Quiróz that her personal papers were all lost when her home was destroyed by fire after the earthquake.

49 Smith 2005, 138.

50 Burgess 1992, 143-5, 155-7, 173-4.

51 Burgess 1992, 175. The following appeared in the *Oakland Tribune* society page for Sept. 30, 1905: "Miss Chabot is going out more this winter, since she is no longer in mourning. She was very good indeed to her stepmother, the late Mrs. Chabot, and many years of her life were devoted to her care. The late Mrs. Chabot never cared to travel and she had a special aversion to a Pullman car. So she and her daughter for many years lived quietly in their home by the lake, Miss Chabot devoting herself to her books and music. She has a charming voice, beautifully cultivated. . . . Miss Chabot is very pretty, very well gowned, very charming, very accomplished, and an heiress. So one awaits with interest her social season in Washington, which is really a belated debut." On Feb. 8, 1906, Ellen was the guest of President and Mrs. Roosevelt at the White House (Washington Post, Feb. 9, 1906). In 1907, Ellen and a friend spent a month in Santa Barbara (Oakland Tribune, Jan. 19, 1907, Feb. 16, 1907).

52 U. S. Federal Census, Oakland, California, 1910.

53 Berkeley city directory, 1915.

54 Collins 2003, 327-32.

55 Collins 2003, 289-90.

56 Emelie Chabot was active in the Women's Congress of the Pacific Coast, an organization that campaigned for women's voting rights. Burgess 1992, 173.

57 According to Hymowitz and Weissman (1978, 234), "By 1900 there were five million female wage earners in the United States, making up one-fifth of the nation's total work force. Two million of these women were domestics — maids, cooks, nurses, and laundresses, who worked in private homes. . . . After 1880. . . factory work became the second most common kind of employment for women."

Chapter III- Return to San Francisco

58 San Francisco city directory, 1919.

59 San Francisco city directories, 1920, 1921.

60 Author's interviews with Sigi Pineda and Jaime Quiróz.

61 San Francisco city directory, 1919.

62 San Francisco city directories, 1920-27.

63 San Francisco city directories, 1923, 1924. "Miss Margt Clayes pres Lola C Monroe v-pres R L Simpson sec Edith L Kesler treas. Art goods 214 Post. Clayes Marget pres Miss Clayes r2623 Divisadero." Lola C. Monroe was Margaret Clayes's sister. In the 1910 and 1920 census, Margaret Clayes was living in San Francisco with her mother, Flora B. Clayes. She was teaching in 1910 and, by 1920, she was "a merchant of goods." Margaret was peripatetic, moving to a different address almost every year. In 1919, Margaret was living at 999 Bush, and in 1920, at 1355 Taylor. Her address for the years 1921-1922 was 2134 Green and in 1924 it was 2623 Divisadero. She gave two different addresses in 1916, 2325 Sacramento in April, and 545 Post in October. Thus, it would seem possible that Margaret Clayes could have lived briefly with Albina at 1008 Larkin (U. S. Federal Census for 1900, 1910, and 1920; Voter Registration records from 1912-1938; San Francisco city directory, 1944).

64 *Directory of Manufacturers of San Francisco, California,* 1922. In the 1920 federal census, Margaret Clayes lists herself as a merchant of art goods.

65 *Official Catalogue of Exhibitors 1915*, 19, 20.

66 San Francisco city directories, 1926, 1927.

67 The business card reads, "Miss Clayes of San Francisco at 284 Post Street, below Union Square, Kearny 686. For Gifts unusual. Oriental Applied Art. Del Monte: Hotel Del Monte; Los Angeles: 2116 W. Seventh Street, opposite Westlake Park; Santa Barbara: 17 E. Carrillo Street."

In February 1920, Miss Clayes received a patent for "The Chinese Launderer." This ceramic flower holder, when produced, was signed "Clayes" and marked "California Faience." California Faience was the trade name of a company, The Tile Shop, in Berkeley, that probably produced Miss Clayes's flowerpot (U. S. Patent Office, patent no. D54379). Also see Stern 2001, 12, 31, 108.

68 San Francisco city directories, 1926, 1927.

69 U. S. Federal Census statistics.

70 U. S. Department of Labor, Women's Bureau, statistics.

71 U. S. Federal Census, San Francisco, 1930. According to the 1931 Berkeley Blue and Gold Yearbook (p. 172), Evans Carr was a member of the Class of 1931 in the School of Dentistry.

72 For more on the history of O'Connor, Moffatt, & Company, see Macy's, Inc.'s website, http://www.fds.com/pressroom/History/MacysAHistory.aspx; and Time Magazine, Monday, July 16, 1945.

73 San Francisco city directory, 1939; Hotel Taxqueño register, May 10, 1938.

74 San Francisco city directory, 1929.

75 San Francisco city directories, 1929, 1930.

Chapter IV- Albina's Death

76 *San Francisco Examiner*, Feb. 20, 1931, p. 1.

77 *San Francisco Examiner,* Feb. 18, 1931, p. 1.

78 *San Francisco Examiner*, Feb. 20, 1931. Most

of the story comes from this edition of the *Examiner*, with embellishments on Feb. 21, 22, and 24. This story was picked up by papers all over the United States, including the *Nevada State Journal*, the *Circleville* [Ohio] *Herald*, *The Chronicle-Telegram* of Elyria, Ohio, the *Coshocton* [Ohio] *Tribune*, the *San Antonio* [Texas] *Express*, and *The Daily Northwestern* of Oshkosh, Wisconsin.

79 *San Francisco Examiner*, Feb. 20, 1931. The Examiner identifies Charles Simpson's father as an old friend: "[The murderer] is the son of Charles Simpson, plumber, old friend of Mrs. Voorhies and owner of the little store which she rented and operated rather than be a burden to her children." Charles Simpson stated in his confession, "I had known Mrs. Voorhies quite awhile. She was a queer old lady, but seemed kind to everybody. She lived alone there in the same building where she kept a store, and everybody thought she had a lot of money hidden somewhere in the place. I thought so, too." And again, "The old lady was a little queer, but she always seemed to sort of like me, so I knew I could get in. Of course, with her knowing me for so many years, I knew I'd have to kill her to cover up, but I thought it'd be all OK if I burned the place down."

80 *San Francisco Examiner,* Feb. 18, 1931.

81 *San Francisco Examiner,* Feb. 20, 1931.

82 State of California, City and County of San Francisco, Department of Public Health, Certificate of Vital Record, Feb. 18, 1931. Certificate of Death for Albina Durand Voorhies. Date of death February 17, 1931; cause of death, murder by general burning of body by fire.

83 The bill from Martin & Brown, San Francisco for funeral services has an attached copy of the obituary. This document was filed with the Certificate of Death.

84 *San Francisco Examiner*, Feb. 18, 1931, p. 19.

85 Superior Court of California, San Francisco, Criminal Division, Criminal Records. Conviction date, April 14, 1931, Superior Judge Louis H. Ward, presiding; Death Warrant and Execution data, July 17, 1931.

86 Probate file for Albina Durand Voorhies, number 58072. Superior Court of California, in City and State of San Francisco, a Petition for Letters of Administration in the Matter of the Estate of Albina Durand Voorhies, deceased. The petitioner was Margot Carr.

87 "Que todas las noches de tu vida sean siempre. . . buenas. . . que todos tus días estén llenos de felicidad."

88 San Francisco city directories, 1932-1937. *San Francisco Examiner*, July 1, 1936, p. 15, col. 5, "Divorce Complaints Filed."

Chapter V- From San Francisco to Mexico

89 Author's interview of Jaime Quiróz, 2008.

90 Hotel Taxqueño Register. On August 25, 1937, Margot Carr signed the register and gave her address as Chopo #139. She was accompanied by Leona and Aurora Eckstrom of San Francisco and Marion Gregory of Los Angeles. Author's interview of María Castillo, 1993.

91 Brenner 1932, 310; 1935, 357. Also in *Terry's Guide to Mexico* (1935), Brehme ran a full-page ad illustrated with one of his photographs. See , *Artes de México,* "La Tarjeta Postal" Number 48 (December 1999).

92 The photograph, *Toluca Market*, was found in 2009, hanging on the wall of the Casa Delicias, Spratling's former residence in Taxco, now owned by the Ulrichts.

93 While the photograph of Santa Prisca is not signed with the artist's name, the lettering of the title closely resembles Margot's handwriting.

94 Hotel Taxqueño Registers, 1937-1938. Dorothy Dunnicliff's daughter, Maryanne Richardson Brady, verifies that her mother traveled with Margot and Consuelo Barreto to Taxco in 1938 (Finch interview, 2009). Margaret Aileen Thomson was married to Carl R. Thomson who was in Japan in 1918-19 (passport applications for Carl and Margaret Aileen Thomson, 1918-19).

95 Hotel Taxqueño Register, 1938. The Eckstrom sisters are listed as Annette and Leona in the San Francisco city directories for 1939-40.

96 Author's interviews with Antonio Castillo and María Castillo, 1992.

Chapter VI- Castillos en el Aire

97 Author's interview with Antonio Castillo. Antonio and Margot went to San Francisco in 1940 and apparently set up a temporary workshop in a basement room. Antonio relates that one day he looked out of the basement window and noticed that people were no longer walking by; instead, all the shoes were facing the same direction. He came out of the basement to discover a group of San Franciscans pointing towards the sun as it broke through the clouds. At that moment, he made a decision that he and Margot would return to Taxco. (Story told to Marc Navarro.)

98 Spratling 1967, 77-8.

99 Spratling 1967, 78-9.

100 Author's interview with Antonio Castillo 1992.

101 Author's interview with Antonio Castillo.

102 Spratling 1967, 82.

103 Spratling 1967, 77.

104 Spratling published his first catalogue in 1938.

105 I would like to thank Elizabeth Finch for this insight.

106 Read Christie Romero's essay on Mexican metalsmithing (Morrill 2002, 166-68, 170-71) for descriptions of *repoussage* and how beads were made.

107 Also see the ivory and silver dress clip in Morrill and Berk 2007, 85, Plate VI-9.

108 San Francisco city directories, 1924, 1930, 1935, 1936, 1938. Miss Margaret Clayes lists herself as a merchant, with her residence at 2623 Divisadero until 1935. In 1936 and 1938, she simply lists her residence.

109 According to her 1921 passport, Margaret Clayes was going to travel to the following countries in 1922 on commercial business: China, Hong Kong, Saigon, Dutch East Indies, French Indochina, Siam, Ceylon, Burma, and India. See also her passports for 1917 and 1919. Records of the Immigration and Naturalization Service, RG 85; National Archives, Washington, D.C. Passenger lists indicate that on Sept. 22, 1927 Margaret Clayes returned from Yokohama to San Francisco and, on Aug. 7, 1929, from Kobe to San Francisco. Ancestry.com. California Passenger and Crew Lists, 1893-1957 [database on-line]. Provo, UT, USA: The Generations Network, Inc., 2008. Passenger Lists of Vessels Arriving at San Francisco, 1893-1953; (National Archives Microfilm Publication M1410, 429 rolls).

110 Information concerning the ivory figures was provided by Dr. Robert De Caroli, specialist in Southeast Asian Art and Director of the Art History Department at George Mason University.

111 Dr. James Ulak is Deputy Director of the Smithsonian's Freer and Arthur M. Sackler Galleries, where he previously served as Director of Collections and Research.

112 Author's interview with Ana Brilanti, who said of Margot, "Ella fue el corazon de la cosa."

113 Author's interviews of Sigi Pineda and Antonio Castillo. The silversmiths of Los Castillo in a 1942 photograph of the workshop are working in repoussage (Morrill and Berk 2007, Plate VI-6, p. 83, and list of silversmiths on p. 85).

114 Spratling 1967, 80.

115 Author's interview of Antonio Castillo.

116 Author's interview of Sigi Pineda. See Morrill 2004, 2-9.

117 Natalie Scott Papers, Special Collections Division, Howard Tilton Library, Tulane University, Jennison to Scott, Dec. 31, 1943.

118 Interestingly, the company Heller-Deltah that owned La Tausca, held the La Tausca Art Competition annually in New York beginning in 1946, and displayed the works of the entrants and prize-winners. Robert M. Coates. "Review of art show at the Riverside Museum, sponsored by Heller-Deltah Company, makers of the La Tausca Pearls." *The New Yorker* (Feb. 8, 1947); Dr. Howard E. Wooden, Jr. Introduction to "The Neglected Generation of American Realist Painters: 1930-1948." Exhibit catalog, Wichita Art Museum (May-June 1981).

119 Author's interviews with Jaime Quiróz and Sigi Pineda.

120 Author's interview with Antonio Castillo, 1992.

121 The information concerning Rafael Castillo comes from the author's interviews with Sigi Pineda, Melecio Rodríguez, Jaime Quiróz, and María Pineda.

122 Author's interviews of María Pineda and Sigi Pineda, 2006.

Chapter VII Margot de Taxco and Enameling

123 Author's interview with María Castillo, 1992.

124 Author's interview of Jaime Quiróz, 2009. Article #33 of the Mexican Constitution allowed for the forced extradition of foreigners considered a threat to the country. According to Quiróz, Margot went every year to Mexico City to file her papers. She had a friend in the government office, who always assisted her.

125 Sigi Pineda and Melecio Rodríguez informed the author about the early years in the enamel workshop.

126 Author's interview of Jaime Quiróz, 2009.

127 Carol Leake, Assistant Professor of painting and drawing, College of Music and Fine Arts, Loyola University, New Orleans, Louisiana, identified Margot's design drawings as pencil and gouache sketches.

128 The description of the enameling technique was supplied by Sigi Pineda, Melecio Rodríguez, and Jaime Quiróz.

129 Author's interview of Jaime Quiróz, 1993.

130 Pack 1947, 138.

131 See Christie Romero's excellent text and photographic essay on Mexican metalsmithing (Morrill 2002, 172-74) concerning die-stamping and enameling.

132 Melecio Rodríguez states that all the workshops, including those of Spratling, Héctor Aguilar, and Los Castillo, when demand dropped, would let go workers who had most recently arrived. The fluctuation of demand seemed to follow a pattern of three good months and three bad months.

133 Information concerning the enameling process came from author's interviews of Jaime Quiróz, Melecio Rodríguez, and Sigi Pineda.

134 I am indebted to Christie Romero for answers to these technical questions. Also see her essay,

"Mexican Metalsmithing Techniques," Morrill 2002, 172-4.

135 Author's interview of Jaime Quiróz.

136 Jim Tracy and Jim Black have discovered an elegant enamel on copper buckle in pale pink and green, with a large "S' curve in black (5794).

137 Morrill and Berk 2007, 200-3.

138 Author's interview of Jaime Quiróz.

139 Passenger lists for Air France from Mexico City to New York, April 27, 1954; June 23-24, 1954 on Air France from Paris to New York; Oct. 2, 1955 arrival in New York.

140 Jaime Quiróz and Sigi Pineda provided this information to the author.

141 Carlos Sánchez's culpability is still debated in Taxco today.

142 Author's interview with Melecio Rodríguez.

143 Author's interview of Jaime Quiróz, 2009. According to Marcial Chávez (author's interview, 2009), William Spratling warned Margot about her manager, advice she obviously did not heed.

144 Morrill 1996, 211, footnote 26.

145 Author's interview with Marcial Chávez, 2009.

146 Author's interviews with Jaime Quiróz and Melecio Rodríguez.

147 Spratling 1955, 89.

148 In a conversation with the author, Christine Till described being present as a tourist in Taxco and watching the citizens put out the fire.

149 Author's interviews of Betty Hays and Jaime Quiróz, 1992-93.

Chapter VIII The Margot de Taxco Style

150 Author's interview of Jaime Quiróz, 2009.

151 Dunlap 1997, 15-21.

152 Dunlap 1997, 135.

153 This identification was made by Dr. Robert De Caroli.

154 The identification of the opal was made by Melanie A. Marts, G. G. (G.I.A.).

155 See Elizabeth Kennedy Easby and John F. Scott, *Before Cortés: Sculpture of Middle America*. Exhibit catalog. Metropolitan Museum of Art (New York, 1970): Plate 174. For a more recent description, see Mary Miller and Simon Martin, *Courtly Art of the Ancient Maya*. Exhibit catalog. Fine Arts Museum of San Francisco (New York: Thames and Hudson, 2004): Plate 1, pp. 28-9.

156 See Morrill 2002, Fig. 24, p. 26.

157 According the Jaime Quiróz, Margot produced this design intentionally to capture the flowing form of the chrysanthemum in a circular format at the request of a client.

158 Author's interview of Jaime Quiróz.

159 Author's interview of Jaime Quiróz. Margot brought back a St. Francis figure from Italy, which she gave to Jaime Quiróz.

1610Hall 1979, 154.

161 Hall 1979, 143.

162 This information was provided by Linda Thrift, Administrator, and Patricia H. Svoboda, Research Coordinator, at the Center for Electronic Research and Outreach Services, National Portrait Gallery, Smithsonian Institution.

Chapter IX- The Last Years

166 The directions for design number 5389 from Margot's notebook in the Spratling-Taxco Collection at Tulane's Latin American Library read as follows, "Collar con dos piezas al centro y 10 engarses unidas con la misma pieza y las centro con tapitas. Todas las piezas llevan una sobrepuesta atrornillada con cuerda y tuerca. Pulsera con 6 piezas unidas con la misma pieza con las mismas piezas sobrepuestas lleva caja y cadena de guardia. Aretes con dos piezas sacadas de la pulsera con las piezas sobrepuestas con tornillo. Prendedor con las dos piezas del centro con las piezas sobrepuestas con juego de broches."

164 Author's interviews of Jaime Quiróz and María Pineda.

165 Elizabeth Finch interview with Elizabeth Freeman, 2008.

166 Morrill 1996, 63-5, 175-7; Stromberg 1985, 52-9.

167 Stromberg 1985, 60.

168 Author's interview of Jaime Quiróz.

169 Morrill and Berk 2007, 114-6; author's interview of Benjamín Santarriaga in 1993.

170 Author's interviews of Jaime Quiróz and Melecio Rodríguez.

171 The sad story of the demise of Margot de Taxco and Margot's own last years was told to the author by María Pineda, Melecio Rodríguez, Sigi Pineda, and Jaime Quiróz. Margot remained youthful to the end, claiming to be 68 years of age just before she died. Also see Morrill and Berk 2007, 176-89.

Chapter X-Contemporary Enamelists in Mexico

172 Author's interview of Adrian Brun, 2008.

173 Author's interview of Miguel Pineda, 2008.

174 Jazzar and Nelson 2007, 267.

175 Fastag 1991.

176 Fastag 1991.

177 Author's interview of Miguel Pineda, 2008.

178 Author's interview of Priscila Canales, 2008.

179 Terán 1994, 36.

180 Terán 1994, 69-73.

181 Author's interview with Jaime Quiróz, 2009.

Bibliography

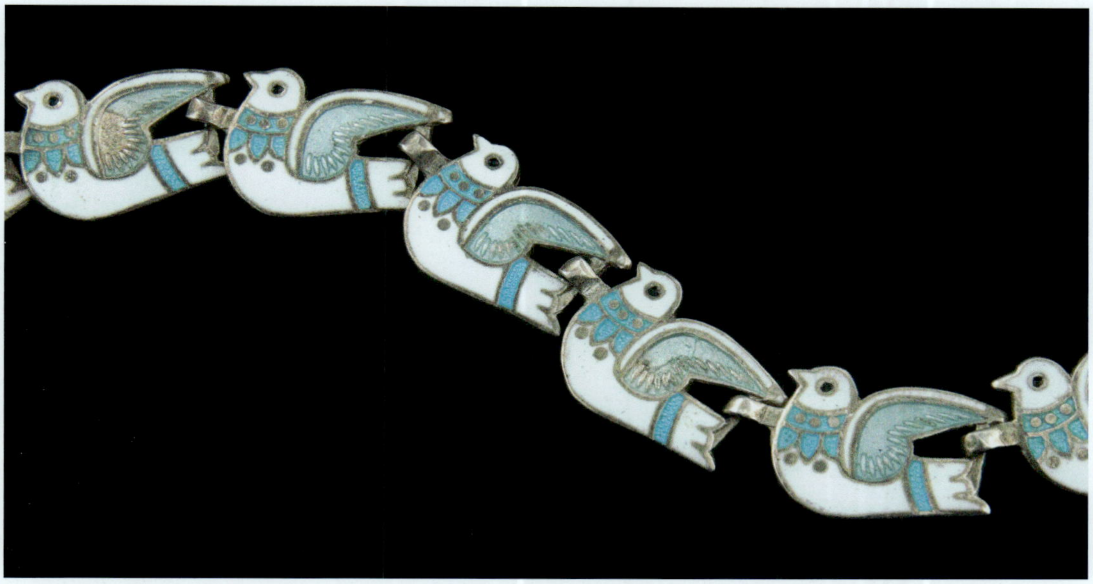

Margot de Taxco 5868. Folk Art Doves necklace, detail

Aguilar Ricketson, Héctor. *Artesanía de la Plata*. Sección 6: Tecnología. Mexico: Manuales Union Tipográfica Editorial Hispano Americana, 1968.

Blewett, Mary H. *The Last Generation: Work and Life in the Textile Mills of Lowell, 1910-1960*. Amherst, MA: University of Massachusetts Press, 1990.

Brenner, Anita. *Your Mexican Holiday: A Modern Guide*. New York and London: G. P. Putnam's Sons, 1932, and revised edition, 1935.

Burgess, Sherwood D. *The Water King Anthony Chabot: His Life and Times*. Davis, CA: Panorama West Publishing, 1992.

Christoph, Florence A. *Van Voorhees Family in America: The Seventh and Eighth Generations*. 2 volumes. Rockport, ME: Penobscot Press, 2003.

Cox, Anna Sprague. Letter to Uncle Wales (Wales L. Knox). Knox Family Papers: Earthquake Selection: Manuscript letter addressed to Uncle Wales, 1906 (BANC MSS 85/163 c). Bancroft Library, University of California, Berkeley, California.

Davis, Mary L., and Greta Pack. *Mexican Jewelry*. Austin, TX: University of Texas Press, 1963.

Directory of Manufacturers of San Francisco, California. San Francisco Chamber of Commerce. San Francisco, CA: The Hansen Co., 1922.

Dunlap, Richard A. *The Golden Ratio and Fibonacci Numbers*. New Jersey and London: World Scientific, 2008 reprint of 1997 edition.

Fastag, Miriam. "Inview: Miguel Angel Pineda." *Glass on Metal Magazine*. Vol. 10, Number 1 (February 1991).

Hall, James. *Dictionary of Subjects and Symbols in Art*. New York: Harper and Row, Publishers, Inc., 1979.

Hofstadter, Richard, William Miller, and Daniel Aaron. *The United States: The History of a Republic*. Englewood Cliffs, NJ: Prentice-Hall, Inc., 1967.

Hotel Taxqueño Registers. Family of Dorothy Sutherland Chittim, Eagle Pass, Texas.

Hymowitz, Carol, and Michaele Weissman. *A History of Women in America*. New York: Bantam Books, 1978.

Jazzar, Bernard N., and Harold B. Nelson. *Painting with Fire: Masters of Enameling in America, 1930-1980*. Long Beach, CA: Long Beach Museum of Art, 2007.

Louisiana: A Guide to the State. Writers' Program of the Work Projects Administration in the State of Louisiana. American Guide Series. New York: Hastings House Publishers, 1945.

Matthews, Glenice Lesley. *Enamels, Enameling, Enamelists*. Radnor, PA: Chilton Book Company, 1984.

Milbank, Caroline Reynolds. *New York Fashion: The Evolution of American Style*. New York, NY: Harry N. Abrams, Inc., 1989.

Mills, Charles. *Treasure Legends of the Civil War.* BookSurge Publishing, 2002.

Morison, Samuel Eliot, and Henry Steele Commager. *The Growth of the American Republic*. 2 Volumes. New York: Oxford University Press, 1962.

Morrill, Penny C. *Hecho en México*. Catalogue for exhibit, Carole A. Berk Gallery. Washington, D. C.: Colortone Press, 1990.

Morrill, Penny C. *Silver Masters of Mexico: Héctor Aguilar and the Taller Borda*. Atglen, PA: Schiffer Publishing, Ltd., 1996.

Morrill, Penny C., author, editor. *Maestros de Plata: William Spratling and the Mexican Silver Renaissance*. New York: Harry Abrams, Inc., 2002.

Morrill, Penny C. "William Spratling and his Marketing Ways." *Silver Magazine* 36, 3, (May-June 2004): 2-9.

Morrill, Penny C., and Carole A. Berk. *Mexican Silver: Modern Hand-wrought Jewelry and Metalwork*. Atglen, PA: Schiffer Publishing, Ltd., 1994, 4th edition 2007.

Official Catalogue of Exhibitors: Panama-Pacific International Exposition, San Francisco, California, 1915. San Francisco, CA: The Wahlgreen Company, 1915.

Ormond, Richard, and Elaine Kilmurray. *John Singer Sargent: The Early Portraits*. Volume I of *Complete Paintings*. New Haven, CT and London: Yale University Press, with The Paul Mellon Centre for Studies in British Art, 1998.

Pack, Greta. *Jewelry and Enameling*. New York: D. Van Nostrand Company, Inc., 1941, sixth printing, 1947.

Smith, Dennis. *San Francisco is Burning: The Untold Story of the 1906 Earthquake and Fires*. New York, NY: The Penguin Group, 2005.

Spratling, William. "25 Years of Mexican Silverware." *Artes de México*. Year 111, No. 10 (Mexico, D.F., December 1955): 87-90.

Spratling, William. *File on Spratling*. Boston and Toronto: Little, Brown and Company 1967.

Stern, Bill. *California Pottery: From Missions to Modernism*. San Francisco, CA: Chronicle Books, 2001.

Stromberg, Gobi. *El Juego del Coyote: Platería y Arte en Taxco*. Mexico: Fondo de Cultura Económica, 1985.

Terán, Silvia. *La Platería en Yucatán*. Mexico: Dirección General de Culturas Populares with Casa de las Artesanías del Estado de Yucatán, 2nd edition, 1994.

Terry, T. Philip. *Terry's Guide to Mexico*. Boston and New York: Houghton Mifflin Company, 1935.

Tulane University, Latin American Library, Spratling-Taxco Collection.

Untracht, Oppi. *Enameling on Metal*. New York: Greenberg, 1957.

Untracht, Oppi. *Metal Techniques for Craftsmen: A Basic Manual on the Methods of Forming and Decorating Metals*. New York: Doubleday, 1975.

Van Voorhies Carr, Margot. *Notes and Instructions Concerning Enamel Jewelry*. 1948-1974.

Watkins, Mrs. James T. "The 1906 San Francisco Earthquake: A Personal Account." *California Geology*, Vol. 34, No. 12 (December 1961).

ARCHIVES

Border Crossings: From Mexico to the United States, 1903-1957. Records of the Immigration and Naturalization Service, Record Group 85, National Archives, Washington, D.C.

California Passenger and Crew Lists, 1893-1957. Database on-line. Provo, UT, USA: The Generations Network, Inc., 2008. Passenger Lists of Vessels Arriving at San Francisco, 1893-1953; (National Archives Microfilm Publication M1410, 429 rolls).

California Voter Registrations, 1900-1968. Database on-line. Provo, UT: The Generations Network, Inc., 2008. Original data: State of California. Great Register of Voters. Sacramento, CA: California State Library.

Census of the United States. Bureau of the Census. Washington, D.C.: National Archives and Record Administration.

Hunting for Bears Genealogy Society, compiler. *Louisiana Marriages, 1718-1925*. Database online.

Natalie Scott Papers, Special Collections Division, Howard Tilton Library, Tulane University.

New Orleans, Louisiana Marriage Records Index, 1831-1925. Database on-line. Provo, UT: The Generations Network, Inc., 2002. Original data from State of Louisiana, Secretary of State, Division of Archives, Records Management, and History. Vital Records Indices, Baton Rouge, LA.

Quebec Vital and Church Records (Drouin Collection), 1621-1967. Provo, UT: The Generations Network, Inc., 2008. *Actes d'état civil et registres d'église du Québec (Collection Drouin), 1621 a 1967*. Original data from Gabriel Drouin, compiler. Drouin Collection. Montreal, Quebec, Canada: Institut Généalogique Drouin.

Records of the Immigration and Naturalization Service, RG 85; National Archives, Washington, D.C.

San Francisco Chronicle website, *The Great Quake: 1906-2006* (http://sfchronicle.us/greatquake/), Carle Nolte, Chronicle Staff Writer, 2006.

Superior Court of California, San Francisco, Criminal Division, Criminal Records. Conviction date, April 14, 1931, Superior Judge Louis H. Ward, presiding; Death Warrant and Execution data, July 17, 1931.

Superior Court of California, in City and State of San Francisco, a Petition for Letters of Administration in the Matter of the Estate of Albina Durand Voorhies, deceased. Probate file for Albina Durand Voorhies, number 58072.

The Generations Network: Ancestry.com. Provo, UT, 2004. Database on-line.

Tulane University, Latin American Library. The Spratling-Taxco Collection.

United States City Directories. Library of Congress Microfilm Reading Room.

United States Department of Labor, Women's Bureau, statistics.

United States Patent and Trademark Office: http://www.uspto.gov/index.html. Design no. D54379. M. Clayes. Flower Holder. Application filed July 14, 1919. Patented Feb. 10, 1920.

United States Passport Applications, January 2, 1906-March 31, 1925. National Archives Microfilm. General Records of the Department of State. Record Group 59. National Archives, Washington, D. C.

WorldVitalRecords.com. Database on-line. University of California, *Berkeley Blue and Gold Yearbook*, 1931. Archived at the Bancroft Library, University of California, Berkeley, California.

PERIODICALS

The Chronicle-Telegram, Elyria, Ohio. February 24, 1931, "Held in Torch Murder."

The Circleville Herald, Circleville, Ohio. Forty-eighth year, No. 43, February 20, 1931, "Youth Held for Gruesome Crime."

Coshocton Tribune, Coshocton, Ohio. Vol. XXII, No. 169, February 20, 1931, "Brutal Crime Confessed by Calif. Youth."

The Daily Northwestern, Oshkosh, Wisconsin. February 20, 1931, "Confesses he Beat and Burned Woman to Death in Store." July 18, 1931, "Two Slayers Hanged."

Nevada State Journal, Reno, Nevada. Vol. LVI, No. 289, February 18, 1931, "Woman Storekeeper Killed in Bay City." No. 293, February 22, 1931, "'Three Dollar' Woman Slayer Only Sorry for Method and Sweetheart" and "Love Blighted by Tragedy" [photo]. No. 294, February 23, 1931, "Insanity Plea Plan of 'Torch' Murderer." No. 295, February 24, 1931, "Mother Visits San Francisco Torch Slayer: Youthful Killer Says 'Bad Booze' Caused Murder." No. 296, February 25, 1931, "Simpson Indicted for S.F. Slaying." No. 351, April 21, 1931, Guilty Plea by S. F. Torch Killer Fails to Bring Clemency of Judge" and "Sneers at Sentence of Death" [photo].

Oakland Tribune, Oakland, California. Vol. CXIV, February 17, 1931, "Aged Woman Burned Alive by Robbers: Body of S.F. Shop Keeper Found Still Smouldering in Living Quarters of Store; Place Ransacked; 'Most Brutal Murder in History of San Francisco,' Says Police Official, as Hunt for Killers Begins." February 21, 1931, "'My Boy is Insane,' Says Mother of Torch Killer: Mrs. Simpson Blames Bump Slayer Received on Head as a Child." "Hang Me Now, Cries Young Torch Killer." "Girl Friend Denies Troth: Anna Recalls Boy Slayer Obsessed With Idea of Old Woman's Wealth." "Torch Murderer Tells Own Life Story." and "Parents, Brothers and Sisters of Torch Fiend"[photo]. February 22, 1931, "Pictures Tell Life Story of S. F. 'Torch Murderer'." February 24, 1931, "Torch Killer is Moved to County Jail: Grand Jury to Meet Tonight With Quick Indictment Forecast for Slayer; Attorney Makes Plea." February 25, 1931, "Torch Killer Indicted at Brief Hearing: Counsel for Boy Charged with Murder of Aged Woman Says He Will Make Insanity Defense." February 27, 1931, "Quick Trial Awaiting Torch Killer: Simpson May Face Jury in Two Weeks as Judge Speeds Assignment of Case to Criminal Office." March 6, 1931, "Torch Slayer's Trial is Set for April 14." April 16, 1931, "'Blood Taint' Plea Made to Save Son: Mother of Charles Simpson, Killer of Aged Woman, Tells Court of Near Relationship of his Parents; Acts of Destruction and Rage on Part of Torch Slayer Related in Claim That He Never Has Been Normal." Vol. CXV, July 2, 1931, "S. F. 'Torch Murderer' Loses Hope: Governor Refuses to Save Youthful Slayer of Aged Storekeeper, Branding Crime as Brutal Act." July 17, 1931 "Torch Killer Pays Price on Gallows: Charles Simpson, Slayer of Aged Woman, Requests Guards 'Let's Make it Snappy' As Noose Falls."

San Antonio Express, San Antonio, Texas. February 20, 1931, "Youth Admits Killing and Burning Woman." July 18, 1931, "California Youth Dies for Woman's Murder."

San Francisco Chronicle. various editions

San Francisco Examiner. various editions

INTERVIEWS by the Author

Adrian Brun, 2008
Priscila Canales, 2008
Antonio Castillo, 1992-2000
María Castillo, 1992-1993
Marcial Chávez, 2009
María Pineda, 2000-2009
Miguel Angel Pineda, 2008
Suigifrido (Sigi) Pineda, 1992-2009
Jaime Quiróz Duque, 1992-2009
Melecio Rodríguez, 2008
Benjamín Santarriaga, 1993

INTERVIEWS by Elizabeth Finch

Maryanne Richardson Brady 2009
Elizabeth Freeman 2008-2009
Ann Nunn 2008-2009

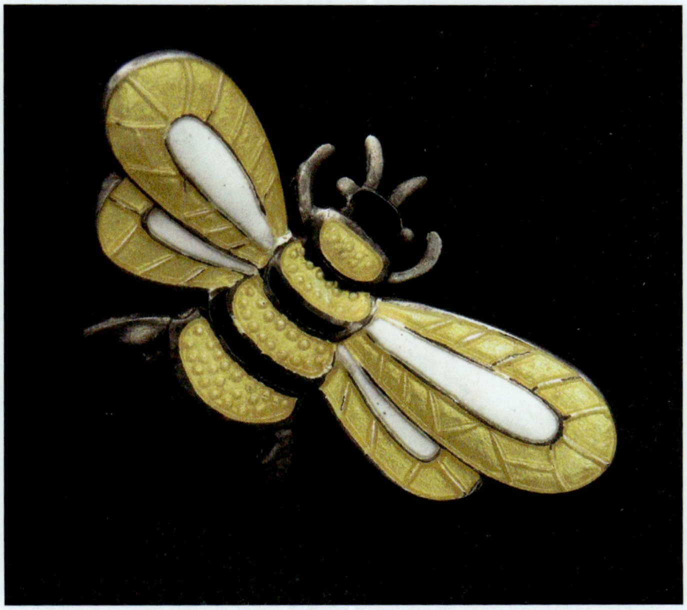

Margot de Taxco 5661. *Bumblebee* brooch 0.75 x 1.5 in.
Stella and Fred Krieger Collection

Index

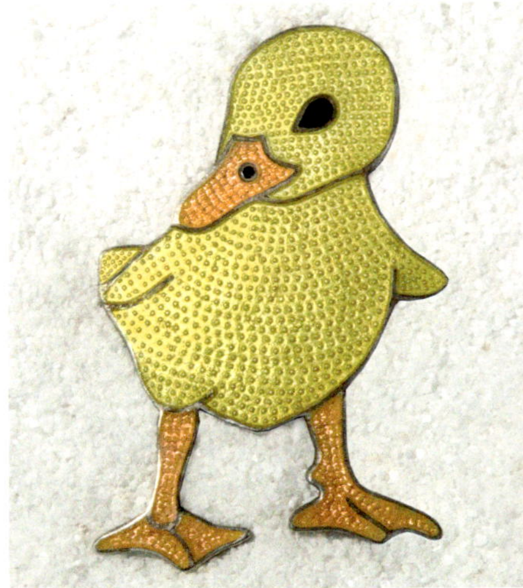

Margot de Taxco 5687. Duck brooch
2 x 1.25 in. Katherine Armitre Collection